ELECTRIC CIRCUITS FUNDAMENTALS

THIRD EDITION

ELECTRIC CIRCUITS FUNDAMENTALS

Thomas L. Floyd

Prentice Hall

Englewood Cliffs, New Jersey ■ **Columbus, Ohio**

Cover art by Todd Yarrington and Thomas Mack
Editor: Dave Garza
Developmental Editor: Carol Hinklin Robison
Production Editor: Rex Davidson
Text Designer: Thomas Mack
Cover Designer: Thomas Mack
Production Buyer: Patricia A. Tonneman
Illustrations: Precision Graphics

This book was set in Times Roman by The Clarinda Company and was printed and bound by R. R. Donnelley
& Sons Company. The cover was printed by Lehigh Press, Inc.

 © 1995 by Prentice-Hall, Inc.
A Simon & Schuster Company
Englewood Cliffs, New Jersey 07632

Second Edition copyright © 1991 by Macmillan Publishing Company.
First Edition copyright © 1987 by Merrill Publishing Company.

Photo credits for chapter openers and part openers: American Stock (Chapter 2); Courtesy of Battelle (Chapter
11); Comstock (Part 2); Courtesy of DeVry Institutes (Chapters 4, 14, 15); FPG International-The Stock
Directory (Chapter 13); Honeywell Home and Building Control, Golden Valley, Minn. (Chapter 7); The Stock
Market (Part 1, Chapter 10); Tony Stone Images (Chapters 3, 5); and Tektronix Inc. (Chapters 1, 6, 8, 9, 12).

Printed in the United States of America

10 9 8 7 6 5 4 3 2 1

ISBN 0-02-338646-0

Prentice-Hall International (UK) Limited, *London*
Prentice-Hall of Australia Pty. Limited, *Sydney*
Prentice-Hall Canada Inc., *Toronto*
Prentice-Hall Hispanoamericana, S. A., *Mexico*
Prentice-Hall of India Private Limited, *New Delhi*
Prentice-Hall of Japan, Inc., *Tokyo*
Simon & Schuster Asia Pte. Ltd., *Singapore*
Editora Prentice-Hall do Brasil, Ltda., *Rio de Janeiro*

PREFACE

Electric Circuits Fundamentals, Third Edition provides thorough, comprehensive, and practical coverage of basic electrical and electronic concepts and circuits with special emphasis on troubleshooting and applications. This third edition has been carefully reviewed and special efforts have been made to ensure a high level of accuracy and clarity.

This book is divided into two basic parts. Chapters 1 through 7 cover dc topics and Chapters 8 through 15 generally cover ac topics.

New and Improved Features

In addition to the many popular features of the second edition, several new features, as well as improvements to existing features, have been incorporated into this third edition.

- [ ] An innovative section at the end of each chapter (except Chapter 1) called Application Assignment
- [ ] An Application Assignment preview in each chapter opener (except Chapter 1)
- [ ] Distinctive full-color format used throughout
- [ ] An introductory message and a list of performance-based objectives at the beginning of each section
- [ ] An exercise in each worked example that is related to the example
- [ ] More section review questions with a complete answer section at the end of each chapter
- [ ] Improved consistency in accuracy (number of significant figures) of numerical results and in the use of metric prefixes
- [ ] An increased emphasis on troubleshooting and applications

An improved and expanded ancillary package for this edition includes

- [ ] Instructor's Edition
- [ ] Test Item File
- [ ] Prentice-Hall Test Manager 2.0
- [ ] Transparency Masters
- [ ] Transparencies
- [ ] Lab Manual: *Experiments in Electric Circuits Fundamentals, Third Edition* by Dave Buchla
- [ ] Student Study Guide by Wally McIntyre
- [ ] Video library of DC/AC topics

Illustration of Features Within Each Chapter

Chapter Opener   As shown in Figure P–1, each chapter begins with a two-page opener. The left page includes a listing of the sections within the chapter and a brief overview

Chapter overview

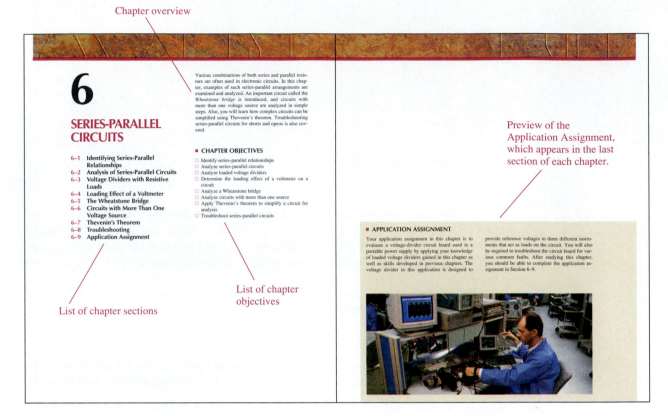

Preview of the
Application Assignment,
which appears in the last
section of each chapter.

List of chapter
objectives

List of chapter sections

FIGURE P–1
Chapter opener.

and introduction as well as a list of main objectives. The right page presents a preview of the Application Assignment section that is found at the end of the chapter.

Section Opener and Section Review Each section within a chapter begins with a brief introduction that either highlights the material to be covered or provides a general overview and includes a list of performance-based objectives for the section with a main objective and subobjectives. Each section ends with review questions that focus on the main concepts presented in the section. Solutions/answers to these review questions are given at the end of the chapter. Figure P–2 illustrates these two features.

Examples Frequent worked examples help to illustrate and clarify basic concepts. Each example concludes with a related exercise that reinforces or expands on the concept in some way. Some related exercises require the student to repeat the procedure demonstrated in the example but with a different set of values or conditions. Others focus on a more limited part of the example or encourage further thought. Answers to all related exercises are given at the end of the chapter. A typical example problem and related exercise are shown in Figure P–3.

Application Assignment The last section of each chapter (except Chapter 1) provides a practical application of material presented in the chapter. A series of steps generally require the student to compare circuit layouts with a schematic, analyze circuits using concepts or theories learned in the chapter, and evaluate and/or troubleshoot circuits. A typical Application Assignment section is shown in Figure P–4.

Results and answers for the steps in the application assignments are provided in the *Instructor's Edition*. Optional worksheets are provided in Dave Buchla's coordinated lab manual.

FIGURE P–2
Section review and section opener.

The Application Assignment is a very versatile feature which provides both motivation and practical experiences in the classroom and can be used as

☐ An integral part of the chapter for illustrating how the concepts and theories can be applied in a practical situation. All or selected activities can be assigned and discussed in class or turned in for a grade.

☐ Separate out-of-class assignments to be turned in for extra credit.

☐ In-class activities to promote and stimulate discussion and interaction among students and between students and the instructor.

☐ Examples to help answer the question that most students ask: "Why do I have to learn this?"

Chapter End Matter At the end of most chapters is a summary, a glossary, a formula list, a multiple-choice self-test, and sectionalized basic and advanced problem sets. Also, the answers to section reviews and to related exercises are found at the end of each chapter. Terms that appear boldface in the text are defined in the glossary.

Examples are
contained within
a ruled box.

Each example contains an
exercise related to the
example.

FIGURE P–3
An example problem and related exercise.

Suggestions for Use

Electric Circuits Fundamentals, Third Edition can be used to accommodate a variety of scheduling and program requirements.

A two-term dc/ac sequence can cover most, if not all, of the book, although some selectivity may be necessary in some cases. Chapters 1 through 7 can be covered in the first term. The topics of capacitors and inductors (Chapters 9 and 10) can be studied at the end of the first term, if desired, by delaying coverage of the ac topics in Sections 9–6, 9–7, 10–6, and 10–7 until the second term.

To the Student

The material in this preface is intended to help both you and your instructor make the most effective use of this textbook as a teaching and learning tool. Although you should certainly read everything in this preface, this part is especially for you, the student.

FIGURE P–4
An Application Assignment section.

I am sure that you realize that knowledge and skills are not obtained easily or without effort. Much hard work is required to properly prepare yourself for any career, and electronics is no exception. You should use this book as more than just a reference. You must really dig in by reading, thinking, and doing. Don't expect every concept or procedure to become immediately clear. Some topics may take several readings, working many problems, and much help from your instructor before you really understand them.

Work through each example step-by-step and then do the related exercise. Answer the review questions at the end of each section. If you don't understand an example or if you can't answer a question, review the section. Check your answers at the end of the chapter. The multiple-choice self-tests at the end of each chapter are a good way to check your overall comprehension and retention of the subjects covered. You should do the self-test before you start the problems. Check your answers to the self-tests at the end of the book.

The problem sets at the end of each chapter provide problems with varying degrees of difficulty. In any technical field, it is important that you work lots of problems. Working through a problem gives you a level of insight and understanding that reading or classroom lectures alone do not provide. Never think that you fully understand a concept or procedure by simply watching or listening to someone else. In the final analysis, you must do it yourself and you must do it to the best of your ability.

Careers in Electronics

The field of electronics is very diverse, and career opportunities are available in many areas. Because electronics is currently found in so many different applications and new technology is being developed at a fast rate, its future appears limitless. There is hardly an area of our lives that is not enhanced to some degree by electronics technology. Those who acquire a sound, basic knowledge of electrical and electronic principles and are willing to continue learning will always be in demand.

The importance of obtaining a thorough understanding of the basic principles contained in this text cannot be overemphasized. Most employers prefer to hire people who have both a thorough grounding in the basics and the ability and eagerness to grasp new concepts and techniques. If you have a good training in the basics, an employer will train you in the specifics of the job to which you are assigned.

There are many types of job classifications for which a person with training in electronics technology may qualify. A few of the most common job functions are discussed briefly in the following paragraphs.

Service Shop Technician   Technical personnel in this category are involved in the repair or adjustment of both commercial and consumer electronic equipment that is returned to the dealer or manufacturer for service. Specific areas include TVs, VCRs, CD players, stereo equipment, CB radios, and computer hardware. This area also offers opportunities for self-employment.

Industrial Manufacturing Technician   Manufacturing personnel are involved in the testing of electronic products at the assembly-line level or in the maintenance and troubleshooting of electronic and electromechanical systems used in the testing and manufacturing of products. Virtually every type of manufacturing plant, regardless of its product, uses automated equipment that is electronically controlled.

Laboratory Technician   These technicians are involved in breadboarding, prototyping, and testing new or modified electronic systems in research and development laboratories. They generally work closely with engineers during the development phase of a product.

Field Service Technician   Field service personnel service and repair electronic equipment—for example, computer systems, radar installations, automatic banking equipment, and security systems—at the user's location.

Engineering Assistant/Associate Engineer   Personnel in this category work closely with engineers in the implementation of a concept and in the basic design and development of electronic systems. Engineering assistants are frequently involved in a project from its initial design through the early manufacturing stages.

Technical Writer   Technical writers compile technical information and then use the information to write and produce manuals and audiovisual materials. A broad knowledge of a particular system and the ability to clearly explain its principles and operation are essential.

Technical Sales   Technically trained people are in demand as sales representatives for high-technology products. The ability both to understand technical concepts and to communicate the technical aspects of a product to a potential customer is very valuable. In this area, as in technical writing, competency in expressing yourself orally and in writing is essential. Actually, being able to communicate well is very important in any technical job category because you must be able to record data clearly and explain procedures, conclusions, and actions taken so that others can readily understand what you are doing.

A Look Back

Now, before you begin your study of electric circuits, let's briefly look at the beginnings of electronics and some of the important developments that have led to the electronics technology that we have today. It is always good to have a sense of the history of your career field. The names of many of the early pioneers in electricity and electromagnetics still live on in the names of familiar units and quantities. Names such as Ohm, Ampere,

Volta, Farad, Henry, Coulomb, Oersted, and Hertz are some of the better known examples. More widely known names such as Franklin and Edison are also very significant in the history of electricity and electronics because of their tremendous contributions.

The Beginning of Electronics   The early experiments in electronics involved electric currents in glass vacuum tubes. One of the first to conduct such experiments was a German named Heinrich Geissler (1814–1879). Geissler removed most of the air from a glass tube and found that the tube glowed when there was an electric current through it. Around 1878, British scientist Sir William Crookes (1832–1919) experimented with tubes similar to those of Geissler. In his experiments, Crookes found that the current in the vacuum tubes seemed to consist of particles.

Thomas Edison (1847–1931), experimenting with the carbon-filament light bulb he had invented, made another important finding. He inserted a small metal plate in the bulb. When the plate was positively charged, there was a current from the filament to the plate. This device was the first thermionic diode. Edison patented it but never used it.

The electron was discovered in the 1890s. The French physicist Jean Baptiste Perrin (1870–1942) demonstrated that the current in a vacuum tube consisted of the movement of negatively charged particles in a given direction. Some of the properties of these particles were measured by Sir Joseph Thomson (1856–1940), a British physicist, in experiments he performed between 1895 and 1897. These negatively charged particles later became known as electrons. The charge on the electron was accurately measured by an American physicist, Robert A. Millikan (1868–1953), in 1909. As a result of these discoveries, electrons could be controlled, and the electronic age was ushered in.

Putting the Electron to Work   In 1904 a British scientist, John A. Fleming, constructed a vacuum tube that allowed electrical current in only one direction. The tube was used to detect electromagnetic waves. Called the Fleming valve, it was the forerunner of the more recent vacuum diode tubes. Major progress in electronics, however, awaited the development of a device that could boost, or amplify, a weak electromagnetic wave or radio signal. This device was the audion, patented in 1907 by Lee deForest, an American. It was a triode vacuum tube capable of amplifying small electrical ac signals.

Two other Americans, Harold Arnold and Irving Langmuir, made great improvements in the triode vacuum tube between 1912 and 1914. About the same time, deForest and Edwin Armstrong, an electrical engineer, used the triode tube in an oscillator circuit. In 1914, the triode was incorporated in the telephone system and made the transcontinental telephone network possible. In 1916 Walter Schottky, a German, invented the tetrode tube. The tetrode, along with the pentode (invented in 1926 by Dutch engineer Tellegen), greatly improved the triode. The first television picture tube, called the kinescope, was developed in the 1920s by Vladimir Sworykin, an American researcher.

During World War II, several types of microwave tubes were developed that made possible modern microwave radar and other communications systems. In 1939, the magnetron was invented in Britain by Henry Boot and John Randall. In the same year, the klystron microwave tube was developed by two Americans, Russell Varian and his brother Sigurd Varian. The traveling-wave tube (TWT) was invented in 1943 by Rudolf Komphner, an Austrian-American.

Solid-State Electronics   The crystal detectors used in early radios were the forerunners of modern solid-state devices. However, the era of solid-state electronics began with the invention of the transistor in 1947 at Bell Labs. The inventors were Walter Brattain, John Bardeen, and William Shockley.

In the early 1960s, the integrated circuit (IC) was developed. It incorporated many transistors and other components on a single small chip of semiconductor material. Integrated circuit technology has been continuously developed and improved, allowing increasingly more complex circuits to be built on smaller chips.

Around 1965, the first integrated general-purpose operational amplifier was introduced. This low-cost, highly versatile device incorporated nine transistors and twelve resistors in a small package. It proved to have many advantages over comparable discrete component circuits in terms of reliability and performance. Since this introduction, the IC operational amplifier has become a basic building block for a wide variety of linear systems.

Acknowledgments

I wish to express thanks to all the people who had a part in making this revision a reality. In particular, the efforts of David Garza, Carol Robison, Pete Robison, Rex Davidson, Pat Tonneman, and Tom Mack are greatly appreciated. While space does not permit acknowledgment of all those who have submitted suggestions or comments for this new edition, I do wish to thank the following reviewers for their invaluable input: Richard Bacon, Santa Rosa Junior College; Mort Broad, Pinellas Vocational-Technical Institute; Dave Buchla, Yuba College; David Fridenmaker, ITT Technical Institute; Jim Gee, ITT Technical Institute; Jim Hart, DeVry Institute of Technology–Chicago; Carl Jensen, DeVry Institute of Technology–Woodbridge; Osama Maarouf, Microcomputer Technical Institute; Raymond Rich, ITT Technical Institute; David Soldner, ITT Technical Institute; Peter Westray, Los Angeles Valley College; and Jerry Wilson, Lamar University–Beaumont. In addition, I am grateful to Gary Snyder for a superb job of checking the manuscript for accuracy and to Lois Porter for another excellent job of editing the manuscript.

Tom Floyd

CONTENTS

xiii

PART TWO ■ AC CIRCUITS

15 Pulse Response of Reactive Circuits 626

Appendices

ELECTRIC CIRCUITS FUNDAMENTALS

PART ONE

DC CIRCUITS

1

COMPONENTS, QUANTITIES, AND UNITS

The topics in this chapter present a basic introduction to the field of electronics. An overview of electrical and electronic components and instruments gives you a preview of the types of things you will study throughout this book.

You must be familiar with the units used in electronics and know how to express electrical quantities in various ways using metric prefixes. Scientific notation is an indispensable tool whether you use a computer, a calculator, or do computations the old-fashioned way.

■ CHAPTER OBJECTIVES

☐ Recognize some common electrical components and measuring instruments
☐ State the electrical and magnetic quantities and their units
☐ Use scientific notation (powers of ten)
☐ Use metric prefixes to express large and small quantities
☐ Convert from one metric unit to another

■ APPLICATION ASSIGNMENT

At the beginning of each chapter starting with Chapter 2, you will find an application assignment that relates to that chapter. These assignments present a variety of practical job situations that a technician might encounter in industry.

As you study each chapter, think about how to approach the application assignment. When you have completed each chapter, you should have a sufficient knowledge of the topics covered to enable you to complete the assignment that appears as the last section in each chapter.

1–1 ■ CIRCUIT COMPONENTS AND ELECTRONIC MEASURING INSTRUMENTS

In this book you will study many types of electrical components and measuring instruments. A thorough background in dc and ac circuit fundamentals provides the foundation for understanding electronic devices and circuits. A preview of the basic types of electrical and electronic components and instruments that you will be studying in detail in this and in other courses is provided in this section.

After completing this section, you should be able to

■ **Recognize some common electrical components and measuring instruments**
   □ State the function of a resistor
   □ State the function of a capacitor
   □ State the function of an inductor
   □ State the function of a transformer
   □ List the basic types of electronic measuring instruments

Resistors

Resistors resist, or limit, electrical current in a circuit. Several common types of resistors are shown in Figure 1–1.

(a)   (b)

(c)   (d)

FIGURE 1–1

Typical fixed and variable resistors. (a) Carbon-composition resistors with standard power ratings of ⅛ W, ¼ W, ½ W, 1 W, and 2 W (courtesy of Allen-Bradley Co.). (b) Wirewound resistors (courtesy of Dale Electronics, Inc.). (c) Variable resistors or potentiometers (courtesy of Bourns, Inc.). (d) Resistor networks (courtesy of Bourns, Inc.).

Capacitors

Capacitors store electrical charge; they used to be called *condensers* and are found in a variety of applications. Figure 1–2 shows several typical capacitors.

FIGURE 1–2
Typical capacitors. Parts (a, c, d, and g) courtesy of Mepco/Centralab. Part (b) courtesy of Murata Erie North America. Parts (e) and (f) courtesy of Sprague Electric Co.

Inductors

Inductors, also known as *coils,* are used to store energy in an electromagnetic field; they serve many useful functions in an electrical circuit. Figure 1–3 shows several typical inductors.

FIGURE 1–3
Typical inductors. Parts (a) and (b) courtesy of Dale Electronics, Inc. Part (c) courtesy of Delevan.

Transformers

Transformers are sometimes used to couple ac voltages from one point in a circuit to another, or to increase or decrease the ac voltage. Several types of transformers are shown in Figure 1–4.

Semiconductor Devices

Several varieties of diodes, transistors, and integrated circuits are shown in Figure 1–5.

FIGURE 1–4
Typical transformers (courtesy of Dale Electronics, Inc.).

(a) Diodes and transistors

(b) Integrated circuits

FIGURE 1–5
A grouping of typical semiconductor devices (courtesy of Motorola, Inc. Used by permission.).

(a)   (b)   (c)   (d)

FIGURE 1–6
*Typical instruments (a) DC power supply (courtesy of B & K Precision). (b) Analog multimeter
(courtesy of B & K Precision). (c) Digital multimeter (courtesy of B & K Precision).
(d) Oscilloscope (courtesy of Tektronix, Inc.).*

Electronic Instruments

Figure 1–6 shows a variety of electronic instruments that are discussed throughout the text. Typical instruments include the power supply, for providing voltage and current; the voltmeter, for measuring voltage; the ammeter, for measuring current; the ohmmeter, for measuring resistance; and the oscilloscope for observing and measuring ac voltages.

SECTION 1–1 REVIEW	**1.** Name three types of common electrical components. **2.** What instrument is used for measuring electrical current? **3.** What instrument is used for measuring voltage? **4.** What instrument is used for measuring resistance?

1–2 ■ ELECTRICAL AND MAGNETIC UNITS

In electronics work, you must deal with measurable quantities. For example, you must be able to express how many volts are measured at a certain test point in a circuit, how much current is flowing through a wire, or how much power a certain amplifier produces. In this section you will learn the units and symbols for many of the electrical and magnetic quantities that are used throughout the book. Definitions of the quantities are presented as they are needed in later chapters.

After completing this section, you should be able to

■ **State the electrical and magnetic quantities and their units**
   ☐ Specify the symbol for each quantity
   ☐ Specify the symbol for each unit

Symbols are used in electronics to represent both quantities and their units. One symbol is used to represent the name of the quantity, and another is used to represent the unit of measurement of that quantity. For example, P stands for *power*, and W stands for *watt*, which is the unit of power. Table 1–1 lists the most important electrical quantities, along

TABLE 1–1
Electrical quantities and units with SI symbols

QUANTITY	SYMBOL	UNIT	SYMBOL
capacitance	C	farad	F
charge	Q	coulomb	C
conductance	G	siemens	S
current	I	ampere	A
energy	W	joule	J
frequency	f	hertz	Hz
impedance	Z	ohm	Ω
inductance	L	henry	H
power	P	watt	W
reactance	X	ohm	Ω
resistance	R	ohm	Ω
time	t	second	s
voltage	V	volt	V

TABLE 1–2
Magnetic quantities and units with SI symbols

QUANTITY	SYMBOL	UNIT	SYMBOL
flux density	B	tesla	T
magnetic flux	ϕ	weber	Wb
magnetizing force	H	ampere-turns/meter	At/m
magnetomotive force	F_m	ampere-turn	At
permeability	μ	webers/ampere-turns-meter	Wb/Atm
reluctance	$\mathcal{R}$	ampere-turns/weber	At/Wb

with their SI units and symbols. Table 1–2 lists magnetic quantities with their SI units and symbols. The term *SI* is the French abbreviation for *International System (Système International* in French).

SECTION 1–2 REVIEW	1. What does *SI* stand for? 2. Without referring to Table 1–1, list as many electrical quantities as possible, including their symbols, units, and unit symbols. 3. Without referring to Table 1–2, list as many magnetic quantities as possible, including their symbols, units, and unit symbols.

1–3 ▪ SCIENTIFIC NOTATION

As an electronics technician, you will encounter both very small and very large quantities. For example, it is common to have electrical current values of only a few thousandths or even a few millionths of an ampere. On the other hand, you will find resistance values of several thousand or several million ohms. This range of values is typical of many other electrical quantities also.

After completing this section, you should be able to

▪ **Use scientific notation (powers of ten)**
  ☐ Express any number using a power of ten
  ☐ Do calculations with powers of ten

Powers of Ten

Scientific notation uses powers of ten, a method that makes it much easier to express large and small numbers and to do calculations involving such numbers.

Table 1–3 lists some powers of ten, both positive and negative. The power of ten is expressed as an exponent of the base 10 in each case. The exponent indicates the number of decimal places to the right or left of the decimal point in the expanded number. If the power is positive, the decimal point is moved to the right. For example,

$$10^4 = 1 \times 10^4 = 1.0000. = 10,000.$$

If the power is negative, the decimal point is moved to the left. For example,

$$10^{-4} = 1 \times 10^{-4} = .0001. = 0.0001$$

TABLE 1–3
Some positive and negative powers of ten

$10^6 = 1{,}000{,}000$	$10^{-6} = 0.000001$
$10^5 = 100{,}000$	$10^{-5} = 0.00001$
$10^4 = 10{,}000$	$10^{-4} = 0.0001$
$10^3 = 1{,}000$	$10^{-3} = 0.001$
$10^2 = 100$	$10^{-2} = 0.01$
$10^1 = 10$	$10^{-1} = 0.1$
$10^0 = 1$	

EXAMPLE 1–1

Express each number as a number between 1 and 10 times a positive power of ten.
(a) 200   **(b)** 5000   **(c)** 85,000   **(d)** 3,000,000

Solution   In each case, the decimal point is moved an appropriate number of places to the left.
(a) $200 = 2 \times 10^2$ **(b)** $5000 = 5 \times 10^3$
(c) $85{,}000 = 8.5 \times 10^4$ **(d)** $3{,}000{,}000 = 3 \times 10^6$

Related Exercise   Express 750,000,000 as a number between 1 and 10 times a positive power of ten.

EXAMPLE 1–2

Express each number as a number between 1 and 10 times a negative power of ten.
(a) 0.2   **(b)** 0.005   **(c)** 0.00063   **(d)** 0.000015

Solution   In each case, the decimal point is moved an appropriate number of places to the right.
(a) $0.2 = 2 \times 10^{-1}$ **(b)** $0.005 = 5 \times 10^{-3}$
(c) $0.00063 = 6.3 \times 10^{-4}$ **(d)** $0.000015 = 1.5 \times 10^{-5}$

Related Exercise   Express 0.00000093 as a number between 1 and 10 times a negative power of ten.

EXAMPLE 1–3

Express each of the following powers of ten numbers as a regular decimal number:
(a) 1×10^5 **(b)** 2×10^3 **(c)** 3.2×10^{-2} **(d)** 250×10^{-6}

Solution   The decimal point is moved to the right or left a number of places indicated by the positive or the negative power of ten respectively.
(a) $1 \times 10^5 = 100{,}000$ **(b)** $2 \times 10^3 = 2000$
(c) $3.2 \times 10^{-2} = 0.032$ **(d)** $250 \times 10^{-6} = 0.000250$

Related Exercise   Express 8.2×10^8 as a regular decimal number.

Calculating with Powers of Ten

The advantage of scientific notation is in addition, subtraction, multiplication, and division of very small or very large numbers.

Rules for Addition   The rules for adding numbers in powers of ten are as follows:

1. Express the numbers to be added in the same power of ten.
2. Add the numbers without their powers of ten to get the sum.
3. Bring down the common power of ten, which is the power of ten of the sum.

EXAMPLE 1–4   Add 2×10^6 and 5×10^7.

Solution
1. Express both numbers in the same power of ten: $(2 \times 10^6) + (50 \times 10^6)$
2. Add $2 + 50 = 52$.
3. Bring down the common power of ten (10^6) and the sum is 52×10^6.

Related Exercise   Add 41×10^3 and 79×10^2.

Rules for Subtraction   The rules for subtracting numbers in powers of ten are as follows:

1. Express the numbers to be subtracted in the same power of ten.
2. Subtract the numbers without their powers of ten to get the difference.
3. Bring down the common power of ten, which is the power of ten of the difference.

EXAMPLE 1–5   Subtract 25×10^{-12} from 75×10^{-11}.

Solution
1. Express each number in the same power of ten: $(75 \times 10^{-11}) - (2.5 \times 10^{-11})$
2. Subtract $75 - 2.5 = 72.5$.
3. Bring down the common power of ten (10^{-11}) and the difference is 72.5×10^{-11}.

Related Exercise   Subtract 3.5×10^{-6} from 2.2×10^{-5}.

Rules for Multiplication   The rules for multiplying numbers in powers of ten are as follows:

1. Multiply the numbers directly without their powers of ten.
2. Add the powers of ten algebraically (the powers do not have to be the same).

EXAMPLE 1–6   Multiply 5×10^{12} by 3×10^{-6}.

Solution   Multiply the numbers, and algebraically add the powers:

$$(5 \times 10^{12})(3 \times 10^{-6}) = 15 \times 10^{12+(-6)} = 15 \times 10^6$$

Related Exercise   Multiply 12×10^3 by 4×10^2.

Rules for Division   The rules for dividing numbers in powers of ten are as follows:

1. Divide the numbers directly without their powers of ten.

2. Subtract the power of ten in the denominator from the power of ten in the numerator (the powers do not have to be the same).

EXAMPLE 1–7   Divide 50×10^8 by 25×10^3.

Solution   The division problem is written with a numerator and denominator:

$$\frac{50 \times 10^8}{25 \times 10^3}$$

Divide the numbers and subtract 3 from 8:

$$\frac{50 \times 10^8}{25 \times 10^3} = 2 \times 10^{8-3} = 2 \times 10^5$$

Related Exercise   Divide 16×10^{-6} by 8×10^{-10}.

SECTION 1–3 REVIEW

1. Scientific notation uses powers of ten (True or False).
2. Express 100 as a power of ten.
3. Do the following operations:
   (a) $(1 \times 10^5) + (2 \times 10^5)$ (b) $(3 \times 10^6)(2 \times 10^4)$
   (c) $(8 \times 10^3) \div (4 \times 10^2)$ (d) $(25 \times 10^{-6}) - (130 \times 10^{-7})$

1–4 ■ METRIC PREFIXES

In electrical and electronics applications, certain powers of ten are used more often than others. It is common practice to use metric prefixes to represent these quantities. You can think of the metric prefix as a shorthand way to express a large or small number.

After completing this section, you should be able to

■ **Use metric prefixes to express large and small quantities**
    ☐ List the metric prefixes
    ☐ Change a power of ten to a metric prefix
    ☐ Use a calculator for metric prefixed numbers
    ☐ Use engineering notation

Table 1–4 lists the metric prefix for each of the commonly used powers of ten.

TABLE 1–4
Commonly used metric prefixes and their symbols

METRIC PREFIX	METRIC SYMBOL	POWER OF TEN	VALUE
tera	T	10^{12}	one trillion
giga	G	10^9	one billion
mega	M	10^6	one million
kilo	k	10^3	one thousand
milli	m	10^{-3}	one-thousandth
micro	μ	10^{-6}	one-millionth
nano	n	10^{-9}	one-billionth
pico	p	10^{-12}	one-trillionth

Use of Metric Prefixes

Now some examples will illustrate the use of metric prefixes. The number 2000 can be expressed in scientific notation as 2×10^3. Suppose we wish to represent 2000 W with a metric prefix. Since $2000 = 2 \times 10^3$, the metric prefix *kilo* (k) is used for 10^3. So we can express 2000 W as 2 kW (2 kilowatts).

    As another example, 0.015 A can be expressed as 15×10^{-3} A. The metric prefix *milli* (m) is used for 10^{-3}. So 0.015 becomes 15 mA (15 milliamperes).

EXAMPLE 1–8

Express each quantity using a metric prefix.
(a) 50,000 V   **(b)** 25,000,000 Ω   **(c)** 0.000036 A

Solution
(a) 50,000 V $= 50 \times 10^3$ V $= 50$ kV
(b) 25,000,000 Ω $= 25 \times 10^6$ Ω $= 25$ MΩ
(c) 0.000036 A $= 36 \times 10^{-6}$ A $= 36\ \mu$A

Related Exercise   Express using metric prefixes:
(a) 56,000,000 Ω   **(b)** 0.000470 A

Entering Numbers with Metric Prefixes on the Calculator

To enter a number expressed in scientific notation on the calculator, use the EXP key (the EE key on some calculators). The following example shows how to enter numbers with metric prefixes on a typical scientific calculator.

EXAMPLE 1–9

(a) Enter 3.3 kΩ (3.3×10^3 Ω) on your calculator.
(b) Enter 450 μA (450×10^{-6} A) on your calculator.

Solution

(a) **Step 1:** Enter ⟨3⟩ ⟨·⟩ ⟨3⟩. The display shows 3.3.

Step 2: Press ⟨EXP⟩. The display shows 3.3 00.

Step 3: Enter ⟨3⟩. The display shows 3.3 03.

(b) **Step 1:** Enter ⟨4⟩ ⟨5⟩ ⟨0⟩. The display shows 450.

Step 2: Press ⟨EXP⟩. The display shows 450 00.

Step 3: Press ⟨+/−⟩. Enter ⟨6⟩. The display shows 450 – 06.

Related Exercise Enter 259 mA on your calculator.

Engineering Notation

Engineering notation is a term that refers to the application of scientific notation in which the powers of ten are limited to multiples of three, such as 10^3, 10^{-3}, 10^6, 10^{-6}, 10^9, 10^{-9}, 10^{12}, and 10^{-12}. The reason for this is that all units used in the fields of engineering and technology, such as ohms, amps, volts, and watts are expressed with prefixes of k (kilo, 10^3), m (milli, 10^{-3}), M (mega, 10^6), μ (micro, 10^{-6}), G (giga, 10^9), n (nano, 10^{-9}), T (tera, 10^{12}), p (pico, 10^{-12}).

SECTION 1–4 REVIEW

1. List the metric prefix for each of the following powers of ten: 10^6, 10^3, 10^{-3}, 10^{-6}, 10^{-9}, and 10^{-12}.
2. Use an appropriate metric prefix to express 0.000001 A.
3. Use an appropriate metric prefix to express 250,000 W.

1–5 ■ METRIC UNIT CONVERSIONS

It is often necessary or convenient to convert a quantity from one metric prefixed unit to another, such as from milliamperes (mA) to microamperes (μA). A metric prefix conversion is accomplished by moving the decimal point in the number an appropriate number of places to the left or to the right, depending on the particular conversion.

After completing this section, you should be able to

■ **Convert from one metric unit to another**
   □ Convert between milli, micro, nano, and pico
   □ Convert between kilo and mega

The following basic rules apply to metric unit conversions.

1. When converting from a larger unit to a smaller unit, move the decimal point to the right.

2. When converting from a smaller unit to a larger unit, move the decimal point to the left.

3. Determine the number of places that the decimal point is moved by finding the difference in the powers of ten of the units being converted.

For example, when converting from milliamperes (mA) to microamperes (μA), move the decimal point three places to the right because there is a three-place difference between the two units (mA is 10^{-3} A and μA is 10^{-6} A). The following examples illustrate a few conversions.

EXAMPLE 1–10   Convert 0.15 milliampere (0.15 mA) to microamperes (μA).

Solution   Move the decimal point three places to the right.
$$0.15 \text{ mA} = 0.15 \times 10^{-3} \text{ A} = 150 \times 10^{-6} \text{ A} = 150 \ \mu\text{A}$$

Related Exercise   Convert 1 mA to microamperes.

EXAMPLE 1–11   Convert 4500 microvolts (4500 μV) to millivolts (mV).

Solution   Move the decimal point three places to the left.
$$4500 \ \mu\text{V} = 4500 \times 10^{-6} \text{ V} = 4.5 \times 10^{-3} \text{ V} = 4.5 \text{ mV}$$

Related Exercise   Convert 1000 μV to millivolts.

EXAMPLE 1–12   Convert 5000 nanoamperes (5000 nA) to microamperes (μA).

Solution   Move the decimal point three places to the left.
$$5000 \text{ nA} = 5000 \times 10^{-9} \text{ A} = 5 \times 10^{-6} \text{ A} = 5 \ \mu\text{A}$$

Related Exercise   Convert 893 nA to microamperes.

EXAMPLE 1–13   Convert 47,000 picofarads (47,000 pF) to microfarads (μF).

Solution   Move the decimal point six places to the left.
$$47,000 \text{ pF} = 47,000 \times 10^{-12} \text{ F} = 0.047 \times 10^{-6} \text{ F} = 0.047 \ \mu\text{F}$$

Related Exercise   Convert 0.0022 μF to picofarads.

EXAMPLE 1–14   Convert 0.00022 microfarad (0.00022 μF) to picofarads (pF).

Solution   Move the decimal point six places to the right.

$$0.00022 \ \mu F = 0.00022 \times 10^{-6} \ F = 220 \times 10^{-12} \ F = 220 \ pF$$

Related Exercise   Convert 10,000 pF to microfarads.

EXAMPLE 1–15   Convert 1800 kilohms (1800 kΩ) to megohms (MΩ).

Solution   Move the decimal point three places to the left.

$$1800 \ k\Omega = 1800 \times 10^3 \ \Omega = 1.8 \times 10^6 \ \Omega = 1.8 \ M\Omega$$

Related Exercise   Convert 2.2 kΩ to megohms.

When adding (or subtracting) quantities with different metric prefixes, first convert one of the quantities to the same prefix as the other as the next example shows.

EXAMPLE 1–16   Add 15 mA and 8000 μA and express the result in milliamperes.

Solution   Convert 8000 μA to 8 mA and add.

$$15 \ mA + 8000 \ \mu A = 15 \times 10^{-3} \ A + 8000 \times 10^{-6} \ A$$
$$= 15 \times 10^{-3} \ A + 8 \times 10^{-3} \ A = 15 \ mA + 8 \ mA = 23 \ mA$$

Related Exercise   Add 2873 mA to 10,000 μA.

SECTION 1–5 REVIEW

1. Convert 0.01 MV to kilovolts (kV).
2. Convert 250,000 pA to milliamperes (mA).
3. Add 0.05 MW and 75 kW.

▪ **SUMMARY**

- Resistors limit electrical current.
- Capacitors store electrical charge.
- Inductors store energy electromagnetically.
- Inductors are also known as *coils*.
- Transformers magnetically couple ac voltages.
- Semiconductor devices include diodes, transistors, and integrated circuits.
- Power supplies provide current and voltage.
- Voltmeters measure voltage.
- Ammeters measure current.
- Ohmmeters measure resistance.

■ **SELF-TEST**

1. Which of the following is not an electrical quantity?
   (a) current (b) voltage (c) time (d) power

2. The unit of current is
   (a) volt (b) watt (c) ampere (d) joule

3. The unit of voltage is
   (a) ohm (b) watt (c) volt (d) farad

4. The unit of resistance is
   (a) ampere (b) henry (c) hertz (d) ohm

5. 15,000 W is the same as
   (a) 15 mW (b) 15 kW (c) 15 MW (d) 15 μW

6. The quantity 4.7×10^3 is the same as
   (a) 470 (b) 4700 (c) 47,000 (d) 0.0047

7. The quantity 56×10^{-3} is the same as
   (a) 0.056 (b) 0.560 (c) 560 (d) 56,000

8. The number 3,300,000 can be expressed as
   (a) 3300×10^3 (b) 3.3×10^{-6} (c) 3.3×10^6 (d) either (a) or (c)

9. Ten milliamperes can be expressed as
   (a) 10 MA (b) 10 μA (c) 10 kA (d) 10 mA

10. Five thousand volts can be expressed as
    (a) 5000 V (b) 5 MV (c) 5 kV (d) either (a) or (c)

11. Twenty million ohms can be expressed as
    (a) 20 mΩ (b) 20 MW (c) 20 MΩ (d) 20 $\mu\Omega$

12. Hertz is the unit of
    (a) power (b) inductance (c) frequency (d) time

■ **PROBLEMS**

SECTION 1–3 Scientific Notation

1. Express each of the following numbers as a power of ten:
   (a) 3000 (b) 75,000 (c) 2,000,000

2. Express each number as a power of ten.
   (a) 1/500 (b) 1/2000 (c) 1/5,000,000

3. Express each of the following numbers in three ways, using 10^3, 10^4, and 10^5:
   (a) 8400 (b) 99,000 (c) 0.2×10^6

4. Express each of the following numbers in three ways, using 10^{-3}, 10^{-4}, and 10^{-5}:
   (a) 0.0002 (b) 0.6 (c) 7.8×10^{-2}

5. Express each power of ten in regular decimal form.
   (a) 2.5×10^{-6} (b) 50×10^2 (c) 3.9×10^{-1}

6. Express each power of ten in regular decimal form.
   (a) 45×10^{-6} (b) 8×10^{-9} (c) 40×10^{-12}

7. Add the following numbers:
   (a) $(92 \times 10^6) + (3.4 \times 10^7)$ (b) $(5 \times 10^3) + (85 \times 10^{-2})$ (c) $(560 \times 10^{-8}) + (460 \times 10^{-9})$

8. Perform the following subtractions:
   (a) $(3.2 \times 10^{12}) - (1.1 \times 10^{12})$ (b) $(26 \times 10^8) - (1.3 \times 10^9)$ (c) $(150 \times 10^{-12}) - (8 \times 10^{-11})$

9. Perform the following multiplications:
   (a) $(5 \times 10^3)(4 \times 10^5)$ (b) $(12 \times 10^{12})(3 \times 10^2)$ (c) $(2.2 \times 10^{-9})(7 \times 10^{-6})$

10. Divide the following:
    (a) $(10 \times 10^3) \div (2.5 \times 10^2)$ (b) $(250 \times 10^{-6}) \div (50 \times 10^{-8})$ (c) $(4.2 \times 10^8) \div (2 \times 10^{-5})$

11. Convert each of the following numbers to another number having a multiplier of 10^{-6}:
    (a) 2.37×10^{-3} (b) 0.001856×10^{-2} (c) 5743.89×10^{-12} (d) 100×10^3

12. Perform the following indicated operations:
    (a) $(2.8 \times 10^3)(3 \times 10^2)/2 \times 10^2$ (b) $(46)(10^{-3})(10^5)/10^6$
    (c) $(7.35)(0.5 \times 10^{12})/[(2)(10 \times 10^{10})]$ (d) $(30)^2(5)^3/10^{-2}$

13. Perform each operation:
(a) $49 \times 10^6 / 4 \times 10^{-8}$ (b) $(3 \times 10^3)^2 / 1.8 \times 10^3$
(c) $(1.5 \times 10^{-6})^2 (4.7 \times 10^6)$ (d) $1/[(5 \times 10^{-3})(0.01 \times 10^{-6})]$

SECTION 1–4 Metric Prefixes

14. Express each of the following as a quantity having a metric prefix:
(a) 31×10^{-3} A (b) 5.5×10^3 V (c) 20×10^{-12} F

15. Express the following using metric prefixes:
(a) 3×10^{-6} F (b) 3.3×10^6 Ω (c) 350×10^{-9} A

16. Express each quantity as a power of ten:
(a) $5\ \mu$A (b) 43 mV (c) 275 kΩ (d) 10 MW

SECTION 1–5 Metric Unit Conversions

17. Perform the indicated conversions:
(a) 5 mA to microamperes (b) 3200 μW to milliwatts
(c) 5000 kV to megavolts (d) 10 MW to kilowatts

18. Determine the following:
(a) The number of microamperes in 1 milliampere
(b) The number of millivolts in 0.05 kilovolt
(c) The number of megohms in 0.02 kilohm
(d) The number of kilowatts in 155 milliwatts

■ **ANSWERS TO SECTION REVIEWS**

Section 1–1

1. Resistors, capacitors, inductors, and transformers

2. Ammeter

3. Voltmeter

4. Ohmmeter

Section 1–2

1. SI is the abbreviation for Système International.

2. Refer to Table 1–1 after you have compiled your list.

3. Refer to Table 1–2 after you have compiled your list.

Section 1–3

1. True **2.** 10^2

3. (a) 3×10^5 (b) 6×10^{10} (c) 2×10^1 (d) 12×10^{-6}

Section 1–4

1. Mega (M), kilo (k), milli (m), micro (μ), nano (n), and pico (p)

2. 1 μA (one microampere) **3.** 250 kW (250 kilowatts)

Section 1–5

1. 0.01 MV = 10 kV

2. 250,000 pA = 0.00025 mA

3. 125 kW

■ **ANSWERS TO RELATED EXERCISES FOR EXAMPLES**

1–1	7.5×10^8	1–2	9.3×10^{-7}
1–3	820,000,000	1–4	48.9×10^3
1–5	1.85×10^{-5}	1–6	4.8×10^6
1–7	2×10^4	1–8	(a) 56 MΩ (b) 470 μA
1–9	[2] [5] [9] [EXP] [+/−] [3]	1–10	1000 μA
1–11	1 mV	1–12	0.893 μA
1–13	2200 pF	1–14	0.01 μF
1–15	0.0022 MΩ	1–16	2883 mA

2

VOLTAGE, CURRENT, AND RESISTANCE

The three basic electrical quantities presented in this chapter are voltage, current, and resistance. No matter what type of electrical or electronic equipment you may work with, these quantities will always be of primary importance.

To help you understand voltage, current, and resistance, the basic structure of the atom is discussed and the concept of charge is introduced. The basic electric circuit is studied, along with techniques for measuring voltage, current, and resistance.

■ CHAPTER OBJECTIVES

☐ Describe the basic structure of an atom
☐ Explain the concept of electrical charge
☐ Define *voltage* and discuss its characteristics
☐ Define *current* and discuss its characteristics
☐ Define *resistance* and discuss its characteristics
☐ Describe a basic electrical circuit
☐ Make basic circuit measurements

■ APPLICATION ASSIGNMENT

As technician for special effects, you are asked to hook up and check out a circuit for use in the special lighting effects for an outdoor play. The requirements are

1. There will be six lamps but only one lamp will be turned on at a time.
2. The sequence in which the lamps are turned on and off will vary.
3. It will be necessary to vary the brightness of each lamp.
4. The lamps must operate from a 12 V battery for portability, and the circuit must be protected by a fuse.

   After studying this chapter, you should be able to complete the application assignment in Section 2–8.

2–1 ■ ATOMS

The atomic structure of the materials used in electronics determines, among other things, the ease with which current is established.

After completing this section, you should be able to

■ **Describe the basic structure of an atom**
    ☐ Define *electron*
    ☐ Define *proton* and *neutron*
    ☐ Define *atomic weight* and *atomic number*
    ☐ Define *free electron*
    ☐ Discuss shells

An **atom** is the smallest particle of an **element** that still retains the characteristics of that element. Different elements have different types of atoms. In fact, every element has a unique atomic structure.

According to the Bohr theory, atoms have a planetary type of structure, consisting of a central nucleus surrounded by orbiting electrons. The nucleus consists of positively charged particles called **protons** and uncharged particles called **neutrons.** The **electrons** are the basic particles of negative charge.

Each type of atom has a certain number of electrons and protons that distinguishes the atom from all other atoms of other elements. For example, the simplest atom is that of hydrogen. It has one proton and one electron, as pictured in Figure 2–1(a). The helium atom, shown in Figure 2–1(b), has two protons and two neutrons in the nucleus, which is orbited by two electrons.

One electron orbiting the nucleus

2 electrons

Nucleus with one proton

Nucleus with 2 protons and 2 neutrons

(a) Hydrogen

(b) Helium

FIGURE 2–1
Hydrogen and helium atoms.

Atomic Number and Weight

All elements are arranged in the periodic table of the elements in order according to their **atomic number,** which is the number of protons in the nucleus. The elements can also be arranged by their **atomic weight,** which is approximately the number of protons and neutrons in the nucleus. For example, hydrogen has an atomic number of one and an atomic weight of one. The atomic number of helium is two, and its atomic weight is four.

In their normal (or neutral) state, all atoms of a given element have the same number of electrons as protons. So the positive charges cancel the negative charges, and the atom has a net charge of zero.

The Copper Atom

Since copper is the most commonly used metal in **electrical** applications, let's examine its atomic structure. The copper atom has 29 electrons in orbit around the nucleus. They do not all occupy the same orbit, however. They move in orbits at varying distances from the nucleus.

The orbits in which the electrons revolve are called **shells.** The number of electrons in each shell follows a predictable pattern according to the formula, $2N^2$, where N is the number of the shell. The first shell of any atom can have up to 2 electrons, the second shell up to 8 electrons, the third shell up to 18 electrons, and the fourth shell up to 32 electrons.

A copper atom is shown in Figure 2–2. Notice that the outermost shell, called the **valence** shell, has only one electron, called the **valence electron.**

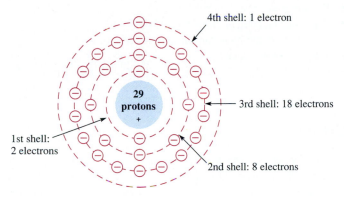

FIGURE 2–2
The copper atom.

Free Electrons

When the valence electron in the outer shell of the copper atom gains sufficient **energy** from the surrounding medium, it can break away from the parent atom and become what is called a **free electron.** Materials with fewer valence electrons in their atoms have more free electrons because the valence electrons are more loosely bound to the atoms. The free electrons in the copper material are capable of moving from one atom to another in the material. In other words, they drift randomly from atom to atom within the copper. As you will see, the free electrons make electrical current possible.

Three categories of materials are used in electronics: conductors, semiconductors, and insulators.

Conductors

Conductors are materials that allow current to flow easily. They have a large number of free electrons and are characterized by one to three valence electrons in their atomic structures. Most metals are good conductors. Silver is the best conductor, and copper is next. Copper is the most widely used conductive material because it is less expensive than silver. Copper wire is commonly used as a conductor in electric circuits.

Semiconductors

Semiconductors are classed below the conductors in their ability to carry current because they have fewer free electrons than do conductors. Semiconductors are characterized by four valence electrons in their atomic structures. However, because of their unique characteristics, certain semiconductor materials are the basis for modern **electronic** devices such as the diode, transistor, and integrated circuit. Silicon and germanium are common semiconductor materials.

Insulators

Insulating materials are poor conductors of electric current. In fact, **insulators** are used to prevent current where it is not wanted. Compared to conductive materials, insulators have very few free electrons and are characterized by more than four valence electrons in their atomic structures.

2–2 ■ ELECTRICAL CHARGE

As you learned in the last section, the two types of charge are the positive charge and the negative charge. The electron is the smallest particle that exhibits negative electrical charge. When an excess of electrons exists in a material, there is a net negative electrical charge. When a deficiency of electrons exists, there is a net positive electrical charge.

After completing this section, you should be able to

■ **Explain the concept of electrical charge**
  □ Name the unit of charge
  □ Name the types of charge
  □ Describe the forces between charges
  □ Determine the amount of charge on a given number of electrons

The **charge** of an electron and that of a proton are equal in magnitude. Electrical charge is symbolized by Q. Static electricity is the presence of a net positive or negative charge in a material. Everyone has experienced the effects of static electricity from time to time, for example, when attempting to touch a metal surface or another person or when the clothes in a dryer cling together.

Materials with charges of opposite polarity are attracted to each other; materials with charges of the same polarity are repelled, as indicated symbolically in Figure 2–3. A force acts between charges, as evidenced by the attraction or repulsion. This force, called an *electric field,* consists of invisible lines of force as shown in Figure 2–4.

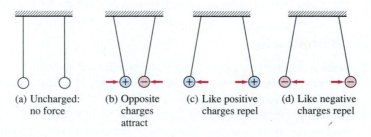

(a) Uncharged: no force (b) Opposite charges attract (c) Like positive charges repel (d) Like negative charges repel

FIGURE 2–3
Attraction and repulsion of electrical charges.

FIGURE 2–4
Electric field between oppositely charged surfaces.

Coulomb: The Unit of Charge

Electrical charge is measured in **coulombs,** abbreviated C.

> **One coulomb is the total charge possessed by 6.25×10^{18} electrons.**

A single electron has a charge of 1.6×10^{-19} C. This unit is named for Charles Coulomb (1736–1806), a French scientist. The total charge, Q, in a given number of electrons is found by the following formula:

$$Q = \frac{\text{number of electrons}}{6.25 \times 10^{18} \text{ electrons/Coulomb}} \qquad (2\text{–}1)$$

Positive and Negative Charge

Consider a neutral atom—that is, one that has the same number of electrons and protons and thus has no net charge. If a valence electron is pulled away from the atom by the application of energy, the atom is left with a net positive charge (more protons than electrons) and becomes a positive ion. If an atom acquires an extra electron in its outer shell, it has a net negative charge and becomes a negative ion.

The amount of energy required to free a valence electron is related to the number of electrons in the outer shell. An atom can have up to eight valence electrons. The more complete the outer shell, the more stable the atom and thus the more energy is required to release an electron. Figure 2–5 illustrates the creation of a positive and a negative ion

Sodium atom
(11 protons, 11 electrons)

Chlorine atom
(17 protons, 17 electrons)

(a) The sodium atom has a single valence electron.

(b) The atoms combine by sharing the valence electron to form sodium chloride (table salt).

Positive sodium ion
(11 protons, 10 electrons)

Negative chlorine ion
(17 protons, 18 electrons)

(c) When dissolved, the sodium atom gives up the valence electron to become a positive ion, and the chlorine atom retains the extra valence electron to become a negative ion.

FIGURE 2–5
Example of the formation of positive and negative ions.

when sodium chloride dissolves and the sodium atom gives up its single valence electron to the chlorine atom.

EXAMPLE 2–1   How many coulombs of charge do 93.8×10^{16} electrons represent?

Solution

$$Q = \frac{\text{number of electrons}}{6.25 \times 10^{18} \text{ electrons/Coulomb}} = \frac{93.8 \times 10^{16} \text{ electrons}}{6.25 \times 10^{18} \text{ electrons/C}}$$

$$= 15 \times 10^{-2} \text{ C} = 0.15 \text{ C}$$

Related Exercise   How many electrons does it take to have 3 C of charge?

SECTION 2–2 REVIEW

1. What is the symbol used for charge?
2. What is the unit of charge, and what is the unit symbol?
3. What are the two types of charge?
4. How much charge, in coulombs, is there in 10×10^{12} electrons?

2–3 ▪ VOLTAGE

As you have seen, a force of attraction exists between a positive and a negative charge. A certain amount of energy in the form of work is required to overcome the force and move the charges a given distance apart. All opposite charges possess a certain potential energy because of the separation between them. The difference in potential energy of the charges is the potential difference or voltage. Voltage is the driving force in electric circuits and is what produces current.

After completing this section, you should be able to

▪ **Define** *voltage* **and discuss its characteristics**
  □ State the formula for voltage
  □ Name and define the unit of voltage
  □ Describe the basic sources of voltage

Consider a water tank that is supported several feet above the ground. A given amount of energy (work) is required to pump water up to fill the tank. Once the water is stored in the tank, it has a certain potential energy which, if released, can be used to perform work. For example, the water can be allowed to fall down a chute to turn a water wheel.

Potential difference in electrical terms is more commonly called **voltage** (*V*) and is expressed as energy (*W*) divided by charge (*Q*).

$$V = \frac{W}{Q} \qquad (2\text{–}2)$$

where *W* is expressed in **joules** (J) and *Q* is in coulombs (C).

Volt: The Unit of Voltage

The unit of voltage is the **volt,** symbolized by V.

> **One volt is the potential difference (voltage) between two points when one joule of energy is used to move one coulomb of charge from one point to the other.**

EXAMPLE 2–2

If 50 J of energy are available for every 10 C of charge, what is the voltage?

Solution

$$V = \frac{W}{Q} = \frac{50 \text{ J}}{10 \text{ C}} = 5 \text{ V}$$

Related Exercise   How much energy is used to move 50 C from one point in a circuit to another when the voltage between the two points is 12 V?

A Voltage Analogy

You can think of voltage as the pressure difference created by a pump that causes the water to flow through the pipe in a water system.

Sources of Voltage

The Battery   A voltage source is a **source** of potential energy that is also called *electromotive force* (emf). The **battery** is one type of voltage source that converts chemical energy into electrical energy. A voltage exists between the electrodes (terminals) of a battery, as shown by a voltaic cell in Figure 2–6. One electrode is positive and the other negative as a result of the separation of charges caused by the chemical action when two different conducting materials are dissolved in the electrolyte.

FIGURE 2–6

A voltaic cell converts chemical energy into electrical energy.

Battery cells are generally classified as *primary cells,* which cannot be recharged, and *secondary cells,* which can be recharged by reversal of the chemical action. Examples of primary cells are carbon-zinc cells and alkaline cells. Examples of secondary cells are lead-acid cells and nickel-cadmium (Ni-Cd) cells. The amount of voltage provided by a cell varies depending on the materials used in the construction of the cell. For example, a carbon-zinc cell (flashlight battery) produces 1.5 V, whereas a lead-acid cell used in automobile batteries produces 2 V. Individual cells are combined to produce higher voltages. A 9 V battery is made up of six 1.5 V cells and a 12 V car battery consists of six 2 V cells. For additional coverage of batteries, see Appendix B.

The Electronic Power Supply   These voltage sources convert the ac voltage from the wall outlet to a constant (dc) voltage which is available across two output terminals, as indicated in Figure 2–7(a). Typical commercial **power supplies** are shown in part (b).

(a)

(b)

FIGURE 2–7
Electronic power supplies. Part (b) courtesy of B & K Precision.

The Solar Cell The operation of solar cells is based on the principle of *photovoltaic action,* which is the process whereby light energy is converted directly into electrical energy. A basic solar cell consists of two layers of different semiconductive materials joined together to form a junction. When one layer is exposed to light, many electrons acquire enough energy to break away from their parent atoms and cross the junction. This process forms negative ions on one side of the junction and positive ions on the other, and thus a potential difference (voltage) is developed. Basic construction of a solar cell is shown in Figure 2–8.

FIGURE 2–8
Basic construction of solar cell.

The Generator **Generators** convert mechanical energy into electrical energy using a principle called *electromagnetic induction* (see Chapter 7). A conductor is rotated through a magnetic field, and a voltage is produced across the conductor. A typical generator is pictured in Figure 2–9.

FIGURE 2–9
Cutaway view of dc generator.

2–4 ▪ CURRENT

Voltage makes electrons move through a circuit, and this movement of electrons is the current. Current does the work in an electric circuit.

After completing this section, you should be able to

- **Define *current* and discuss its characteristics**
  - ☐ Explain the movement of electrons
  - ☐ State the formula for current
  - ☐ Name and define the unit of current

As you have learned, there are free electrons available in all conductive and semiconductive materials. These electrons drift randomly in all directions, from atom to atom, within the structure of the material, as indicated in Figure 2–10.

FIGURE 2–10
Random motion of free electrons in a material.

Now, if a voltage is placed across the conductive or semiconductive material, one end becomes positive and the other negative, as indicated in Figure 2–11. The repulsive force between the negative voltage at the left end causes the free electrons (negative charges) to

FIGURE 2–11
Electrons flow from negative to positive when a voltage is applied across a conductive or semiconductive material.

move toward the right. The attractive force between the positive voltage at the right end pulls the free electrons to the right. The result is a net movement of the free electrons from the negative end of the material to the positive end, as shown in Figure 2–11.

The movement of the free electrons from the negative end of the material to the positive end is the electrical **current,** symbolized by *I*.

Electrical current is the rate of flow of electrons.

Current is measured by the number of electrons (amount of charge, *Q*) that flow past a point in a unit of time. Current *(I)* equals charge *(Q)* divided by time *(t)*.

$$I = \frac{Q}{t} \qquad (2\text{–}3)$$

where *I* is current in amperes, *Q* is the charge of the electrons in coulombs, and *t* is the time in seconds.

Ampere: The Unit of Current

Current is measured in a unit called the **ampere** or *amp* for short, symbolized by A. It is named after André Ampère (1775–1836), a French physicist whose work contributed to the understanding of electrical current and its effects.

One ampere (1 A) is the amount of current when a number of electrons having one coulomb (1 C) of charge move past a given point *(P)* in one second (1 s).

One coulomb is the charge carried by 6.25×10^{18} electrons. See Figure 2–12.

FIGURE 2–12
Illustration of 1 A of current in a material (1 C/s).

EXAMPLE 2-3   Ten coulombs of charge flow past a given point in a wire in 2 s. How many amperes of current are flowing?

Solution
$$I = \frac{Q}{t} = \frac{10\text{ C}}{2\text{ s}} = 5\text{ A}$$

Related Exercise   If 8 A of current flow through the filament of a light bulb in 1.5 s, how many coulombs have moved through the filament?

A Current Analogy

You can think of current as the water flowing through the pipe in a water system when pressure (voltage) is applied by the pump (voltage source).

SECTION 2-4 REVIEW
1. Define *current* and state its unit.
2. How many electrons make up one coulomb of charge?
3. What is the current in amperes when 20 C flow past a point in a wire in 4 s?

2-5 ■ RESISTANCE

*When there is current in a material, the free electrons move through the material and occasionally collide with atoms. These collisions cause the electrons to lose some of their energy, and thus their movement is restricted. The more collisions, the more the flow of electrons is restricted. This restriction varies and is determined by the type of material. The property of a material that restricts the flow of electrons is called **resistance**, designated with an R.*

After completing this section, you should be able to

■ **Define *resistance* and discuss its characteristics**
   □ Name and define the unit of resistance
   □ Describe the basic types of resistors
   □ Determine resistance value by color code or labeling

Resistance is the opposition to current.

The schematic symbol for resistance is shown in Figure 2-13.
   When current flows through any material that has resistance, heat is produced by the collisions of electrons and atoms. Therefore, wire, which typically has a very small resistance, becomes warm when there is current through it.

FIGURE 2-13
Resistance/resistor symbol.   —⩗⩘⩗⩘— R

Ohm: The Unit of Resistance

Resistance, R, is expressed in the unit of **ohms,** named after Georg Simon Ohm and is symbolized by the Greek letter omega (Ω).

> **One ohm (1 Ω) of resistance exists when one ampere (1 A) of current flows in a material when one volt (1 V) is applied across the material.**

Resistors

Components that are specifically designed to have a certain amount of resistance are called **resistors.** The principal applications of resistors are to limit current, divide voltage, and, in certain cases, generate heat. Although there are a variety of different types of resistors that come in many shapes and sizes, they can all be placed in one of two main categories: fixed or variable.

Fixed Resistors   The fixed resistor is available with a large selection of resistance values that are set during manufacturing and cannot be changed easily. Fixed resistors are constructed using various methods and materials. Several common types are shown in Figure 2–14.

One common fixed resistor is the carbon-composition type, which is made with a mixture of finely ground carbon, insulating filler, and a resin binder. The ratio of carbon to insulating filler sets the resistance value. The mixture is formed into rods, and lead connections are made. The entire resistor is then encapsulated in an insulated coating for protection. The chip resistor is another type of fixed resistor and is in the category of SMT (surface mount technology) components. It has the advantage of a very small size

(a)

(b)

(c)

FIGURE 2–14

Typical fixed resistors. Parts (a) and (b) courtesy of Stackpole Carbon Co. (c) Resistor networks (courtesy of Bourns, Inc.).

(a) Cutaway view of a carbon-composition resistor

(b) Cutaway view of a chip resistor

FIGURE 2–15
Two types of fixed resistors.

for compact assemblies. Figure 2–15(a) shows the construction of a typical carbon-composition resistor and Figure 2–15(b) shows the construction of a chip resistor.

Other types of fixed resistors include carbon film, metal film, and wirewound. In film resistors, a resistive material is deposited evenly onto a high-grade ceramic rod. The resistive film may be carbon (carbon film) or nickel chromium (metal film). In these types of resistors, the desired resistance value is obtained by removing part of the resistive material in a helical pattern along the rod using a spiraling technique as shown in Figure 2–16(a). Very close **tolerance** can be achieved with this method. Film resistors are also available in the form of resistor networks as shown in Figure 2–16(b).

(a) Film resistor showing spiraling technique

(b) Resistor network

FIGURE 2–16
Construction views of typical film resistors.

Wirewound resistors are constructed with resistive wire wound around an insulating rod and then sealed. Normally, wirewound resistors are used because of their relatively high power ratings. Some typical fixed resistors are shown in Figure 2–17.

Resistor Color Codes   Many types of fixed resistors with value tolerances of 5%, 10%, or 20% are color coded with four bands to indicate the resistance value and the tolerance. This color-code band system is shown in Figure 2–18, and the color code is listed in Table 2–1.

The color code is read as follows:

1. Beginning at the banded end, the first band is the first digit of the resistance value. If it is not clear which is the banded end, start from the end that does not begin with a gold or silver band.

2. The second band is the second digit.

FIGURE 2–17
Typical wirewound power resistors.

FIGURE 2–18
Color-code bands on a resistor.

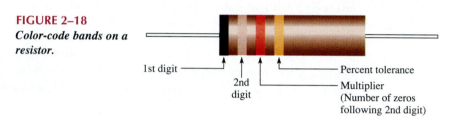

1st digit ——————
2nd digit ——
——— Percent tolerance
——— Multiplier
(Number of zeros following 2nd digit)

TABLE 2–1
Resistor color code

	DIGIT	COLOR
Resistance value, first three bands: First band—1st digit Second band—2nd digit Third band—number of zeros following the 2nd digit	0	Black
	1	Brown
	2	Red
	3	Orange
	4	Yellow
	5	Green
	6	Blue
	7	Violet
	8	Gray
	9	White
Tolerance, fourth band	5%	Gold
	10%	Silver
	20%	No band

3. The third band is the number of zeros following the second digit, or the *multiplier*.

4. The fourth band indicates the tolerance and is usually gold or silver.

For example, a 5% tolerance means that the *actual* resistance value is within ±5% of the color-coded value. Thus, a 100 Ω resistor with a tolerance of ±5% can have acceptable values as low as 95 Ω and as high as 105 Ω.

For resistance values less than 10 Ω, the third band is either gold or silver. Gold represents a multiplier of 0.1, and silver represents 0.01. For example, a color code of red, violet, gold, and silver represents 2.7 Ω with a tolerance of ±10%.

EXAMPLE 2–4   Find the resistance values in ohms and the percent tolerance for each of the color-coded resistors shown in Figure 2–19.

FIGURE 2–19

Solution

(a) First band is red = 2, second band is violet = 7, third band is orange = 3 zeros, fourth bank is silver = 10% tolerance.

$$R = 27,000 \ \Omega \pm 10\%$$

(b) First band is brown = 1, second band is black = 0, third band is brown = 1 zero, fourth band is silver = 10% tolerance.

$$R = 100 \ \Omega \pm 10\%$$

(c) First band is green = 5, second band is blue = 6, third band is green = 5 zeros, fourth band is gold = 5% tolerance.

$$R = 5,600,000 \ \Omega \pm 5\%$$

Related Exercise   A certain resistor has a yellow first band, a violet second band, a red third band, and no fourth band. Determine the value in ohms.

Certain precision resistors with tolerances of 1% or 2% are color coded with five bands. Beginning at the banded end, the first band is the first digit of the resistance value, the second band is the second digit, the third band is the third digit, the fourth band is the multiplier, and the fifth band indicates the tolerance. Table 2–1 applies, except that gold in the fourth band indicates a multiplier of 0.1 and silver indicates a multiplier of 0.01. Brown in the fifth band indicates 1% tolerance and red indicates 2% tolerance.

Numerical labels are also commonly used on certain types of resistors where the resistance value and tolerance are stamped on the body of the resistor. For example, a common system uses R to designate the decimal point and letters to indicate tolerance as follows:

$$F = \pm 1\%, \quad G = \pm 2\%, \quad J = \pm 5\%, \quad K = \pm 10\%, \quad M = \pm 20\%$$

For values above 100 Ω, three digits are used to indicate resistance value, followed by a fourth digit that specifies the number of zeros. For values less than 100 Ω, R indicates the decimal point.

Some examples are as follows: 6R8M is a 6.8 Ω ±20% resistor; 3301F is a 3300 Ω ±1% resistor; and 2202J is a 22,000 Ω ±5% resistor.

Resistor Reliability Band A fifth band on some color-coded resistors indicates the resistor's reliability in percent of failures per 1000 hours (1000 h) of use. The fifth-band reliability color code is listed in Table 2–2. For example, a brown fifth band means that if a group of like resistors are operated under standard conditions for 1000 h, 1% of the resistors in that group will fail.

TABLE 2–2
Fifth-band reliability color code

COLOR	FAILURES DURING 1000 H OF OPERATION
Brown	1.0%
Red	0.1%
Orange	0.01%
Yellow	0.001%

Variable Resistors Variable resistors are designed so that their resistance values can be changed easily with a manual or an automatic adjustment.

Two basic uses for variable resistors are to divide voltage and to control current. The variable resistor used to divide voltage is called a **potentiometer.** The variable resistor used to control current is called a **rheostat.** Schematic symbols for these types are shown in Figure 2–20. The potentiometer is a three-terminal device, as indicated in part (a). Terminals 1 and 2 have a fixed resistance between them, which is the total resistance. Terminal 3 is connected to a moving contact (**wiper).** You can vary the resistance between 3 and 1 or between 3 and 2 by moving the contact up or down.

Figure 2–20(b) shows the rheostat as a two-terminal variable resistor. Part (c) shows how you can use a potentiometer as a rheostat by connecting terminal 3 to either terminal 1 or terminal 2. Part (d) shows a basic simplified construction diagram of a potentiometer. Some typical potentiometers are pictured in Figure 2–21.

(a) Potentiometer   (b) Rheostat   (c) Potentiometer connected as a rheostat

(d) Basic construction

FIGURE 2–20
Potentiometer and rheostat symbols and basic construction of one type of potentiometer.

FIGURE 2–21
(a) Typical potentiometers (courtesy of Allen-Bradley Co.). (b) Trimmer potentiometers with construction views (courtesy of Bourns Trimpot).

Potentiometers and rheostats can be classified as linear or tapered, as shown in Figure 2–22, where a potentiometer with a total resistance of 100 Ω is used as an example. As shown in part (a), in a linear potentiometer, the resistance between either terminal and the moving contact varies linearly with the position of the moving contact. For example, one-half of a turn results in one-half the total resistance. Three-quarters of a turn results in three-quarters of the total resistance between the moving contact and one terminal, or one-quarter of the total resistance between the other terminal and the moving contact.

In the **tapered** potentiometer, the resistance varies nonlinearly with the position of the moving contact, so that one-half of a turn does not necessarily result in one-half the total resistance. This concept is illustrated in Figure 2–22(b), where the nonlinear values are arbitrary.

The potentiometer is used as a voltage-control device because when a fixed voltage is applied across the end terminals, a variable voltage is obtained at the wiper contact with respect to either end terminal. The rheostat is used as a current-control device because the current can be changed by changing the wiper position. You will learn the reasons for these functions later.

FIGURE 2–22

Examples of (a) linear and (b) tapered potentiometers.

1. Quarter turn   2. Half turn   3. Three-quarter turn

(a) Linear

1. Quarter turn   2. Half turn   3. Three-quarter turn

(b) Tapered (nonlinear)

Two Types of Automatically Variable Resistors   A **thermistor** is a type of variable resistor that is temperature-sensitive. Its resistance changes inversely with temperature. That is, it has a negative temperature coefficient. When temperature increases, the resistance decreases, and vice versa.

The resistance of a **photoconductive cell** changes with a change in light intensity. This cell also has a negative temperature coefficient. Symbols for both of these devices are shown in Figure 2–23.

FIGURE 2–23

Resistive devices with sensitivities to temperature and light.

(a) Thermistor   (b) Photoconductive cell

A Resistor Analogy

You can think of a resistor as a partially open valve in a water system that restricts the amount of water flowing through the pipe. If the valve is opened more (less resistance), the water flow (current) increases. If the valve is closed more (more resistance), the water flow (current) decreases.

SECTION 2–5 REVIEW

1. Define *resistance* and name its unit.
2. What are the two main categories of resistors? Briefly explain the difference between them.
3. In the resistor color code, what does each band represent?
4. Determine the resistance and tolerance for each of the following color codes:
   (a) yellow, violet, red, gold   (b) blue, red, orange, silver
   (c) brown, gray, black, gold
5. What is the basic difference between a rheostat and a potentiometer?
6. What is a thermistor?

2–6 ■ THE ELECTRIC CIRCUIT

A basic electric circuit is an arrangement of physical components which utilize voltage, current, and resistance to perform some useful function.

After completing this section, you should be able to

■ **Describe a basic electric circuit**
  □ Relate a schematic to a physical circuit
  □ Define *open circuit* and *closed circuit*
  □ Describe various types of protective devices
  □ Describe various types of switches

Basically, an electric **circuit** consists of a voltage source, a load, and a path for current between the source and the **load.** Figure 2–24 shows an example of a simple electric circuit: a battery connected to a lamp with two wires. The battery is the voltage source, the lamp is the load on the battery because it draws current from the battery, and the two wires provide the current path from the negative terminal of the battery to the lamp and back to the positive terminal of the battery, as shown in part (b). Current flows through the filament of the lamp (which has a resistance), causing it to emit visible light. Current flows through the battery by chemical action. In many practical cases, one terminal of the battery is connected to a *ground* point. For example, in automobiles, the negative battery terminal is connected to the metal chassis of the car. The chassis is the ground for the automobile electrical system. The concept of circuit ground is covered in detail in Chapter 4.

(a)   (b)

FIGURE 2–24
A simple electric circuit.

The Electrical Schematic

An electric circuit can be represented by a **schematic** using standard symbols for each element, as shown in Figure 2–25 for the simple circuit in Figure 2–24. The purpose of a schematic is to show in an organized manner how the various components in a given circuit are interconnected so that the operation of the circuit can be determined.

FIGURE 2–25
A schematic for the circuit in Figure 2–24.

Closed and Open Circuits

The example circuit in Figure 2–24 illustrated a **closed circuit**—that is, a circuit in which the current has a complete path. When the current path is broken so that current cannot flow, the circuit is called an **open circuit.**

Switches **Switches** are commonly used for controlling the opening or closing of circuits by either mechanical or electronic means. For example, a switch is used to turn a lamp on or off as illustrated in Figure 2–26. Each circuit pictorial is shown with its associated schematic. The type of switch indicated is a *single-pole–single-throw* (SPST) toggle switch.

Figure 2–27 shows a somewhat more complicated circuit using a *single-pole–double-throw* (SPDT) type of switch to control the current to two different lamps. When

FIGURE 2–26
Basic closed and open circuits using an SPST switch for control.

(a) Current flows in a *closed* circuit (switch is ON or in the *closed* position).

(b) No current flows in an *open* circuit (switch is OFF or in the *open* position).

FIGURE 2–27
An example of an SPDT switch controlling two lamps.

(a) Pictorial

(b) A schematic showing Lamp 1 on and Lamp 2 off

(c) A schematic showing Lamp 2 on and Lamp 1 off

one lamp is on, the other is off, and vice versa, as illustrated by the two schematics that represent each of the switch positions.

The term *pole* refers to the movable arm in a switch, and the term *throw* indicates the number of contacts that are affected (either opened or closed) by a single switch action (a single movement of a pole).

In addition to the SPST and the SPDT switches (symbols are shown in Figure 2–28 (a) and (b)), several other types of switches are important.

☐ *Double-pole–single-throw (DPST).* The DPST switch permits simultaneous opening or closing of two sets of contacts. The symbol is shown in Figure 2–28(c). The dashed line indicates that the contact arms are mechanically linked so that both move with a single switch action.

☐ *Double-pole–double-throw (DPDT).* The DPDT switch provides connection from one set of contacts to either of two other sets. The schematic symbol is shown in Figure 2–28(d).

☐ *Push button (PB).* In the normally open push-button switch (NOPB), shown in Figure 2–28(e), connection is made between two contacts when the button is depressed, and connection is broken when the button is released. In the normally closed push-button switch (NCPB), shown in Figure 2–28(f), connection between the two contacts is broken when the button is depressed.

☐ *Rotary.* In a rotary switch, a knob is turned to make connection between one contact and any one of several others. A symbol for a simple six-position rotary switch is shown in Figure 2–28(g).

FIGURE 2–28
Switch symbols.

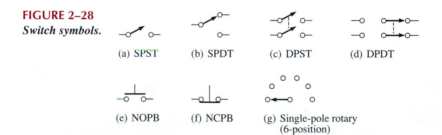

(a) SPST   (b) SPDT   (c) DPST   (d) DPDT

(e) NOPB   (f) NCPB   (g) Single-pole rotary (6-position)

Several varieties of switches are shown in Figure 2–29.

Protective Devices **Fuses** and **circuit breakers** are placed in the current path and are used to deliberately create an open circuit when the current exceeds a specified number of amperes due to a malfunction or other abnormal condition in a circuit. For example, a fuse or circuit breaker with a 20 A rating will open a circuit when the current exceeds 20 A.

The major difference between a fuse and a circuit breaker is that when a fuse is "blown," it must be replaced; but when a circuit breaker opens, it can be reset and reused repeatedly. Both of these devices protect against damage to a circuit due to excess current or prevent a hazardous condition created by the overheating of wires and other components when the current is too great. Because fuses cut off excess current more quickly than circuit breakers, fuses are used whenever delicate electronic equipment needs to be protected. Several typical fuses and circuit breakers, along with their schematic symbols, are shown in Figure 2–30.

Wires

Wires are a common form of conductive material used in electrical circuits. They vary in diameter size and are arranged according to standard gage numbers, called **American Wire Gage (AWG)** sizes. The larger the gage number is, the smaller the wire diameter is. The AWG sizes are listed in Table 2–3 on page 41.

(a) Toggle switches

(b) Rocker switches

(c) Rocker switches in dual-in-line packages

(d) Push-button switches

(e) Rotary switches

FIGURE 2–29
Switches. Parts (a), (b), and (d) courtesy of Eaton Corp. Parts (c) and (e) courtesy of Grayhill, Inc.

(a) Power fuses

(b) Fuses and fuse holders

(c) Fuse symbol   (d) Circuit breakers

(e) Remote control circuit breaker

(f) Circuit breaker symbol

FIGURE 2–30
Fuses and circuit breakers. Parts (a), (b), and (d) courtesy of Bussmann, Cooper Industries. Part (e) courtesy of Eaton Corp.

TABLE 2–3

American Wire Gage (AWG) sizes and resistances for solid round copper

AWG #	AREA (CM)	Ω/1000 FT AT 20°C	AWG #	AREA (CM)	Ω/1000 FT AT 20°C
0000	211,600	0.0490	19	1,288.1	8.051
000	167,810	0.0618	20	1,021.5	10.15
00	133,080	0.0780	21	810.10	12.80
0	105,530	0.0983	22	642.40	16.14
1	83,694	0.1240	23	509.45	20.36
2	66,373	0.1563	24	404.01	25.67
3	52,634	0.1970	25	320.40	32.37
4	41,742	0.2485	26	254.10	40.81
5	33,102	0.3133	27	201.50	51.47
6	26,250	0.3951	28	159.79	64.90
7	20,816	0.4982	29	126.72	81.83
8	16,509	0.6282	30	100.50	103.2
9	13,094	0.7921	31	79.70	130.1
10	10,381	0.9989	32	63.21	164.1
11	8,234.0	1.260	33	50.13	206.9
12	6,529.0	1.588	34	39.75	260.9
13	5,178.4	2.003	35	31.52	329.0
14	4,106.8	2.525	36	25.00	414.8
15	3,256.7	3.184	37	19.83	523.1
16	2,582.9	4.016	38	15.72	659.6
17	2,048.2	5.064	39	12.47	831.8
18	1,624.3	6.385	40	9.89	1049.0

As the table shows, the size of a wire is also specified in terms of its cross-sectional area, as illustrated also in Figure 2–31. The unit of cross-sectional area is the **circular mil** abbreviated CM. One circular mil is the area of a wire with a diameter of 0.001 inch (0.001 in., or 1 mil). Find the cross-sectional area by expressing the diameter in thousandths of an inch (mils) and squaring it, as follows:

$$A = d^2 \tag{2–4}$$

where A is the cross-sectional area in circular mils and d is the diameter in mils.

FIGURE 2–31
Cross-sectional area of a wire.

Cross-sectional area, A

0.001 in.

$A = 1$ CM

EXAMPLE 2–5

What is the cross-sectional area of a wire with a diameter of 0.005 inch?

Solution

$d = 0.005$ in. $= 5$ mils
$A = d^2 = (5 \text{ mils})^2 = 25$ CM

Related Exercise What is the cross-sectional area of a 0.0201 in. diameter wire? What is the AWG # for this wire?

Wire Resistance

Although copper wire conducts electricity extremely well, it still has some resistance, as do all conductors. The resistance of a wire depends on four factors: (1) type of material, (2) length of wire, (3) cross-sectional area, and (4) temperature.

Each type of conductive material has a characteristic called its *resistivity, ρ*. For each material, ρ is a constant value at a given temperature. The formula for the resistance of a wire of length l and cross-sectional area A is

$$R = \frac{\rho l}{A}$$

(2–5)

This formula shows that resistance increases with resistivity and length and decreases with cross-sectional area. For resistance to be calculated in ohms, the length must be in feet (ft), the cross-sectional area in circular mils, and the resistivity in CM-Ω/ft.

EXAMPLE 2–6

Find the resistance of a 100 ft length of copper wire with a cross-sectional area of 810.1 CM. The resistivity of copper is 10.4 CM-Ω/ft.

Solution

$$R = \frac{\rho l}{A} = \frac{(10.4 \text{ CM-}\Omega/\text{ft})(100 \text{ ft})}{810.1 \text{ CM}} = 1.284 \ \Omega$$

Related Exercise Determine the resistance of a 1000 ft length of #22 AWG copper wire.

Table 2–3 lists the resistance of the various standard wire sizes in ohms per 1000 ft at 20°C. For example, a 1000 ft length of 14-gage copper wire has a resistance of 2.525 Ω. A 1000 ft length of 22-gage wire has a resistance of 16.14 Ω. For a given length, the smaller wire has more resistance. Thus, for a given voltage, larger wires can carry more current than smaller ones.

SECTION 2–6 REVIEW

1. What are the basic elements of an electric circuit?
2. Define *open circuit.*
3. Define *closed circuit.*
4. What is the purpose of a fuse?
5. What is the difference between a fuse and a circuit breaker?
6. Which wire is larger in diameter, AWG #3 or AWG #22?

2–7 ▪ BASIC CIRCUIT MEASUREMENTS

An electronics technician cannot function without knowing how to measure voltage, current, and resistance.

After completing this section, you should be able to

▪ **Make basic circuit measurements**
  □ Properly place a voltmeter in a circuit
  □ Properly place an ammeter in a circuit
  □ Properly connect an ohmmeter to measure resistance
  □ Set up and read basic meters

Voltage, current, and resistance measurements are commonly required in electronics work. Special types of instruments are used to measure these basic electrical quantities.

The instrument used to measure voltage is a **voltmeter,** the instrument used to measure current is an **ammeter,** and the instrument used to measure resistance is an **ohmmeter.** Commonly, all three instruments are combined into a single instrument known as a **multimeter,** or VOM (volt-ohm-milliammeter), in which you can choose what specific quantity to measure by selecting the switch setting.

Typical multimeters are shown in Figure 2–32. Part (a) shows an analog meter, that is, with a needle pointer, and part (b) shows a digital multimeter (DMM), which provides a digital readout of the measured quantity.

(a)   (b)

FIGURE 2–32
Typical multimeters. (a) Analog and (b) digital (courtesy of B & K Precision).

Meter Symbols

Throughout this book, certain symbols will be used in circuits to represent the different meters, as shown in Figure 2–33. You may see any of three types of symbols for volt-

FIGURE 2–33
Meter symbols used in this book.

Analog   Digital   General   Analog   Digital   General

(a) Voltmeter   (b) Ammeter

Analog   Digital   General

(c) Ohmmeter

meters, ammeters, and ohmmeters, depending on which symbol most effectively conveys the information required. Although the digital meter is much more widely used than the analog meter in industry, we will use the analog meter symbol in certain situations to illustrate better the operation of a circuit when relative measurements or changes in quantities need to be depicted by the position or movement of the needle. The digital meter symbol is used when fixed values are to be indicated in a circuit. The general schematic symbol is used to indicate placement of meters in a circuit when no values or value changes need to be shown.

How to Measure Current with an Ammeter

Figure 2–34 illustrates how to measure current with an ammeter. Part (a) shows the simple circuit in which the current through the resistor is to be measured. Connect the amme-

FIGURE 2–34

Example of an ammeter connection in a simple circuit.

(a) Circuit in which the current is to be measured

(b) Open the circuit either between the resistor and the positive terminal or between the resistor and the negative terminal of source.

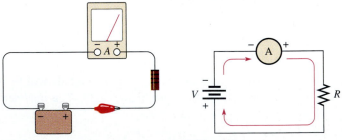

(c) Install the ammeter in the current path with polarity as shown (negative to negative, positive to positive).

ter in the current path by first opening the circuit, as shown in part (b). Then insert the meter as shown in part (c). Such a connection is a *series* connection. The polarity of the meter must be such that the current flows in at the negative terminal and out at the positive terminal.

How to Measure Voltage with a Voltmeter

To measure voltage, connect the voltmeter across the component for which the voltage is to be found. Such a connection is a *parallel* connection. The negative terminal of the

FIGURE 2–35
Example of a voltmeter connection in a simple circuit.

meter must be connected to the negative side of the circuit, and the positive terminal of the meter to the positive side of the circuit. Figure 2–35 shows a voltmeter connected to measure the voltage across the resistor.

How to Measure Resistance with an Ohmmeter

To measure resistance, connect the ohmmeter across the resistor. *The resistor must first be removed or disconnected from the circuit.* This procedure is shown in Figure 2–36.

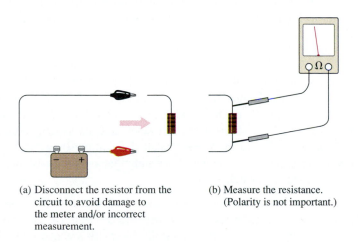

(a) Disconnect the resistor from the circuit to avoid damage to the meter and/or incorrect measurement.

(b) Measure the resistance. (Polarity is not important.)

FIGURE 2–36
Example of using an ohmmeter.

Reading Analog Multimeters

Functions   The face of a typical analog meter is represented in Figure 2–37. This particular instrument can be used to measure both direct current (dc) and alternating current (ac) quantities as well as resistance values. It has four selectable functions: dc volts (DC VOLTS), dc milliamperes (DC mA), ac volts (AC VOLTS), and OHMS. Most analog multimeters are similar to this one, although range selections and scales may vary.

Ranges   Within each function there are several ranges, as indicated by the brackets around the selector switch. For example, the DC VOLTS function has 0.3 V, 3 V, 12 V, 60 V, 300 V, and 600 V ranges. Thus, dc voltages from 0.3 V full-scale to 600 V full-scale can be measured. On the DC mA function, direct currents from 0.06 mA full-scale to 120 mA full-scale can be measured. On the ohm scale, the range settings are ×1, ×10, ×100, ×1000, and ×100,000.

FIGURE 2–37

A typical analog multimeter.

The Ohm Scale Ohms are read on the top scale of the meter. This scale is nonlinear; that is, the values represented by each division (large or small) vary as you go across the scale. In Figure 2–37, notice how the scale becomes more compressed as you go from right to left.

To read the actual value in ohms, multiply the number on the scale as indicated by the pointer by the factor selected by the switch. For example, when the switch is set at ×100 and the pointer is at 20, the reading is 20 × 100 = 2000 Ω.

As another example, assume that the switch is at ×10 and the pointer is at the seventh small division between the 1 and 2 marks, indicating 17 Ω (1.7 × 10). Now, if the meter remains connected to the same resistance and the switch setting is changed to ×1, the pointer will move to the second small division between the 15 and 20 marks. This, of course, is also a 17 Ω reading, illustrating that a given resistance value can often be read at more than one switch setting.

The AC-DC Scales The second, third, and fourth scales from the top (labeled "AC" and "DC") are used in conjunction with the DC VOLTS, DC mA, and AC VOLTS functions. The upper ac-dc scale ends at the 300 mark and is used with the range settings that are multiples of three, such as 0.3, 3, and 300. For example, when the switch is at 3 on the DC VOLTS function, the 300 scale has a full-scale value of 3 V. At the range setting of 300, the full-scale value is 300 V, and so on.

The middle ac-dc scale ends at 60. This scale is used in conjunction with range settings that are multiples of 6, such as 0.06, 60, and 600. For example, when the switch is at 60 on the DC VOLTS function, the full-scale value is 60 V.

The lower ac-dc scale ends at 12 and is used in conjunction with switch settings that are multiples of 12, such as 1.2, 12, and 120.

The remaining scales are for ac current and for decibels, which are discussed later in the book.

EXAMPLE 2–7 In Figure 2–38, determine the quantity (voltage, current, or resistance) that is being measured and its value.

FIGURE 2–38

(a)

(b)

(c)

Solution

(a) The switch in Figure 2–38(a) is set on the DC VOLTS function and the 60 V range. The reading taken from the middle dc scale is 18 V.

(b) The switch in Figure 2–38(b) is set on the DC mA function and the 12 mA range. The reading taken from the upper dc scale is 7.2 mA.

(c) The switch in Figure 2–38(c) is set on the ohm (OHMS) function and the ×1000 range. The reading taken from the ohm scale (top scale) is 7 kΩ.

Related Exercise In Figure 2–38(c) the switch is moved to the ×100 setting. Assuming that the same resistance is being measured, what will the needle do?

Digital Multimeters (DMMs)

DMMs are the most widely used type of electronic measuring instrument. Generally, DMMs provide more functions, better accuracy, greater ease of reading, and greater reliability than do many analog meters. Analog meters have at least one advantage over DMMs, however. They can track short-term variations and trends in a measured quantity that many DMMs are too slow to respond to. Three typical DMMs are shown in Figure 2–39. Many DMMs are autoranging types in which the proper range is automatically selected by internal circuitry.

DMM Functions   The basic functions found on most DMMs include the following:

☐ Ohms

☐ DC voltage and current

☐ AC voltage and current

Some DMMs provide special functions such as transistor or diode tests, power measurement, and decibel measurement for audio amplifier tests.

DMM Displays   DMMs are available with either LCD (liquid-crystal display) or LED (light-emitting diode) readouts. The LCD is the most commonly used readout in battery-powered instruments because it requires only very small amounts of current. A typical battery-powered DMM with an LCD readout operates on a 9 V battery that will last from a few hundred hours to 2000 hours and more. The disadvantages of LCD readouts are

(a)   (b)

(c)

FIGURE 2–39

Typical digital multimeters (DMMs). Part (a) courtesy of Triplett Corp. Parts (b) and (c) courtesy of John Fluke Manufacturing Co., Inc.

that (a) they are difficult or impossible to see in low-light conditions and (b) they are relatively slow to respond to measurement changes. LEDs, on the other hand, can be seen in the dark and respond quickly to changes in measured values. LED displays require much more current than LCDs, and, therefore, battery life is shortened when they are used in portable equipment.

Both LCD and LED DMM displays are in a seven-segment format. Each digit in the display consists of seven separate segments, as shown in Figure 2–40(a). Each of the ten decimal digits is formed by activation of appropriate segments, as illustrated in Figure 2–40(b). In addition to the seven segments, there is also a decimal point.

(a)　　　　　　　(b)

FIGURE 2–40
Seven-segment display.

Resolution　The **resolution** of a meter is the smallest increment of a quantity that the meter can measure. The smaller the increment, the better the resolution. One factor that determines the resolution of a meter is the number of digits in the display.

Because many meters have 3½ digits in their display, we will use this case for illustration. A 3½-digit multimeter has three digit positions that can indicate from 0 through 9, and one digit position that can indicate only a value of 1. This latter digit, called the *half-digit,* is always the most significant digit in the display. For example, suppose that a DMM is reading 0.999 V, as shown in Figure 2–41(a). If the voltage increases by 0.001 V to 1 V, the display correctly shows 1.000 V, as shown in part (b). The "1" is the half-digit. Thus, with 3½ digits, a variation of 0.001 V, which is the resolution, can be observed.

(a) Resolution: 0.001 V　　　　　　(b) Resolution: 0.001 V

FIGURE 2–41
A 3½-digit DMM illustrates how the resolution changes with the number of digits in use.

(c) Resolution: 0.001 V   (d) Resolution: 0.01 V

FIGURE 2–41 *Continued*

Now, suppose that the voltage increases to 1.999 V. This value is indicated on the meter as shown in Figure 2–41(c). If the voltage increases by 0.001 V to 2 V, the half-digit cannot display the "2," so the display shows 2.00. The half-digit is blanked and only three digits are active, as indicated in part (d). With only three digits active, the resolution is 0.01 V rather than 0.001 V as it is with 3½ active digits. The resolution remains 0.01 V up to 19.99 V. The resolution goes to 0.1 V for readings of 20.0 V to 199.9 V. At 200 V, the resolution goes to 1 V, and so on.

The resolution capability of a DMM is also determined by the internal circuitry and the rate at which the measured quantity is sampled. DMMs with displays of 4½ through 8½ digits are also available.

Accuracy   The accuracy of a DMM is established strictly by its internal circuitry. For typical meters, accuracies range from 0.01% to 0.5%, with some precision laboratory-grade meters going to 0.002%.

SECTION 2–7 REVIEW

1. Name the meters for measurement of (a) current, (b) voltage, and (c) resistance.
2. Place two ammeters in the circuit of Figure 2–27 to measure the current through either lamp (be sure to observe the polarities). How can the same measurements be accomplished with only one ammeter?
3. Show how to place a voltmeter to measure the voltage across lamp 2 in Figure 2–27.
4. The multimeter in Figure 2–37 is set on the 3 V range to measure dc voltage. Assume the pointer is at 150 on the upper ac-dc scale. What voltage is being measured?
5. How do you set up the meter in Figure 2–37 to measure 275 V dc, and on what scale do you read the voltage?
6. If you expect to measure a resistance in excess of 20 kΩ with the meter in Figure 2–37, where do you set the switch?

7. List two common types of DMM displays, and discuss the advantages and disadvantages of each.
8. Define *resolution* in a DMM.

2–8 ▪ APPLICATION ASSIGNMENT

The requirements for a lighting system were stated in the chapter opener. You will go through a step-by-step process of working through this assignment using things that you've learned in this chapter. Recall from the opener that the lighting system will consist of six lamps that are controlled so that only one is turned on at a time and the sequence can be in any order. Also, the system must be fused and operate from a 12 V battery and a method must be provided for varying the brightness of the lamp that is on.

Step 1: The Circuit

Select the circuit from the four configurations in Figure 2–42 that meets the requirements. Explain why the rejected circuits will not meet the requirements. Explain the purpose of each component in your circuit.

FIGURE 2–42

(a) (b) (c) (d)

Step 2: Hook Up the Circuit

From the selection of components in Figure 2–43, specify the quantity of each component required for the lighting circuit. Hook up the lighting circuit by developing a point-to-point (From–To) wiring list using component designations and pin numbers.

FIGURE 2–43
Component selection.

Lamp (L)   12 V battery (V)   Fuse (F)

SPST toggle switch (SW)   Rotary switch (SW)   Potentiometer (R)

Step 3: Determine the Current Rating of the Fuse

The fuse protects the circuit by burning open when there is excessive current. The type of lamp to be used draws approximately 5 A when 12 V are applied. Choose a fuse based on its ampere rating that is the most appropriate for your circuit from the following ratings: 1 A, 2 A, 3 A, 4 A, 5 A, 6 A, 7 A, 8 A, 9 A, 10 A, 20 A, and 30 A. Explain your choice.

Step 4: Determine the Capacity of the Battery

A battery can provide a certain amount of current for a certain period of time at its rated voltage. The amount of amps that a battery can deliver to a load for a given period of time at its rated voltage is its *ampere-hour rating*. Determine the minimum ampere-hour rating of the battery if the lighting circuit must operate for four hours before the battery is recharged.

Step 5: Troubleshoot the Circuit

Once you have connected all the components, it is time to check out the operation to verify that everything is functioning properly.

☐ Write a brief description of the procedure you would use to thoroughly test the circuit. This is called a *test procedure*.

☐ For each of the following cases where the circuit is not working properly, determine the possible fault or faults and specify how you would correct it:
   1. Each lamp can be turned on except one.
   2. None of the lamps can be turned on.
   3. Each lamp is too dim and cannot be brightened by adjusting the rheostat.
   4. Each lamp is too dim; however, the amount of light can be varied with the rheostat, but not to full brightness.

SECTION 2–8 REVIEW

1. If the resistance of the rheostat in your lighting circuit is reduced, does the brightness of the selected lamp increase or decrease?
2. Does removing any one of the lamps affect the operation of any of the other lamps?
3. What is the current from the battery when no lamps are on?
4. What will happen if you try to turn on two lamps at the same time? Explain.

■ SUMMARY

- An atom is the smallest particle of an element that retains the characteristics of that element.
- The electron is the basic particle of negative electrical charge.
- The proton is the basic particle of positive charge.
- An ion is an atom that has gained or lost an electron and is no longer neutral.
- When electrons in the outer orbit of an atom (valence electrons) break away, they become free electrons.
- Free electrons make current possible.
- Like charges repel each other, and opposite charges attract cach other.
- Voltage must be applied to a circuit before current can flow.
- Resistance limits the current.
- Basically, an electric circuit consists of a source, a load, and a current path.
- An open circuit is one in which the current path is broken.
- A closed circuit is one which has a complete current path.
- An ammeter is connected in line with the current path.
- A voltmeter is connected across the current path.
- An ohmmeter is connected across a resistor after removal from circuit.
- Figure 2–44 shows the electrical symbols introduced in this chapter.

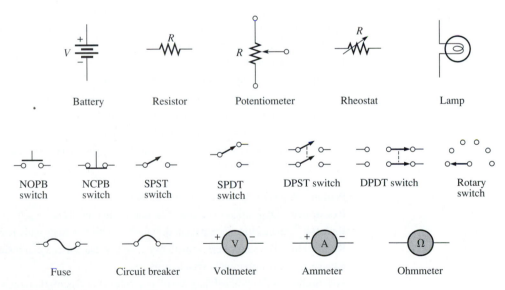

FIGURE 2–44

■ GLOSSARY

Ammeter   An electrical instrument used to measure current.

Ampere (A)   The unit of electrical current.

Atom   The smallest particle of an element possessing the unique characteristics of that element.

Atomic number   The number of electrons in a neutral atom.

Atomic weight   The number of protons and neutrons in the nucleus of an atom.

AWG   American wire gage. A standardization based on wire diameter.

Battery An energy source that uses a chemical reaction to convert chemical energy into electrical energy.

Charge An electrical property of matter that exists because of an excess or a deficiency of electrons. Charge can be either positive or negative.

Circuit An interconnection of electrical components designed to produce a desired result. A basic circuit consists of a source, a load, and an interconnecting current path.

Circuit breaker A resettable protective device used for interrupting excessive current in an electric circuit.

Circular mil (CM) The unit of the cross-sectional area of a wire.

Closed circuit A circuit with a complete current path.

Conductor A material in which electrical current can flow with relative ease. An example is copper.

Coulomb (C) The unit of electrical charge.

Current The rate of flow of electrons.

Electrical Related to the use of electrical voltage and current to achieve desired results.

Electron The basic particle of electrical charge in matter. The electron possesses negative charge.

Electronic Related to the movement and control of free electrons in semiconductors or vacuum devices.

Element One of the unique substances that make up the known universe. Each element is characterized by a unique atomic structure.

Energy The ability to do work.

Free electron A valence electron that has broken away from its parent atom and is free to move from atom to atom within the atomic structure of a material.

Fuse A protective device that burns open when excessive current flows in a circuit.

Generator An energy source that produces electrical signals.

Insulator A material that does not allow current under normal conditions.

Joule (J) The unit of energy.

Load The device in a circuit upon which work is done.

Multimeter An instrument that measures voltage, current, and resistance.

Neutron An atomic particle having no electrical charge.

Ohm (Ω) The unit of resistance.

Ohmmeter An instrument for measuring resistance.

Open circuit A circuit in which there is not a complete current path.

Photoconductive cell A type of variable resistor that is light-sensitive.

Potentiometer A three-terminal variable resistor.

Power supply An electronic instrument that produces voltage, current, and power from the ac power line or batteries in a form suitable for use in various applications to power electronic equipment.

Proton A positively charged atomic particle.

Resistance Opposition to current. The unit is the ohm (Ω).

Resistor An electrical component designed specifically to provide resistance.

Resolution The smallest increment of a quantity that a meter can measure.

Rheostat A two-terminal variable resistor.

Schematic A symbolized diagram of an electrical or electronic circuit.

Semiconductor A material that has a conductance value between that of a conductor and that of an insulator. Silicon and germanium are examples.

Shell The orbit in which an electron revolves.

Source Any device that produces energy.

Switch An electrical device for opening and closing a current path.

Tapered Nonlinear, such as a tapered potentiometer.

Thermistor A type of variable resistor that is temperature-sensitive.

Tolerance The limits of variation in the value of a component.

Valence Related to the outer shell or orbit of an atom.

Valence electron An electron that is present in the outermost shell of an atom.

Volt The unit of voltage or electromotive force.

Voltage The amount of energy available to move a certain number of electrons from one point to another in an electrical circuit.

Voltmeter An instrument used to measure voltage.

Wiper The sliding contact in a potentiometer.

■ FORMULAS

(2–1) $$Q = \frac{\text{number of electrons}}{6.25 \times 10^{18} \text{ electrons/Coulomb}}$$ Charge

(2–2) $$V = \frac{W}{Q}$$ Voltage equals energy divided by charge.

(2–3) $$I = \frac{Q}{t}$$ Current equals charge divided by time.

(2–4) $$A = d^2$$ Cross-sectional area equals the diameter squared.

(2–5) $$R = \frac{\rho l}{A}$$ Resistance is resistivity times length divided by cross-sectional area.

■ SELF-TEST

1. A neutral atom with an atomic number of three has how many electrons?
   (a) 1 (b) 3 (c) none (d) depends on the type of atom

2. Electron orbits are called
   (a) shells (b) nuclei (c) waves (d) valences

3. Materials in which no current can flow are called
   (a) filters (b) conductors (c) insulators (d) semiconductors

4. When placed close together, a positively charged material and a negatively charged material will
   (a) repel (b) become neutral (c) attract (d) exchange charges

5. The charge on a single electron is
   (a) 6.25×10^{-18} C (b) 1.6×10^{-19} C (c) 1.6×10^{-19} J (d) 3.14×10^{-6} C

6. *Potential difference* is another term for
   (a) energy (b) voltage (c) distance of an electron from the nucleus (d) charge

7. The unit of energy is the
   (a) watt (b) coulomb (c) joule (d) volt

8. Which one of the following is not a type of energy source?
   (a) battery (b) solar cell (c) generator (d) potentiometer

9. Which one of the following is not a possible condition in an electric circuit?
   (a) voltage and no current (b) current and no voltage
   (c) voltage and current (d) no voltage and no current

10. Electrical current is defined as
    (a) free electrons (b) the rate of flow of free electrons
    (c) the energy required to move electrons (d) the charge on free electrons

11. Current will not flow in a circuit when
    (a) a switch is closed (b) a switch is open (c) there is no voltage
    (d) both (a) and (c) (e) both (b) and (c)

12. The primary purpose of a resistor is to
    (a) increase current (b) limit current (c) produce heat (d) resist current change

13. Potentiometers and rheostats are types of
    (a) voltage sources (b) variable resistors (c) fixed resistors (d) circuit breakers

14. The current in a given circuit is not to exceed 22 A. Which value of fuse is best?
    (a) 10 A (b) 25 A (c) 20 A (d) a fuse is not necessary

■ BASIC PROBLEMS

SECTION 2–2 Electrical Charge

1. How many coulombs of charge do 50×10^{31} electrons possess?
2. How many electrons does it take to make 80 μC of charge?

SECTION 2–3 Voltage

3. Determine the voltage in each of the following cases:
   (a) 10 J/C (b) 5 J/2 C (c) 100 J/25 C
4. Five hundred joules of energy are used to move 100 C of charge through a resistor. What is the voltage across the resistor?
5. What is the voltage of a battery that uses 800 J of energy to move 40 C of charge through a resistor?
6. How much energy does a 12 V battery in your car use to move 2.5 C through the electrical circuit?

SECTION 2–4 Current

7. Determine the current in each of the following cases:
   (a) 75 C in 1 s (b) 10 C in 0.5 s (c) 5 C in 2 s
8. Six-tenths coulomb passes a point in 3 s. What is the current in amperes?
9. How long does it take 10 C to flow past a point if the current is 5 A?
10. How many coulombs pass a point in 0.1 s when the current is 1.5 A?

SECTION 2–5 Resistance

11. Figure 2–45(a) shows color-coded resistors. Determine the resistance value and the tolerance of each.
12. Find the minimum and the maximum resistance within the tolerance limits for each resistor in Figure 2–45(a).
13. (a) If you need a 270 Ω resistor, what color bands should you look for?
    (b) From the selection of resistors in Figure 2–45(b), choose the following values: 330 Ω, 2.2 kΩ, 56 kΩ, 100 kΩ, and 39 kΩ.

(a)

(b)

FIGURE 2–45

14. The adjustable contact of a linear potentiometer is set at the mechanical center of its adjustment. If the total resistance is 1000 Ω, what is the resistance between each end terminal and the adjustable contact?

SECTION 2–6 The Electric Circuit

15. Trace the current path in the lamp circuit of Figure 2–27(a) with the switch in position 2.

16. With the switch in either position, redraw the lamp circuit in Figure 2–27(b) with a fuse connected to protect the circuit against excessive current.

SECTION 2–7 Basic Circuit Measurements

17. Show the placement of an ammeter and a voltmeter to measure the current and the source voltage in Figure 2–46.

18. Show how you would measure the resistance of R_2 in Figure 2–46.

19. In Figure 2–47, how much voltage does each meter indicate when the switch (SW) is in position 1? In position 2?

20. In Figure 2–47, show how to connect an ammeter to measure the current from the voltage source regardless of the switch (SW) position.

21. What are the voltage readings of the multimeters in Figures 2–48(a) and 2–48(b)?

FIGURE 2–46   **FIGURE 2–47**

(a)

(b)

FIGURE 2–48

FIGURE 2–49

ANALOG MULTIMETER

22. How much resistance is the multimeter in Figure 2–49 measuring?

23. Determine the resistance indicated by each of the following ohmmeter readings and range settings:
    (a) pointer at 2, range setting at R ×100
    (b) pointer at 15, range setting at R ×10M
    (c) pointer at 45, range setting at R ×100

24. A multimeter has the following ranges: 1 mA, 10 mA; 100 mA, 100 mV, 1 V, 10 V; R ×1, R ×10, R ×100. Indicate schematically how you would connect the multimeter in Figure 2–50 to measure the following quantities:

    (a) I_1 (b) V_1 (c) R_1

    In each case indicate the *function* to which you would set the meter and the *range* that you would use.

FIGURE 2–50

R_1
4.7 kΩ

10 V

R_2
2.2 kΩ

R_3
8.2 kΩ

■ **ADVANCED PROBLEMS**

25. A resistor with a current of 2 A through it in an amplifier circuit converts 1000 J of electrical energy to heat energy in 15 s. What is the voltage across the resistor?

26. If 574×10^{15} electrons flow through a speaker wire in 250 ms, what is the current in amperes?

27. A 120 V source is to be connected to a 1500 Ω resistive load by two lengths of wire as shown in Figure 2–51. The voltage source is to be located 50 ft from the load. Using Table 2–3, determine the gage number of the *smallest* wire that can be used if the total resistance of the two lengths of wire is not to exceed 6 Ω.

FIGURE 2–51

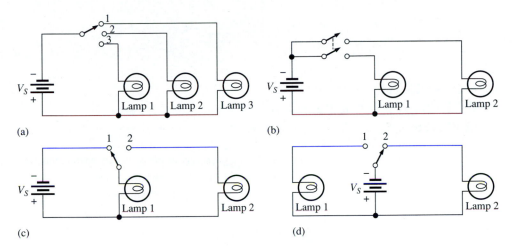

28. Determine the resistance and tolerance of each resistor labeled as follows:
    (a) 4R7J (b) 5602M (c) 1501F

29. There is only one circuit in Figure 2–52 in which it is possible to have all lamps on at the same time. Determine which circuit it is.

(a)

(b)

(c)

(d)

FIGURE 2–52

30. Through which resistor in Figure 2–53 does current always flow, regardless of the position of the switches?

31. In Figure 2–53, show the proper placement of ammeters to measure the current through each resistor and the current out of the battery.

32. Show the proper placement of voltmeters to measure the voltage across each resistor in Figure 2–53.

FIGURE 2–53

33. Devise a switch arrangement whereby two voltage sources (V_{S1} and V_{S2}) can be connected simultaneously to either of two resistors (R_1 and R_2) as follows:

$$V_{S1} \text{ connected to } R_1 \text{ and } V_{S2} \text{ connected to } R_2$$

or $\qquad V_{S1}$ connected to R_2 and V_{S2} connected to R_1

34. The different sections of a stereo system are represented by the blocks in Figure 2–54. Show how a single switch can be used to connect the phonograph, the CD (compact disk) player, the tape deck, the AM tuner, or the FM tuner to the amplifier by a single knob control. Only one section can be connected to the amplifier at any time.

FIGURE 2–54

▪ **ANSWERS TO SECTION REVIEWS**

Section 2–1

1. The basic particle of negative charge is the electron.

2. An atom is the smallest particle of an element that retains the unique characteristics of the element.

3. An atom consists of a positively charged nucleus surrounded by orbiting electrons.

4. No, all elements do not have the same type of atoms.

5. A free electron is an outer-shell electron that has drifted away from the parent atom.

Section 2–2

1. Q is the symbol for charge.

2. Coulomb is the unit of charge and C is the symbol for Coulomb.

3. The two types of charge are positive and negative.

4. $Q = \dfrac{10 \times 10^{12} \text{ electrons}}{6.25 \times 10^{18} \text{ electrons/C}} = 1.6 \times 10^{-6} \text{C}.$

Section 2–3

1. Voltage is energy per unit charge.

2. The unit of voltage is the volt.

3. $V = 24 \text{ J}/10 \text{ C} = 2.4 \text{ V}$

4. Sources of voltage are the battery, power supply, solar cell, and generator.

Section 2–4

1. Current is the rate of flow of electrons; the unit of current is the ampere (A).

2. There are 6.25×10^{18} electrons in one coulomb.

3. $I = 20 \text{ C}/4 \text{ s} = 5 \text{ A}$

Section 2–5

1. Resistance is opposition to current and its unit is the ohm (Ω).

2. Two resistor categories are fixed and variable. The value of a fixed resistor cannot be changed, but that of a variable resistor can.

3. *First band:* first digit of resistance value. *Second band:* second digit of resistance value. *Third band:* number of zeros following the 2nd digit. *Fourth band:* tolerance.

4. **(a)** Yellow, violet, red, gold = 4700 Ω ±5%
  (b) Blue, red, orange, silver = 62,000 Ω ±10%
  (c) Brown, gray, black, gold = 18 Ω ±5%

5. A rheostat has two terminals; a potentiometer has three terminals.

6. A thermistor is a temperature-sensitive resistor.

Section 2–6

1. A basic electric circuit consists of source, load, and current path between source and load.

2. An open circuit is one that has no path for current.

3. A closed circuit is one that has a complete path for current.

4. A fuse protects a circuit against excessive current.

5. A fuse must be replaced once blown. A circuit breaker can be reset once tripped.

6. AWG #3 is larger than AWG #22.

Section 2–7

1. **(a)** Ammeter measures current. **(b)** Voltmeter measures voltage. **(c)** Ohmmeter measures resistance.

2. See Figure 2–55.

3. See Figure 2–56.

(a) Two ammeters   (b) One ammeter

FIGURE 2–55   **FIGURE 2–56**

4. The voltage being measured is 1.5 V.

5. DC volts, 300 setting; 275 V is read on the 300 scale.

6. OHMS ×1000

7. Two types of DMM displays are LED and LCD. The LCD requires little current, but it is difficult to see in low light and is slow to respond. The LED can be seen in the dark, and it responds quickly; however, it requires much more current than does the LCD.

8. Resolution is the smallest increment that the meter can measure.

Section 2–8

1. The brightness increases.

2. No

3. There is no current when all lights are off.

4. The total current would increase, possibly causing the fuse to blow.

■ **ANSWERS TO RELATED EXERCISES FOR EXAMPLES**		
	2–1	1.88×10^{19} electrons
	2–2	600 J
	2–3	12 C
	2–4	$4.7 \text{ k}\Omega \pm 20\%$
	2–5	404.01 CM; #24
	2–6	16.14 Ω
	2–7	The needle will move left to the "70" mark on the top scale.

3

OHM'S LAW, ENERGY, AND POWER

Georg Simon Ohm (1787–1854) experimentally found that voltage, current, and resistance are all related in a specific way. This basic relationship, known as *Ohm's law,* is one of the most fundamental and important laws in the fields of electricity and electronics. In this chapter, Ohm's law is examined, and its use in practical circuit applications is discussed and demonstrated by numerous examples.

In addition to Ohm's law, the concepts and definitions of energy and power in electric circuits are introduced and the Watt's law power formulas are given.

■ CHAPTER OBJECTIVES

☐ Explain Ohm's law
☐ Use Ohm's law to determine voltage, current, or resistance
☐ Define *energy* and *power*
☐ Calculate power in a circuit
☐ Properly select resistors based on power rating
☐ Explain energy loss and voltage drop
☐ Discuss power supplies and their characteristics

■ APPLICATION ASSIGNMENT

As a technician working in the engineering development lab of an electronics company, you are assigned the task of modifying an existing test fixture for use in a new application. The test fixture is a resistor box with an array of switch-selectable resistors of various values. Your job will be to determine and specify the changes to be made in the existing circuit and develop a procedure to test the box once the modifications have been made. To complete the assignment, you will

1. Determine color-coded resistor values.

2. Determine resistor power ratings.

3. Draw the schematic of the existing circuit.

4. Draw the schematic to meet the new requirements.

5. Use Watt's law power formulas to determine required power ratings.

6. Use Ohm's law to determine resistor values to meet specifications.

After studying this chapter, you should be able to complete the application assignment in Section 3–8.

3–1 ■ OHM'S LAW

Ohm's law describes how voltage, current, and resistance in a circuit are related. Ohm's law is used in three equivalent forms depending on which quantity you need to determine. In this section, you will learn each of these forms.

After completing this section, you should be able to

■ **Explain Ohm's law**
  □ Describe how V, I, and R are related
  □ Express I as a function of V and R
  □ Express V as a function of I and R
  □ Express R as a function of V and I

Ohm determined experimentally that if the voltage across a resistor is increased, the current through the resistor will also increase; and, likewise, if the voltage is decreased, the current will decrease. For example, if the voltage is doubled, the current will double. If the voltage is halved, the current will also be halved. This relationship is illustrated in Figure 3–1, with meter indications of voltage and current.

Ohm's law also states that if the voltage is kept constant, less resistance results in more current, and, also, more resistance results in less current. For example, if the resistance is halved, the current doubles. If the resistance is doubled, the current is halved. This concept is illustrated by the meter indications in Figure 3–2, where the resistance is increased and the voltage is held constant.

FIGURE 3–1

Effect on the current of changing the voltage with the resistance at a constant value.

(a) Less V, less I (b) More V, more I

FIGURE 3–2

Effect on the current of changing the resistance with the voltage at a constant value.

(a) Less R, more I (b) More R, less I

Formula for Current

Ohm's law can be stated as follows:

$$I = \frac{V}{R}$$

(3–1)

Equation (3–1) states the relationship indicated by the action in the circuits of Figures 3–1 and 3–2.

For a constant resistance, if the voltage in a circuit is increased, more current will flow; and if the voltage is decreased, less current will flow.

$$I = V / R \qquad\qquad I = V / R \qquad\qquad R \text{ constant}$$

Increase V, I increases

Decrease V, I decreases

For a constant voltage, if the resistance in a circuit is increased, less current will flow; and if the resistance is decreased, more current will flow.

$$I = V / R \qquad\qquad I = V / R \qquad\qquad V \text{ constant}$$

Increase R, I decreases   Decrease R, I increases

Using Equation (3–1), you can calculate the current if the values of voltage and resistance are known.

EXAMPLE 3–1

Using the Ohm's law formula in Equation (3–1), verify that the current through a 10 Ω resistor increases when the voltage is increased from 5 V to 20 V.

Solution The following calculations show that the current increases from 0.5 A to 2 A. The calculator sequence for each calculation is also shown.

For $V = 5$ V,

$$I = \frac{V}{R} = \frac{5\text{ V}}{10\ \Omega} = 0.5\text{ A} \qquad \boxed{5}\ \boxed{\div}\ \boxed{1}\ \boxed{0}\ \boxed{=}$$

For $V = 20$ V,

$$I = \frac{V}{R} = \frac{20\text{ V}}{10\ \Omega} = 2\text{ A} \qquad \boxed{2}\ \boxed{0}\ \boxed{\div}\ \boxed{1}\ \boxed{0}\ \boxed{=}$$

Related Exercise Show that the current decreases when the resistance is increased from 5 Ω to 20 Ω and the voltage is a constant 10 V.

Formula for Voltage

Ohm's law can also be stated another way. By multiplying both sides of Equation (3–1) by R and transposing terms, you can obtain an equivalent form of Ohm's law, as follows:

$$V = IR \qquad\qquad (3\text{–}2)$$

With this equation, you can calculate voltage if the current and resistance are known.

EXAMPLE 3–2   Use the Ohm's law formula in Equation (3–2) to calculate the voltage across a 100 Ω resistor when the current is 2 A.

Solution $V = IR = (2 \text{ A})(100 \text{ Ω}) = 200 \text{ V}$ `2` `×` `1` `0` `0` `=`

Related Exercise   Find the voltage across a 1 kΩ resistor when the current is 1 mA.

Formula for Resistance

There is a third equivalent way to state Ohm's law. By dividing both sides of Equation (3–2) by *I* and transposing terms, you obtain the following formula:

$$R = \frac{V}{I}$$

(3–3)

This form of Ohm's law is used to determine resistance if voltage and current values are known.

The three formulas—Equations (3–1), (3–2), and (3–3)—are all equivalent. They are simply three different ways of expressing Ohm's law.

EXAMPLE 3–3   Use the Ohm's law formula in Equation (3–3) to calculate the resistance in a circuit when the voltage is 12 V and the current is 0.5 A.

Solution $R = \frac{V}{I} = \frac{12 \text{ V}}{0.5 \text{ A}} = 24 \text{ Ω}$ `1` `2` `÷` `0` `.` `5` `=`

Related Exercise   Find the resistance when the voltage is 9 V and the current is 10 mA.

The Linear Relationship of Current and Voltage

When we say that the current and voltage are **linearly** proportional, we mean that if one is increased or decreased by a certain percentage, the other will increase or decrease by the same percentage, assuming that the resistance is constant in value. For example, if the voltage across a resistor is tripled, the current will triple.

EXAMPLE 3–4   Show that if the voltage in the circuit of Figure 3–3 is increased to three times its present value, the current will triple in value.

FIGURE 3–3

Solution With 10 V, the current is

$$I = \frac{V}{R} = \frac{10 \text{ V}}{100 \text{ }\Omega} = 0.1 \text{ A}$$

If the voltage is increased to 30 V, the current will be

$$I = \frac{V}{R} = \frac{30 \text{ V}}{100 \text{ }\Omega} = 0.3 \text{ A}$$

The current went from 0.1 A to 0.3 A when the voltage was tripled to 30 V.

Related Exercise Show that if the voltage in Figure 3–3 is quadrupled, the current will also quadruple?

A Graph of Current Versus Voltage

Let's take a constant value of resistance, for example, 10 Ω, and calculate the current for several values of voltage ranging from 10 V to 100 V. The current values obtained are shown in Figure 3–4(a). The graph of the I values versus the V values is shown in Figure 3–4(b). Note that it is a straight line graph. This graph shows that a change in voltage results in a linearly proportional change in current. No matter what value R is, assuming that R is constant, the graph of I versus V will always be a straight line.

FIGURE 3–4
Graph of current versus voltage for $R = 10 \text{ }\Omega$.

V	I
10 V	1 A
20 V	2 A
30 V	3 A
40 V	4 A
50 V	5 A
60 V	6 A
70 V	7 A
80 V	8 A
90 V	9 A
100 V	10 A

$$I = \frac{V}{10 \text{ }\Omega}$$

(a)

(b)

A Graphic Aid for Ohm's Law

You may find the graphic aid in Figure 3–5 of value for applying any of the three forms of Ohm's law. It is a way to help you remember the formulas.

FIGURE 3–5
A graphic aid for the Ohm's law formulas.

$$I = \frac{V}{R} \qquad V = IR \qquad R = \frac{V}{I}$$

3–2 ▪ APPLICATION OF OHM'S LAW

This section provides examples of the application of Ohm's law for calculating voltage, current, and resistance in electric circuits. You will also see how to use quantities expressed with metric prefixes in circuit calculations.

After completing this section, you should be able to

▪ **Use Ohm's law to determine voltage, current, or resistance**
 □ Use Ohm's law to find current when voltage and resistance are known
 □ Use Ohm's law to find voltage when current and resistance are known
 □ Use Ohm's law to find resistance when voltage and current are known
 □ Use quantities with metric prefixes

How to Determine *I* When *V* and *R* Are Known

In these examples you will learn to determine current values when you know the values of voltage and resistance. In these problems, the formula $I = V/R$ is used. In order to get current in amperes, you must express the value of V in volts and the value of R in ohms.

EXAMPLE 3–5

How many amperes of current are in the circuit of Figure 3–6?

FIGURE 3–6

Solution Use the formula $I = V/R$ and substitute 100 V for V and 20 Ω for R:

$$I = \frac{V}{R} = \frac{100 \text{ V}}{20 \text{ Ω}} = 5 \text{ A}$$

There are 5 A of current in this circuit.

Related Exercise What is the current in Figure 3–6 if the voltage is reduced to 50 V?

In electronics, resistance values of thousands or millions of **ohms** are common. As you learned in Chapter 1, large values of resistance are indicated by the metric system prefixes *kilo* (k) and *mega* (M). Thus, thousands of ohms are expressed in *kilohms* (kΩ), and millions of ohms in *megohms* (MΩ). The following examples illustrate how to use kilohms and megohms when you use Ohm's law to calculate current.

EXAMPLE 3–6

Calculate the current in the circuit of Figure 3–7.

FIGURE 3–7

Solution   Remember that 1 kΩ is the same as 1×10^3 Ω. Use the formula $I = V/R$ and substitute 50 V for V and 1×10^3 Ω for R:

$$I = \frac{V}{R} = \frac{50 \text{ V}}{1 \text{ k}\Omega} = \frac{50 \text{ V}}{1 \times 10^3 \ \Omega} = 50 \times 10^{-3} \text{ A} = 50 \text{ mA}$$

The calculator sequence is

Related Exercise   If the resistance in Figure 3–7 is increased to 10 kΩ, what is the current?

In Example 3–6, the current is expressed as 50 mA. Thus, when volts are divided by kilohms, the current will be in milliamperes.

If volts are used when resistance values are in megohms, the current is in microamperes (μA) as Example 3–7 illustrates.

EXAMPLE 3–7

Determine the amount of current in the circuit of Figure 3–8.

FIGURE 3–8

Solution   Recall that 5 MΩ equals 5×10^6 Ω. Use the formula $I = V/R$ and substitute 25 V for V and 5×10^6 Ω for R:

$$I = \frac{V}{R} = \frac{25 \text{ V}}{5 \text{ M}\Omega} = \frac{25 \text{ V}}{5 \times 10^6 \ \Omega} = 5 \times 10^{-6} \text{ A} = 5 \ \mu\text{A}$$

The calculator sequence is

[2] [5] [÷] [5] [EXP] [6] [=]

Related Exercise   If the resistance in Figure 3–8 is decreased to 1 MΩ, what is the current?

Small voltages, usually less than 50 V, are common in electronic circuits. Occasionally, however, large voltages are encountered. For example, the high-voltage supply in a television receiver is around 20,000 V (20 kV), and transmission voltages generated by the power companies may be as high as 345,000 V (345 kV).

EXAMPLE 3–8   How much current is there through 100 MΩ when 50 kV are applied?

Solution   In this case, divide 50 kV by 100 MΩ to get the current. Substitute 50×10^3 V for 50 kV and 100×10^6 Ω for 100 MΩ:

$$I = \frac{V}{R} = \frac{50 \text{ kV}}{100 \text{ M}\Omega} = \frac{50 \times 10^3 \text{ V}}{100 \times 10^6 \text{ }\Omega} = 0.5 \times 10^{-3} \text{ A} = 0.5 \text{ mA}$$

The calculator sequence is

[5] [0] [EXP] [3] [÷] [1] [0] [0] [EXP] [6] [=]

Related Exercise   How much current is there through 10 MΩ when 2 kV are applied?

How to Determine *V* When *I* and *R* Are Known

In these examples you will learn to determine voltage values when the current and resistance are known using the formula $V = IR$. To obtain voltage in volts, you must express the value of I in amperes and the value of R in ohms.

EXAMPLE 3–9   In the circuit of Figure 3–9, how much voltage is needed to produce 5 A of current?

FIGURE 3–9

Solution   Substitute 5 A for I and 100 Ω for R into the formula $V = IR$:

$$V = IR = (5 \text{ A})(100 \text{ }\Omega) = 500 \text{ V}$$

Thus, 500 V are required to produce 5 A of current through a 100 Ω resistor.

Related Exercise How much voltage is required to produce 8 A in the circuit of Figure 3–9?

EXAMPLE 3–10

How much voltage will be measured across the resistor in Figure 3–10?

FIGURE 3–10

Solution Note that 5 mA equals 5×10^{-3} A. Substitute the values for I and R into the formula $V = IR$:

$$V = IR = (5 \text{ mA})(50 \ \Omega) = (5 \times 10^{-3} \text{ A})(50 \ \Omega) = 250 \text{ mV}$$

When milliamperes are multiplied by ohms, the result is millivolts.
   The calculator sequence is

5 EXP +/− 3 × 5 0 =

Related Exercise Change the resistor in Figure 3–10 to 22 Ω and determine the voltage required to produce 10 mA.

EXAMPLE 3–11

The circuit in Figure 3–11 has a current of 10 mA. What is the voltage?

FIGURE 3–11

Solution Note that 10 mA equals 10×10^{-3} A and that 3.3 kΩ equals 3.3×10^{3} Ω. Substitute these values into the formula $V = IR$:

$$V = IR = (10 \text{ mA})(3.3 \text{ k}\Omega) = (10 \times 10^{-3} \text{ A}) (3.3 \times 10^{3} \ \Omega) = 33 \text{ V}$$

When milliamperes and kilohms are multiplied, the result is volts.
   The calculator sequence is

1 0 EXP +/− 3 × 3 • 3 EXP 3 =

Related Exercise What is the voltage in Figure 3–11 if the current is 5 mA?

How to Determine *R* When *V* and *I* Are Known

In these examples you will learn to determine resistance values when the voltage and current are known using the formula $R = V/I$. To find resistance in ohms, you must express the value of *V* in volts and the value of *I* in amperes.

EXAMPLE 3–12   In the circuit of Figure 3–12 how much resistance is needed to draw 3 A of current from the battery?

FIGURE 3–12

Solution   Substitute 12 V for *V* and 3 A for *I* into the formula $R = V/I$:

$$R = \frac{V}{I} = \frac{12\text{ V}}{3\text{ A}} = 4\ \Omega$$

Related Exercise   How much resistance is required to draw 3 mA from the battery in Figure 3–12?

EXAMPLE 3–13   Suppose that the ammeter in Figure 3–13 indicates 5 mA of current and the voltmeter reads 150 V. What is the value of *R?*

FIGURE 3–13

Solution   Note that 5 mA equals 5×10^{-3} A. Substitute the voltage and current values into the formula $R = V/I$:

$$R = \frac{V}{I} = \frac{150\text{ V}}{5\text{ mA}} = \frac{150\text{ V}}{5 \times 10^{-3}\text{ A}} = 30 \times 10^3\ \Omega = 30\text{ k}\Omega$$

Thus, if volts are divided by milliamperes, the resistance will be in kilohms.

Related Exercise   Determine the value of *R* if $V = 50$ V and $I = 500$ mA in Figure 3–13.

SECTION 3–2 REVIEW

1. $V = 10$ V and $R = 5$ Ω. Find I.
2. If a 5 MΩ resistor has 20 kV across it, how much current flows?
3. How much current will 10 kV across 2 kΩ produce?
4. $I = 1$ A and $R = 10$ Ω. Find V.
5. What voltage do you need to produce 3 mA of current in a 3 kΩ resistance?
6. A battery produces 2 A of current into a 6 Ω resistive load. What is the battery voltage?
7. $V = 10$ V and $I = 2$ A. Find R.
8. In a stereo amplifier circuit there is a resistor across which you measure 25 V, and your ammeter indicates 50 mA of current through the resistor. What is the resistor's value in kilohms? In ohms?

3–3 ■ ENERGY AND POWER

Energy is the fundamental ability to do work and produce action or results. When there is current through a resistance, energy is released in the form of heat. A common example of this is a light bulb that becomes too hot to touch. The current through the filament that produces light also produces unwanted heat because the filament has resistance. Power is a measure of how fast energy is being used; electrical components must be able to dissipate a certain amount of energy in a given period of time.

After completing this section, you should be able to

■ **Define energy and power**
  □ Express power in terms of energy
  □ State the unit of power
  □ State the common units of energy
  □ Perform energy and power calculations

Power is the rate at which energy is used.

In other words, **power,** symbolized by P, is a certain amount of **energy** used in a certain length of time, expressed as follows:

$$\text{Power} = \frac{\text{energy}}{\text{time}}$$

$$P = \frac{W}{t} \tag{3–4}$$

Energy is measured in **joules** (J), time is measured in seconds (s), and power is measured in **watts** (W). Energy in joules divided by time in seconds gives power in watts. For example, if 50 J of energy are used in 2 s, the power is 50 J/2 s = 25 W. By definition.

One watt is the amount of power when one joule of energy is consumed in one second.

Thus, the number of joules consumed in 1 s is always equal to the number of watts. For example, if 75 J are used in 1 s, the power is 75 W. The following example illustrates use of these units. To avoid confusion between the symbols for energy and watts, remember that an italic W stands for energy and a nonitalic W stands for watts.

EXAMPLE 3–14   An amount of energy equal to 100 J is used in 5 s. What is the power in watts?

Solution

$$P = \frac{\text{energy}}{\text{time}} = \frac{W}{t} = \frac{100 \text{ J}}{5 \text{ s}} = 20 \text{ W}$$

Related Exercise   If 100 W of power occurs for 30 s, how much energy in joules is consumed?

Amounts of power much less than one watt are common in certain areas of electronics. As with small current and voltage values, metric prefixes are used to designate small amounts of power. Thus, milliwatts (mW) and microwatts (μW) are commonly found in some applications.

In the electrical utilities field, kilowatts (kW) and megawatts (MW) are common units. Radio and television stations also use large amounts of power to transmit signals.

EXAMPLE 3–15   Express the following powers using appropriate metric prefixes:
(a) 0.045 W   (b) 0.000012 W   (c) 3500 W   (d) 10,000,000 W

Solution
(a) $0.045 \text{ W} = 45 \times 10^{-3} \text{ W} = 45 \text{ mW}$ (b) $0.000012 \text{ W} = 12 \times 10^{-6} \text{ W} = 12 \ \mu\text{W}$
(c) $3500 \text{ W} = 3.5 \times 10^3 \text{ W} = 3.5 \text{ kW}$ (d) $10,000,000 \text{ W} = 10 \times 10^6 \text{ W} = 10 \text{ MW}$

Related Exercise   Express the following amounts of power in watts without metric prefixes:
(a) 1 mW   (b) 1800 μW   (c) 3 MW   (d) 10 kW

The Kilowatt-hour (kWh) Unit of Energy

Since power is the rate of energy usage, power utilized over a period of time represents energy consumption. If we multiply power and time, we have energy, symbolized by W.

$$\text{Energy} = \text{power} \times \text{time}$$

$$W = Pt \tag{3–5}$$

Earlier, the joule was defined as a unit of energy. However, there is another way to express energy. Since power is expressed in watts and time in seconds, we can use units of energy called the watt-second (Ws), watt-hour (Wh), and kilowatt-hour (kWh).

When you pay your electric bill, you are charged on the basis of the amount of energy you use. Because power companies deal in huge amounts of energy, the most practical unit is the **kilowatt-hour.** *You use a kilowatt-hour of energy when you use the equivalent of 1000 W of power for 1 h.* For example, a 100 W light bulb burning for 10 h uses 1 kWh of energy.

EXAMPLE 3–16

Determine the number of kilowatt-hours for each of the following energy consumptions:

(a) 1400 W for 1 hr   (b) 2500 W for 2 h   (c) 100,000 W for 5 h

Solution

(a) 1400 W = 1.4 kW
    Energy = power × time = (1.4 kW)(1 h) = 1.4 kWh
(b) 2500 W = 2.5 kW
    Energy = (2.5 kW)(2 h) = 5 kWh
(c) 100,000 W = 100 kW
    Energy = (100 kW)(5 h) = 500 kWh

Related Exercise How many kilowatt-hours of energy are used by a 250 W light bulb burning for 8 h?

SECTION 3–3 REVIEW

1. Define *power*.
2. Write the formula for power in terms of energy and time.
3. Define *watt*.
4. Express each of the following values of power in the most appropriate units:
    (a) 68,000 W   (b) 0.005 W   (c) 0.000025 W
5. If you use 100 W of power for 10 h, how much energy (in kilowatt-hours) have you consumed?
6. Convert 2000 Wh to kilowatt-hours.
7. Convert 360,000 Ws to kilowatt-hours.

3–4 ■ POWER IN AN ELECTRIC CIRCUIT

The loss of energy in an electric circuit is generally an unwanted by-product of current through the resistance in the circuit. In some cases, however, the generation of heat is the primary purpose of a circuit as, for example, in an electric resistive heater. In any case, you must always deal with power in electrical and electronic circuits.

After completing this section, you should be able to

■ **Calculate power in a circuit**
    ☐ Determine power knowing I and R
    ☐ Determine power knowing V and I
    ☐ Determine power knowing V and R

When there is current through a resistance, the collisions of the electrons give off heat, resulting in a loss of energy as indicated in Figure 3–14. There is always a certain amount of power in an electric circuit, and it is dependent on the amount of resistance and on the amount of current, expressed as follows:

$$P = I^2 R$$

(3–6)

FIGURE 3–14

Power in an electric circuit is seen as heat given off by the resistance.

Heat produced by current through resistance is energy loss.

We can produce an equivalent expression for power by substituting V for IR (I^2 is $I \times I$).

$$P = I^2R = (I \times I)R = I(IR) = (IR)I$$

$$\boxed{P = VI} \qquad (3\text{–}7)$$

We obtain another equivalent expression by substituting V/R for I (Ohm's law) as follows:

$$P = VI = V\left(\frac{V}{R}\right)$$

$$\boxed{P = \frac{V^2}{R}} \qquad (3\text{–}8)$$

The three power formulas in Equations (3–6), (3–7), and (3–8) are also known as **Watt's law.**

How to Use the Appropriate Power Formula

To calculate the power in a resistance, you can use any one of the three Watt's law power formulas, depending on what information you have. For example, assume that you know the values of current and voltage; in this case, calculate the power with the formula $P = VI$. If you know I and R, use the formula $P = I^2R$. If you know V and R, use the formula $P = V^2/R$.

EXAMPLE 3–17   Calculate the power in each of the three circuits of Figure 3–15.

(a)   (b)   (c)

FIGURE 3–15

Solution   In circuit (a), V and I are known. The power is determined as follows with calculator sequences also shown:

$$P = VI = (10 \text{ V})(2 \text{ A}) = 20 \text{ W}$$

In circuit (b), I and R are known. Therefore,

$$P = I^2R = (2 \text{ A})^2(50 \text{ } \Omega) = 200 \text{ W}$$

$$\boxed{2} \boxed{x^2} \boxed{\times} \boxed{5} \boxed{0} \boxed{=}$$

In circuit (c), V and R are known. Therefore,

$$P = \frac{V^2}{R} = \frac{(5 \text{ V})^2}{10 \text{ } \Omega} = 2.5 \text{ W}$$

$$\boxed{5} \boxed{x^2} \boxed{\div} \boxed{1} \boxed{0} \boxed{=}$$

Related Exercise Determine the power in each circuit of Figure 3–15 for the following changes:
(a) I doubled with V the same (b) R doubled with I the same
(c) V halved with R the same

EXAMPLE 3–18

A 100 W light bulb operates on 120 V. How much current does it require?

Solution Use the formula $P = VI$ and solve for I as follows. Transpose the terms to get I on the left of the equation:

$$VI = P$$

Divide both sides of equation by V to get I by itself:

$$\frac{\cancel{V}I}{\cancel{V}} = \frac{P}{V}$$

The Vs cancel on the left, leaving

$$I = \frac{P}{V}$$

Substitute 100 W for P and 120 V for V:

$$I = \frac{P}{V} = \frac{100 \text{ W}}{120 \text{ V}} = 0.833 \text{ A} = 833 \text{ mA}$$

Related Exercise A light bulb draws 545 mA from a 110 V source. What is the power?

**SECTION 3–4
REVIEW**

1. If there are 10 V across a resistor and a current of 3 A through it, what is the power?
2. If a current of 5 A is flowing through a 50 Ω resistor, what is the power?
3. How much power is produced by 20 mA through a 5 kΩ resistor?
4. Five volts are applied to a 10 Ω resistor. What is the power?
5. How much power does a 2 kΩ resistor with 8 V across it produce?
6. What is the resistance of a 75 W bulb that draws 0.5 A?

3–5 ■ THE POWER RATING OF RESISTORS

As you know, a resistor gives off heat when there is current through it. There is a limit to the amount of heat that a resistor can give off, which is specified by its power rating.

After completing this section, you should be able to

■ **Properly select resistors based on power rating**
   □ Define *power rating*
   □ Explain how physical characteristics of resistors determine their power rating
   □ Check for resistor failure with an ohmmeter

The **power rating** is the maximum amount of power that a resistor can dissipate without being damaged by excessive heat buildup. The power rating is not related to the ohmic value (resistance) but rather is determined mainly by the physical size and shape of the resistor. The larger the surface area of a resistor, the more power it can dissipate. *The surface area of a cylindrically shaped resistor is equal to the length (l) times the circumference (c),* as indicated in Figure 3–16.

Length (*l*)

Circumference (*c*)

Surface area = *l* × *c*

FIGURE 3–16
The power rating of a resistor is directly related to its surface area.

Carbon-composition resistors are available in standard power ratings from ⅛ W to 2 W, as shown in Figure 3–17. Available power ratings for other types of resistors vary. For example, carbon-film and metal-film resistors have ratings up to 10 W, and wire-wound resistors have ratings up to 225 W or greater. Figure 3–18 shows some of these resistors.

FIGURE 3–17
Carbon-composition resistors with standard power ratings of ⅛ W, ¼ W, ½ W, 1 W, and 2 W (courtesy of Allen-Bradley Co.).

(a) Vitreous-enameled resistor   (b) "Corrib" corrugated ribbon resistor   (c) Molded vitreous-enameled wire-wound axial lead resistor

(d) Thin resistors   (e) Metal-film precision resistors

FIGURE 3–18
Typical resistors with high power ratings (courtesy of Ohmite Manufacturing Co.).

How to Select the Proper Power Rating for an Application

When a resistor is used in a circuit, its power rating must be greater than the maximum power that it will have to handle. For example, if a carbon-composition resistor is to dissipate 0.75 W in a circuit application, its rating should be at least the next higher standard value which is 1 W. *Ideally, a rating that is approximately twice the actual power should be used when possible.*

EXAMPLE 3–19   Choose an adequate power rating for each of the carbon-composition resistors represented in Figure 3–19 ($\frac{1}{8}$ W, $\frac{1}{4}$ W, $\frac{1}{2}$ W, 1 W, or 2 W).

FIGURE 3–19

(a)   (b)

Solution   For the circuit in Figure 3–19(a), the actual power is

$$P = \frac{V^2}{R} = \frac{(10 \text{ V})^2}{62 \text{ }\Omega} = \frac{100 \text{ V}^2}{62 \text{ }\Omega} = 1.6 \text{ W}$$

Select a resistor with a power rating higher than the actual power. In this case, a 2 W resistor should be used.

    For the circuit in Figure 3–19(b), the actual power is

$$P = I^2R = (10 \text{ mA})^2(1000 \text{ }\Omega) = 0.1 \text{ W}$$

A $\frac{1}{8}$ W (0.125 W) resistor should be used in this case.

Related Exercise   A certain resistor is required to dissipate 0.25 W ($\frac{1}{4}$ W). What minimum standard rating should be used?

Resistor Failures

When the power in a resistor is greater than its rating, the resistor will become excessively hot. As a result, either the resistor will burn open or its resistance value will be greatly altered.

A resistor that has been damaged because of overheating can often be detected by the charred or altered appearance of its surface. If there is no visual evidence, a resistor that is suspected of being damaged can be checked with an ohmmeter for an open or increased resistance value.

EXAMPLE 3–20   Determine whether the resistor in each circuit of Figure 3–20 has possibly been damaged by overheating.

(a)   (b)   (c)

FIGURE 3–20

Solution   For the circuit in Figure 3–20(a),

$$P = \frac{V^2}{R} = \frac{(9\ V)^2}{100\ \Omega} = 0.81\ W$$

The rating of the resistor is ¼ W (0.25 W), which is insufficient to handle the power. The resistor has been overheated and may be burned out.

For the circuit in Figure 3–20(b),

$$P = \frac{V^2}{R} = \frac{(24\ V)^2}{1.5\ k\Omega} = 0.384\ W$$

The rating of the resistor is ½ W (0.5 W), which is sufficient to handle the power.

For the circuit in Figure 3–20(c),

$$P = \frac{V^2}{R} = \frac{(5\ V)^2}{10\ \Omega} = 2.5\ W$$

The rating of the resistor is 1 W, which is insufficient to handle the power. The resistor has been overheated and may be burned out.

Related Exercise   A ¼ W, 1 kΩ resistor is connected across a 12 V battery. Will it overheat?

How to Check a Resistor with an Ohmmeter

A typical analog multimeter (VOM) and a digital multimeter are shown in Figures 3–21(a) and 3–21(b), respectively. The large switch on the analog meter is called a *range*

(a)   (b)

FIGURE 3–21
Typical portable multimeters. (a) Analog (courtesy B & K Precision).
(b) Digital (courtesy John Fluke Manufacturing Co.).

switch. Notice the resistance (OHMS) settings on both meters. For the analog meter in part (a), each setting indicates the amount by which the ohms scale (top scale) on the meter is to be multiplied. For example, if the pointer is at 50 on the ohms scale and the range switch is set at ×100, the resistance being measured is $50 \times 100 \ \Omega = 5000 \ \Omega$. *If the resistor is open, the pointer will stay at full left scale (∞ means infinite) regardless of the range switch setting.*

For the digital meter in Figure 3–21(b), you use the range switch to select the appropriate setting for the value of resistance being measured. You do not have to multiply to get the correct reading because you have a direct digital readout of the resistance value.

The Basic Analog Ohmmeter   A basic ohmmeter, shown in Figure 3–22(a), contains a battery and a variable resistor in series with the meter movement. The range and function selection circuitry is omitted for simplicity. To measure resistance, connect the leads

FIGURE 3–22
Basic analog ohmmeter circuit.

(a)

Test leads

(b)

across the external resistor to be measured, as shown in part (b). This connection completes the circuit, allowing the internal battery to produce current through the movement, in turn causing a deflection of the pointer proportional to the value of the external resistance being measured.

Zero Adjustment of an Analog Ohmmeter

When the ohmmeter leads are open, as in Figure 3–23(a), the pointer is at full left scale, indicating infinite (∞) resistance (open circuit). When the leads are shorted, as in Figure 3–23(b), the pointer is at full right scale, indicating zero resistance.

The variable resistor adjusts the current so that the pointer is at exactly zero when the leads are shorted. It is used to compensate for changes in the internal battery voltage due to aging. Only analog ohmmeters need to be zeroed and they should be rezeroed each time the ohms scale is changed.

Because of the ohmmeter's internal battery, any voltage source must be disconnected from a circuit before a resistance measurement is made. If the voltage is left on, there is a risk of damaging the meter.

The following steps should be used in measuring resistance:

1. Disconnect the voltage from the circuit.

2. To prevent other parts of the circuit from altering the measurement, disconnect the resistor to be checked from its circuit.

3. Select the proper range switch setting. On the analog meter, it is best to select a setting that produces a reading on the right half of the scale for better accuracy. On the digital meter, set the switch at the lowest setting that is greater than the value you expect to measure.

4. Touch the test leads together, and use the ohms adjust knob to set the pointer at 0 Ω (analog meter only).

5. Connect the meter leads across the resistor. Polarity is not important in resistance measurements. Be careful not to touch the test lead contacts with your fingers because your body resistance will affect the reading.

FIGURE 3–23
Zero adjustment.

(a) Open (infinite resistance)   (b) Short (zero resistance)

SECTION 3–5 REVIEW

1. Name two important parameters associated with a resistor.
2. How does the physical size of a resistor determine the amount of power that it can handle?
3. List the standard power ratings of carbon-composition resistors.

4. A resistor must handle 0.3 W. What standard size carbon-composition resistor should be used to dissipate the energy properly?

5. Why is an internal battery required in an ohmmeter?

6. If the pointer indicates 8 on the ohmmeter scale and the range switch is set at R × 1k, what resistance is being measured?

3–6 ■ ENERGY LOSS AND VOLTAGE DROP IN RESISTANCE

As you have seen, when there is current through a resistance, energy is dissipated in the form of heat. This heat loss is caused by collisions of the free electrons within the atomic structure of the resistive material. When a collision occurs, heat is given off; and the electron loses some of its acquired energy.

After completing this section, you should be able to

■ **Explain energy loss and voltage drop**
   □ Discuss the cause of energy loss in a circuit
   □ Define *voltage drop*
   □ Explain the relationship between energy loss and voltage drop

In Figure 3–24, electrons are flowing out of the negative terminal of the battery. They have acquired energy from the battery and are at their highest energy level at the negative side of the circuit. As the electrons move through the resistor, they lose energy. The electrons emerging from the upper end of the resistor are at a lower energy level than those entering the lower end. The drop in energy level through the resistor creates a potential difference, or **voltage drop,** across the resistor having the polarity shown in Figure 3–24.

The upper end of the resistor in Figure 3–24 is less negative (more positive) than the lower end. When we follow the flow of electrons, this change corresponds to a voltage drop from a more negative potential to a less negative potential.

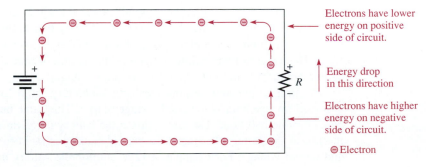

FIGURE 3–24
Illustration of electron flow.

SECTION 3–6 REVIEW

1. What is the basic reason for energy loss in a resistor?

2. What is a voltage drop?

3. What is the polarity of a voltage drop in relation to current direction?

3–7 ■ POWER SUPPLIES

A power supply is a device that provides power to a load. A load is any electrical device or circuit that is connected to the output of the power supply and draws current from the supply.

After completing this section, you should be able to

■ **Discuss power supplies and their characteristics**
   □ Define *ampere-hour rating*
   □ Discuss power supply efficiency

Figure 3–25 shows a block diagram of a power supply with a loading device connected to it. The load can be anything from a light bulb to a computer. The power supply produces a voltage across its two output terminals and provides current through the load, as indicated in the figure. The product IV_{OUT} is the amount of power produced by the supply and consumed by the load. For a given output voltage (V_{OUT}), more current drawn by the load means more power from the supply.

FIGURE 3–25
Block diagram of power supply and load.

Power supplies range from simple batteries to regulated electronic circuits where an accurate output voltage is automatically maintained. A battery is a dc power supply that converts chemical energy into electrical energy. Electronic power supplies normally convert 110 V ac (alternating current) from a wall outlet into a regulated dc (direct current) or ac voltage.

Ampere-hour Ratings of Batteries

Batteries convert chemical energy into electrical energy. Because of their limited source of chemical energy, batteries have a certain capacity that limits the amount of time over which they can produce a given power level. This capacity is measured in ampere-hours (Ah). The **ampere-hour rating** determines the length of time that a battery can deliver a certain amount of current to a load at the rated voltage.

A rating of one ampere-hour means that a battery can deliver one ampere of current to a load for one hour at the rated voltage output. This same battery can deliver two amperes for one-half hour. The more current the battery is required to deliver, the shorter is the life of the battery. In practice, a battery usually is rated for a specified current level and output voltage. For example, a 12 V automobile battery may be rated for 70 Ah at 3.5 A. This means that it can produce 3.5 A for 20 h.

EXAMPLE 3–21   For how many hours can a battery deliver 2 A if it is rated at 70 Ah?

Solution   The ampere-hour rating is the current times the hours.

$$70 \text{ Ah} = (2 \text{ A})(x \text{ h})$$

Solving for x h,

$$x = \frac{70 \text{ Ah}}{2 \text{ A}} = 35 \text{ h}$$

Related Exercise A certain battery delivers 10 A for 6 h. What is its Ah rating?

Power Supply Efficiency

An important characteristic of electron power supplies is efficiency. **Efficiency** is the ratio of the output power to the input power.

$$\text{Efficiency} = \frac{\text{output power}}{\text{input power}}$$

$$\text{Efficiency} = \frac{P_{\text{OUT}}}{P_{\text{IN}}} \qquad (3\text{--}9)$$

Efficiency is often expressed as a percentage. For example, if the input power is 100 W and the output power is 50 W, the efficiency is $(50 \text{ W}/100 \text{ W}) \times 100\% = 50\%$.

All power supplies require that power be put into them. For example, an electronic power supply might use the ac power from a wall outlet as its input. Its output may be regulated dc or ac. The output power is *always* less than the input power because some of the total power must be used internally to operate the power supply circuitry. This amount is normally called the *power loss*.

The output power is the input power minus the amount of internal power loss. High efficiency means that little power is lost and there is a higher proportion of output power for a given input power.

EXAMPLE 3–22 A certain power supply unit requires 25 W of input power. It can produce an output power of 20 W. What is its efficiency, and what is the power loss?

Solution

$$\text{Efficiency} = \left(\frac{P_{\text{OUT}}}{P_{\text{IN}}}\right) 100\% = \left(\frac{20 \text{ W}}{25 \text{ W}}\right) 100\% = 80\%$$

$$P_{\text{LOSS}} = P_{\text{IN}} - P_{\text{OUT}} = 25 \text{ W} - 20 \text{ W} = 5 \text{ W}$$

Related Exercise A power supply has an efficiency of 92%. If P_{IN} is 50 W, what is P_{OUT}?

SECTION 3–7 REVIEW
1. When a loading device draws an increased amount of current from a power supply, does this change represent a greater or a smaller load on the supply?
2. A power supply produces an output voltage of 10 V. If the supply provides 0.5 A to a load, what is the power output?

3. If a battery has an ampere-hour rating of 100 Ah, how long can it provide 5 A to a load?

4. If the battery in Question 3 is a 12 V device, what is its power output for the specified value of current?

5. An electronic power supply used in the lab operates with an input power of 1 W. It can provide an output power of 750 mW. What is its efficiency?

3–8 ■ APPLICATION ASSIGNMENT

In this assignment, an existing resistor box that is to be used as part of a test setup in the lab is to be checked out and modified. Your task is to modify the circuit so that it will meet the requirements of the new application. You will have to apply your knowledge of Ohm's law and Watt's law in order to complete this assignment.

The specifications are as follows:

1. Each resistor is switch selectable and only one resistor is selected at a time.

2. The lowest resistor value is to be 10 Ω.

3. Each successively higher resistance in the switch sequence must be a decade (10 times) increase over the previous one.

4. The maximum resistor value must be 1 MΩ.

5. The maximum voltage across any resistor in the box will be 4 V.

6. Two additional resistors are required, one to limit the current to 10 mA ± 10% and the other to limit the current to 5 mA ± 10% with a 4 V drop.

Step 1: Inspect the Existing Resistor Box

The existing resistor box is shown in both top and bottom views in Figure 3–26. The switch is a rotary type. The resistors all have a power rating of 1 W.

Step 2: Draw the Existing Schematic

From Figure 3–26, determine the resistor values and draw the schematic for the existing circuit so that you will know what you have to work with. Determine the resistor numbering from the *R* labels on the top view.

(a) Top view

(b) Bottom view

FIGURE 3–26

Step 3: Develop the Schematic for the New Requirements

Draw the schematic for a circuit that will accomplish the following:

1. One resistor at a time is to be connected by the switch between terminals 1 and 2 of the box.

2. Provide switch selectable resistor values beginning with 10 Ω and increasing in decade increments to 1 MΩ.

3. Each of the resistors from Step 1 must be selectable by a sequence of adjacent switch positions in ascending order.

4. In addition to the resistors in Step 1, there must be two switch-selectable resistors, one is in switch position 1 (shown in Figure 3–26, bottom view) and must limit the current to 10 mA ± 10% with a 4 V drop and the other is in switch position 8 and must limit the current to 5 mA ± 10% with a 4 V drop.

5. All the resistors must be standard values with 10% tolerance and have a sufficient power rating for the 4 V operating requirement.

Step 4: Compare the Schematics

Determine the modifications that must be made to the existing box to meet the specifications and develop a detailed list of the changes including resistance values, power ratings, wiring, and new components. You should number each point in the schematic for easy reference.

Step 5: Develop a Test Procedure

After the box has been modified to meet the new specifications, it must be tested to see if it is working properly. Determine how you would test the resistor box and what instruments you would use and then detail your test procedure in a step-by-step format.

Step 6: Troubleshoot the Circuit

When an ohmmeter is connected across terminals 1 and 2 of the resistor box, determine the most likely fault in each of the following cases:

1. The ohmmeter shows an infinitely high resistance when the switch is in position 3.

2. The ohmmeter shows an infinitely high resistance in all switch positions.

3. The ohmmeter shows an incorrect value of resistance when the switch is in position 6.

SECTION 3–8 REVIEW	1. Explain how you applied Watt's law to this application assignment. 2. Explain how you applied Ohm's law to this application assignment. 3. How did you determine the power ratings of the resistors in the existing box?

■ SUMMARY

- Voltage and current are linearly proportional.
- Ohm's law gives the relationship of voltage, current, and resistance.
- Current is directly proportional to voltage.
- Current is inversely proportional to resistance.
- A kilohm (kΩ) is one thousand ohms.
- A megohm (MΩ) is one million ohms.

- A microampere (μA) is one-millionth of an ampere.
- A milliampere (mA) is one-thousandth of an ampere.
- Use $I = V/R$ when calculating the current.
- Use $V = IR$ when calculating the voltage.
- Use $R = V/I$ when calculating the resistance.
- One watt equals one joule per second.
- Watt is the unit of power, joule is a unit of energy, and second is a unit of time.
- The power rating of a resistor determines the maximum power that it can handle safely.
- Resistors with a larger physical size can dissipate more power in the form of heat than smaller ones.
- A resistor should have a power rating higher than the maximum power that it is expected to handle in the circuit.
- Power rating is not related to resistance value.
- A resistor normally opens when it burns out.
- Energy is equal to power multiplied by time.
- The kilowatt-hour is a unit of energy.
- One kilowatt-hour is one thousand watts used for one hour.
- A power supply is an energy source used to operate electrical and electronic devices.
- A battery is one type of power supply that converts chemical energy into electrical energy.
- An electronic power supply converts commercial energy (ac from the power company) to regulated dc or ac at various voltage levels.
- The output power of a supply is the output voltage times the load current.
- A load is a device that draws current from the power supply.
- The capacity of a battery is measured in ampere-hours (Ah).
- One ampere-hour equals one ampere used for one hour, or any other combination of amperes and hours that has a product of one.
- A power supply with a high efficiency wastes less power than one with a lower efficiency.
- The formula wheel in Figure 3–27 gives the Ohm's law and Watt's law relationships.

FIGURE 3–27

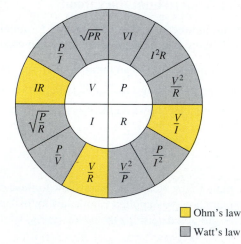

■ Ohm's law
■ Watt's law

■ GLOSSARY

Ampere-hour rating   A number given in ampere-hours (Ah) determined by multiplying the current (A) times the length of time (h) a battery can deliver that current to a load.

Efficiency   The ratio of the output power to the input power.

Energy   The fundamental ability to do work and produce action or results..

Joule (J)   The unit of energy.

Kilowatt-hour (kWh)   A common unit of energy used mainly by utility companies.

Linear   Characterized by a straight-line relationship.

Ohm (Ω)   The unit of resistance.

Ohm's law   A law stating that current is directly proportional to voltage and inversely proportional to resistance.

Power   The rate of energy consumption.

Power rating   The maximum amount of power that a resistor can dissipate without being damaged by excessive heat buildup.

Voltage drop   The potential difference in voltage between two points when there is a drop in energy level due to resistance.

Watt (W)   The unit of power.

Watt's law   A law that states the relationships of power to current, voltage, and resistance.

■ FORMULAS

(3–1)	$I = \dfrac{V}{R}$	Ohm's law
(3–2)	$V = IR$	Ohm's law
(3–3)	$R = \dfrac{V}{I}$	Ohm's law
(3–4)	$P = \dfrac{W}{t}$	Power equals energy divided by time.
(3–5)	$W = Pt$	Energy equals power multiplied by time.
(3–6)	$P = I^2R$	Power equals current squared times resistance.
(3–7)	$P = VI$	Power equals voltage times current.
(3–8)	$P = \dfrac{V^2}{R}$	Power equals voltage squared divided by resistance.
(3–9)	$\text{Efficiency} = \dfrac{P_{OUT}}{P_{IN}}$	Power supply efficiency

---

■ SELF-TEST

1. Ohm's law states that
   (a) current equals voltage times resistance
   (b) voltage equals current times resistance
   (c) resistance equals current divided by voltage
   (d) voltage equals current squared times resistance

2. When the voltage across a resistor is doubled, the current will
   (a) triple   (b) halve   (c) double   (d) not change

3. When 10 V are applied across a 20 Ω resistor, the current is
   (a) 10 A   (b) 0.5 A   (c) 200 A   (d) 2 A

4. When 10 mA flow through a 1 kΩ resistor, the voltage across the resistor is
   (a) 100 V   (b) 0.1 V   (c) 10 kV   (d) 10 V

5. If 20 V are applied across a resistor and there are 6.06 mA of current, the resistance is
   (a) 3.3 kΩ (b) 33 kΩ (c) 330 Ω (d) 3.03 kΩ

6. A current of 250 μA through a 4.7 kΩ resistor produces a voltage drop of
   (a) 53.2 V   (b) 1.175 mV   (c) 18.8 V   (d) 1.175 V

7. A resistance of 2.2 MΩ is connected across a 1 kV source. The resulting current is approximately
   (a) 2.2 mA   (b) 0.455 mA   (c) 45.5 μA   (d) 0.455 A

8. Power can be defined as
   (a) energy   (b) heat
   (c) the rate at which energy is used   (d) the time required to use energy

9. For 10 V and 50 mA, the power is
   (a) 500 mW   (b) 0.5 W   (c) 500,000 μW   (d) answers (a), (b), and (c)

10. When the current through a 10 kΩ resistor is 10 mA, the power is
    (a) 1 W   (b) 10 W   (c) 100 mW   (d) 1000 μW

11. A 2.2 kΩ resistor dissipates 0.5 W. The current is
    (a) 15.1 mA   (b) 0.227 mA   (c) 1.1 mA   (d) 4.4 mA

12. A 330 Ω resistor dissipates 2 W. The voltage is
    (a) 2.57 V   (b) 660 V   (c) 6.6 V   (d) 25.7 V

13. The power rating of a carbon-composition resistor that is to handle up to 1.1 W should be
    (a) 0.25 W   (b) 1 W   (c) 2 W   (d) 5 W

14. A 22 Ω half-watt resistor and a 220 Ω half-watt resistor are connected across a 10 V source. Which one(s) will overheat?
    (a) 22 Ω   (b) 220 Ω   (c) both   (d) neither

15. When the needle of an analog ohmmeter indicates infinity, the resistor being measured is
    (a) overheated   (b) shorted   (c) open   (d) reversed

▪ BASIC
PROBLEMS

SECTION 3–1   Ohm's Law

1. The current in a circuit is 1 A. Determine what the current will be when
   (a) the voltage is tripled
   (b) the voltage is reduced by 80%
   (c) the voltage is increased by 50%

2. The current is a circuit is 100 mA. Determine what the current will be when
   (a) the resistance is increased by 100%
   (b) the resistance is reduced by 30%
   (c) the resistance is quadrupled

3. The current in a circuit is 10 mA. What will the current be if the voltage is tripled and the resistance is doubled?

SECTION 3–2   Application of Ohm's Law

4. Determine the current in each case.
   (a) $V = 5$ V, $R = 1$ Ω   (b) $V = 15$ V, $R = 10$ Ω
   (c) $V = 50$ V, $R = 100$ Ω   (d) $V = 30$ V, $R = 15$ kΩ
   (e) $V = 250$ V, $R = 5$ MΩ

5. Determine the current in each case.
   (a) $V = 9$ V, $R = 2.7$ kΩ   (b) $V = 5.5$ V, $R = 10$ kΩ
   (c) $V = 40$ V, $R = 68$ kΩ   (d) $V = 1$ kV, $R = 2$ kΩ
   (e) $V = 66$ kV, $R = 10$ MΩ

6. A 10 Ω resistor is connected across a 12 V battery. How much current is there through the resistor?

7. A resistor is connected across the terminals of a dc voltage source in each part of Figure 3–28. Determine the current in each resistor.

8. Calculate the voltage for each value of I and R.
   (a) $I = 2$ A, $R = 18$ Ω   (b) $I = 5$ A, $R = 50$ Ω
   (c) $I = 2.5$ A, $R = 600$ Ω   (d) $I = 0.6$ A, $R = 47$ Ω
   (e) $I = 0.1$ A, $R = 500$ Ω

9. Calculate the voltage for each value of I and R.
   (a) $I = 1$ mA, $R = 10$ Ω   (b) $I = 50$ mA, $R = 33$ Ω
   (c) $I = 3$ A, $R = 5$ kΩ   (d) $I = 1.6$ mA, $R = 2.2$ kΩ
   (e) $I = 250$ μA, $R = 1$ kΩ   (f) $I = 500$ mA, $R = 1.5$ MΩ
   (g) $I = 850$ μA, $R = 10$ MΩ   (h) $I = 75$ μA, $R = 50$ Ω

10. Three amperes of current are measured through a 27 Ω resistor connected across a voltage source. How much voltage does the source produce?

11. Assign a voltage value to each source in the circuits of Figure 3–29 to obtain the indicated amounts of current.

12. Calculate the resistance for each value of V and I.
    (a) $V = 10$ V, $I = 2$ A   (b) $V = 90$ V, $I = 45$ A
    (c) $V = 50$ V, $I = 5$ A   (d) $V = 5.5$ V, $I = 10$ A
    (e) $V = 150$ V, $I = 0.5$ A

(a)   (b)   (c)

FIGURE 3–28

FIGURE 3–29

27 kΩ

V

3 mA

(a)

100 MΩ

V

5 μA

(b)

50 Ω

V

2.5 A

(c)

13. Calculate R for each set of V and I values.
   (a) $V = 10$ kV, $I = 5$ A   (b) $V = 7$ V, $I = 2$ mA
   (c) $V = 500$ V, $I = 250$ mA   (d) $V = 50$ V, $I = 500$ μA
   (e) $V = 1$ kV, $I = 1$ mA

14. Six volts are applied across a resistor. A current of 2 mA is measured. What is the value of the resistor?

15. Choose the correct value of resistance to get the current values indicated in each circuit of Figure 3–30.

FIGURE 3–30

R

8 V

2 A

(a)

R

12 V

4 mA

(b)

R

30 V

150 μA

(c)

SECTION 3–3   Energy and Power

16. What is the power when energy is consumed at the rate of 350 J/s?

17. How many watts are used when 7500 J of energy are consumed in 5 h?

18. Convert the following to kilowatts:
    (a) 1000 W (b) 3750 W (c) 160 W (d) 50,000 W

19. Convert the following to megawatts:
    (a) 1,000,000 W (b) 3×10^6 W (c) 15×10^7 W (d) 8700 kW

20. Convert the following to milliwatts:
    (a) 1 W (b) 0.4 W (c) 0.002 W (d) 0.0125 W

21. Convert the following to microwatts:
    (a) 2 W (b) 0.0005 W (c) 0.25 mW (d) 0.00667 mW

22. Convert the following to watts:
    (a) 1.5 kW (b) 0.5 MW (c) 350 mW (d) 9000 μW

SECTION 3–4   Power in an Electric Circuit

23. If a resistor has 5.5 V across it and 3 mA flowing through it, what is the power?

24. An electric heater works on 115 V and draws 3 A of current. How much power does it use?

25. How much power is produced by 500 mA of current through a 4.7 kΩ resistor?

26. Calculate the power handled by a 10 kΩ resistor carrying 100 μA.

27. If there are 60 V across a 600 Ω resistor, what is the power?

28. A 50 Ω resistor is connected across the terminals of a 1.5 V battery. What is the power dissipation in the resistor?

29. If a resistor is to carry 2 A of current and handle 100 W of power, how many ohms must it be? Assume that the voltage can be adjusted to any required value.

30. Convert 5×10^6 Wmin to kWh.

31. Convert 6700 Ws to kWh.

32. How many watt-hours does 50 W used for 12 h equal? How many kilowatt-hours?

SECTION 3–5   The Power Rating of Resistors

33. A 6.8 kΩ resistor has burned out in a circuit. You must replace it with another resistor with the same resistance value. If the resistor carries 10 mA, what should its power rating be? Assume that you have available carbon-composition resistors in all the standard power ratings.

34. A certain type of power resistor comes in the following ratings: 3 W, 5 W, 8 W, 12 W, 20 W. Your particular application requires a resistor that can handle approximately 8 W. Which rating would you use? Why?

SECTION 3–6   Energy Loss and Voltage Drop in Resistance

35. For each circuit in Figure 3–31, assign the proper polarity for the voltage drop across the resistor.

FIGURE 3–31

(a)   (b)   (c)

SECTION 3–7   Power Supplies

36. A 50 Ω load consumes 1 W of power. What is the output voltage of the power supply?

37. A battery can provide 1.5 A of current for 24 h. What is its ampere-hour rating?

38. How much continuous current can be drawn from an 80 Ah battery for 10 h?

39. If a battery is rated at 650 mAh, how much current will it provide for 48 h?

40. If the input power is 500 mW and the output power is 400 mW, how much power is lost? What is the efficiency of this power supply?

41. To operate at 85% efficiency, how much output power must a source produce if the input power is 5 W?

42. A certain power supply provides a continuous 2 W to a load. It is operating at 60% efficiency. In a 24 h period, how many kilowatt-hours does the power supply consume?

■ **ADVANCED PROBLEMS**

43. The filament of a light bulb in the circuit of Figure 3–32(a) has a certain amount of resistance, represented by an equivalent resistance in Figure 3–32(b). If the bulb operates with 120 V and 0.8 A of current, what is the resistance of its filament?

(a)

(b)

FIGURE 3–32

44. A certain electrical device has an unknown resistance. You have available a 12 V battery and an ammeter. How would you determine the value of the unknown resistance? Draw the necessary circuit connections.

45. A variable voltage source is connected to the circuit of Figure 3–33. Start at 0 V and increase the voltage in 10 V steps up to 100 V. Determine the current at each voltage value, and plot a graph of V versus I. Is the graph a straight line? What does the graph indicate?

46. In a certain circuit, $V = 1$ V and $I = 5$ mA. Determine the current for each of the following voltages in the same circuit:
(a) $V = 1.5$ V (b) $V = 2$ V (c) $V = 3$ V (d) $V = 4$ V (e) $V = 10$ V

47. Figure 3–34 is a graph of voltage versus current for three resistance values. Determine R_1, R_2, and R_3.

48. You are measuring the current in a circuit that is operated on a 10 V battery. The ammeter reads 50 mA. Later, you notice that the current has dropped to 30 mA. Eliminating the possibility of a resistance change, you must conclude that the voltage has changed. How much has the voltage of the battery changed, and what is its new value?

49. If you wish to increase the amount of current in a resistor from 100 mA to 150 mA by changing the 20 V source, by how many volts should you change the source? To what new value should you set it?

FIGURE 3–33

FIGURE 3–34

50. By varying the rheostat (variable resistor) in the circuit of Figure 3–35, you can change the amount of current. The setting of the rheostat is such that the current is 750 mA. What is the ohmic value of this setting? To adjust the current to 1 A, to what resistance value must you set the rheostat?

51. A certain resistor has the following color code: orange, orange, red, gold. Determine the maximum and minimum currents you should expect to measure when a 12 V source is connected across the resistor.

52. A 6 V source is connected to a 100 Ω resistor by two 12 ft lengths of 18-gage copper wire. Refer to Table 2–3 to determine the following:
    (a) current (b) resistor voltage (c) voltage across each length of wire

53. A certain appliance uses 300 W. If it is allowed to run continuously for 30 days, how many kilowatt-hours of energy does it consume?

54. At the end of a 31-day period, your utility bill shows that you have used 1500 kWh. What is your average daily power?

55. A certain type of power resistor comes in the following ratings: 3 W, 5 W, 8 W, 12 W, 20 W. Your particular application requires a resistor that can handle approximately 10 W. Which rating would you use? Why?

56. A 12 V source is connected across a 10 Ω resistor for 2 min.
    (a) What is the power?
    (b) How much energy is used?
    (c) If the resistor remains connected for an additional minute, does the power increase or decrease?

57. The rheostat in Figure 3–36 is used to control the current to a heating element. When the rheostat is adjusted to a value of 8 Ω or less, the heating element can burn out. What is the rated value of the fuse needed to protect the circuit if the voltage across the heating element at the point of maximum current is 100 V?

FIGURE 3–35

FIGURE 3–36

■ **ANSWERS TO SECTION REVIEWS**

Section 3–1

1. Ohm's law states that current varies directly with voltage and inversely with resistance.
2. $I = V/R$ 3. $V = IR$ 4. $R = V/I$
5. The current increases; three times when V is tripled
6. Doubling R cuts I in half to 5 mA.
7. No change in I if V and R are doubled.

Section 3–2

1. $I = 10 \text{ V}/5 \text{ Ω} = 2 \text{ A}$ 2. $I = 20 \text{ kV}/5 \text{ MΩ} = 4 \text{ mA}$
3. $I = 10 \text{ kV}/2 \text{ kΩ} = 5 \text{ A}$ 4. $V = (1 \text{ A})(10 \text{ Ω}) = 10 \text{ V}$
5. $V = (3 \text{ mA})(3 \text{ kΩ}) = 9 \text{ V}$ 6. $V = (2 \text{ A})(6 \text{ Ω}) = 12 \text{ V}$
7. $R = 10 \text{ V}/2 \text{ A} = 5 \text{ Ω}$ 8. $R = 25 \text{ V}/50 \text{ mA} = 0.5 \text{ kΩ} = 500 \text{ Ω}$

Section 3–3

1. Power is the rate at which energy is used.
2. $P = W/t$
3. Watt is the unit of power. One watt is the power when 1 J of energy is used in 1 s.

4. **(a)** 68,000 W = 68 kW   **(b)** 0.005 W = 5 mW   **(c)** 0.000025 W = 25 μW

5. (100 W)(10 h) = 1 kWh

6. 2000 Wh = 2 kWh

7. 360,000 Ws = 360 kWs/3600 s/h = 0.1 kWh

Section 3–4

1. $P = (10 \text{ V})(3 \text{ A}) = 30 \text{ W}$

2. $P = (5 \text{ A})^2(50 \text{ }\Omega) = 1250 \text{ W}$

3. $P = (20 \text{ mA})^2(5 \text{ k}\Omega) = 2 \text{ W}$

4. $P = (5 \text{ V})^2/10 \text{ }\Omega = 2.5 \text{ W}$

5. $P = (8 \text{ V})^2/2 \text{ k}\Omega = 32 \text{ mW}$

6. $R = 75 \text{ W}/(0.5 \text{ A})^2 = 300 \text{ }\Omega$

Section 3–5

1. Two resistor parameters are resistance and power rating.

2. A larger resistor surface area dissipates more energy.

3. Standard ratings of carbon-composition resistors are 0.125 W, 0.25 W, 0.5 W, 1 W, 2 W.

4. At least a 0.5 W rating to handle 0.3 W

5. An internal ohmmeter battery is required to produce current through the resistance being measured.

6. $8 \times 1\text{k} = 8 \text{ k}\Omega$

Section 3–6

1. Energy loss in a resistor is caused by collisions of free electrons with the atoms in the material.

2. Voltage drop is the potential difference between two points due to energy loss.

3. Voltage drop is negative to positive in the direction of current.

Section 3–7

1. An increased amount of current represents a greater load.

2. $P_{OUT} = (10 \text{ V})(0.5 \text{ A}) = 5 \text{ W}$

3. 100 Ah/5 A = 20 h

4. $P_{OUT} = (12 \text{ V})(5 \text{ A}) = 60 \text{ W}$

5. Efficiency = (750 mW/1000 mW) 100% = 75%

Section 3–8

1. Watt's law was used to find power ratings ($P = V^2/R$).

2. Ohm's law was used to find the two additional resistors ($R = V/I$).

3. Power ratings were determined by the relative sizes of the resistors, given the range of values.

▪ **ANSWERS TO RELATED EXERCISES FOR EXAMPLES**

3–1	$I_1 = 10 \text{ V}/5 \text{ }\Omega = 2 \text{ A}; I_2 = 10 \text{ V}/20 \text{ }\Omega = 0.5 \text{ A}$	**3–2**	1 V
3–3	900 Ω	**3–4**	Yes
3–5	2.5 A	**3–6**	5 mA
3–7	25 μA	**3–8**	200 μA
3–9	800 V	**3–10**	220 mV
3–11	16.5 V	**3–12**	4 kΩ
3–13	100 Ω	**3–14**	3000 J
3–15	**(a)** 0.001 W **(b)** 0.0018 W **(c)** 3,000,000 W **(d)** 10,000 W	**3–16**	2 kWh
3–17	**(a)** 40 W **(b)** 400 W **(c)** 0.625 W	**3–18**	60 W
3–19	0.5 W (½ W)	**3–20**	No
3–21	60 Ah	**3–22**	46 W

4

SERIES CIRCUITS

Resistive circuits can be of two basic forms: series or parallel. In this chapter we discuss series circuits. Parallel circuits are covered in Chapter 5, and combinations of series and parallel circuits are examined in Chapter 6. In this chapter you will see how Ohm's law is used in series circuits; and you will study another important law, Kirchhoff's voltage law. Also, several important applications of series circuits are presented.

■ CHAPTER OBJECTIVES

☐ Identify a series circuit
☐ Determine the current in a series circuit
☐ Determine the total series resistance
☐ Apply Ohm's law in series circuits
☐ Determine the total effect of voltage sources in series
☐ Apply Kirchhoff's voltage law
☐ Use a series circuit as a voltage divider
☐ Determine power in a series circuit
☐ Determine and identify ground in a circuit
☐ Troubleshoot series circuits

■ APPLICATION ASSIGNMENT

For the application assignment in this chapter, your supervisor has given you a voltage-divider board to evaluate and modify if necessary. You will use it to obtain five different voltage levels from a 12 V battery. The voltage divider will be used to provide positive reference voltages to an electronic circuit in an analog-to-digital converter. Your task will be to check the circuit to see if it provides the required voltages and, if not, you will modify it so that it does. Also, you must make sure that the power ratings of the resistors are adequate for the application and determine how long the battery will last with the voltage divider connected to it. After studying this chapter, you should be able to complete the application assignment in Section 4–11.

4–1 ■ RESISTORS IN SERIES

When connected in series, resistors form a "string" in which there is only one path for current.

After completing this section, you should be able to

- ■ **Identify a series circuit**
  - ☐ Translate a physical arrangement of resistors into a schematic

Figure 4–1(a) shows two resistors connected in **series** between point *A* and point *B*. Part (b) shows three in series, and part (c) shows four in series. Of course, there can be any number of resistors in series.

FIGURE 4–1
Resistors in series.

(a)

(b)

(c)

The only way for electrons to get from point *A* to point *B* in any of the connections of Figure 4–1 is to go through each of the resistors. The following statement identifies a series circuit:

> **A series circuit provides only one path for current between two points in a circuit so that the current is the same through each series resistor.**

Identifying Series Circuits

In an actual circuit diagram, a series circuit may not always be as easy to identify as those in Figure 4–1. For example, Figure 4–2 shows series resistors drawn in other ways. Remember, if there is only one current path between two points, the resistors between those two points are in series, no matter how they appear in a diagram.

(a)   (b)   (c)   (d)   (e)

FIGURE 4–2
Some examples of series circuits. Notice that the current must be the same at all points.

EXAMPLE 4–1   Five resistors are positioned on a circuit board as shown in Figure 4–3. Wire them together in series so that, starting from the negative (–) terminal, R_1 is first, R_2 is second, R_3 is third, and so on. Draw a schematic showing this connection.

FIGURE 4–3

Solution   The wires are connected as shown in Figure 4–4(a), which is the assembly diagram. The schematic is shown in Figure 4–4(b). Note that the schematic does not necessarily show the actual physical arrangement of the resistors as does the assembly diagram. The schematic shows how components are connected electrically; the assembly diagram shows how components are arranged physically.

(a) Assembly diagram

(b) Schematic

FIGURE 4–4

Related Exercise   (a) Show how you would rewire the circuit board in Figure 4–4(a) so that all the odd-numbered resistors come first followed by the even-numbered ones. (b) Determine the value of each resistor.

EXAMPLE 4–2

Describe how the resistors on the printed circuit (PC) board in Figure 4–5 are related electrically. Determine the resistance value of each resistor.

FIGURE 4–5

Solution Resistors R_1 through R_7 are in series with each other. This series combination is connected between pins 1 and 2 on the PC board.

Resistors R_8 through R_{13} are in series with each other. This series combination is connected between pins 3 and 4 on the PC board.

The values of the resistors are $R_1 = 2.2$ kΩ, $R_2 = 3.3$ kΩ, $R_3 = 1$ kΩ, $R_4 = 1.2$ kΩ, $R_5 = 3.3$ kΩ, $R_6 = 4.7$ kΩ, $R_7 = 5.6$ kΩ, $R_8 = 12$ kΩ, $R_9 = 68$ kΩ, $R_{10} = 27$ kΩ, $R_{11} = 12$ kΩ, $R_{12} = 82$ kΩ, and $R_{13} = 270$ kΩ.

Related Exercise How is the circuit in Figure 4–5 changed when pin 2 and pin 3 are connected?

SECTION 4–1 REVIEW

1. How are the resistors connected in a series circuit?
2. How can you identify a series circuit?
3. Complete the schematics for the circuits in each part of Figure 4–6 by connecting the resistors in series in numerical order from A to B.
4. Now connect each group of series resistors in Figure 4–6 in series.

FIGURE 4–6

4–2 ■ CURRENT IN A SERIES CIRCUIT

The same amount of current is through all points in a series circuit. The current through each resistor in a series circuit is the same as the current through all the other resistors that are in series with it.

After completing this section, you should be able to

■ **Determine the current in a series circuit**
  ☐ Show that the current is the same at all points in a series circuit.

Figure 4–7 shows three resistors connected in series to a voltage source. *At any point in this circuit, the current into that point must equal the current out of that point,* as illustrated by the directional arrows. Notice that the current out of each of the resistors must equal the current in, because it is essentially forced through the resistor and out the other end. Since there are no paths other than the single series path into which the current can branch off, the current in each section of the circuit must be the same as the current in all other sections. *Although resistance reduces the current, the reduction is felt at all points equally.*

FIGURE 4–7

Current entering any point in a series circuit is the same as the current leaving that point.

(a) Pictorial   (b) Schematic

FIGURE 4–8

Current is the same at all points in a series circuit.

(a) Pictorial   (b) Schematic

In Figure 4–8, the battery supplies 1 A of current to the series resistors. One ampere is flowing out of the battery's negative terminal. When ammeters are connected in series at several points as shown, each meter reads 1 A.

SECTION 4–2 REVIEW

1. What statement can you make about the amount of current at any point in a series circuit?
2. In a series circuit with a 10 Ω and a 5 Ω resistor in series, 2 A flow through the 10 Ω resistor. How much current flows through the 5 Ω resistor?
3. A milliammeter is connected between points *A* and *B* in Figure 4–9. It measures 50 mA. If you move the meter and connect it between points *C* and *D,* how much current will it indicate? Between *E* and *F?*
4. In Figure 4–10, how much current does ammeter 1 indicate? Ammeter 2?

FIGURE 4–9

4–3 ▪ TOTAL SERIES RESISTANCE

The total resistance of a series circuit is equal to the sum of the resistances of each individual resistor.

After completing this section, you should be able to

▪ **Determine total series resistance**
  □ Explain why resistance values add in series
  □ Apply the series resistance formula

It is understandable that resistor values add in series because each of the resistors in series offers opposition to the current in direct proportion to its resistance. A greater number of resistors connected in series creates more opposition to current. More opposition to current implies a higher value of resistance. So every time a resistor is added in series, the total resistance increases.

Series Resistor Values Add

Figure 4–11 illustrates how series resistances add to increase the total resistance (R_T). Part (a) of the figure has a single 10 Ω resistor. Part (b) shows another 10 Ω resistor connected in series with the first one, making a total resistance of 20 Ω. If a third 10 Ω resistor is connected in series with the first two, as shown in part (c), the total resistance becomes 30 Ω.

(a)　　　　　　　(b)　　　　　　　(c)

FIGURE 4–11
Total resistance increases with each additional series resistor.

Series Resistance Formula

For any number of individual resistors connected in series, the total resistance is the sum of each of the individual values:

$$R_T = R_1 + R_2 + R_3 + \cdots + R_n \qquad \text{(4–1)}$$

where R_T is the total resistance and R_n is the last resistor in the series string (n can be any positive integer equal to the number of resistors in series). For example, if there are four resistors in series ($n = 4$), the total resistance formula is

$$R_T = R_1 + R_2 + R_3 + R_4$$

If there are six resistors in series ($n = 6$), the total resistance formula is

$$R_T = R_1 + R_2 + R_3 + R_4 + R_5 + R_6$$

To illustrate the calculation of total series resistance, let's determine R_T of the circuit of Figure 4–12, where V_S is the source voltage. This circuit has five resistors in series. To find the total resistance, simply add the values:

$$R_T = 56 \ \Omega + 100 \ \Omega + 27 \ \Omega + 10 \ \Omega + 5.6 \ \Omega = 198.6 \ \Omega$$

Note in Figure 4–12 that the order in which the resistances are added does not matter. You can physically change the positions of the resistors in the circuit without affecting the total resistance.

FIGURE 4–12
Example of five resistors in series.

EXAMPLE 4–3   Connect the resistors in Figure 4–13 in series, and determine the total resistance, R_T.

FIGURE 4–13

Solution   The resistors are connected as shown in Figure 4–14. Find the total resistance by adding all the values:

$$R_T = 33 \ \Omega + 68 \ \Omega + 100 \ \Omega + 47 \ \Omega + 10 \ \Omega = 258 \ \Omega$$

FIGURE 4–14

(a) Circuit assembly   (b) Schematic

Related Exercise   Determine the total resistance in Figure 4–14(a) if the positions of R_2 and R_4 are interchanged.

EXAMPLE 4–4   Calculate R_T for each circuit in Figure 4–15.

FIGURE 4–15

(a)   (b)

Solution   For circuit (a),

$$R_T = 1 \text{ k}\Omega + 2.2 \text{ k}\Omega + 4.7 \text{ k}\Omega + 10 \text{ k}\Omega = 17.9 \text{ k}\Omega$$

For circuit (b),

$$R_T = 4.7 \text{ k}\Omega + 10 \text{ k}\Omega + 1 \text{ k}\Omega + 2.2 \text{ k}\Omega = 17.9 \text{ k}\Omega$$

Notice that the total resistance does not depend on the position of the resistors. Both circuits are identical in terms of total resistance.

The calculator sequence for R_T in circuit (a) is

[1] [EXP] [3] [+] [2] [•] [2] [EXP] [3] [+] [4] [•] [7] [EXP] [3] [+] [1] [0] [EXP] [3] [=]

Related Exercise   What is the total resistance for the following series resistors: 1 kΩ, 2.2 kΩ, 3.3 kΩ, and 5.6 kΩ?

EXAMPLE 4–5   Determine R_4 in the circuit of Figure 4–16.

FIGURE 4–16

Solution   From the ohmmeter reading, $R_T = 17.9 \text{ k}\Omega$.

$$R_T = R_1 + R_2 + R_3 + R_4$$

Solving for R_4 yields

$$R_4 = R_T - (R_1 + R_2 + R_3)$$
$$= 17.9 \text{ k}\Omega - (1 \text{ k}\Omega + 2.2 \text{ k}\Omega + 4.7 \text{ k}\Omega) = 10 \text{ k}\Omega$$

Related Exercise   Determine R_4 in Figure 4–16 if the ohmmeter reading is 14,700 Ω.

Equal-Value Series Resistors

When a circuit has more than one resistor of the same value in series, there is a shortcut method to obtain the total resistance: Simply multiply the resistance value of the resistors having the same value by the number of equal-value resistors that are in series. This method is essentially the same as adding the values. For example, five 100 Ω resistors in series have an R_T of 5(100 Ω) = 500 Ω. In general, the formula is expressed as

$$R_T = nR \tag{4–2}$$

where n is the number of equal-value resistors and R is the resistance value.

EXAMPLE 4–6   Find the R_T of eight 22 Ω resistors in series.

Solution   Find R_T by adding the values:

$$R_T = 22 \ \Omega + 22 \ \Omega + 22 \ \Omega + 22 \ \Omega + 22 \ \Omega + 22 \ \Omega + 22 \ \Omega + 22 \ \Omega = 176 \ \Omega$$

However, it is much easier to multiply.

$$R_T = 8(22 \ \Omega) = 176 \ \Omega$$

Related Exercise   Find R_T for three 1 kΩ and two 720 Ω resistors in series.

SECTION 4–3 REVIEW

1. Calculate R_T between points A and B for each circuit in Figure 4–17.

(a)   (b)   (c)

FIGURE 4–17

2. The following resistors are in series: one 100 Ω, two 47 Ω, four 12 Ω, and one 330 Ω. What is the total resistance?
3. Suppose that you have one resistor each of the following values: 1 kΩ, 2.7 kΩ, 3.3 kΩ, and 1.8 kΩ. To get a total resistance of 10 kΩ, you need one more resistor. What should its value be?
4. What is the R_T for twelve 47 Ω resistors in series?

4–4 ▪ OHM'S LAW IN SERIES CIRCUITS

The use of Ohm's law and the basic concepts of series circuits are applied in several examples.

After completing this section, you should be able to

▪ **Apply Ohm's law in series circuits**
  ☐ Find the current in a series circuit
  ☐ Find the voltage across each resistor in series

EXAMPLE 4–7   Some resistors are not color coded with bands but have the values stamped on the resistor body. When the portion of the circuit board shown in Figure 4–18 was assembled, someone mounted the resistors with the labels turned down, and there is no documentation showing the resistor values. Assume that there are a voltmeter, ammeter, and power supply immediately available, but no ohmmeter. Without removing the resistors from the board, use Ohm's law to determine the resistance of each one.

FIGURE 4–18

Solution   The resistors are all in series, so the current is the same through each one. Measure the current by connecting a 12 V source (arbitrary value) and an ammeter as shown in Figure 4–19. Measure the voltage across each resistor by placing the voltmeter across the first resistor. Then repeat this measurement for the other three resistors. For illustration, the voltage values indicated in color are assumed to be the measured values.

FIGURE 4–19
The voltmeter readings across each resistor are indicated.

Determine the resistance of each resistor by substituting the measured values of current and voltage into the Ohm's law formula as follows:

$$R_1 = \frac{V_1}{I} = \frac{2.5 \text{ V}}{25 \text{ mA}} = 100 \ \Omega$$

$$R_2 = \frac{V_2}{I} = \frac{3 \text{ V}}{25 \text{ mA}} = 120 \ \Omega$$

$$R_3 = \frac{V_3}{I} = \frac{4.5 \text{ V}}{25 \text{ mA}} = 180 \ \Omega$$

$$R_4 = \frac{V_4}{I} = \frac{2 \text{ V}}{25 \text{ mA}} = 80 \ \Omega$$

The calculator sequence for R_1 is

[2] [•] [5] [÷] [2] [5] [EXP] [+/−] [3] [=]

Related Exercise What is an easier way to determine the resistance values?

EXAMPLE 4–8

Find the current in the circuit of Figure 4–20.

FIGURE 4–20

Solution The current is determined by the voltage and the total resistance. First, calculate the total resistance:

$$R_T = R_1 + R_2 + R_3 + R_4$$
$$= 82 \ \Omega + 18 \ \Omega + 15 \ \Omega + 10 \ \Omega = 125 \ \Omega$$

Next, use Ohm's law to calculate the current:

$$I = \frac{V_S}{R_T} = \frac{25 \text{ V}}{125 \text{ }\Omega} = 0.2 \text{ A} = 200 \text{ mA}$$

Remember, the same current flows at all points in the circuit. Thus, each resistor has 200 mA through it.

Related Exercise   What is the current in Figure 4–20 if R_4 is changed to 100 Ω?

EXAMPLE 4–9   The current in the circuit of Figure 4–21 is 1 mA. For this amount of current, what must the source voltage V_S be?

FIGURE 4–21

Solution   In order to calculate V_S, first determine R_T:

$$R_T = 1.2 \text{ k}\Omega + 5.6 \text{ k}\Omega + 1.2 \text{ k}\Omega + 1.5 \text{ k}\Omega = 9.5 \text{ k}\Omega$$

Now use Ohm's law to get V_S:

$$V_S = IR_T = (1 \text{ mA})(9.5 \text{ k}\Omega) = 9.5 \text{ V}$$

Related Exercise   If the 5.6 kΩ resistor is changed to 3.9 kΩ, what value of V_S is necessary to keep the current at 1 mA?

EXAMPLE 4–10   Calculate the voltage across each resistor in Figure 4–22, and find the value of V_S. To what value can V_S be raised before the 5 mA fuse blows?

FIGURE 4–22

Solution By Ohm's law, the voltage across each resistor is equal to its resistance multiplied by the current through it. Using the Ohm's law formula $V = IR$, determine the voltage across each of the resistors. Keep in mind that the same current flows through each series resistor. The voltage across R_1 is

$$V_1 = IR_1 = (1 \text{ mA})(1 \text{ k}\Omega) = 1 \text{ V}$$

The voltage across R_2 is

$$V_2 = IR_2 = (1 \text{ mA})(3.3 \text{ k}\Omega) = 3.3 \text{ V}$$

The voltage across R_3 is

$$V_3 = IR_3 = (1 \text{ mA})(4.7 \text{ k}\Omega) = 4.7 \text{ V}$$

The source voltage V_S is equal to the current times the total resistance.

$$R_T = 1 \text{ k}\Omega + 3.3 \text{ k}\Omega + 4.7 \text{ k}\Omega = 9 \text{ k}\Omega$$
$$V_S = (1 \text{ mA})(9 \text{ k}\Omega) = 9 \text{ V}$$

Notice that if you add the voltage drops of the resistors, they total 9 V, which is the same as the source voltage.

The fuse can handle a maximum current of 5 mA. The maximum value of V_S is therefore

$$V_S = IR_T = (5 \text{ mA})(9 \text{ k}\Omega) = 45 \text{ V}$$

Related Exercise Determine all the voltages in the circuit of Figure 4–22 if R_3 is changed to 2.2 kΩ and I is maintained at 1 mA.

SECTION 4–4 REVIEW

1. A 10 V battery is connected across three 100 Ω resistors in series. What is the current through each resistor?
2. How much voltage is required to produce 5 A through the circuit of Figure 4–23?

FIGURE 4–23

3. How much voltage is dropped across each resistor in Figure 4–23 with 5 A?
4. Four equal-value resistors are connected in series with a 5 V source. The measured current is 4.63 mA. What is the value of each resistor?

4–5 ▪ VOLTAGE SOURCES IN SERIES

A voltage source is an energy source that provides a constant voltage to a load. Batteries and power supplies are practical examples. When two or more voltage sources are in series, the total voltage is equal to the algebraic sum of the individual source voltages.

After completing this section, you should be able to

▪ **Determine the total effect of voltage sources in series**
  ☐ Determine the total voltage of series sources with the same polarities
  ☐ Determine the total voltage of series sources with opposite polarities

When batteries are placed in a flashlight, they are connected series-aiding to produce a larger voltage, as illustrated in Figure 4–24. In this example, three 1.5 V batteries are placed in series to produce a total voltage (V_{ST}) as follows:

$$V_{ST} = V_{S1} + V_{S2} + V_{S3} = 1.5 \text{ V} + 1.5 \text{ V} + 1.5 \text{ V} = 4.5 \text{ V}$$

FIGURE 4–24
Example of series-aiding voltage sources.

(a) Flashlight with series batteries

(b) Schematic of flashlight circuit

Series voltage sources are added when their polarities are in the same direction and are subtracted when their polarities are in opposite directions, or series-opposing. For example, if one of the batteries in the flashlight is turned around, as indicated in the schematic of Figure 4–25, its voltage subtracts, because it has a negative value, and reduces the total voltage:

$$V_{ST} = V_{S1} - V_{S2} + V_{S3} = 1.5 \text{ V} - 1.5 \text{ V} + 1.5 \text{ V} = 1.5 \text{ V}$$

FIGURE 4–25
Opposite polarities subtract.

EXAMPLE 4–11   What is the total source voltage (V_{ST}) in Figure 4–26?

FIGURE 4–26

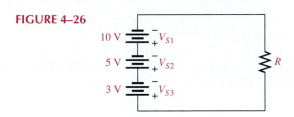

Solution   The polarity of each source is the same (the sources are connected in the same direction in the circuit). Therefore, sum the three voltages to get the total:

$$V_{ST} = V_{S1} + V_{S2} + V_{S3} = 10 \text{ V} + 5 \text{ V} + 3 \text{ V} = 18 \text{ V}$$

The three individual sources can be replaced by a single equivalent source of 18 V with its polarity as shown in Figure 4–27.

FIGURE 4–27

Related Exercise   If the battery V_{S3} in Figure 4–26 is reversed, what is the total voltage?

EXAMPLE 4–12   Determine V_{ST} in Figure 4–28.

FIGURE 4–28

Solution   These sources are connected in opposing directions. If you go clockwise around the circuit, you go from plus to minus through V_{S1}, and minus to plus through V_{S2}. The total voltage is the difference of the two source voltages (algebraic sum of oppositely signed values). The total voltage has the same polarity as the larger-value source. Here we will choose V_{S2} to be positive.

$$V_{ST} = V_{S2} - V_{S1} = 25 \text{ V} - 15 \text{ V} = 10 \text{ V}$$

The two sources in Figure 4–28 can be replaced by a 10 V equivalent one with polarity as shown in Figure 4–29.

FIGURE 4–29

Related Exercise   If an 8 V source in the same direction as V_{S1} is connected in series in Figure 4–28, what is the total voltage?

SECTION 4–5 REVIEW

1. How many 12 V batteries must be connected in series to produce 60 V? Sketch a schematic that shows the battery connections.
2. Four 1.5 V flashlight batteries are connected in series positive to negative. What is the total voltage across the bulb?
3. The resistive circuit in Figure 4–30 is used to bias a transistor amplifier. Show how to connect the resistors to two 15 V power supplies in order to get 30 V across the series resistors.

FIGURE 4–30

4. Determine the total source voltage in each circuit of Figure 4–31.
5. Sketch the equivalent single source circuit for each circuit of Figure 4–31.

(a)   (b)

FIGURE 4–31

4–6 ■ KIRCHHOFF'S VOLTAGE LAW

Kirchhoff's voltage law is a fundamental circuit law that states that the algebraic sum of all the voltages around a closed path is zero or, in other words, the sum of the voltage drops equals the total source voltage.

After completing this section, you should be able to

▪ **Apply Kirchhoff's voltage law**
  ☐ State Kirchhoff's voltage law
  ☐ Determine the source voltage by adding the voltage drops
  ☐ Determine an unknown voltage drop

In an electric circuit, the voltages across the resistors (voltage drops) *always* have polarities opposite to the source voltage polarity. For example, in Figure 4–32, follow a clockwise loop around the circuit and note that the source polarity is plus-to-minus and each voltage drop is minus-to-plus.

Also notice in Figure 4–32 that the current is out of the negative side of the source and through the resistors as indicated by the arrows. The current is into the negative side of each resistor and out the positive side. As the electrons flow through a resistor, they lose energy and are therefore at a lower energy level when they emerge. The lower energy side is less negative (more positive) than the higher energy side. The drop in energy level across a resistor creates a potential difference, or voltage drop, with a minus-to-plus polarity in the direction of the current.

Notice that the voltage from point A to point B in the circuit of Figure 4–32 equals the source voltage, V_S. Also, the voltage from A to B is the sum of the series resistor voltage drops. Therefore, the source voltage is equal to the sum of the three voltage drops.

This discussion is an example of **Kirchhoff's voltage law,** which is generally stated as:

> **The sum of all the voltage drops around a single closed loop in a circuit is equal to the total source voltage in that loop.**

FIGURE 4–32

Kirchhoff's voltage law: The sum of the voltage drops equals the source voltage.

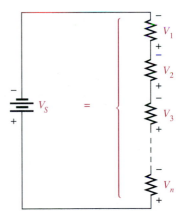

FIGURE 4–33

Sum of n voltage drops equals the source voltage.

The general concept of Kirchhoff's voltage law is illustrated in Figure 4–33 and expressed by Equation (4–3):

$$V_S = V_1 + V_2 + V_3 + \cdots + V_n \qquad (4\text{–}3)$$

where *n* represents the number of voltage drops.

Another Way to State Kirchhoff's Voltage Law

If all the voltage drops around a closed loop are added and then this total is subtracted from the source voltage, the result is zero. This result occurs because the sum of the voltage drops always equals the source voltage.

>**The algebraic sum of all voltages (both source and drops) around a closed path is zero.**

Therefore, another way of expressing Kirchhoff's voltage law is as follows:

$$V_S - V_1 - V_2 - V_3 = 0 \qquad (4\text{–}4)$$

You can verify Kirchhoff's voltage law by connecting a circuit and measuring each resistor voltage and the source voltage as illustrated in Figure 4–34. When the resistor voltages are added together, their sum will equal the source voltage. Any number of resistors can be added.

FIGURE 4–34

Experimental verification of Kirchhoff's voltage law.

EXAMPLE 4–13 Determine the source voltage V_S in Figure 4–35 where the two voltage drops are given.

FIGURE 4–35

Solution By Kirchhoff's voltage law (Eq. 4–3), the source voltage (applied voltage) must equal the sum of the voltage drops. Adding the voltage drops gives the value of the source voltage:

$$V_S = 5 \text{ V} + 10 \text{ V} = 15 \text{ V}$$

Related Exercise If V_S is increased to 30 V in Figure 4–35, what are the two voltage drops?

EXAMPLE 4–14   Determine the unknown voltage drop, V_3, in Figure 4–36.

FIGURE 4–36

Solution   By Kirchhoff's voltage law (Eq. 4–4), the algebraic sum of all the voltages around the circuit is zero (the signs of the voltage drops are opposite the sign of the source).

$$V_S - V_1 - V_2 - V_3 = 0$$

The value of each voltage drop except V_3 is known. Substitute these values into the equation:

$$50 \text{ V} - 12 \text{ V} - 25 \text{ V} - V_3 = 0 \text{ V}$$

Next combine the known values:

$$13 \text{ V} - V_3 = 0 \text{ V}$$

Transpose 13 V to the right side of the equation, and cancel the minus signs:

$$-V_3 = -13 \text{ V}$$
$$V_3 = 13 \text{ V}$$

The voltage drop across R_3 is 13 V, and its polarity is as shown in Figure 4–36.

Related Exercise   Determine V_3 in Figure 4–36 if the source is changed to 25 V.

EXAMPLE 4–15   Find the value of R_4 in Figure 4–37.

FIGURE 4–37

Solution   In this problem you must use both Ohm's law and Kirchhoff's voltage law. Follow this procedure carefully.

First, use Ohm's law to find the voltage drop across each of the known resistors:

$$V_1 = IR_1 = (200 \text{ mA})(10 \text{ }\Omega) = 2 \text{ V}$$
$$V_2 = IR_2 = (200 \text{ mA})(47 \text{ }\Omega) = 9.4 \text{ V}$$
$$V_3 = IR_3 = (200 \text{ mA})(100 \text{ }\Omega) = 20 \text{ V}$$

Next, use Kirchhoff's voltage law to find V_4, the voltage drop across the unknown resistor:

$$V_S - V_1 - V_2 - V_3 - V_4 = 0 \text{ V}$$
$$100 \text{ V} - 2 \text{ V} - 9.4 \text{ V} - 20 \text{ V} - V_4 = 0 \text{ V}$$
$$68.6 \text{ V} - V_4 = 0 \text{ V}$$
$$V_4 = 68.6 \text{ V}$$

Now that you know V_4, you can use Ohm's law to calculate R_4 as follows:

$$R_4 = \frac{V_4}{I} = \frac{68.6 \text{ V}}{200 \text{ mA}} = 343 \text{ }\Omega$$

R_4 is most likely a 330 Ω resistor because 343 Ω is within a standard tolerance range (+5%) of 330 Ω.

Related Exercise Determine R_4 in Figure 4–37 for $V_S = 150$ V and $I = 200$ mA.

SECTION 4–6 REVIEW

1. State Kirchhoff's voltage law in two ways.
2. A 50 V source is connected to a series resistive circuit. What is the total of the voltage drops in this circuit?
3. Two equal-value resistors are connected in series across a 10 V battery. What is the voltage drop across each resistor?
4. In a series circuit with a 25 V source, there are three resistors. One voltage drop is 5 V, and the other is 10 V. What is the value of the third voltage drop?
5. The individual voltage drops in a series circuit are 1 V, 3 V, 5 V, 7 V, and 8 V. What is the total voltage applied across the series circuit?

4–7 ▪ VOLTAGE DIVIDERS

A series circuit acts as a voltage divider. You will learn what this term means and why voltage dividers are an important application of series circuits.

After completing this section, you should be able to

▪ **Use a series circuit as a voltage divider**
  □ Apply the voltage-divider formula
  □ Use the potentiometer as an adjustable voltage divider
  □ Describe some voltage-divider applications

To illustrate how a series string of resistors acts as a voltage divider, we will examine Figure 4–38(a), where there are two resistors in series. As you already know, there are two voltage drops: one across R_1 and one across R_2. These voltage drops are V_1 and V_2, respectively, as indicated in the diagram.

The source voltage is applied between these two terminals. This notation is equivalent to a battery symbol.

(a)   (b)

FIGURE 4–38
A two-resistor voltage divider.

Since the same current flows through each resistor, the voltage drops are proportional to the values of the resistors. For example, if the value of R_2 is twice that of R_1, then the value of V_2 is twice that of V_1. In other words, the total voltage drop divides among the series resistors in amounts directly proportional to the resistance values. The smallest resistance has the least voltage, and the largest resistance has the most voltage ($V = IR$). For example, in Figure 4–38(b), if V_S is 10 V, R_1 is 50 Ω, and R_2 is 100 Ω, then V_1 is one-third the total voltage, or 3.33 V, because R_1 is one-third the total resistance. Likewise, V_2 is two-thirds V_S, or 6.67 V, because R_2 is two-thirds the total resistance.

Voltage-Divider Formula

With a few calculations, a formula for determining how the voltages divide among series resistors can be developed. Let's assume that we have several resistors in series as shown in Figure 4–39. This figure shows five resistors, but there can be any number.

Let's call the voltage drop across any one of the resistors V_x, where x represents the number of a particular resistor (1, 2, 3, and so on). By Ohm's law, the voltage drop across any of the resistors in Figure 4–39 can be written as follows:

$$V_x = IR_x$$

where $x = 1, 2, 3, 4,$ or 5.

FIGURE 4–39
Five-resistor voltage divider.

The current is equal to the source voltage divided by the total resistance ($I = V_S/R_T$). For our example circuit of Figure 4–39, the total resistance is $R_1 + R_2 + R_3 + R_4 + R_5$. Substituting V_S/R_T for I in the expression for V_x,

$$V_x = \left(\frac{V_S}{R_T}\right)R_x$$

By rearranging,

$$V_x = \left(\frac{R_x}{R_T}\right)V_S \qquad\qquad (4\text{–}5)$$

Equation (4–5) is the general voltage-divider formula. It can be stated as follows:

The voltage drop across any resistor or combination of resistors in a series circuit is equal to the ratio of that resistance value to the total resistance, multiplied by the source voltage.

The following three examples illustrate use of the voltage-divider formula.

EXAMPLE 4–16 Determine V_1, the voltage across R_1, and V_2, the voltage across R_2, in the voltage divider in Figure 4–40.

FIGURE 4–40

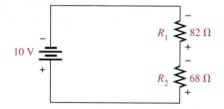

Solution Use the voltage-divider formula, $V_x = (R_x/R_T)V_S$. In this problem you are looking for V_1; so $V_x = V_1$ and $R_x = R_1$. The total resistance is

$$R_T = R_1 + R_2 = 82\ \Omega + 68\ \Omega = 150\ \Omega$$

R_1 is 82 Ω and V_S is 10 V. Substitute these values into the voltage-divider formula:

$$V_1 = \left(\frac{R_1}{R_T}\right)V_S = \left(\frac{82\ \Omega}{150\ \Omega}\right)10\text{ V} = 5.47\text{ V}$$

The calculator sequence is

8 2 ÷ 1 5 0 × 1 0 =

There are two ways to find the value of V_2 in this problem: Kirchhoff's voltage law or the voltage-divider formula. First, use Kirchhoff's voltage law, $V_S = V_1 + V_2$. By substituting the values for V_S and V_1, you can solve for V_2 as follows:

$$V_2 = 10\text{ V} - 5.47\text{ V} = 4.53\text{ V}$$

A second way is to use the voltage-divider formula to find V_2.

$$V_2 = \left(\frac{R_2}{R_T}\right)V_S = \left(\frac{68\ \Omega}{150\ \Omega}\right)10\text{ V} = 4.53\text{ V}$$

The calculator sequence is

6 8 ÷ 1 5 0 × 1 0 =

Related Exercise Find the voltage drops across R_1 and R_2 if R_2 is changed to 180 Ω in Figure 4–40.

EXAMPLE 4–17   Calculate the voltage drop across each resistor in the voltage divider of Figure 4–41.

FIGURE 4–41

Solution   Look at the circuit for a moment and consider the following: The total resistance is 1000 Ω. Ten percent of the total voltage is across R_1 because it is 10% of the total resistance (100 Ω is 10% of 1000 Ω). Likewise, 22% of the total voltage is dropped across R_2 because it is 22% of the total resistance (220 Ω is 22% of 1000 Ω). Finally, R_3 drops 68% of the total voltage (680 Ω is 68% of 1000 Ω).

Because of the convenient values in this problem, it is easy to figure the voltage drops mentally (0.10×100 V = 10 V, 0.22×100 V = 22 V, and 0.68×100 V = 68 V). Such is not always the case, but sometimes a little thinking will produce a result more efficiently and eliminate some calculating.

Although you have already reasoned through this problem, the calculations are

$$V_1 = \left(\frac{R_1}{R_T}\right)V_S = \left(\frac{100\ \Omega}{1000\ \Omega}\right)100\ \text{V} = 10\ \text{V}$$

$$V_2 = \left(\frac{R_2}{R_T}\right)V_S = \left(\frac{220\ \Omega}{1000\ \Omega}\right)100\ \text{V} = 22\ \text{V}$$

$$V_3 = \left(\frac{R_3}{R_T}\right)V_S = \left(\frac{680\ \Omega}{1000\ \Omega}\right)100\ \text{V} = 68\ \text{V}$$

Notice that the sum of the voltage drops is equal to the source voltage, in accordance with Kirchhoff's voltage law. This check is a good way to verify your results.

Related Exercise   If R_1 and R_2 are both changed to 680 Ω in Figure 4–41, what are all the voltage drops?

EXAMPLE 4–18

Determine the voltages between the following points in the voltage divider of Figure 4–42: **(a)** *A* to *B*   **(b)** *A* to *C*   **(c)** *B* to *C*   **(d)** *B* to *D*   **(e)** *C* to *D*

FIGURE 4–42

Solution   First determine R_T.

$$R_T = 1 \text{ k}\Omega + 8.2 \text{ k}\Omega + 3.3 \text{ k}\Omega = 12.5 \text{ k}\Omega$$

Now apply the voltage-divider formula to obtain each required voltage.

(a) The voltage *A* to *B* is the voltage drop across R_1.

$$V_{AB} = \left(\frac{R_1}{R_T} \right) V_S = \left(\frac{1 \text{ k}\Omega}{12.5 \text{ k}\Omega} \right) 25 \text{ V} = 2 \text{ V}$$

(b) The voltage from *A* to *C* is the combined voltage drop across both R_1 and R_2. In this case, R_x in the general formula given in Equation (4–5) is $R_1 + R_2$.

$$V_{AC} = \left(\frac{R_1 + R_2}{R_T} \right) V_S = \left(\frac{9.2 \text{ k}\Omega}{12.5 \text{ k}\Omega} \right) 25 \text{ V} = 18.4 \text{ V}$$

(c) The voltage from *B* to *C* is the voltage drop across R_2.

$$V_{BC} = \left(\frac{R_2}{R_T} \right) V_S = \left(\frac{8.2 \text{ k}\Omega}{12.5 \text{ k}\Omega} \right) 25 \text{ V} = 16.4 \text{ V}$$

(d) The voltage from *B* to *D* is the combined voltage drop across both R_2 and R_3. In this case, R_x in the general formula is $R_2 + R_3$.

$$V_{BD} = \left(\frac{R_2 + R_3}{R_T} \right) V_S = \left(\frac{11.5 \text{ k}\Omega}{12.5 \text{ k}\Omega} \right) 25 \text{V} = 23 \text{ V}$$

(e) Finally, the voltage from *C* to *D* is the voltage drop across R_3.

$$V_{CD} = \left(\frac{R_3}{R_T} \right) V_S = \left(\frac{3.3 \text{ k}\Omega}{12.5 \text{ k}\Omega} \right) 25 \text{ V} = 6.6 \text{ V}$$

If you connect this voltage divider in the lab, you can verify each of the calculated voltages by connecting a voltmeter between the appropriate points in each case.

Related Exercise   Determine each of the previously calculated voltages if V_S is doubled.

The Potentiometer As an Adjustable Voltage Divider

Recall from Chapter 2 that a potentiometer is a variable resistor with three terminals. A potentiometer connected to a voltage source is shown in Figure 4–43. Notice that the two end terminals are labeled 1 and 2. The adjustable terminal or wiper is labeled 3. The potentiometer acts as a voltage divider, which can be illustrated by separating the total resistance into two parts as shown in Figure 4–43(c). The resistance between terminal 1 and terminal 3 (R_{13})

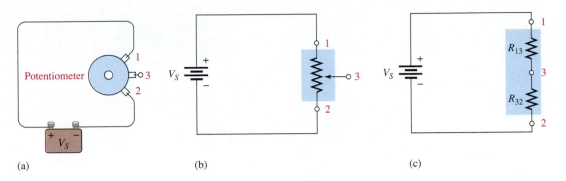

FIGURE 4–43

The potentiometer as a voltage divider.

is one part, and the resistance between terminal 3 and terminal 2 (R_{32}) is the other part. So this potentiometer actually is a two-resistor voltage divider that can be manually adjusted.

Figure 4–44 shows what happens when the wiper terminal (3) is moved. In part (a), the wiper is exactly centered, making the two resistances equal. If we measure the voltage across terminals 3 to 2 as indicated by the voltmeter symbol, we have one-half of the total source voltage. When the wiper is moved up, as in part (b), the resistance between terminals 3 and 2 increases, and the voltage across it increases proportionally. When the wiper is moved down, as in part (c), the resistance between terminals 3 and 2 decreases, and the voltage decreases proportionally.

FIGURE 4–44

Adjusting the voltage divider.

Applications of Voltage Dividers

The volume control of radio or TV receivers is a common application of a potentiometer used as a voltage divider. Since the loudness of the sound is dependent on the amount of voltage associated with the audio signal, you can increase or decrease the volume by adjusting the potentiometer, that is, by turning the knob of the volume control on the set. The block diagram in Figure 4–45 shows how a potentiometer can be used for volume control in a typical receiver.

FIGURE 4–45

A voltage divider used for volume control.

Another application of a voltage divider is illustrated in Figure 4–46, which depicts a potentiometer voltage divider as a fuel-level sensor in an automobile gas tank. As shown in part (a), the float moves up as the tank is filled and moves down as the tank empties.

FIGURE 4–46

A potentiometer voltage divider used as an automotive fuel-level sensor.

(a) Fuel tank

(b) Detail of fuel-level sensor

(c) Schematic of fuel-level sensor

The float is mechanically linked to the wiper arm of a potentiometer, as shown in part (b). The output voltage varies proportionally with the position of the wiper arm. As the fuel in the tank decreases, the sensor output voltage also decreases. The output voltage goes to the indicator circuitry, which controls the fuel gauge or digital readout to show the fuel level. The schematic of this system is shown in part (c).

Still another application for voltage dividers is in setting the dc operating voltage (bias) in transistor amplifiers. Figure 4–47 shows a voltage divider used for this purpose. You will study transistor amplifiers and biasing later, so it is important that you understand the basics of voltage dividers at this point.

These examples are only three out of many possible applications of voltage dividers.

FIGURE 4–47

The voltage divider as a bias circuit for a transistor amplifier.

SECTION 4–7 REVIEW

1. What is a voltage divider?
2. How many resistors can there be in a series voltage divider?
3. Write the general voltage-divider formula.
4. If two series resistors of equal value are connected across a 20 V source, how much voltage is there across each resistor?
5. A 56 Ω resistor and an 82 Ω resistor are connected as a voltage divider. The source voltage is 100 V. Sketch the circuit, and determine the voltage across each of the resistors.
6. The circuit of Figure 4–48 is an adjustable voltage divider. If the potentiometer is linear, where would you set the wiper in order to get 5 V from B to A and 5 V from C to B?

FIGURE 4–48

4–8 ▪ POWER IN A SERIES CIRCUIT

The power dissipated by each individual resistor in a series circuit contributes to the total power in the circuit. The individual powers are additive.

After completing this section, you should be able to

▪ **Determine power in a series circuit**

The total amount of power in a series resistive circuit is equal to the sum of the powers in each resistor in series:

$$P_T = P_1 + P_2 + P_3 + \cdots + P_n \qquad (4\text{–}6)$$

where n is the number of resistors in series, P_T is the total power, and P_n is the power in the last resistor in series. In other words, the powers are additive.

The power formulas that you learned in Chapter 3 are, of course, directly applicable to series circuits. Since the same current flows through each resistor in series, the following formulas are used to calculate the total power:

$$P_T = V_S I$$
$$P_T = I^2 R_T$$
$$P_T = \frac{V_S^2}{R_T}$$

where V_S is the source voltage across the series circuit and R_T is the total resistance. Example 4–19 illustrates how to calculate total power in a series circuit.

EXAMPLE 4–19   Determine the total amount of power in the series circuit in Figure 4–49.

FIGURE 4–49

Solution   The source voltage is 15 V. The total resistance is

$$R_T = 10\ \Omega + 12\ \Omega + 56\ \Omega + 22\ \Omega = 100\ \Omega$$

The easiest formula to use is $P_T = V_S^2/R_T$, since you know both V_S and R_T.

$$P_T = \frac{V_S^2}{R_T} = \frac{(15\ \text{V})^2}{100\ \Omega} = \frac{225\ \text{V}^2}{100\ \Omega} = 2.25\ \text{W}$$

If you determine the power of each resistor separately and add all these powers, you obtain the same result. Another calculation will illustrate. First, find the current:

$$I = \frac{V_S}{R_T} = \frac{15\ \text{V}}{100\ \Omega} = 0.15\ \text{A}$$

Next, calculate the power for each resistor using $P = I^2R$:

$$P_1 = (0.15\ \text{A})^2(10\ \Omega) = 0.225\ \text{W}$$
$$P_2 = (0.15\ \text{A})^2(12\ \Omega) = 0.270\ \text{W}$$
$$P_3 = (0.15\ \text{A})^2(56\ \Omega) = 1.260\ \text{W}$$
$$P_4 = (0.15\ \text{A})^2(22\ \Omega) = 0.495\ \text{W}$$

Now, add these powers to get the total power:

$$P_T = 0.225\ \text{W} + 0.270\ \text{W} + 1.260\ \text{W} + 0.495\ \text{W} = 2.25\ \text{W}$$

This result shows that the sum of the individual powers is equal to the total power as determined by one of the power formulas.

Related Exercise   What is the total power in the circuit of Figure 4–49 if V_S is increased to 30 V?

SECTION 4–8 REVIEW

1. If you know the power in each resistor in a series circuit, how can you find the total power?
2. The resistors in a series circuit have the following powers: 1 W, 2 W, 5 W, and 8 W. What is the total power in the circuit?
3. A circuit has a 100 Ω, a 330 Ω, and a 680 Ω resistor in series. A current of 1 A flows through the circuit. What is the total power?

4–9 ■ CIRCUIT GROUND

Voltage is relative. That is, the voltage at one point in a circuit is always measured relative to another point. For example, if we say that there are +100 V at a certain point in a circuit, we mean that the point is 100 V more positive than some reference point in the circuit. This reference point in a circuit is usually called ground.

After completing this section, you should be able to

■ **Determine and identify ground in a circuit**
  □ Measure voltage with respect to ground
  □ Define the terms *circuit ground* and *earth ground*

The term *ground* derives from the method used in ac power lines, in which one side of the line is neutralized by connecting it to a water pipe or a metal rod driven into the ground. This method of grounding is called **earth ground.**

In most electronic equipment, a large conductive area on a printed circuit board or the metal chassis that houses the assembly is used as the common or reference point, called the **circuit ground** or *chassis ground,* as illustrated in Figure 4–50. This ground provides a convenient way of connecting all common points within the circuit back to one side of the battery or other energy source. The circuit or chassis ground does not necessarily have to be connected to earth ground. However, in many cases it is earth grounded in order to prevent a shock hazard due to a potential difference between circuit and earth ground.

FIGURE 4–50
Simple illustration of circuit ground.

In summary, **ground** is the reference point in electronic circuits. It has a potential of zero volts (0 V) with respect to all other points in the circuit that are referenced to it, as illustrated in Figure 4–51. In part (a), the negative side of the source is grounded, and all voltages indicated are positive with respect to ground. In part (b), the positive side of the source is ground. The voltages at all other points are therefore negative with respect to ground.

(a) Negative ground   (b) Positive ground

FIGURE 4–51
Examples of negative and positive grounds.

Measurement of Voltages with Respect to Ground

When voltages are measured in a circuit, one meter lead is connected to the circuit ground, and the other to the point at which the voltage is to be measured. In a negative ground circuit, the negative meter terminal is connected to the circuit ground. The positive terminal of the voltmeter is then connected to the positive voltage point. Measurement of positive voltage is illustrated in Figure 4–52, where the meter reads the voltage at point *A* with respect to ground.

For a circuit with a positive ground, the positive voltmeter lead is connected to ground, and the negative lead is connected to the negative voltage point, as indicated in Figure 4–53. Here the meter reads the voltage at point *A* with respect to ground.

FIGURE 4–52
Measuring a voltage with respect to negative ground.

FIGURE 4–53
Measuring a voltage with respect to positive ground.

When voltages must be measured at several points in a circuit, the ground lead can be clipped to ground at one point in the circuit and left there. The other lead is then moved from point to point as the voltages are measured. This method is illustrated in Figure 4–54.

Measurement of Voltage across an Ungrounded Resistor

Voltage can normally be measured across a resistor, as shown in Figure 4–55, even though neither side of the resistor is connected to circuit ground. In some cases, when the meter is not isolated from power line ground, the negative lead of the meter will ground one side of the resistor and alter the operation of the circuit. In this situation, another method must be used, as illustrated in Figure 4–56. The voltages on each side of the resistor are measured with respect to ground. The difference of these two measurements is the voltage drop across the resistor.

(a) (b) (c)

FIGURE 4–54
Measuring voltages at several points in a circuit.

FIGURE 4–55
Measuring voltage across a resistor.

FIGURE 4–56
Measuring voltage across a resistor with two separate measurements to ground.

EXAMPLE 4–20   Determine the voltages at each of the indicated points in each circuit of Figure 4–57. Assume that 25 V are dropped across each resistor.

(a)   (b)   (c)

FIGURE 4–57

Solution   For the circuit in Figure 4–57(a), the voltage polarities are as shown. Point E is ground. Single-letter subscripts denote voltage at a point with respect to ground. The voltages with respect to ground are as follows:

$$V_E = 0 \text{ V}$$
$$V_D = +25 \text{ V}$$
$$V_C = +50 \text{ V}$$
$$V_B = +75 \text{ V}$$
$$V_A = +100 \text{ V}$$

For the circuit in Figure 4–57(b), the voltage polarities are as shown. Point D is ground. The voltages with respect to ground are as follows:

$$V_E = -25 \text{ V}$$
$$V_D = 0 \text{ V}$$
$$V_C = +25 \text{ V}$$
$$V_B = +50 \text{ V}$$
$$V_A = +75 \text{ V}$$

For the circuit in Figure 4–57(c), the voltage polarities are as shown. Point C is ground. The voltages with respect to ground are as follows:

$$V_E = -50 \text{ V}$$
$$V_D = -25 \text{ V}$$
$$V_C = 0 \text{ V}$$
$$V_B = +25 \text{ V}$$
$$V_A = +50 \text{ V}$$

Related Exercise   If the ground is at point A in Figure 4–57, what are the voltages at each of the other points with respect to ground?

SECTION 4–9 REVIEW

1. What is the common point in a circuit called?
2. Most voltages in a circuit are referenced to ground (T or F).
3. The housing or chassis is often used as circuit ground (T or F).
4. What is the symbol for ground?
5. What does *earth ground* mean?

4–10 ■ TROUBLESHOOTING

Open resistors or contacts and one point shorted to another are common problems in all circuits including series circuits.

After completing this section, you should be able to

■ **Troubleshoot series circuits**
   ☐ Check for an open circuit
   ☐ Check for a short circuit
   ☐ Identify primary causes of shorts

Open Circuit

The most common failure in a series circuit is an open. For example, when a resistor or a lamp burns out, it causes a break in the current path and creates an **open circuit** as illustrated in Figure 4–58.

An open in a series circuit prevents current.

FIGURE 4–58
An open circuit prevents current.

(a) Complete circuit

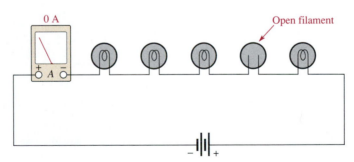

(b) Open circuit

How to Check for an Open Element Sometimes a visual check will reveal a charred resistor or an open lamp filament. However, it is possible for a resistor to open without showing visible signs of damage. In this situation, a voltage check of the series circuit is required. The general procedure is as follows: *Measure the voltage across each resistor in series. The voltage across all of the good resistors will be zero. The voltage across the open resistor will equal the total voltage across the series combination.*

This condition occurs because an open resistor will prevent current through the series circuit. With no current, there can be no voltage drop across any of the good resistors. Since $IR = (0\ A)R = 0\ V$, in accordance with Ohm's law, the voltage on each side of a good resistor is the same. The total voltage must then appear across the open resistor in accordance with Kirchhoff's voltage law, as illustrated in Figure 4–59. To fix the circuit, replace the open resistor.

(a) Voltage check shows that R_3 is open.

$V_T = V_1 + V_2 + V_3 + V_4$
$V_3 = V_T - V_1 - V_2 - V_4$
$\quad = 10\,V - 0\,V - 0\,V - 0\,V$
$\quad = 10\,V$

Kirchhoff's voltage law requires that the total voltage appear across an open.

(b)

FIGURE 4–59
Troubleshooting a series circuit for an open element.

Short Circuit

Sometimes a **short circuit** occurs when two conductors touch or a foreign object such as solder or a wire clipping accidentally connects two sections of a circuit together. This situation is particularly common in circuits with a high component density. Several potential causes of short circuits are illustrated on the PC board in Figure 4–60.

When there is a short, a portion of the series resistance is bypassed (all of the current goes through the short), thus reducing the total resistance as illustrated in Figure 4–61. Notice that the current increases as a result of the short.

A short in a series circuit causes more current.

FIGURE 4–60
Examples of shorts on a PC board.

FIGURE 4–61
The effect of a short in a series circuit.

(a) Before short$\qquad\qquad$(b) After short

SECTION 4–10 REVIEW

1. Define *open circuit*.
2. Define *short circuit*.
3. What happens when a series circuit opens?
4. Name two general ways in which an open circuit can occur in practice. Why might a short circuit occur?
5. When a resistor fails, it will normally open (T or F).
6. The total voltage across a string of series resistors is 24 V. If one of the resistors is open, how much voltage is there across it? How much is there across each of the good resistors?

4–11 ■ APPLICATION ASSIGNMENT

For this assignment, your supervisor has given you a voltage-divider board to evaluate and modify if necessary. It will be used to obtain five different voltage levels from a 12 V battery that has a 6.5 Ah rating. The voltage divider is to be used to provide positive reference voltages to an electronic circuit in an analog-to-digital converter. Your task will be to check the circuit to see if it provides the following voltages within a tolerance of ±5% with respect to the negative side of the battery: 10.4 V, 8.0 V, 7.3 V, 6.0 V, and 2.7 V. If the existing circuit does not provide the specified voltages, you will modify it so that it does. Also, you must make sure that the power ratings of the resistors are adequate for the application and determine how long the battery will last with the voltage divider connected to it.

Step 1: Draw the Schematic of the Circuit

Use Figure 4–62 to determine the resistor values and draw the schematic of the voltage-divider circuit so you will know what you are working with. All the resistors on the board are ¼ W.

FIGURE 4–62

Step 2: Determine the Voltages

Determine each output voltage on the existing circuit board when the positive side of the 12 V battery is connected to pin 3 and the negative side is connected to pin 1. Compare the existing output voltages to the following specifications:

Pin 1: negative terminal of 12 V battery

Pin 2: 2.7 V ± 5%

Pin 3: positive terminal of 12 V battery

Pin 4: 10.4 V ± 5%

Pin 5: 8.0 V ± 5%

Pin 6: 7.3 V + 5%

Pin 7: 6.0 V ± 5%

Step 3: Modify The Existing Circuit (if necessary)

If the output voltages of the existing circuit are not the same as those stated in the specifications of Step 2, make the necessary changes in the circuit to meet the specifications. Draw a schematic of the modified circuit showing resistor values and adequate power ratings.

Step 4: Determine the Life of the Battery

Find the total current drawn from the 12 V battery when the voltage-divider circuit is connected and determine how many days the 6.5 Ah battery will last.

Step 5: Develop a Test Procedure

Determine how you would test the voltage-divider board and what instruments you would use and then detail your test procedure in a step-by-step format.

Step 6: Troubleshoot the Circuit

Determine the most likely fault for each of the following cases (voltages are with respect to the negative battery terminal (pin 1 on the circuit board):

1. No voltage at any of the pins on the circuit board
2. 12 V at pins 3 and 4. All other pins have 0 V.
3. 12 V at all pins except 0 V at pin 1
4. 12 V at pin 6 and 0 V at pin 7
5. 3.3 V at pin 2

SECTION 4–11 REVIEW

1. What is the total power dissipated by the voltage-divider circuit with a 12 V battery?
2. What are the output voltages from the voltage divider if a 6 V battery is used?
3. When the voltage-divider board is connected to the electronic circuit to which it is providing positive reference voltages, which pin on the board should be connected to the ground of the electronic circuit?

▪ SUMMARY

- The same amount of current flows at all points in a series circuit.
- The total series resistance is the sum of all resistors in the series circuit.
- The total resistance between any two points in a series circuit is equal to the sum of all resistors connected in series between those two points.
- If all of the resistors in a series circuit are of equal value, the total resistance is the number of resistors multiplied by the resistance value.

- Voltage sources in series add algebraically.
- First statement of Kirchhoff's voltage law: The sum of the voltage drops equals the total source voltage.
- Second statement of Kirchhoff's voltage law: The sum of all the voltages around a closed path is zero.
- The voltage drops in a circuit are always opposite in polarity to the total source voltage.
- Current is out of the negative side of a source and into the positive side.
- Current is into the negative side of each resistor and out of the positive side.
- A voltage divider is a series arrangement of resistors.
- A voltage divider is so named because the voltage drop across any resistor in the series circuit is divided down from the total voltage by an amount proportional to that resistance value in relation to the total resistance.
- A potentiometer can be used as an adjustable voltage divider.
- The total power in a resistive circuit is the sum of all the individual powers of the resistors making up the series circuit.
- All voltages in a circuit are referenced to ground unless otherwise specified.
- Ground is zero volts with respect to all points referenced to it in the circuit.
- *Negative ground* is the term used when the negative side of the source is grounded.
- *Positive ground* is the term used when the positive side of the source is grounded.
- The voltage across an open series element equals the source voltage.

■ GLOSSARY

Circuit ground   A method of grounding whereby a large conducive area on a printed circuit board or the metal chassis that houses the assembly is used as the common or reference point; also called *chassis ground*.

Earth ground   A method of grounding whereby one side of a power line is neutralized by connecting it to a water pipe or a metal rod driven into the ground.

Ground   In electronic circuits, the common or reference point.

Kirchhoff's voltage law   A law stating that (1) the sum of the voltage drops around a closed loop equals the source voltage or (2) the sum of all the voltages (drops and sources) around a closed loop is zero.

Open circuit   A circuit in which the current path is broken.

Series   In an electrical circuit, a relationship of components in which the components are connected such that they provide a single current path between two points.

Short circuit   A zero or abnormally low resistance between two points; usually an inadvertant condition.

■ FORMULAS

(4–1) $R_T = R_1 + R_2 + R_3 + \cdots + R_n$ Total resistance of n resistors in series

(4–2) $R_T = nR$ Total resistance of n equal-value resistors in series

(4–3) $V_S = V_1 + V_2 + V_3 + \cdots V_n$ Kirchhoff's voltage law

(4–4) $V_S - V_1 - V_2 - V_3 = 0$ Kirchhoff's voltage law

(4–5) $V_x = \left(\dfrac{R_x}{R_T}\right)V_S$ Voltage-divider formula

(4–6) $P_T = P_1 + P_2 + P_3 + \cdots + P_n$ Total power

■ SELF-TEST

1. Five resistors are connected in series and there is a current of 2 A into the first resistor. The amount of current out of the second resistor is
   (a) 2 A   (b) 1 A   (c) 4 A   (d) 0.4 A

2. To measure the current out of the third resistor in a circuit consisting of four series resistors, an ammeter can be placed
   (a) between the third and fourth resistors   (b) between the second and third resistors
   (c) at the positive terminal of the source   (d) at any point in the circuit

3. When a third resistor is connected in series with two series resistors, the total resistance
   (a) remains the same   (b) increases   (c) decreases   (d) increases by one-third

4. When one of four series resistors is removed from a circuit and the circuit reconnected, the current
   (a) decreases by the amount of current through the removed resistor
   (b) decreases by one-fourth
   (c) quadruples
   (d) increases

5. A series circuit consists of three resistors with values of 100 Ω, 220 Ω, and 330Ω. The total resistance is
   (a) less than 100 Ω (b) the average of the values   (c) 550 Ω (d) 650 Ω

6. A 9 V battery is connected across a series combination of 68 Ω, 33 Ω, 100 Ω, and 47 Ω resistors. The amount of current is
   (a) 36.3 mA   (b) 27.6 A   (c) 22.3 mA   (d) 363 mA

7. While putting four 1.5 V batteries in a flashlight, you accidentally put one of them in backward. The light will be
   (a) brighter than normal   (b) dimmer than normal   (c) off   (d) the same

8. If you measure all the voltage drops and the source voltage in a series circuit and add them together, taking into consideration the polarities, you will get a result equal to
   (a) the source voltage   (b) the total of the voltage drops
   (c) zero   (d) the total of the source voltage and the voltage drops

9. There are six resistors in a given series circuit and each resistor has 5 V dropped across it. The source voltage is
   (a) 5 V   (b) 30 V   (c) dependent on the resistor values   (d) dependent on the current

10. A series circuit consists of a 4.7 kΩ, a 5.6 kΩ, and a 10 kΩ resistor. The resistor that has the most voltage across it is
    (a) the 4.7 kΩ (b) the 5.6 kΩ
    (c) the 10 kΩ (d) impossible to determine from the given information

11. Which of the following series combinations dissipates the most power when connected across a 100 V source?
    (a) one 100 Ω resistor   (b) two 100 Ω resistors
    (c) three 100 Ω resistors   (d) four 100 Ω resistors

12. The total power in a certain circuit is 10 W. Each of the five equal-value series resistors making up the circuit dissipates
    (a) 10 W   (b) 50 W   (c) 5 W   (d) 2 W

13. When you connect an ammeter in a series resistive circuit and turn on the source voltage, the meter reads zero. You should check for
    (a) a broken wire   (b) a shorted resistor   (c) an open resistor   (d) both (a) and (c)

14. While checking out a series resistive circuit, you find that the current is higher than it should be. You should look for
    (a) an open circuit   (b) a short   (c) a low resistor value   (d) both (b) and (c)

■ **BASIC PROBLEMS**

SECTION 4–1   Resistors in Series

1. Connect each set of resistors in Figure 4–63 in series between points A and B.

2. Determine which resistors in Figure 4–64 are in series. Show how to interconnect the pins to put all the resistors in series.

FIGURE 4–63

(a)   (b)   (c)

FIGURE 4–64

SECTION 4–2   Current in a Series Circuit

3. What is the current through each of four resistors in a series circuit if the source voltage is 12 V and the total resistance is 120 Ω?

4. The current from the source in Figure 4–65 is 5 mA. How much current does each milliammeter in the circuit indicate?

FIGURE 4–65

SECTION 4–3   Total Series Resistance

5. An 82 Ω resistor and a 56 Ω resistor are connected in series. What is the total resistance?

6. Find the total resistance of each group of series resistors shown in Figure 4–66.

FIGURE 4–66

(a)

(b)

(c)

(d)

7. Determine R_T for each circuit in Figure 4–67. Show how to measure R_T with an ohmmeter.

8. What is the total resistance of twelve 5.6 kΩ resistors in series?

9. Six 47 Ω resistors, eight 100 Ω resistors, and two 22 Ω resistors are in series. What is the total resistance?

10. The total resistance in Figure 4–68 is 20 kΩ. What is the value of R_5?

11. Determine the resistance between each of the following sets of pins on the PC board in Figure 4–64.

    (a) pin 1 and pin 8   (b) pin 2 and pin 3   (c) pin 4 and pin 7   (d) pin 5 and pin 6

12. If all the resistors in Figure 4–64 are connected in series, what is the total resistance?

FIGURE 4–67   **FIGURE 4–68**

SECTION 4–4   Ohm's Law in Series Circuits

13. What is the current in each circuit of Figure 4–69? Show how to connect an ammeter in each case.

14. Determine the voltage across each resistor in Figure 4–69.

15. Three 470 Ω resistors are in series with a 500 V source. How much current is there?

16. Four equal-value resistors are in series with a 5 V source, and a current of 1 mA is measured. What is the value of each resistor?

FIGURE 4–69

SECTION 4–5   Voltage Sources in Series

17. A 5 V battery and a 9 V battery are connected in series with their polarities in the same direction. What is the total voltage?

18. Determine the total source voltage in each circuit of Figure 4–70.

FIGURE 4–70

SECTION 4–6   Kirchhoff's Voltage Law

19. The following voltage drops are measured across each of three resistors in series: 5.5 V, 8.2 V, and 12.3 V. What is the value of the source voltage to which these resistors are connected?

20. Five resistors are in series with a 20 V source. The voltage drops across four of the resistors are 1.5 V, 5.5 V, 3 V, and 6 V. How much voltage is across the fifth resistor?

21. Determine the unspecified voltage drop(s) in each circuit of Figure 4–71. Show how to connect a voltmeter to measure each unknown voltage drop.

(a)   (b)

FIGURE 4–71

SECTION 4–7   Voltage Dividers

22. The total resistance of a series circuit is 500 Ω. What percentage of the total voltage appears across a 22 Ω resistor in the series circuit?

23. Find the voltage between A and B in each voltage divider of Figure 4–72.

24. What is the voltage across each resistor in Figure 4–73? R is the lowest value and all others are multiples of that value as indicated.

25. What is the voltage across each resistor in Figure 4–74?

(a)   (b)

FIGURE 4–72

FIGURE 4–73

(b) Circuit board with meter leads and power
supply leads (red and gray) connected

(a) Meter with leads going to circuit board

FIGURE 4–74

SECTION 4–8   Power in a Series Circuit

26. Five series resistors each dissipate 50 mW of power. What is the total power?

27. Use the results of Problem 25 to find the total power in Figure 4–74.

SECTION 4–9   Circuit Ground

28. Determine the voltage at each point with respect to ground in Figure 4–75.

29. In Figure 4–76, how would you determine the voltage across R_2 by measurement, without connecting a meter directly across the resistor?

30. Determine the voltage at each point with respect to ground in Figure 4–76.

FIGURE 4–75

FIGURE 4–76

SECTION 4–10 Troubleshooting

31. By observing the meters in Figure 4–77, determine the types of failures in the circuits and which components have failed.

32. Is the ohmmeter reading in Figure 4–78 correct? If not, what is wrong?

FIGURE 4–77

(a)

(b)

FIGURE 4–78

(a) Meter with leads going to circuit board

(b) Circuit board with meter leads connected

▪ **ADVANCED**
PROBLEMS

33. Determine the unknown resistance (R_3) in the circuit of Figure 4–79.

34. You have the following resistor values available to you in the lab in unlimited quantities: 10 Ω, 100 Ω, 470 Ω, 560 Ω, 680 Ω, 1 kΩ, 2.2 kΩ, and 5.6 kΩ. All of the other standard values are out of stock. A project that you are working on requires an 18 kΩ resistance. What combination of available values can you use to obtain the needed value?

35. Determine the voltage at each point in Figure 4–80 with respect to the negative side of the battery.

FIGURE 4–79   **FIGURE 4–80**

36. Find all the unknown quantities (in color) in Figure 4–81.

FIGURE 4–81

37. There are 250 mA in a series circuit with a total resistance of 1.5 kΩ. The current must be reduced by 25%. Determine how much resistance must be added in order to accomplish this reduction in current.

38. Four ½ W resistors are in series: 47 Ω, 68 Ω, 100 Ω, and 120 Ω. To what maximum value can the current be raised before the power rating of one of the resistors is exceeded? Which resistor will burn out first if the current is increased above the maximum?

39. A certain series circuit is made up of a ⅛ W resistor, a ¼ W resistor, and a ½ W resistor. The total resistance is 2400 Ω. If each of the resistors is operating at its maximum power level, determine the following:
    (a) I (b) V_S (c) the value of each resistor

40. Using 1.5 V batteries, a switch, and three lamps, devise a circuit to apply 4.5 V across one lamp, two lamps in series, or three lamps in series with a single-control switch. Draw the schematic.

41. Develop a variable voltage divider to provide output voltages ranging from a minimum of 10 V to a maximum of 100 V using a 120 V source. The maximum voltage must be at the maximum resistance setting of the potentiometer. The minimum voltage must be at the minimum resistance (zero ohms) setting. The maximum current is to be 10 mA.

42. Using the standard resistor values given in Appendix A, build a voltage divider to provide the following approximate voltages with respect to the negative terminal of a 30 V source: 8.18 V, 14.73 V, and 24.55 V. The current drain on the source must be limited to no more than 1 mA. The number of resistors, their resistance values, and their power ratings must be specified. Draw a schematic showing the circuit with all resistor values indicated.

FIGURE 4–82

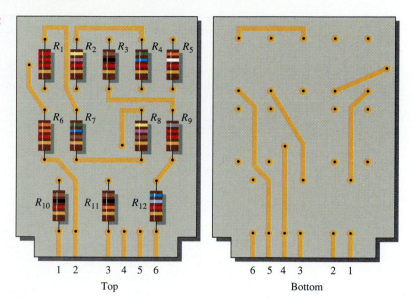

Top   Bottom

43. On the double-sided PC board in Figure 4–82, identify each group of series resistors and determine its total resistance. Note that many of the interconnections feed through the board from the top side to the bottom side.

44. What is the total resistance from A to B for each switch position in Figure 4–83?

45. Determine the current measured by the meter in Figure 4–84 for each switch position.

46. Determine the current measured by the meter in Figure 4–85 for each position of the ganged switch.

47. Determine the voltage across each resistor in Figure 4–86 for each switch position.

FIGURE 4–83

FIGURE 4–84

FIGURE 4–85

FIGURE 4–86

48. Table 4–1 shows the results of resistance measurements on the PC circuit board in Figure 4–82. Are these results correct? If not, identify the possible problems.

TABLE 4–1

BETWEEN PINS	RESISTANCE
1 and 2	∞
1 and 3	∞
1 and 4	4.23 kΩ
1 and 5	∞
1 and 6	∞
2 and 3	23.6 kΩ
2 and 4	∞
2 and 5	∞
2 and 6	∞
3 and 4	∞
3 and 5	∞
3 and 6	∞
4 and 5	∞
4 and 6	∞
5 and 6	19.9 kΩ

49. You measure 15 kΩ between pins 5 and 6 on the PC board in Figure 4–82. Does this indicate a problem? If so, identify it.

50. In checking out the PC board in Figure 4–82, you measure 17.83 kΩ between pins 1 and 2. Also, you measure 13.6 kΩ between pins 2 and 4. Does this indicate a problem on the PC board? If so, identify the fault.

51. The three groups of series resistors on the PC board in Figure 4–82 are connected in series with each other to form a single series circuit by connecting pin 2 to pin 4 and pin 3 to pin 5. A voltage source is connected across pins 1 and 6 and an ammeter is placed in series. As you increase the source voltage, you observe the corresponding increase in current. Suddenly, the current drops to zero and you smell smoke. All resistors are $\frac{1}{2}$ W.

(a) What has happened?

(b) Specifically, what must you do to fix the problem?

(c) At what voltage did the failure occur?

■ **ANSWERS TO SECTION REVIEWS**

Section 4–1

1. Series resistors are connected end-to-end in a "string."

2. There is a single current path in a series circuit.

3. See Figure 4–87.   **4.** See Figure 4–88.

FIGURE 4–87

FIGURE 4–88

Section 4–2

1. The current is the same at all points in a series circuit.

2. There are 2 A through the 5 Ω resistor.

3. 50 mA between C and D, 50 mA between E and F

4. Ammeter 1 indicates 1.79 A; Ammeter 2 indicates 1.79 A.

Section 4–3

1. (a) $R_T = 33\ \Omega + 100\ \Omega + 10\ \Omega = 143\ \Omega$
   (b) $R_T = 39\ \Omega + 56\ \Omega + 10\ \Omega = 105\ \Omega$
   (c) $R_T = 820\ \Omega + 2200\ \Omega + 1000\ \Omega = 4020\ \Omega$

2. $R_T = 100\ \Omega + 2(47\ \Omega) + 4(12\ \Omega) + 330\ \Omega = 572\ \Omega$

3. $10\ k\Omega - 8.8\ k\Omega = 1.2\ k\Omega$

4. $R_T = 12(47\ \Omega) = 564\ \Omega$

Section 4–4

1. $I = 10\ \text{V}/300\ \Omega = 0.033\ \text{A} = 33\ \text{mA}$

2. $V = (5\ \text{A})(43\ \Omega) = 215\ \text{V}$

3. $V_1 = (5\ \text{A})(10\ \Omega) = 50\ \text{V}$, $V_2 = (5\ \text{A})(15\ \Omega) = 75\ \text{V}$, $V_3 = (5\ \text{A})(18\ \Omega) = 90\ \text{V}$

4. $R = 125\ \text{V}/4.63\ \text{mA} = 270\ \Omega$

Section 4–5

1. Five $(5)(12\ \text{V}) = 60\ \text{V}$; see Figure 4–89.

FIGURE 4–89

2. $V_T = (4)(1.5\ \text{V}) = 6.0\ \text{V}$

3. See Figure 4–90.

4. (a) $V_{ST} = 100\ \text{V} + 50\ \text{V} - 75\ \text{V} = 75\ \text{V}$
   (b) $V_{ST} = 20\ \text{V} + 10\ \text{V} - 10\ \text{V} - 5\ \text{V} = 15\ \text{V}$

5. See Figure 4–91.

FIGURE 4–90   **FIGURE 4–91**

Section 4–6

1. Kirchhoff's voltage law states:
   (a) The sum of the voltages around a closed path is zero.
   (b) The sum of the voltage drops equals the total source voltage.

2. $V_{RT} = V_{ST} = 50\ \text{V}$

3. $V_R = 10\ \text{V}/2 = 5\ \text{V}$

4. $V_{R3} = 25\ \text{V} - 5\ \text{V} - 10\ \text{V} = 10\ \text{V}$

5. $V_T = 1\ \text{V} + 3\ \text{V} + 5\ \text{V} + 7\ \text{V} + 8\ \text{V} = 24\ \text{V}$

Section 4–7

1. A voltage divider is a series circuit with two or more resistors with an output taken across one or more of the resistors.

2. There can be any number of resistors in a voltage divider.

3. $V_x = (R_x/R_T) V_S$ is the general voltage-divider formula.

4. $V_R = 20\ V/2 = 10\ V$

5. $V_{56\Omega} = (56\ \Omega/138\ \Omega)100\ V = 40.6\ V$, $V_{82\Omega} = (82\ \Omega/138\Omega)100\ V = 59.4\ V$. See Figure 4–92.

6. Set the potentiometer at the midpoint.

FIGURE 4–92

100 V   56 Ω   82 Ω

Section 4–8

1. Add the powers in each resistor to get total power.

2. $P_T = 1\ W + 2\ W + 5\ W + 8\ W = 16\ W$

3. $P_T = (1\ A)^2(100\ \Omega + 330\ \Omega + 680\ \Omega) = 1110\ W$

Section 4–9

1. Ground   **2.** True   **3.** True   **4.** See Figure 4–93.

5. Earth ground is obtained by driving a conductive rod into the ground.

FIGURE 4–93

Section 4–10

1. An open is a break in the current path.

2. A short is a zero resistance path that bypasses a portion of a circuit.

3. Current ceases when a series circuit opens.

4. An open can be created by a switch or a component failure. A short can be created by a switch or by wire clippings, solder splashes, etc.

5. True

6. 24 V across the open R; 0 V across the other Rs.

Section 4–11

1. $P_T = (12\ V)^2/16.6\ k\Omega = 8.67\ mW$

2. Pin 2: 4.59 V; Pin 6: 2.35 V; Pin 5: 1.99 V; Pin 4: 795 mV; Pin 7: 2.89 V

3. Pin 3 connects to ground.

■ **ANSWERS TO RELATED EXERCISES FOR EXAMPLES**

4–1   (a) Left end of R_1 to "minus," right end of R_1 to top end of R_3, bottom end of R_3 to right end of R_5, left end of R_5 to left end of R_2, right end of R_2 to right end of R_4, left end of R_4 to "plus".

      (b) $R_1 = 1\ k\Omega$, $R_2 = 33\ k\Omega$, $R_3 = 39\ k\Omega$, $R_4 = 470\ \Omega$, $R_5 = 22\ k\Omega$

4–2   The two series circuits are connected in series.

4–3   258 Ω (No change)

4–4   12.1 kΩ

4–5   6.8 kΩ

4–6 $4.44 \text{ k}\Omega$

4–7 Measure each R with an ohmmeter.

4–8 116 mA

4–9 7.8 V

4–10 $V_1 = 1 \text{ V}, V_2 = 3.3 \text{ V}, V_3 = 2.2 \text{ V}, V_S = 6.5 \text{ V}$

4–11 12 V

4–12 2 V

4–13 $10 \text{ V}, 20 \text{ V}$

4–14 6.5 V

4–15 $593 \ \Omega$

4–16 $V_1 = 3.13 \text{ V}; V_2 = 6.87 \text{ V}$

4–17 33.3 V

4–18 $V_{AB} = 4 \text{ V}; V_{AC} = 36.8 \text{ V}; V_{BC} = 32.8 \text{ V}; V_{BD} = 46 \text{ V}; V_{CD} = 13.2 \text{ V}$

4–19 9 W

4–20 $V_A = 0 \text{ V}; V_B = -25 \text{ V}; V_C = -50 \text{ V}; V_D = -75 \text{ V}; V_E = -100 \text{ V}$

5

PARALLEL CIRCUITS

In this chapter, parallel circuits are introduced. Parallel circuits are found in many applications, such as lighting systems, sound systems, appliances, and most electronic equipment.

■ CHAPTER OBJECTIVES

☐ Identify a parallel circuit
☐ Determine the voltage across each parallel branch
☐ Apply Kirchhoff's current law
☐ Determine total parallel resistance
☐ Apply Ohm's law in a parallel circuit
☐ Use a parallel circuit as a current divider
☐ Determine power in a parallel circuit
☐ Describe some basic applications of parallel circuits
☐ Troubleshoot parallel circuits

■ APPLICATION ASSIGNMENT

As a line technician in an electronic instruments company, your job is to troubleshoot any defective instrument that fails the routine line check. In this particular assignment, you must determine the problem(s) with a defective five-range milliammeter so that it can be repaired. The knowledge of parallel circuits and of basic ammeters that you will acquire in this chapter plus your understanding of Ohm's law, current dividers, and the resistor color code will be put to good use. After studying this chapter, you should be able to complete the application assignment in Section 5–10.

5–1 ■ RESISTORS IN PARALLEL

When two or more resistors are individually connected between the same two points, they are in parallel with each other. A parallel circuit provides more than one path for current.

After completing this section, you should be able to

■ **Identify a parallel circuit**
   □ Translate a physical arrangement of resistors into a schematic

Each parallel path in a circuit is called a **branch.** Two resistors connected in parallel are shown in Figure 5–1(a). As shown in part (b), the current out of the source divides when it gets to point *A*. Part of it goes through R_1 and part through R_2. The two currents come back together at point *B*. If additional resistors are connected in parallel with the first two, more current paths are provided, as shown in Figure 5–1(c).

(c)

FIGURE 5–1
Resistors in parallel.

Identifying Parallel Circuits

It is obvious that the resistors in Figure 5–1 are connected in **parallel.** However, in actual circuit diagrams, the parallel relationship often is not so clear. It is important that you learn to recognize parallel circuits regardless of how they may be drawn.

A way to identify a parallel resistive circuit is as follows:

> **If there is more than one resistor connected from point *A* to point *B* and if the voltage across each of the resistors is the same, then the resistors are in parallel.**

Figure 5–2 shows parallel resistors drawn in different ways between two points labeled *A* and *B*. Notice that in each case, the current "travels" two paths going from *A* to *B,* and the voltage across each branch is the same. Although these figures show only two parallel paths, there can be any number of resistors in parallel.

(a) (b) (c)

(d) (e) (f)

FIGURE 5–2
Examples of circuits with two parallel paths.

EXAMPLE 5–1   Five resistors are positioned on a circuit board as shown in Figure 5–3. Show the wiring required to connect all the resistors in parallel, and draw a schematic showing this circuit with resistance values.

FIGURE 5–3

Solution   Wires are connected as shown in the assembly wiring diagram of Figure 5–4(a). The schematic is shown in Figure 5–4(b) with resistance values from the color codes. Again, note that the schematic does not necessarily have to show the actual physical arrangement of the resistors. The schematic shows how components are connected electrically.

(a) Assembly wiring diagram

(b) Schematic

FIGURE 5–4

Related Exercise   How would the circuit have to be rewired if R_2 is removed?

EXAMPLE 5–2   Determine the parallel groupings in Figure 5–5 and the value of each resistor.

FIGURE 5–5

Solution   Resistors R_1 through R_4 and R_{11} and R_{12} are all in parallel. This parallel combination is connected to pins 1 and 4. All resistors in this group are 56 kΩ.

Resistors R_5 through R_{10} are all in parallel. This combination is connected to pins 2 and 3. All resistors in this group are 100 kΩ.

Related Exercise   How would you connect all the resistors on the PC board in parallel?

SECTION 5–1 REVIEW

1. How are the resistors connected in a parallel circuit?
2. How do you identify a parallel circuit?
3. Complete the schematics for the circuits in each part of Figure 5–6 by connecting the resistors in parallel between points A and B.
4. Now connect each group of parallel resistors in Figure 5–6 in parallel with each other.

(a) (b) (c)

FIGURE 5–6

5–2 ■ VOLTAGE DROP IN PARALLEL CIRCUITS

As mentioned in the previous section, each current path in a parallel circuit is called a branch. The voltage across any given branch of a parallel circuit is equal to the voltage across each of the other branches in parallel.

After completing this section, you should be able to

■ **Determine the voltage across each parallel branch**
   □ Explain why the voltage is the same across all parallel resistors

To illustrate voltage drop in a parallel circuit, let's examine Figure 5–7(a). Points A, B, C, and D along the left side of the parallel circuit are electrically the same point because the voltage is the same along this line. You can think of all of these points as being connected by a single wire to the negative terminal of the battery. The points E, F, G, and H along the right side of the circuit are all at a voltage equal to that of the positive terminal of the

(a) (b)

FIGURE 5–7
Voltage across parallel branches is the same.

source. Thus, voltage across each parallel resistor is the same, and each is equal to the source voltage.

Figure 5–7(b) is the same circuit as in part (a), drawn in a slightly different way. Here the left side of each resistor is connected to a single point, which is the negative battery terminal. The right side of each resistor is connected to a single point, which is the positive battery terminal. The resistors are still all in parallel across the source.

In Figure 5–8, a 12 V battery is connected across three parallel resistors. When the voltage is measured across the battery and then across each of the resistors, the readings are the same. As you can see, the same voltage appears across each branch in a parallel circuit.

FIGURE 5–8

The same voltage appears across each resistor in parallel.

(a) Pictorial   (b) Schematic

EXAMPLE 5–3   Determine the voltage across each resistor in Figure 5–9.

FIGURE 5–9

Solution   The five resistors are in parallel; so the voltage drop across each one is equal to the applied source voltage. There is no voltage drop across the fuse.

$$V_{ST} = V_1 = V_2 = V_3 = V_4 = V_5 = 25 \text{ V}$$

Related Exercise   If R_4 is removed from the circuit, what is the voltage across R_3?

**SECTION 5–2
REVIEW**

1. A 10 Ω and a 22 Ω resistor are connected in parallel with a 5 V source. What is the voltage across each of the resistors?
2. A voltmeter connected across R_1 in Figure 5–10 measures 118 V. If you move the meter and connect it across R_2, how much voltage will it indicate? What is the source voltage?
3. In Figure 5–11, how much voltage does voltmeter 1 indicate? Voltmeter 2?
4. How are voltages across each branch of a parallel circuit related?

FIGURE 5–10　　　　　　　　**FIGURE 5–11**

5–3 ■ KIRCHHOFF'S CURRENT LAW

In the last chapter, you learned Kirchhoff's voltage law that dealt with voltages in a closed series circuit. Now, you will learn Kirchhoff's current law that deals with currents in a parallel circuit.

After completing this section, you should be able to

■ **Apply Kirchhoff's current law**
　□ State Kirchhoff's current law
　□ Determine the total current by adding the branch currents
　□ Determine an unknown branch current

Kirchhoff's current law is stated as follows:

> **The sum of the currents into a junction (total current in) is equal to the sum of the currents out of that junction (total current out).**

A **junction** is any point in a circuit where two or more circuit paths come together. In a parallel circuit, a junction is a point where the parallel branches connect together.

　For example, in the circuit of Figure 5–12, point A is one junction and point B is another. Let's start at the negative terminal of the source and follow the current. The total current I_T flows from the source and *into* the junction at point A. At this point, the current splits up among the three branches as indicated. Each of the three branch currents (I_1, I_2, and I_3) flows *out of* junction A. Kirchhoff's current law says that the total current into junction A is equal to the total current out of junction A; that is,

$$I_T = I_1 + I_2 + I_3$$

　Now, following the currents in Figure 5–12 through the three branches, you see that they come back together at point B. Currents I_1, I_2, and I_3 flow into junction B, and I_T flows out. Kirchhoff's current law formula at this junction is therefore the same as at junction A:

$$I_T = I_1 + I_2 + I_3$$

FIGURE 5–12
Kirchhoff's current law: The current into a junction equals the current out of that junction.

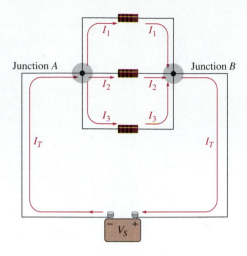

General Formula for Kirchhoff's Current Law

The previous discussion used a specific case to illustrate Kirchhoff's current law, often abbreviated KCL. Now let's look at the general case. Figure 5–13 shows a generalized circuit junction where a number of branches are connected to a point in the circuit. Currents $I_{IN(1)}$ through $I_{IN(n)}$ flow into the junction (n can be any number). Currents $I_{OUT(1)}$ through $I_{OUT(m)}$ flow out of the junction (m can be any number but not necessarily equal to n).

By Kirchhoff's current law, the sum of the currents into a junction must equal the sum of the currents out of the junction. With reference to Figure 5–13, the general formula for Kirchhoff's current law is

$$I_{IN(1)} + I_{IN(2)} + \cdots + I_{IN(n)} = I_{OUT(1)} + I_{OUT(2)} + \cdots + I_{OUT(m)} \qquad (5\text{–}1)$$

If all of the terms on the right side of Equation (5–1) are brought over to the left side, their signs change to negative, and a zero is left on the right side. Kirchhoff's current law can also be stated in this way:

> **The algebraic sum of all the currents entering and leaving a junction is equal to zero.**

FIGURE 5–13
Generalized circuit junction to illustrate Kirchhoff's current law.

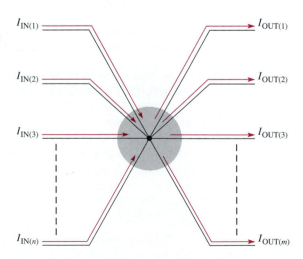

$$I_{IN(1)} + I_{IN(2)} + I_{IN(3)} + \cdots + I_{IN(n)}$$
$$= I_{OUT(1)} + I_{OUT(2)} + I_{OUT(3)} + \cdots I_{OUT(m)}$$

FIGURE 5–14

Experimental verification of Kirchhoff's current law.

You can verify Kirchhoff's current law by connecting a circuit and measuring each branch current and the total current from the source, as illustrated in Figure 5–14. When the branch currents are added together, their sum will equal the total current. This rule applies for any number of branches.

The following three examples illustrate use of Kirchhoff's current law.

EXAMPLE 5–4

The branch currents in the circuit of Figure 5–15 are known. Determine the total current entering junction A and the total current leaving junction B.

FIGURE 5–15

Solution The total current out of junction A is the sum of the two branch currents. So the total current into A is

$$I_T = I_1 + I_2 = 5 \text{ mA} + 12 \text{ mA} = 17 \text{ mA}$$

The total current entering point B is the sum of the two branch currents. So the total current out of B is

$$I_T = I_1 + I_2 = 5 \text{ mA} + 12 \text{ mA} = 17 \text{ mA}$$

The calculator sequence is

Related Exercise If a third resistor is connected in parallel to the circuit of Figure 5–15, and its current is 3 mA, what is the total current into junction A and out of junction B?

EXAMPLE 5–5   Determine the current through R_2 in Figure 5–16.

FIGURE 5–16

Solution   The total current into the junction of the three branches is known, and two of the branch currents are known. The current equation at this junction is

$$I_T = I_1 + I_2 + I_3$$

Solving for I_2,

$$I_2 = I_T - I_1 - I_3 = 100 \text{ mA} - 30 \text{ mA} - 20 \text{ mA} = 50 \text{ mA}$$

Related Exercise   Determine I_T and I_2 if a fourth branch is added to the circuit in Figure 5–16 and it has 12 mA through it.

EXAMPLE 5–6   Use Kirchhoff's current law to find the current measured by ammeters A1 and A2 in Figure 5–17.

FIGURE 5–17

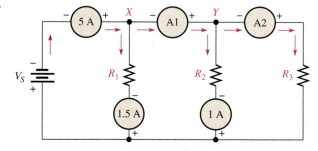

Solution   The total current into junction X is 5 A. Two currents are out of junction X: 1.5 A through resistor R_1 and the current through A1. Kirchhoff's current law at junction X is

$$5 \text{ A} = 1.5 \text{ A} + I_{A1}$$

Solving for I_{A1} yields

$$I_{A1} = 5 \text{ A} - 1.5 \text{ A} = 3.5 \text{ A}$$

The total current into junction Y is $I_{A1} = 3.5$ A. Two currents are out of junction Y: 1 A through resistor R_2 and the current through A2 and R_3. Kirchhoff's current law applied to junction Y gives

$$3.5 \text{ A} = 1 \text{ A} + I_{A2}$$

Solving for I_{A2} yields

$$I_{A2} = 3.5 \text{ A} - 1 \text{ A} = 2.5 \text{ A}$$

Related Exercise How much current will an ammeter measure when it is placed in the circuit in Figure 5–17 right below R_3? Below the positive battery terminal?

SECTION 5–3 REVIEW

1. State Kirchhoff's current law in two ways.
2. A total current of 2.5 A is into the junction of three parallel branches. What is the sum of all three branch currents?
3. In Figure 5–18, 100 mA and 300 mA are into the junction. What is the amount of current out of the junction?
4. Determine I_1 in the circuit of Figure 5–19.
5. Two branch currents enter a junction, and two branch currents leave the same junction. One of the currents entering the junction is 1 A, and one of the currents leaving the junction is 3 A. The total current entering and leaving the junction is 8 A. Determine the value of the unknown current entering the junction and the value of the unknown current leaving the junction.

FIGURE 5–18

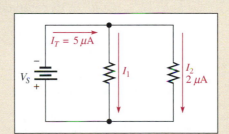

FIGURE 5–19

5–4 ▪ TOTAL PARALLEL RESISTANCE

When resistors are connected in parallel, the total resistance of the circuit decreases. The total resistance of a parallel circuit is always less than the value of the smallest resistor. For example, if a 10 Ω resistor and a 100 Ω resistor are connected in parallel, the total resistance is less than 10 Ω.

After completing this section, you should be able to

▪ **Determine total parallel resistance**
  □ Explain why resistance decreases as resistors are connected in parallel
  □ Apply the parallel-resistance formula

The Number of Current Paths Affects Resistance

As you know, when resistors are connected in parallel, the current has more than one path. The number of current paths is equal to the number of parallel branches.

For example, in Figure 5–20(a), there is only one current path because it is a series circuit. A certain amount of current, I_1, flows through R_1. If resistor R_2 is connected in

FIGURE 5–20

Adding resistors in parallel reduces total resistance and increases total current.

(a)   (b)

parallel with R_1, as shown in Figure 5–20(b), an additional amount of current, I_2, flows through R_2. The total current coming from the source has increased with the addition of the parallel branch. An increase in the total current from the source means that the total resistance has decreased, in accordance with Ohm's law. Additional resistors connected in parallel will further reduce the resistance and increase the total current.

Formula for Total Parallel Resistance

The circuit in Figure 5–21 shows a general case of *n* resistors in parallel (*n* can be any number). From Kirchhoff's current law, the current equation is

$$I_T = I_1 + I_2 + I_3 + \cdots + I_n$$

Since V_S is the voltage across each of the parallel resistors, by Ohm's law, $I_1 = V_S/R_1$, $I_2 = V_S/R_2$, and so on. By substituting into the current equation,

$$\frac{V_S}{R_T} = \frac{V_S}{R_1} + \frac{V_S}{R_2} + \frac{V_S}{R_3} + \cdots + \frac{V_S}{R_n}$$

We factor V_S out of the right side of the equation and cancel it with V_S on the left side, leaving only the resistance terms.

$$\frac{1}{R_T} = \frac{1}{R_1} + \frac{1}{R_2} + \frac{1}{R_3} + \cdots + \frac{1}{R_n} \tag{5–2}$$

We get another useful form of Equation (5–2) by taking the reciprocal of (that is, by inverting) both sides of the equation.

$$R_T = \frac{1}{\left(\dfrac{1}{R_1}\right) + \left(\dfrac{1}{R_2}\right) + \left(\dfrac{1}{R_3}\right) + \cdots + \left(\dfrac{1}{R_n}\right)} \tag{5–3}$$

Equation (5–3) shows that to find the total parallel resistance, add all the 1/*R* terms and then take the reciprocal of the sum. Example 5–7 shows how to use this formula in a specific case, and Example 5–8 shows a simple method of using the calculator to determine parallel resistances. The reciprocal of resistance (1/*R*) is called *conductance*. Conductance is symbolized by *G*, and its unit is the siemens (S).

FIGURE 5–21

Circuit with n resistors in parallel.

EXAMPLE 5–7

Calculate the total parallel resistance between points A and B of the circuit in Figure 5–22.

FIGURE 5–22

Solution First, find the reciprocal (conductance) of each of the three resistors:

$$\frac{1}{R_1} = \frac{1}{100\ \Omega} = 10\ \text{mS}$$

$$\frac{1}{R_2} = \frac{1}{47\ \Omega} = 21.3\ \text{mS}$$

$$\frac{1}{R_3} = \frac{1}{22\ \Omega} = 45.5\ \text{mS}$$

Next, calculate R_T by adding $1/R_1$, $1/R_2$, and $1/R_3$ and taking the reciprocal of the sum:

$$R_T = \frac{1}{10\ \text{mS} + 21.3\ \text{mS} + 45.5\ \text{mS}} = \frac{1}{76.8\ \text{mS}} = 13.0\ \Omega$$

For a quick accuracy check, notice that the value of R_T (13.0 Ω) is smaller than the smallest value in parallel, which is R_3 (22 Ω), as it should be.

Related Exercise If a 33 Ω resistor is connected in parallel in Figure 5–22, what is R_T?

Calculator Solution

The parallel-resistance formula is easily solved on an electronic calculator. The general procedure is to enter the value of R_1 and then take its reciprocal by pressing the [1/x] key ([1/x] is a secondary function on some calculators.). Next press the [+] key; then enter the value of R_2 and take its reciprocal. Repeat this procedure until all of the resistor values have been entered and the reciprocal of each has been added. The final step is to press the [1/x] key to convert $1/R_T$ to R_T. The total parallel resistance is now on the display. This calculator procedure is illustrated in Example 5–8.

EXAMPLE 5–8

Show the steps required for a calculator solution of Example 5–7.

Solution

Step 1: Enter 100. Display shows 100.

Step 2: Press [1/x] key. Display shows 0.01.

Step 3: Press [+] key. Display shows 0.01.

Step 4: Enter 47. Display shows 47.

Step 5: Press $\boxed{1/x}$ key. Display shows 0.0212766.

Step 6: Press $\boxed{+}$ key. Display shows 0.0312766.

Step 7: Enter 22. Display shows 22.

Step 8: Press $\boxed{1/x}$ key. Display shows 0.0454545.

Step 9: Press $\boxed{=}$ key. Display shows 0.0767311.

Step 10: Press $\boxed{1/x}$ key. Display shows 13.0325183.

The number displayed in Step 10 is the total resistance in ohms. Round it to 13.0 Ω.

Related Exercise Show the additional calculator steps for R_T when a 33 Ω resistor is placed in parallel in the circuit of Example 5–7.

Two Resistors in Parallel

Equation (5–3) is a general formula for finding the total resistance of any number of re-sistors in parallel. It is often useful to consider only two resistors in parallel because this situation occurs commonly in practice.

Derived from Equation (5–3), the formula for two resistors in parallel is

$$R_T = \frac{R_1 R_2}{R_1 + R_2} \tag{5–4}$$

Equation (5–4) states

The total resistance of two resistors in parallel is equal to the product of the two resistors divided by the sum of the two resistors.

This equation is sometimes referred to as the "product over the sum" formula. Example 5–9 illustrates how to use it.

EXAMPLE 5–9   Calculate the total resistance between the positive and negative terminals of the source of the circuit in Figure 5–23.

FIGURE 5–23

Solution Use Equation (5–4) as follows:

$$R_T = \frac{R_1 R_2}{R_1 + R_2} = \frac{(680\ \Omega)(330\ \Omega)}{680\ \Omega + 330\ \Omega} = \frac{224{,}000\ \Omega^2}{1{,}010\ \Omega} = 222\ \Omega$$

The calculator sequence is

⑥ ⑧ ⓪ ✕ ③ ③ ⓪ ÷ ⦅ ⑥ ⑧ ⓪ ➕ ③ ③ ⓪ ⦆ ⹀

Related Exercise Find R_T if a 220 Ω resistor is added in parallel in Figure 5–23.

Resistors of Equal Value in Parallel

Another special case of parallel circuits is the parallel connection of several resistors having the same value. The following is a shortcut method of calculating R_T when this case occurs:

$$R_T = \frac{R}{n} \tag{5-5}$$

Equation (5–5) says that when any number of resistors *(n)*, all having the same resistance *(R)*, are connected in parallel, R_T is equal to the resistance divided by the number of resistors in parallel. Examples 5–10 and 5–11 show how to use this formula.

EXAMPLE 5–10 Find the total resistance between points *A* and *B* in Figure 5–24.

FIGURE 5–24

Solution There are five 100 Ω resistors in parallel. Use Equation (5–5) as follows:

$$R_T = \frac{R}{n} = \frac{100 \ \Omega}{5} = 20 \ \Omega$$

Related Exercise Find R_T for three 100 kΩ resistors in parallel.

EXAMPLE 5–11 A stereo amplifier drives two 8 Ω speakers in parallel from each channel as shown in Figure 5–25. What is the total resistance across the output terminals of each channel of the amplifier?

FIGURE 5–25

Solution The total parallel resistance across each channel output with the two 8 Ω speakers is

$$R_T = \frac{R}{n} = \frac{8\,\Omega}{2} = 4\,\Omega$$

Related Exercise If two additional 8 Ω speakers are added to each channel, what is the resistance across each channel output?

Notation for Parallel Resistors

Sometimes, for convenience, parallel resistors are designated by two parallel vertical marks. For example, R_1 in parallel with R_2 can be written as $R_1 \| R_2$. Also, when several resistors are in parallel with each other, this notation can be used. For example,

$$R_1 \| R_2 \| R_3 \| R_4 \| R_5$$

indicates that R_1 through R_5 are all in parallel.

This notation is also used with resistance values. For example,

$$10\text{ k}\Omega \| 5\text{ k}\Omega$$

means that a 10 kΩ resistor is in parallel with a 5 kΩ resistor.

SECTION 5–4 REVIEW

1. Does the total resistance increase or decrease as more resistors are connected in parallel?
2. The total parallel resistance is always less than _____ .
3. Determine R_T (between pin 1 and pin 4) for the circuit in Figure 5–26.

FIGURE 5–26

5–5 ■ OHM'S LAW IN PARALLEL CIRCUITS

In this section, you will see how Ohm's law can be applied to parallel circuit analysis. You will learn by several examples how to determine the total current and branch currents in a parallel circuit. You will also learn how to find the voltage across a parallel circuit and the resistance in a parallel circuit.

After completing this section, you should be able to

■ **Apply Ohm's law in a parallel circuit**
  ☐ Find the total current in a parallel circuit
  ☐ Find each branch current in a parallel circuit
  ☐ Find the voltage across a parallel circuit
  ☐ Find the resistance in a parallel circuit

EXAMPLE 5–12

Find the total current produced by the battery in Figure 5–27.

FIGURE 5–27

Solution The battery "sees" a total parallel resistance that determines the amount of current generated. First, calculate R_T:

$$R_T = \frac{R_1 R_2}{R_1 + R_2} = \frac{(100\ \Omega)(56\ \Omega)}{100\ \Omega + 56\ \Omega} = \frac{5600\ \Omega^2}{156\ \Omega} = 35.9\ \Omega$$

The calculator sequence is

[1][0][0][×][5][6][÷][(][(][1][0][0][+][5][6][)][)][=]

The battery voltage is 100 V. Use Ohm's law to find I_T:

$$I_T = \frac{100\ \text{V}}{35.9\ \Omega} = 2.79\ \text{A}$$ [1][0][0][÷][3][5][•][9][=]

Related Exercise Find the currents through R_1 and R_2 in Figure 5–27.

EXAMPLE 5–13

Determine the current through each resistor in the parallel circuit of Figure 5–28.

FIGURE 5–28

Solution The voltage across each resistor (branch) is equal to the source voltage. That is, the voltage across R_1 is 20 V, the voltage across R_2 is 20 V, and the voltage across R_3 is 20 V. The current through each resistor is determined as follows:

$$I_1 = \frac{V_S}{R_1} = \frac{20 \text{ V}}{1 \text{ k}\Omega} = 20 \text{ mA}$$

$$I_2 = \frac{V_S}{R_2} = \frac{20 \text{ V}}{2.2 \text{ k}\Omega} = 9.1 \text{ mA}$$

$$I_3 = \frac{V_S}{R_3} = \frac{20 \text{ V}}{560 \ \Omega} = 35.7 \text{ mA}$$

Related Exercise If an additional resistor of 910 Ω is connected in parallel to the circuit in Figure 5–28, determine all the branch currents.

EXAMPLE 5–14 Find the voltage across the parallel circuit in Figure 5–29.

FIGURE 5–29

Solution The total current into the parallel circuit is 10 mA. You need to find the total resistance, and then apply Ohm's law to get the voltage. The total resistance is

$$R_T = \frac{1}{\left(\dfrac{1}{R_1}\right) + \left(\dfrac{1}{R_2}\right) + \left(\dfrac{1}{R_3}\right)} = \frac{1}{\left(\dfrac{1}{220 \ \Omega}\right) + \left(\dfrac{1}{560 \ \Omega}\right) + \left(\dfrac{1}{1 \text{ k}\Omega}\right)}$$

$$= \frac{1}{4.55 \text{ mS} + 1.79 \text{ mS} + 1 \text{ mS}} = \frac{1}{7.34 \text{ mS}} = 136 \ \Omega$$

$$V_S = I_T R_T = (10 \text{ mA})(136 \ \Omega) = 1.36 \text{ V}$$

Related Exercise Find the voltage if R_3 is decreased to 680 Ω in Figure 5–29. Assume I_T remains 10 mA.

EXAMPLE 5–15 The circuit board in Figure 5–30 has three resistors in parallel used for bias modification in an instrumentation amplifier. The values of two of the resistors are known from the color codes, but the top resistor is unmarked (maybe the bands are worn off from handling). Determine the value of the unknown resistor without disconnecting it from the circuit board or without using an ohmmeter.

FIGURE 5–30

Solution If you can determine the total resistance of the three resistors in parallel, then you can use the parallel-resistance formula to calculate the unknown resistance. You can use Ohm's law to find the total resistance if voltage and total current are known.

In Figure 5–31, a 10 V source (arbitrary value) is connected across the resistors and the total current is measured. Using these measured values, the total resistance is

$$R_T = \frac{V}{I_T} = \frac{10 \text{ V}}{20.1 \text{ mA}} = 498 \ \Omega$$

FIGURE 5–31

Use Equation (5–2) to find the unknown resistance as follows:

$$\frac{1}{R_T} = \frac{1}{R_1} + \frac{1}{R_2} + \frac{1}{R_3}$$

$$\frac{1}{R_1} = \frac{1}{R_T} - \frac{1}{R_2} - \frac{1}{R_3} = \frac{1}{498 \ \Omega} - \frac{1}{1.8 \text{ k}\Omega} - \frac{1}{1 \text{ k}\Omega} = 453 \ \mu S$$

$$R_1 = \frac{1}{453 \ \mu S} = 2.2 \text{ k}\Omega$$

The calculator sequence for R_1 is

[4] [9] [8] [1/x] [−] [1] [EXP] [3] [1/x] [−] [1] [.] [8] [EXP] [3] [1/x] [=] [1/x]

Related Exercise Explain how to determine the value of R_1 using an ohmmeter and without removing R_1 from the circuit.

SECTION 5–5 REVIEW

1. A 10 V battery is connected across three 68 Ω resistors that are in parallel. What is the total current from the battery?
2. How much voltage is required to produce 2 A of current through the circuit of Figure 5–32?

FIGURE 5–32

3. How much current is there through each resistor of Figure 5–32?
4. There are four equal-value resistors in parallel with a 12 V source, and 6 mA of current from the source. What is the value of each resistor?
5. A 1 kΩ and a 2.2 kΩ resistor are connected in parallel. A total of 100 mA flows through the parallel combination. How much voltage is dropped across the resistors?

5–6 ■ CURRENT DIVIDERS

A parallel circuit acts as a current divider because the current entering the junction of parallel branches "divides" up into several individual branch currents.

After completing this section, you should be able to

■ **Use a parallel circuit as a current divider**
  □ Apply the current-divider formula
  □ Determine an unknown branch current

In a parallel circuit, when the total current flows into the junction of the parallel branches, it divides among the branches. Thus, a parallel circuit acts as a current divider. This current divider principle is illustrated in Figure 5–33 for a two-branch parallel circuit in which part of the total current I_T goes through R_1 and part through R_2.

Since the same voltage is across each of the resistors in parallel, the branch currents are inversely proportional to the values of the resistors. For example, if the value of R_2 is twice that of R_1, then the value of I_2 is one-half that of I_1. In other words,

> **The total current divides among parallel resistors in a manner inversely proportional to the resistance values.**

FIGURE 5–33
Total current divides between two branches.

FIGURE 5–34

The branch with the lowest resistance has the most current, and the branch with the highest resistance has the least current.

The branches with higher resistance have less current, and the branches with lower resistance have more current, in accordance with Ohm's law. If all the branches have the same resistance, the branch currents are all equal.

Figure 5–34 shows specific values to demonstrate how the currents divide according to the branch resistances. Notice that in this case the resistance of the upper branch is one-tenth the resistance of the lower branch, but the upper branch current is ten times the lower branch current.

Current-Divider Formulas for Two Branches

You already know how to use Ohm's law ($I = V/R$) to determine the current in any parallel branch when the voltage and resistance are known. When the voltage is not known but the total current is, you can find both of the branch currents (I_1 and I_2) by using the following formulas:

$$I_1 = \left(\frac{R_2}{R_1 + R_2}\right)I_T \qquad\qquad \text{(5–6)}$$

$$I_2 = \left(\frac{R_1}{R_1 + R_2}\right)I_T \qquad\qquad \text{(5–7)}$$

The current in either branch is equal to the opposite branch resistance divided by the sum of the two resistors and then multiplied by the total current.

EXAMPLE 5–16

Find I_1 and I_2 in Figure 5–35.

FIGURE 5–35

Solution   Use Equation (5–6) to determine I_1:

$$I_1 = \left(\frac{R_2}{R_1 + R_2}\right)I_T = \left(\frac{47\ \Omega}{147\ \Omega}\right)100\ \text{mA} = 32.0\ \text{mA}$$

Use Equation (5–7) to determine I_2:

$$I_2 = \left(\frac{R_1}{R_1 + R_2}\right)I_T = \left(\frac{100 \ \Omega}{147 \ \Omega}\right)100 \text{ mA} = 68.0 \text{ mA}$$

The calculator sequence for I_2 is

Related Exercise   If $R_1 = 56 \ \Omega$ and $R_2 = 82 \ \Omega$ in Figure 5–35 and I_T stays the same, what will each branch current be?

General Current-Divider Formula for Any Number of Branches

A generalized parallel circuit with n branches is shown in Figure 5–36 (n can represent any number).

FIGURE 5–36
Generalized parallel circuit with n branches.

The current in any branch can be determined with the following formula:

$$I_x = \left(\frac{R_T}{R_x}\right)I_T \qquad\qquad\qquad \textbf{(5–8)}$$

where I_x represents any branch current (I_1, I_2, and so on) and R_x represents any resistance (R_1, R_2, and so on). For example, the formula for current in the second branch in Figure 5–36 is

$$I_2 = \left(\frac{R_T}{R_2}\right)I_T$$

Equation (5–8) can be used for a parallel circuit with any number of branches. Notice that you must determine R_T in order to use Equation (5–8) to find a branch current.

EXAMPLE 5–17

Determine the current through each resistor in the circuit of Figure 5–37.

FIGURE 5–37

Solution First calculate the total parallel resistance:

$$R_T = \frac{1}{\left(\dfrac{1}{R_1}\right) + \left(\dfrac{1}{R_2}\right) + \left(\dfrac{1}{R_3}\right)} = \frac{1}{\left(\dfrac{1}{680\ \Omega}\right) + \left(\dfrac{1}{330\ \Omega}\right) + \left(\dfrac{1}{220\ \Omega}\right)} = 111\ \Omega$$

The total current is 10 A. Use Equation (5–8) to calculate each branch current:

$$I_1 = \left(\frac{R_T}{R_1}\right) I_T = \left(\frac{111\ \Omega}{680\ \Omega}\right) 10\ \text{A} = 1.63\ \text{A}$$

$$I_2 = \left(\frac{R_T}{R_2}\right) I_T = \left(\frac{111\ \Omega}{330\ \Omega}\right) 10\ \text{A} = 3.36\ \text{A}$$

$$I_3 = \left(\frac{R_T}{R_3}\right) I_T = \left(\frac{111\ \Omega}{220\ \Omega}\right) 10\ \text{A} = 5.05\ \text{A}$$

Related Exercise Determine the current through R_1 and R_2 in Figure 5–37 if R_3 is removed. Assume the source voltage remains the same.

SECTION 5–6 REVIEW

1. A circuit has the following resistors in parallel: 220 Ω, 100 Ω, 68 Ω, 56 Ω, and 22 Ω. Which resistor has the most current through it? The least current?
2. Determine the current through R_3 in Figure 5–38.
3. Find the currents through each resistor in the circuit of Figure 5–39.

FIGURE 5–38

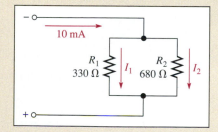

FIGURE 5–39

5–7 ■ POWER IN PARALLEL CIRCUITS

Total power in a parallel circuit is found by adding up the powers of all the individual resistors, the same as you did for series circuits.

After completing this section, you should be able to

■ **Determine power in a parallel circuit**

Equation (5–9) states the formula for finding total power in a concise way for any number of resistors in parallel.

$$P_T = P_1 + P_2 + P_3 + \cdots + P_n \tag{5–9}$$

where P_T is the total power and P_n is the power in the last resistor in parallel. As you can see, the power losses are additive, just as in the series circuit.

The power formulas in Chapter 3 are directly applicable to parallel circuits. The following formulas are used to calculate the total power P_T:

$$P_T = VI_T$$
$$P_T = I_T^2 R_T$$
$$P_T = \frac{V^2}{R_T}$$

where V is the voltage across the parallel circuit, I_T is the total current into the parallel circuit, and R_T is the total resistance of the parallel circuit. Example 5–18 shows how total power can be calculated in a parallel circuit.

EXAMPLE 5–18   Determine the total amount of power in the parallel circuit in Figure 5–40.

FIGURE 5–40

Solution   The total current is 2 A. The total resistance is

$$R_T = \frac{1}{\left(\dfrac{1}{68\ \Omega}\right) + \left(\dfrac{1}{33\ \Omega}\right) + \left(\dfrac{1}{22\ \Omega}\right)} = 11.1\ \Omega$$

The easiest formula to use is $P_T = I_T^2 R_T$ since you know both I_T and R_T. Thus,

$$P_T = I_T^2 R_T = (2\ \text{A})^2 (11.1\ \Omega) = 44.4\ \text{W}$$

To demonstrate that if the power in each resistor is determined and if all of these values are added together, you get the same result, work through another calculation. First, find the voltage across each branch of the circuit:

$$V_S = I_T R_T = (2\ \text{A})(11.1\ \Omega) = 22.2\ \text{V}$$

Remember that the voltage across all branches is the same.

Next, use $P = V^2/R$ to calculate the power for each resistor:

$$P_1 = \frac{(22.2\ \text{V})^2}{68\ \Omega} = 7.25\ \text{W}$$

$$P_2 = \frac{(22.2\ \text{V})^2}{33\ \Omega} = 14.9\ \text{W}$$

$$P_3 = \frac{(22.2\ \text{V})^2}{22\ \Omega} = 22.4\ \text{W}$$

Now, add these powers to get the total power:

$$P_T = 7.25\ \text{W} + 14.9\ \text{W} + 22.4\ \text{W} = 44.6\ \text{W}$$

This calculation shows that the sum of the individual powers is equal (approximately) to the total power as determined by one of the power formulas. Rounding to three significant figures accounts for the difference.

Related Exercise   Find the total power in Figure 5–40 if the total current is doubled.

EXAMPLE 5–19   The amplifier in one channel of a stereo system as shown in Figure 5–41 drives four speakers as shown. If the maximum voltage to the speakers is 15 V, how much power must the amplifier be able to deliver to the speakers?

FIGURE 5–41

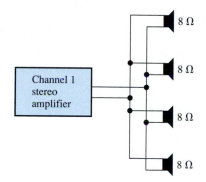

Solution   The speakers are connected in parallel to the amplifier output, so the voltage across each is the same. The maximum power to each speaker is

$$P_{max} = \frac{V_{max}^2}{R} = \frac{(15 \text{ V})^2}{8 \text{ } \Omega} = 28.1 \text{ W}$$

The total power that the amplifier must be capable of delivering to the speaker system is four times the power in an individual speaker because the total power is the sum of the individual powers.

$$P_{T(max)} = P_{(max)} + P_{(max)} + P_{(max)} + P_{(max)} = 4P_{max} = 4(28.1 \text{ W}) = 112.4 \text{ W}$$

Related Exercise   If the amplifier can produce a maximum of 18 V, what is the maximum total power to the speakers?

SECTION 5–7 REVIEW

1. If you know the power in each resistor in a parallel circuit, how can you find the total power?
2. The resistors in a parallel circuit dissipate the following powers: 1 W, 2 W, 5 W, and 8 W. What is the total power in the circuit?
3. A circuit has a 1 kΩ, a 2.7 kΩ, and a 3.9 kΩ resistor in parallel. A total current of 1 A flows into the parallel circuit. What is the total power?

5–8 ■ EXAMPLES OF PARALLEL CIRCUIT APPLICATIONS

Parallel circuits are found in some form in virtually every electronic system. In many of these applications, the parallel relationship of components may not be obvious until you have covered some advanced topics that will come later. For now, we will look at some common and familiar applications of parallel circuits.

After completing this section, you should be able to

■ **Describe some basic applications of parallel circuits**
  □ Discuss the lighting system in automobiles
  □ Discuss residential wiring
  □ Describe how current through a heating element is controlled
  □ Explain basically how a multiple-range ammeter works

Automotive

One advantage of a parallel circuit over a series circuit is that when one branch opens, the other branches are not affected. For example, Figure 5–42 shows a simplified diagram of an automobile lighting system. When one headlight on your car goes out, it does not cause the other lights to go out, because they are all in parallel.

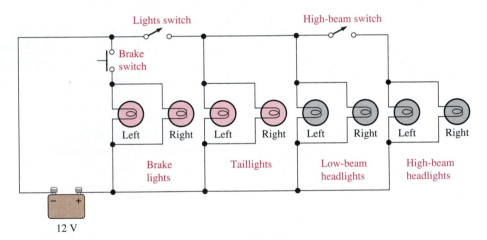

FIGURE 5–42
Simplified diagram of the exterior light system of an automobile. All lights are off with switches in the positions shown.

Notice that the brake lights are switched on independently of the head and taillights. They come on only when the driver closes the brake light switch by depressing the brake pedal.

When the lights switch is closed, both low-beam headlights and both taillights are on. The high-beam headlights are on only when both the lights switch and the high-beam switch are closed. If any one of the lights burns out (opens), there is still current in each of the other lights.

Residential

Another common use of parallel circuits is in residential electrical systems. All the lights and appliances in a home are wired in parallel. Figure 5–43(a) shows a typical room wiring arrangement with two switch-controlled lights and three wall outlets in parallel.

Figure 5–43(b) shows a simplified parallel arrangement of four heating elements on an electric range. The four-position switches in each branch allow the user to control the

FIGURE 5–43

Examples of parallel circuits in residential wiring and appliances.

(a) Simplified diagram of room wiring

(b) Simplified diagram of a four-burner range

amount of current through the heating elements by selecting the appropriate limiting resistor. The lowest resistor value (H setting) allows the highest amount of current for maximum heat. The highest resistor value (L setting) allows the least amount of current for minimum heat; M designates the medium settings.

Ammeters

Another example in which parallel circuits are used is the familiar analog (needle-type) ammeter or milliammeter. Parallel circuits are an important part of the operation of the ammeter because they allow the user to select various ranges in order to measure many different current values.

The mechanism in an ammeter that causes the pointer to move in proportion to the current is called the *meter movement,* which is based on a magnetic principle that you will learn later. Right now, all you need to know is that a meter movement has a certain resistance and a maximum current. This maximum current, called the *full-scale deflection* current, causes the pointer to go all the way to the end of the scale. For example, a certain meter movement has a 50 Ω resistance and a full-scale deflection current of 1 mA. A meter with this particular movement can measure currents of 1 mA or less. Currents greater than 1 mA will cause the pointer to "peg" (or stop) at full scale. Figure 5–44 illustrates a 1 mA meter.

(a) Half-scale deflection

(b) Full-scale deflection

(c) Pegged

FIGURE 5–44

A 1 mA meter.

FIGURE 5–45
A 10 mA meter.

Figure 5–45 shows a simple ammeter with a resistor in parallel with the meter movement; this resistor is called a *shunt resistor*. Its purpose is to bypass any current in excess of 1 mA around the meter movement. The figure specifically shows 9 mA through the shunt resistor and 1 mA through the meter movement. Thus, up to 10 mA can be measured. To find the actual current value, simply multiply the reading on the scale by 10.

A practical ammeter has a range switch that permits the selection of several full-scale current settings. In each switch position, a certain amount of current is bypassed through a parallel resistor as determined by the resistance value. In our example, the current through the movement is never greater than 1 mA.

Figure 5–46 illustrates a meter with three ranges: 1 mA, 10 mA, and 100 mA. When the range switch is in the 1 mA position, all of the current coming into the meter goes through the meter movement. In the 10 mA setting, up to 9 mA goes through R_{SH1} and up to 1 mA through the movement. In the 100 mA setting, up to 99 mA goes through R_{SH2} and the movement can still have only 1 mA for full-scale.

For example, in Figure 5–46, if 50 mA of current are being measured, the needle points to the 0.5 mark on the scale; you must multiply 0.5 by 100 to find the current value. In this situation, 0.5 mA flows through the movement (half-scale deflection), and 49.5 mA flows through R_{SH2}.

Effect of the Ammeter on a Circuit   As you know, an ammeter is connected in series to measure the current in a circuit. Ideally, the meter should not alter the current that it is intended to measure. In practice, however, the meter unavoidably has some effect on the circuit, because its internal resistance is connected in series with the circuit resistance. However, in most cases, the meter's internal resistance is so small compared to the circuit resistance that it can be neglected.

For example, if the meter has a 50 Ω movement (R_M) and a 100 μA full-scale current (I_M), the voltage dropped across the movement is

$$V_M = I_M R_M = (100\ \mu A)(50\ \Omega) = 5\ mV$$

FIGURE 5–46
A milliammeter with three ranges.

The shunt resistance (R_{SH}) for the 10 mA range, for example, is

$$R_{SH} = \frac{V_M}{I_{SH}} = \frac{5 \text{ mV}}{9.9 \text{ mA}} = 0.505 \ \Omega$$

As you can see, the total resistance of the ammeter on the 10 mA range is the resistance of the movement in parallel with the shunt resistance:

$$R_M \| R_{SH} = 50 \ \Omega \| 0.505 \ \Omega = 0.5 \ \Omega$$

EXAMPLE 5–20

How much does an ammeter with a 100 μA, 50 Ω movement affect the current in the circuit of Figure 5–47?

(a) Circuit

(b) Circuit with ammeter connected

FIGURE 5–47

Solution   The true current in the circuit is

$$I = \frac{10 \text{ V}}{1200 \ \Omega} = 8.3333 \text{ mA}$$

The meter is set on the 10 mA range in order to measure this particular amount of current. It was found that the meter's resistance on the 10 mA range is 0.5 Ω. When the meter is connected in the circuit, its resistance is in series with the 1200 Ω resistor. Thus, there is a total of 1200.5 Ω.

The current actually measured by the ammeter is

$$I = \frac{10 \text{ V}}{1200.5 \ \Omega} = 8.3299 \text{ mA}$$

This current differs from the true circuit current by only 0.04%. Therefore, the meter does not significantly alter the current value, a situation which, of course, is necessary because the measuring instrument should not change the quantity that is to be measured accurately.

Related Exercise   How much will the measured current differ from the true current if the circuit resistance is 12 kΩ instead of 1200 Ω?

SECTION 5–8 REVIEW

1. For the ammeter in Figure 5–46, what is the maximum resistance that the meter will have when connected in a circuit? What is the maximum current that can be measured at the setting?
2. Do the shunt resistors have resistance values considerably less than or more than that of the movement? Why?

5–9 ■ TROUBLESHOOTING

In this section, you will see how an open in a parallel branch affects the parallel circuit.

After completing this section, you should be able to

■ **Troubleshoot parallel circuits**
  □ Check for an open circuit

Open Branches

Recall that an open circuit is one in which the current path is interrupted and no current flows. In this section we examine what happens when a branch of a parallel circuit opens.

If a switch is connected in a branch of a parallel circuit, as shown in Figure 5–48, an open or a closed path can be made by the switch. When the switch is closed, as in Figure 5–48(a), R_1 and R_2 are in parallel. The total resistance is 50 Ω (two 100 Ω resistors in parallel). Current flows through both resistors. If the switch is opened, as in Figure 5–48(b), R_1 is effectively removed from the circuit, and the total resistance is 100 Ω. Current is now only through R_2.

In general,

> **When an open circuit occurs in a parallel branch, the total resistance increases, the total current decreases, and the same current continues through each of the remaining parallel paths.**

The decrease in total current equals the amount of current that was previously in the open branch. The other branch currents remain the same.

Consider the lamp circuit in Figure 5–49. There are four bulbs in parallel with a 120 V source. In part (a), there is current through each bulb. Now suppose that one of the bulbs burns out, creating an open path as shown in Figure 5–49(b). This light will go out

(a)   (b)

FIGURE 5–48
When the switch opens, total current decreases and current through R_2 remains unchanged.

(a)   (b)

FIGURE 5–49
When a lamp filament opens, total current decreases and the other branch currents remain unchanged.

because there is no current through the open path. Notice, however, that current continues through all the other parallel bulbs, and they continue to glow. The open branch does not change the voltage across the parallel branches; it remains at 120 V.

You can see that a parallel circuit has an advantage over a series connection in lighting systems because if one or more of the parallel bulbs burn out, the others will stay on. In a series circuit, when one bulb goes out, all of the others go out also because the current path is completely interrupted.

When a resistor in a parallel circuit opens, the open resistor cannot be located by measurement of the voltage across the branches, because the same voltage exists across all the branches. Thus, there is no way to tell which resistor is open by simply measuring voltage. The good resistors will always have the same voltage as the open one, as illustrated in Figure 5–50 (note that the middle resistor is open).

If a visual inspection does not reveal the open resistor, it can be located by current measurements. In practice, measuring current is more difficult than measuring voltage because you must insert the ammeter in series to measure the current. Thus, a wire or a printed circuit connection must be cut or disconnected, or one end of a component must be lifted off the circuit board, in order to connect the ammeter in series. This procedure, of course, is not required when voltage measurements are made, because the meter leads are simply connected across a component.

FIGURE 5–50
All parallel branches (open or not) have the same voltage.

Finding an Open Branch by Current Measurement

In a parallel circuit, the total current should be measured. *When a parallel resistor opens, I_T is always less than its normal value.* Once I_T and the voltage across the branches are known, a few calculations will determine the open resistor when all the resistors are of different values.

Consider the two-branch circuit in Figure 5–51(a). If one of the resistors opens, the total current will equal the current in the good resistor. Ohm's law quickly tells you what the current in each resistor should be:

$$I_1 = \frac{50 \text{ V}}{560 \text{ } \Omega} = 89.3 \text{ mA}$$

$$I_2 = \frac{50 \text{ V}}{100 \text{ } \Omega} = 500 \text{ mA}$$

FIGURE 5–51
Finding an open path by current measurement.

(a)

(b)　　　　　　　　　　(c)

If R_2 is open, the total current is 89.3 mA, as indicated in Figure 5–51(b). If R_1 is open, the total current is 500 mA, as indicated in Figure 5–51(c). This procedure can be extended to any number of branches having unequal resistances.

If the parallel resistances are all equal, the current in each branch must be checked until a branch is found with no current. This branch has the open resistor.

EXAMPLE 5–21   In Figure 5–52, there is a total current of 31.09 mA, and the voltage across the parallel branches is 20 V. Is there an open resistor, and, if so, which one is it?

FIGURE 5–52

Solution   Calculate the current in each branch:

$$I_1 = \frac{V}{R_1} = \frac{20 \text{ V}}{10 \text{ k}\Omega} = 2 \text{ mA}$$

$$I_2 = \frac{V}{R_2} = \frac{20 \text{ V}}{4.7 \text{ k}\Omega} = 4.26 \text{ mA}$$

$$I_3 = \frac{V}{R_3} = \frac{20 \text{ V}}{2.2 \text{ k}\Omega} = 9.09 \text{ mA}$$

$$I_4 = \frac{V}{R_4} = \frac{20 \text{ V}}{1 \text{ k}\Omega} = 2.0 \text{ mA}$$

The total current should be

$$I_T = I_1 + I_2 + I_3 + I_4$$
$$= 2 \text{ mA} + 4.26 \text{ mA} + 9.09 \text{ mA} + 20 \text{ mA} = 35.35 \text{ mA}$$

The actual measured current is 31.09 mA, as stated, which is 4.26 mA less than normal, indicating that the branch carrying 4.26 mA is open. Thus, R_2 must be open.

Related Exercise   What is the total current measured in Figure 5–52 if R_4 and not R_2 is open?

Finding an Open Branch by Resistance Measurement

If the parallel circuit to be checked can be disconnected from its voltage source and from any other circuit to which it may be connected, a measurement of the total resistance can be used to locate an open branch.

Recall that conductance, G, is the reciprocal of resistance ($1/R$) and its unit is the siemens (S). Based on Equation (5–2), the total conductance of a parallel circuit is the sum of the conductances of all the resistors.

$$G_T = G_1 + G_2 + G_3 + \cdots + G_n \tag{5–10}$$

To locate an open branch, do the following steps:

1. Calculate what the total conductance should be using the individual resistor values:

$$G_{T(\text{calc})} = \frac{1}{R_1} + \frac{1}{R_2} + \frac{1}{R_3} + \cdots + \frac{1}{R_n}$$

2. Measure the total resistance with an ohmmeter and calculate the total measured conductance:

$$G_{T(\text{meas})} = \frac{1}{R_{T(\text{meas})}}$$

3. Subtract the measured total conductance (Step 2) from the calculated total conductance (Step 1). The result is the conductance of the open branch and the resistance is obtained by taking its reciprocal ($R = 1/G$).

$$R_{\text{open}} = \frac{1}{G_{T(\text{calc})} - G_{T(\text{meas})}} \qquad (5\text{–}11)$$

EXAMPLE 5–22

Check the printed circuit board in Figure 5–53 for open branches.

FIGURE 5–53

Solution   There are two separate parallel circuits on the board. The circuit between pin 1 and pin 4 is checked as follows (we will assume one of the resistors is open):

1. Calculate what the total conductance should be using the individual resistor values:

$$G_{T(\text{calc})} = \frac{1}{1\text{ k}\Omega} + \frac{1}{1.8\text{ k}\Omega} + \frac{1}{2.2\text{ k}\Omega} + \frac{1}{2.7\text{ k}\Omega} + \frac{1}{3.3\text{ k}\Omega} + \frac{1}{3.9\text{ k}\Omega} = 2.94\text{ mS}$$

2. Measure the total resistance with an ohmmeter and calculate the total measured conductance. Assume that your ohmmeter measures 402 Ω.

$$G_{T(\text{meas})} = \frac{1}{402\ \Omega} = 2.49\text{ mS}$$

3. Subtract the measured total conductance (Step 2) from the calculated total conductance (Step 1). The result is the conductance of the open branch and the resistance is obtained by taking its reciprocal.

$$G_{open} = G_{T(calc)} - G_{T(meas)} = 2.94 \text{ mS} - 2.49 \text{ mS} = 0.45 \text{ mS}$$

$$R_{open} = \frac{1}{G_{open}} = \frac{1}{0.45 \text{ mS}} = 2.2 \text{ k}\Omega$$

Resistor R_3 is open and must be replaced.

Related Exercise Your ohmmeter indicates 9.6 kΩ between pin 2 and pin 3 on the PC board in Figure 5–53. Determine if this is correct and, if not, which resistor is open.

SECTION 5–9 REVIEW

1. If a parallel branch opens, what changes can be detected in the circuit's voltage and the currents, assuming that the parallel circuit is across a constant voltage source?
2. What happens to the total resistance if one branch opens?
3. If several light bulbs are connected in parallel and one of the bulbs opens (burns out), will the others continue to glow?
4. One ampere of current flows in each branch of a parallel circuit. If one branch opens, what is the current in each of the remaining branches?
5. A three-branch circuit normally has the following branch currents: 1 A, 2.5 A, and 1.2 A. If the total current measures 3.5 A, which branch is open?

5–10 ■ APPLICATION ASSIGNMENT

A five-range milliammeter has just come off the assembly line, but has failed the line check. For any current in excess of 1 mA, the needle "pegs" (goes off scale to the right) on all the range settings except 1 mA and 50 mA. In these two ranges, the meter seems to work correctly. As you know, a multiple-range meter circuit is based on shunt resistance placed in parallel with the meter movement to achieve the specified full-scale deflection of the needle. Your task is to troubleshoot the meter and determine what repairs are necessary.

Step 1: Open the Meter and Examine the Circuit

The meter face is shown in Figure 5–54(a). The circuitry is exposed by removing the front of the meter as shown in part (b). The precision resistors are 1% tolerance with a five-band color code. Recall from Chapter 2 that the first three bands indicate resistance value. In the fourth band, gold indicates a multiplier of 0.1 and silver a multiplier of 0.01. The brown fifth band is for 1% tolerance.

Step 2: Compare the Meter Circuit to the Schematic

The meter schematic is shown in Figure 5–55. Carefully check the circuit connections to make sure they agree with the schematic. If there is a wiring problem, specify how to repair it. Resistors R_1, R_2, etc. can be identified by the way they are wired in the circuit. The 100 mA switch position is labeled "1". Resistance values are not shown in the schematic.

(a)   (b)

FIGURE 5–54

FIGURE 5–55

Step 3: Check for Additional Problems

There was a wiring problem in the meter circuit which, let's assume, you have located and corrected. Now, assume that when you connect the meter to the test instrument to check for proper operation you find that the meter works properly on the 1 mA, 5 mA, 10 mA, and 50 mA ranges but the needle still "pegs" on the 100 mA range setting for a current of 50 mA or greater.

Step 4: Analyze the Circuit to Determine the Additional Problem

Since the needle pegs on the 100 mA range, all or most of the current must be going through the meter movement. Recall that a shunt resistor carries most of the current being measured so that the current through the meter movement does not exceed its 1 mA maximum. For this problem to occur, one of the shunt resistors for this range is either not connected, failed open, or has a resistance value that is much too high. Check the circuit in Figure 5–54 again for another wiring error and check the color coded values of R_4 and R_5. For some reason, the schematic does not show the resistor values so you will have to calculate what they should be using your knowledge of Ohm's law and current dividers. Refer to the table of standard resistor values in Appendix A.

SECTION 5–10 REVIEW

1. What is the total resistance between the positive and negative terminals of the milliammeter in Figure 5–54 when the range switch is in the 1 mA position? The 5 mA position? The 10 mA position?
2. When the switch is in the 5 mA position and the needle points to the middle mark on the scale, how much current is being measured?
3. When the meter is on the 5 mA range and is measuring the amount of current determined in Question 2, how much of the total current is through the meter movement and how much through the shunt resistor?

▪ SUMMARY

- Resistors in parallel are connected across the same points.
- A parallel combination has more than one path for current.
- The number of current paths equals the number of resistors in parallel.
- The total parallel resistance is less than the lowest-value resistor.
- The voltages across all branches of a parallel circuit are the same.
- One way to state Kirchhoff's current law: The sum of the currents into a junction (total current in) equals the sum of the currents out of the junction (total current out).
- Another way to state Kirchhoff's current law: The algebraic sum of all the currents entering and leaving a junction is zero.
- A parallel circuit is a current divider, so called because the total current entering the parallel junction divides up into each of the branches.
- If all of the branches of a parallel circuit have equal resistance, the currents through all of the branches are equal.
- The total power in a parallel-resistive circuit is the sum of all of the individual powers of the resistors making up the parallel circuit.
- The total power for a parallel circuit can be calculated with the power formulas using values of total current, total resistance, or total voltage.
- If one of the branches of a parallel circuit opens, the total resistance increases, and therefore the total current decreases.
- If a branch of a parallel circuit opens, there is still current through the remaining branches.

▪ GLOSSARY

Branch   One current path in a parallel circuit.

Junction   A point at which two or more branches are connected.

Kirchhoff's current law   A law stating that the total current into a junction equals the total current out of the junction.

Parallel   The relationship in electric circuits in which two or more current paths are connected between the same two points.

▪ FORMULAS

(5–1)
$$I_{IN(1)} + I_{IN(2)} + I_{IN(3)} + \cdots + I_{IN(n)}$$
$$= I_{OUT(1)} + I_{OUT(2)} + I_{OUT(3)} + \cdots + I_{OUT(m)}$$ Kirchhoff's current law

(5–2) $$\frac{1}{R_T} = \frac{1}{R_1} + \frac{1}{R_2} + \frac{1}{R_3} + \cdots + \frac{1}{R_n}$$ Reciprocal for total parallel resistance

(5–3) $$R_T = \frac{1}{\left(\frac{1}{R_1}\right) + \left(\frac{1}{R_2}\right) + \left(\frac{1}{R_3}\right) + \cdots + \left(\frac{1}{R_n}\right)}$$ Total parallel resistance

(5–4) $$R_T = \frac{R_1 R_2}{R_1 + R_2}$$ Special case for two resistors in parallel

(5–5) $$R_T = \frac{R}{n}$$ Special case for n equal-value resistors in parallel

(5–6) $$I_1 = \left(\frac{R_2}{R_1 + R_2}\right) I_T$$ Two-branch current-divider formula

$$(5\text{–}7) \qquad I_2 = \left(\frac{R_1}{R_1 + R_2} \right) I_T \qquad\qquad \text{Two-branch current-divider for-mula}$$

$$(5\text{–}8) \qquad I_x = \left(\frac{R_T}{R_x} \right) I_T \qquad\qquad \text{General current-divider formula}$$

$$(5\text{–}9) \qquad P_T = P_1 + P_2 + P_3 + \cdots + P_n \qquad\qquad \text{Total power}$$

$$(5\text{–}10) \qquad G_T = G_1 + G_2 + G_3 + \cdots + G_n \qquad\qquad \text{Total conductance}$$

$$(5\text{–}11) \qquad R_{\text{open}} = \frac{1}{G_{T(\text{calc})} - G_{T(\text{meas})}} \qquad\qquad \text{Open branch resistance}$$

■ SELF-TEST

1. In a parallel circuit, each resistor has
   (a) the same current   (b) the same voltage   (c) the same power   (d) all of the above

2. When a 1.2 kΩ resistor and a 100 Ω resistor are connected in parallel, the total resistance is
   (a) greater than 1.2 kΩ
   (b) greater than 100 Ω but less than 1.2 kΩ
   (c) less than 100 Ω but greater than 90 Ω   (d) less than 90 Ω

3. A 330 Ω resistor, a 270 Ω resistor, and a 68 Ω resistor are all in parallel. The total resistance is approximately
   (a) 668 Ω   (b) 47 Ω   (c) 68 Ω   (d) 22 Ω

4. Eight resistors are in parallel. The two lowest-value resistors are both 1 kΩ. The total resistance
   (a) cannot be determined   (b) is greater than 1 kΩ
   (c) is less than 1 kΩ   (d) is less than 500 Ω

5. When an additional resistor is connected across an existing parallel circuit, the total resistance
   (a) decreases   (b) increases
   (c) remains the same   (d) increases by the value of the added resistor

6. If one of the resistors in a parallel circuit is removed, the total resistance
   (a) decreases by the value of the removed resistor   (b) remains the same
   (c) increases   (d) doubles

7. The currents into a junction flow along two paths. One current is 5 A and the other is 3 A. The total current out of the junction is
   (a) 2 A   (b) unknown   (c) 8 A   (d) the larger of the two

8. The following resistors are in parallel across a voltage source: 390 Ω, 560 Ω, and 820 Ω. The resistor with the least current is
   (a) 390 Ω   (b) 560 Ω
   (c) 820 Ω   (d) impossible to determine without knowing the voltage

9. A sudden decrease in the total current into a parallel circuit may indicate
   (a) a short   (b) an open resistor   (c) a drop in source voltage   (d) either (b) or (c)

10. In a four-branch parallel circuit, there are 10 mA of current in each branch. If one of the branches opens, the current in each of the other three branches is
    (a) 13.33 mA   (b) 10 mA   (c) 0 A   (d) 30 mA

11. In a certain three-branch parallel circuit, R_1 has 10 mA through it, R_2 has 15 mA through it, and R_3 has 20 mA through it. After measuring a total current of 35 mA, you can say that
    (a) R_1 is open   (b) R_2 is open
    (c) R_3 is open   (d) the circuit is operating properly

12. If there are a total of 100 mA into a parallel circuit consisting of three branches and two of the branch currents are 40 mA and 20 mA, the third branch current is
    (a) 60 mA   (b) 20 mA   (c) 160 mA   (d) 40 mA

13. A complete short develops across one of five parallel resistors on a PC board. The most likely result is
    (a) the shortest resistor will burn out
    (b) one or more of the other resistors will burn out
    (c) the fuse in the power supply will blow
    (d) the resistance values will be altered

14. The power dissipation in each of four parallel branches is 1 W. The total power dissipation is
    (a) 1 W   (b) 4 W   (c) 0.25 W   (d) 16 W

■ BASIC PROBLEMS

SECTION 5–1 Resistors in Parallel

1. Connect the resistors in Figure 5–56 in parallel across the battery.

2. Determine whether or not all the resistors in Figure 5–57 are connected in parallel on the printed circuit board. Sketch the schematic including resistor values.

FIGURE 5–56

FIGURE 5–57

SECTION 5–2 Voltage Drop in Parallel Circuits

3. Determine the voltage across and the current through each parallel resistor if the total voltage is 12 V and the total resistance is 600 Ω. There are four resistors, all of equal value.

4. The source voltage in Figure 5–58 is 100 V. How much voltage does each of the meters V1–V3 read?

FIGURE 5–58

SECTION 5–3 Kirchhoff's Current Law

5. The following currents are measured in the same direction in a three-branch parallel circuit: 250 mA, 300 mA, and 800 mA. What is the value of the current into the junction of these three branches?

6. Five hundred milliamperes (500 mA) flow into five parallel resistors. The currents through four of the resistors are 50 mA, 150 mA, 25 mA, and 100 mA. How much current flows through the fifth resistor?

7. How much current flows through R_2 and R_3 in Figure 5–59 if R_2 and R_3 have the same resistance? Show how to connect ammeters to measure these currents.

FIGURE 5–59

SECTION 5–4   Total Parallel Resistance

8. The following resistors are connected in parallel: 1 MΩ, 2.2 MΩ, 4.7 MΩ, 12 MΩ, and 22 MΩ. Determine the total resistance.

9. Find the total resistance for each group of parallel resistors in Figure 5–60.

10. Calculate R_T for each circuit in Figure 5–61.

11. What is the total resistance of eleven 22 kΩ resistors in parallel?

12. Five 15 Ω, ten 100 Ω, and two 10 Ω resistors are all connected in parallel. What is the total resistance?

FIGURE 5–60

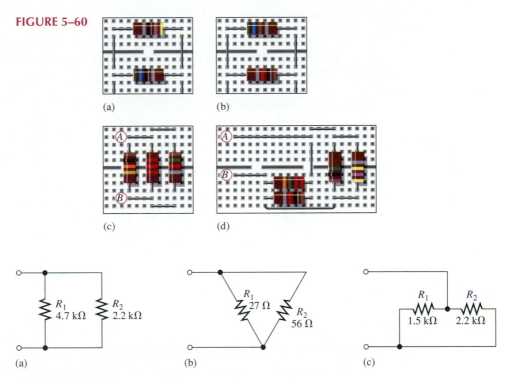

FIGURE 5–61

SECTION 5–5   Ohm's Law in Parallel Circuits

13. What is the total current I_T in each circuit of Figure 5–62?

14. Three 33 Ω resistors are connected in parallel with a 110 V source. How much current is flowing from the source?

15. Which circuit of Figure 5–63 has more total current?

16. Four equal-value resistors are connected in parallel. Five volts are applied across the parallel circuit, and 2.5 mA are measured from the source. What is the value of each resistor?

FIGURE 5–62

FIGURE 5–63

SECTION 5–6 Current Dividers

17. How much branch current should each meter in Figure 5–64 indicate?
18. Using the current-divider formula, determine the current in each branch of the circuits of Figure 5–65.

FIGURE 5–64

FIGURE 5–65

SECTION 5–7 Power in Parallel Circuits

19. Five parallel resistors each handle 40 mW. What is the total power?
20. Determine the total power in each circuit of Figure 5–65.
21. Six light bulbs are connected in parallel across 110 V. Each bulb is rated at 75 W. How much current is through each bulb, and what is the total current?

SECTION 5–9 Troubleshooting

22. If one of the bulbs burns out in Problem 21, how much current will flow through each of the remaining bulbs? What will the total current be?
23. In Figure 5–66, the current and voltage measurements are indicated. Has a resistor opened, and, if so, which one?
24. What is wrong with the circuit in Figure 5–67?

FIGURE 5–66

FIGURE 5–67

▪ ADVANCED PROBLEMS

25. In the circuit of Figure 5–68, determine resistances R_2, R_3, and R_4.

26. The total resistance of a parallel circuit is 25 Ω. If the total current is 100 mA, how much current flows through a 220 Ω resistor that makes up part of the parallel circuit?

27. What is the current through each resistor in Figure 5–69? R is the lowest-value resistor, and all others are multiples of that value as indicated.

FIGURE 5–68

FIGURE 5–69

28. A certain parallel network consists of only ½ W resistors. The total resistance is 1 kΩ, and the total current is 50 mA. If each resistor is operating at one-half its maximum power level, determine the following:
  (a) the number of resistors   **(b)** the value of each resistor
  (c) the current in each branch   **(d)** the applied voltage

29. Find the values of the unspecified labeled quantities in each circuit of Figure 5–70.

FIGURE 5–70

(a)

(b)

(c)

30. What is the total resistance between point A and ground in Figure 5–71 for the following conditions?
  (a) SW$_1$ and SW$_2$ open   **(b)** SW$_1$ closed, SW$_2$ open
  (c) SW$_1$ open, SW$_2$ closed   **(d)** SW$_1$ and SW$_2$ closed

31. To what minimum value can the 100 Ω rheostat in Figure 5–72 be adjusted before the 0.5 A fuse blows?

FIGURE 5–71

FIGURE 5–72

32. Determine the total current from the source and the current through each resistor for each switch position in Figure 5–73.

33. The electrical circuit in a room has a ceiling lamp that draws 1.25 A and four wall outlets. Two table lamps that each draw 0.833 A are plugged into two outlets, and a TV set that draws 1 A is connected to the third outlet. When all of these items are in use, how much current is in the main line serving the room? If the main line is protected by a 5 A circuit breaker, how much current can be drawn from the fourth outlet? Draw a schematic of this wiring.

34. The total resistance of a parallel circuit is 25 Ω. How much current flows through a 220 Ω resistor that makes up part of the parallel circuit if the total current is 100 mA?

35. Identify which groups of resistors are in parallel on the double-sided PC board in Figure 5–74. Determine the total resistance of each group.

FIGURE 5–73

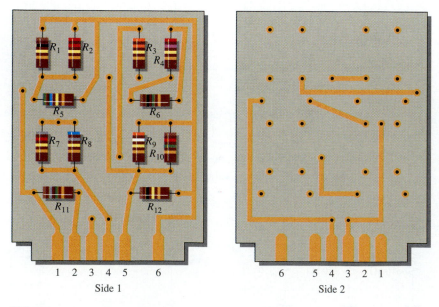

FIGURE 5–74

36. If the total resistance in Figure 5–75 is 200 Ω, what is the value of R_2?

37. Determine the unknown resistances in Figure 5–76.

FIGURE 5–75   FIGURE 5–76

38. A total of 250 mA flows into a parallel circuit with a total resistance of 1.5 kΩ. The current must be increased by 25%. Determine how much resistance to add in parallel to accomplish this increase in current.

39. Sketch the schematic for the setup in Figure 5–77 and determine what is wrong with the circuit, if 25 V are applied across the red and black leads.

(a) Meter with yellow lead going to circuit board and red lead going to the positive terminal of the 25 V power supply.

(b) Circuit board with leads connected. Yellow lead is from meter and gray lead is from 25 V power supply ground. The red meter lead goes to +25 V.

FIGURE 5–77

40. Develop a test procedure to check the circuit in Figure 5–78 to make sure that there are no open components. You must do this test without removing any component from the board. List the procedure in a detailed step-by-step format.

41. A certain parallel circuit consists of five $\frac{1}{2}$ W resistors with the following values: 1.8 kΩ, 2.2 kΩ, 3.3 kΩ, 3.9 kΩ, and 4.7 kΩ. As you slowly increase the voltage across the parallel circuit, the total current slowly increases. Suddenly, the total current drops to a lower value.
   (a) Excluding a power supply failure, what happened?
   (b) What is the maximum voltage you should have applied?
   (c) Specifically, what must be done to repair the fault?

42. For the circuit board shown in Figure 5–79, determine the resistance between the following pins if there is a short between pins 2 and 4.
   (a) pins 1 and 2   (b) pins 2 and 3   (c) pins 3 and 4   (d) pins 1 and 4

43. For the circuit board shown in Figure 5–79, determine the resistance between the following pins if there is a short between pins 3 and 4.
   (a) pins 1 and 2   (b) pins 2 and 3   (c) pins 2 and 4   (d) pins 1 and 4

FIGURE 5–78

FIGURE 5–79

▪ **ANSWERS TO SECTION REVIEWS**

Section 5–1

1. Parallel resistors are connected between the same two points.

2. A parallel connection has more than one current path between two given points.

3. See Figure 5–80.

4. See Figure 5–81.

FIGURE 5–80

(a)   (b)   (c)

FIGURE 5–81

Section 5–2

1. 5 V
2. $V_{R1} = 118$ V; $V_S = 118$ V
3. $V_1 = 50$ V; $V_2 = 50$ V
4. Voltage is the same across all branches.

Section 5–3

1. Kirchhoff's current law states that the algebraic sum of all currents at a junction is zero; the sum of the currents entering a junction equals the sum of the currents leaving that junction.
2. $I_1 + I_2 + I_3 = 2.5$ A
3. 100 mA + 300 mA = 400 mA
4. $I_1 = 5\ \mu A - 2\ \mu A = 3\ \mu A$
5. 8 A − 1 A = 7 A entering, 8 A − 3 A = 5 A leaving

Section 5–4

1. As more resistors are connected in parallel, the total resistance decreases.
2. R_T is always less than the smallest resistance value.
3. $R_T = 2.2\ k\Omega/12 = 183\ \Omega$

Section 5–5

1. $I_T = 10\ V/(68\ \Omega/3) = 441$ mA
2. $V_S = (2\ A)(680\ \Omega \parallel 330\ \Omega) = 444$ V
3. $I_1 = (330\ \Omega \parallel 680\ \Omega/330\ \Omega)2\ A = 653$ mA; $I_2 = (330\ \Omega \parallel 680\ \Omega/680\ \Omega)2\ A = 1.35$ A
4. $R = 4(12\ V/6\ mA) = 8\ k\Omega$
5. $V = (1\ k\Omega \parallel 2.2\ k\Omega)100\ mA = 68.8$ V

Section 5–6

1. The 22 Ω has most current; the 200 Ω has least current.
2. $I_3 = (R_T/R_3)4\ mA = (113.6\ \Omega/470\ \Omega)4\ mA = 967\ \mu A$
3. $I_2 = (R_T/680\ \Omega)10\ mA = 3.27$ mA; $I_1 = (R_T/330\ \Omega)10\ mA = 6.73$ mA

Section 5–7

1. Add the powers in each resistor to get P_T.
2. $P_T = 1\ W + 2\ W + 5\ W + 8\ W = 16$ W
3. $P_T = (1\ A)^2 R_T = 615$ W

Section 5–8

1. $R_{max} = 50\ \Omega$; $I_{max} = 1$ mA
2. Shunt resisters have less resistance than the movement because the shunt resistors must allow currents much greater than the current through the meter movement.

Section 5–9

1. When a parallel branch opens, there is no change in voltage and the total current decreases.
2. When the total resistance is greater than it should be, a branch is open.

3. Yes, all other bulbs continue to glow.

4. The current in each nonopen branch remains 1 A.

5. 1 A + 2.5 A = 3.5 A; therefore, the 1.2 A branch is open.

Section 5–10

1. 50 Ω; 9.94 Ω; 5.08 Ω

2. 2.5 mA

3. $I_M = 0.5$ mA; $I_{R1} = 2$ mA

■ **ANSWERS TO RELATED EXERCISES FOR EXAMPLES**

5–1	No rewiring is necessary.
5–2	Connect pin 1 to pin 2 and pin 3 to pin 4.
5–3	25 V
5–4	20 mA
5–5	$I_T = 112$ mA; $I_2 = 50$ mA
5–6	2.5 A; 5 A
5–7	9.34 Ω
5–8	Replace Step 9 with "Press ⊞ key." Display shows 0.0767311.
	Replace Step 10 with "Enter 33." Display shows 33.
	Step 11: Press ⟨1/x⟩ . Display shows 0.030303.
	Step 12: Press ⊟ key. Display shows 0.1070342.
	Step 13: Press ⟨1/x⟩ key. Display shows 9.3428107.
5–9	111 Ω
5–10	33.3 Ω
5–11	2 Ω
5–12	$I_1 = 1$ A; $I_2 = 1.79$ A
5–13	$I_1 = 20$ mA; $I_2 = 9.1$ mA; $I_3 = 35.7$ mA; $I_4 = 22.0$ mA
5–14	1.28 V
5–15	Use an ohmmeter to measure R_T and then calculate R_1 using the parallel-resistance formula.
5–16	$I_1 = 59.4$ mA; $I_2 = 40.6$ mA
5–17	$I_1 = 1.63$ A; $I_2 = 3.36$ A
5–18	177.6 W
5–19	162 W
5–20	An insignificant amount
5–21	15.4 mA
5–22	Not correct; $R_{10}(68$ kΩ$)$ is open.

6

SERIES-PARALLEL CIRCUITS

Various combinations of both series and parallel resistors are often used in electronic circuits. In this chapter, examples of such series-parallel arrangements are examined and analyzed. An important circuit called the *Wheatstone bridge* is introduced, and circuits with more than one voltage source are analyzed in simple steps. Also, you will learn how complex circuits can be simplified using Thevenin's theorem. Troubleshooting series-parallel circuits for shorts and opens is also covered.

■ CHAPTER OBJECTIVES

☐ Identify series-parallel relationships
☐ Analyze series-parallel circuits
☐ Analyze loaded voltage dividers
☐ Determine the loading effect of a voltmeter on a circuit
☐ Analyze a Wheatstone bridge
☐ Analyze circuits with more than one source
☐ Apply Thevenin's theorem to simplify a circuit for analysis
☐ Troubleshoot series-parallel circuits

■ APPLICATION ASSIGNMENT

Your application assignment in this chapter is to evaluate a voltage-divider circuit board used in a portable power supply by applying your knowledge of loaded voltage dividers gained in this chapter as well as skills developed in previous chapters. The voltage divider in this application is designed to provide reference voltages to three different instruments that act as loads on the circuit. You will also be required to troubleshoot the circuit board for various common faults. After studying this chapter, you should be able to complete the application assignment in Section 6–9.

6–1 ■ IDENTIFYING SERIES-PARALLEL RELATIONSHIPS

A series-parallel circuit consists of combinations of both series and parallel current paths. It is important to be able to identify how the components in a circuit are arranged in terms of their series and parallel relationships.

After completing this section, you should be able to

■ **Identify series-parallel relationships**
  □ Recognize how each resistor in a given circuit is related to the other resistors
  □ Determine series and parallel relationships on a PC board

Figure 6–1(a) shows a simple series-parallel combination of resistors. Notice that the resistance from point *A* to point *B* is R_1. The resistance from point *B* to point *C* is R_2 and R_3 in parallel ($R_2 \| R_3$). The resistance from point *A* to point *C* is R_1 in series with the parallel combination of R_2 and R_3, as indicated in Figure 6–1(b).

FIGURE 6–1

A simple series-parallel circuit.

(a) (b) (c)

When the circuit of Figure 6–1(a) is connected to a voltage source as shown in part (c), the total current flows through R_1 and divides at point *B* into the two parallel paths. These two branch currents then recombine, and the total current flows into the positive source terminal as shown.

Now, to illustrate series-parallel relationships, we will increase the complexity of the circuit in Figure 6–1(a) step-by-step. In Figure 6–2(a), another resistor (R_4) is connected in series with R_1. The resistance between points *A* and *B* is now $R_1 + R_4$, and this combination is in series with the parallel combination of R_2 and R_3, as illustrated in Figure 6–2(b).

FIGURE 6–2

R_4 is added to the circuit in series with R_1.

(a) (b)

In Figure 6–3(a), R_5 is connected in series with R_2. The series combination of R_2 and R_5 is in parallel with R_3. This entire series-parallel combination is in series with the $R_1 + R_4$ combination, as illustrated in Figure 6–3(b).

In Figure 6–4(a), R_6 is connected in parallel with the series combination of R_1 and R_4. The series-parallel combination of R_1, R_4, and R_6 is in series with the series-parallel combination of R_2, R_3, and R_5, as indicated in Figure 6–4(b).

FIGURE 6–3

R_5 is added to the circuit in series with R_2.

FIGURE 6–4

R_6 is added to the circuit in parallel with the series combination of R_1 and R_4.

EXAMPLE 6–1

Identify the series-parallel relationships in Figure 6–5.

FIGURE 6–5

Solution Starting at the negative terminal of the source, follow the current paths. All the current produced by the source must go through R_1, which is in series with the rest of the circuit.

The total current takes two paths when it gets to point *A*. Part of it flows through R_2, and part of it through R_3. R_2 and R_3 are in parallel with each other, and this parallel combination is in series with R_1.

At point *B*, the currents through R_2 and R_3 come together again. Thus, the total current flows through R_4. R_4 is in series with R_1 and the parallel combination of R_2 and R_3. The currents are shown in Figure 6–6, where I_T is the total current.

In summary, R_1 and R_4 are in series with the parallel combination of R_2 and R_3.

FIGURE 6–6

Related Exercise If another resistor, R_5, is connected from point A to the positive side of the source in Figure 6–6, what is its relationship to the other resistors?

EXAMPLE 6–2

Describe the series-parallel combination between points A and D in Figure 6–7.

FIGURE 6–7

Solution Between points B and C, there are two parallel paths. The lower path consists of R_4, and the upper path consists of a series combination of R_2 and R_3. This parallel combination is in series with both R_1 and R_5.

In summary, R_1 and R_5 are in series with the parallel combination of R_4 and $R_2 + R_3$.

Related Exercise If a resistor is connected from point C to point D in Figure 6–7, describe its parallel relationship.

EXAMPLE 6–3

Describe the total resistance between each pair of points in Figure 6–8.

FIGURE 6–8

Solution

1. From point A to point B: R_1 is in parallel with the series combination of R_2 and R_3.
2. From point A to point C: R_3 is in parallel with the series combination of R_1 and R_2.
3. From point B to point C: R_2 is in parallel with the series combination of R_1 and R_3.

Related Exercise Describe the total resistance between each point and ground if a resistor, R_4, is connected from point C to ground in Figure 6–8.

Redrawing a Schematic to Determine the Series-Parallel Relationships

Sometimes it is difficult to see the series-parallel relationships on a schematic because of the way in which it is drawn. In such a situation, it helps to redraw the diagram so that the relationships become clear. Example 6–4 will illustrate.

EXAMPLE 6–4   Identify the series-parallel relationships in Figure 6–9.

FIGURE 6–9

$$R_1 + (R_2 || R_3) + (R_4 || R_5)$$

Solution   Sometimes it is easier to see a particular circuit arrangement if it is drawn in a different way. The circuit schematic is redrawn in Figure 6–10, which better illustrates the series-parallel relationships. Now you can see that R_2 and R_3 are in parallel with each other and also that R_4 and R_5 are in parallel with each other. Both parallel combinations are in series with each other and with R_1.

FIGURE 6–10

Related Exercise If a resistor is connected from the bottom end of R_3 to the top end of R_5 in Figure 6–10, what effect does it have on the circuit?

Determining Relationships on a Printed Circuit Board

Usually, the physical arrangement of components on a PC board bears no resemblance to the actual circuit relationships. By tracing out the circuit on the PC board and rearranging the components on paper into a recognizable form, you can determine the series-parallel relationships. Example 6–5 will illustrate.

EXAMPLE 6–5   Determine the relationships of the resistors on the PC board in Figure 6–11.

FIGURE 6–11

Solution   In Figure 6–12(a), the schematic is drawn approximately in the same arrangement as that of the resistors on the board. In part (b), the resistors are reoriented so that the series-parallel relationships are obvious. Resistors R_1 and R_4 are in series; $R_1 + R_4$ is in parallel with R_2; R_5 and R_6 are in parallel and this combination is in series with R_3. The R_3, R_5, and R_6 series-parallel combination is in parallel with both R_2 and the $R_1 + R_4$ combination. This entire series-parallel combination is in series with R_7. Figure 6–12(c) illustrates these relationships.

FIGURE 6–12

Related Exercise   What happens if there is an open connection between R_1 and R_4 on the printed circuit board in Figure 6–11?

**SECTION 6–1
REVIEW**

1. A certain series-parallel circuit is described as follows: R_1 and R_2 are in parallel. This parallel combination is in series with another parallel combination of R_3 and R_4. Sketch the circuit.
2. In the circuit of Figure 6–13, describe the series-parallel relationships of the resistors.
3. Which resistors are in parallel in Figure 6–14?

FIGURE 6–13

FIGURE 6–14

4. Identify the parallel resistors in Figure 6–15.
5. Are the parallel combinations in Figure 6–15 in series?

FIGURE 6–15

6–2 ▪ ANALYSIS OF SERIES-PARALLEL CIRCUITS

Several quantities are important when you have a circuit that is a series-parallel configuration of resistors.

After completing this section, you should be able to

▪ **Analyze series-parallel circuits**
  ☐ Determine total resistance
  ☐ Determine all the currents
  ☐ Determine all the voltage drops

Total Resistance

In Chapter 4, you learned how to determine total series resistance. In Chapter 5, you learned how to determine total parallel resistance.

To find the total resistance R_T of a series-parallel combination, first identify the series and parallel relationships, and then apply what you have previously learned. The following two examples illustrate the general approach.

EXAMPLE 6–6

Determine R_T between points A and B of the circuit in Figure 6–16.

FIGURE 6–16

Solution   Resistors R_2 and R_3 are in parallel, and this parallel combination is in series with R_1. First find the parallel resistance of R_2 and R_3. Since R_2 and R_3 are equal in value, divide the value by 2:

$$R_{2\text{-}3} = \frac{R}{n} = \frac{100\ \Omega}{2} = 50\ \Omega$$

Now, since R_1 is in series with $R_{2\text{-}3}$, add their values:

$$R_T = R_1 + R_{2\text{-}3} = 10\ \Omega + 50\ \Omega = 60\ \Omega$$

The calculator sequence for R_T is

Related Exercise   Determine R_T in Figure 6–16 if R_3 is changed to 82 Ω.

EXAMPLE 6–7

Find R_T of the circuit in Figure 6–17.

FIGURE 6–17

Solution   In the upper branch, R_2 is in series with R_3. The series combination is designated $R_{2\text{-}3}$ and is equal to $R_2 + R_3$.

$$R_{2\text{-}3} = R_2 + R_3 = 47\ \Omega + 47\ \Omega = 94\ \Omega$$

In the lower branch, R_4 and R_5 are in parallel with each other. This parallel combination is designated $R_{4\text{-}5}$ and is calculated as follows:

$$R_{4\text{-}5} = \frac{R_4 R_5}{R_4 + R_5} = \frac{(68\ \Omega)(39\ \Omega)}{68\ \Omega + 39\ \Omega} = 24.8\ \Omega$$

Also in the lower branch, the parallel combination of R_4 and R_5 is in series with R_6. This series-parallel combination is designated R_{4-5-6} and is calculated as follows:

$$R_{4\text{-}5\text{-}6} = R_6 + R_{4\text{-}5} = 75 \ \Omega + 24.8 \ \Omega = 99.8 \ \Omega$$

Figure 6–18 shows the original circuit in a simplified equivalent form.

FIGURE 6–18

Now you can find the resistance between points A and B. It is $R_{2\text{-}3}$ in parallel with $R_{4\text{-}5\text{-}6}$. The equivalent resistance is calculated as follows:

$$R_{AB} = \cfrac{1}{\left(\cfrac{1}{94 \ \Omega}\right) + \left(\cfrac{1}{99.8 \ \Omega}\right)} = 48.4 \ \Omega$$

Finally, the total resistance is R_1 in series with R_{AB}.

$$R_T = R_1 + R_{AB} = 100 \ \Omega + 48.4 \ \Omega = 148.4 \ \Omega$$

Related Exercise Determine R_T if a 68 Ω resistor is connected to the circuit in Figure 6–17 from point A to point B.

Total Current

Once the total resistance and the source voltage are known, you can find total current in a circuit by applying Ohm's law. Total current is the total source voltage divided by the total resistance:

$$I_T = \frac{V_S}{R_T}$$

For example, let's find the total current in the circuit of Example 6–7 (Figure 6–17). Assume that the source voltage is 30 V. The calculation is

$$I_T = \frac{V_S}{R_T} = \frac{30 \ \text{V}}{148.4 \ \Omega} = 202 \ \text{mA}$$

Branch Currents

Using the current-divider formula, Kirchhoff's current law, Ohm's law, or combinations of these, you can find the current in any branch of a series-parallel circuit. In some cases it may take repeated application of the formula to find a given current.

EXAMPLE 6–8   Determine the current through R_4 in Figure 6–19 if $V_S = 50$ V.

FIGURE 6–19

Solution   First the current (I_2) into the junction of R_3 and R_4 must be found. Once this current is known, you can use the current-divider formula to find I_4.

Notice that there are two main branches in the circuit. The left-most branch consists of only R_1. The right-most branch has R_2 in series with the parallel combination of R_3 and R_4. The voltage across both of these main branches is the same and equal to 50 V. Find the current (I_2) into the junction of R_3 and R_4 by calculating the resistance of the right-most main branch and then applying Ohm's law as follows:

$$R_{2\text{-}3\text{-}4} = R_2 + \frac{R_3 R_4}{R_3 + R_4} = 330\ \Omega + \frac{(330\ \Omega)(560\ \Omega)}{890\ \Omega} = 538\ \Omega$$

$$I_2 = \frac{V_S}{R_{2\text{-}3\text{-}4}} = \frac{50\ \text{V}}{538\ \Omega} = 93\ \text{mA}$$

Using the current-divider formula, calculate I_4 as follows:

$$I_4 = \left(\frac{R_3}{R_3 + R_4}\right)I_2 = \left(\frac{330\ \Omega}{890\ \Omega}\right)93\ \text{mA} = 34.5\ \text{mA}$$

You can find the current in R_3 by applying Kirchhoff's law ($I_2 = I_3 + I_4$) and subtracting I_4 from I_2 ($I_3 = I_2 - I_4$).

You can find the current in R_1 by using Ohm's law ($I_1 = V_S/R_1$).

The calculator sequence for I_4 is

[3] [3] [0] [×] [5] [6] [0] [÷] [(] [(] [3] [3] [0] [+] [5] [6] [0] [)] [)] [+]
[3] [3] [0] [=] [STO] [5] [0] [÷] [RCL] [=] [STO] [3] [3] [0] [÷]
[(] [3] [3] [0] [+] [5] [6] [0] [)] [×] [RCL] [=]

Related Exercise   Find I_1, I_3, and I_T in Figure 6–19.

Voltage Relationships

The circuit in Figure 6–20 illustrates voltage relationships in a series-parallel circuit. Voltmeters are connected to measure each of the resistor voltages, and the readings are indicated.

Some general observations about Figure 6–20 are as follows:

1. V_1 and V_2 are equal because R_1 and R_2 are in parallel. (Recall that voltages across parallel branches are the same.) V_1 and V_2 are the same as the voltage from A to B.

2. V_3 is equal to $V_4 + V_5$ because R_3 is in parallel with the series combination of R_4 and R_5. (V_3 is the same as the voltage from B to C.)

3. V_4 is about one-third of the voltage from B to C because R_4 is about one-third of the resistance $R_4 + R_5$ (by the voltage-divider principle).

FIGURE 6–20
Illustration of voltage relationships.

4. V_5 is about two-thirds of the voltage from B to C because R_5 is about two-thirds of $R_4 + R_5$.

5. $V_1 + V_3$ equals V_S because, by Kirchhoff's voltage law, the sum of the voltage drops must equal the source voltage.

Example 6–9 will verify the meter readings in Figure 6–20.

EXAMPLE 6–9   Verify that the voltmeter readings in Figure 6–20 are correct. The circuit is redrawn as a schematic in Figure 6–21.

FIGURE 6–21

Solution   The resistance from A to B is the parallel combination of R_1 and R_2.

$$R_{AB} = \frac{R_1 R_2}{R_1 + R_2} = \frac{(1\text{ k}\Omega)(2.2\text{ k}\Omega)}{3.2\text{ k}\Omega} = 688\ \Omega$$

The resistance from B to C is R_3 in parallel with the series combination of R_4 and R_5.

$$R_4 + R_5 = 330\ \Omega + 680\ \Omega = 1010\ \Omega = 1.01\ \text{k}\Omega$$

$$R_{BC} = \frac{R_3(R_4 + R_5)}{R_3 + R_4 + R_5} = \frac{(1.8\ \text{k}\Omega)(1.01\ \text{k}\Omega)}{2.81\ \text{k}\Omega} = 647\ \Omega$$

The resistance from A to B is in series with the resistance from B to C, so the total circuit resistance is

$$R_T = R_{AB} + R_{BC} = 688\ \Omega + 647\ \Omega = 1335\ \Omega$$

Using the voltage-divider principle, you can calculate the voltages as follows:

$$V_{AB} = \left(\frac{R_{AB}}{R_T}\right)V_S = \left(\frac{688\ \Omega}{1335\ \Omega}\right)100\ \text{V} = 51.5\ \text{V}$$

$$V_{BC} = \left(\frac{R_{BC}}{R_T}\right)V_S = \left(\frac{647\ \Omega}{1335\ \Omega}\right)100\ \text{V} = 48.5\ \text{V}$$

$$V_1 = V_2 = V_{AB} = 51.5\ \text{V}$$

$$V_3 = V_{BC} = 48.5\ \text{V}$$

$$V_4 = \left(\frac{R_4}{R_4 + R_5}\right)V_{BC} = \left(\frac{330\ \Omega}{1010\ \Omega}\right)48.5\ \text{V} = 15.8\ \text{V}$$

$$V_5 = \left(\frac{R_5}{R_4 + R_5}\right)V_{BC} = \left(\frac{680\ \Omega}{1010\ \Omega}\right)48.5\ \text{V} = 32.7\ \text{V}$$

Related Exercise Determine each voltage drop in Figure 6–21 if the source voltage is doubled.

EXAMPLE 6–10   Determine the voltage drop across each resistor in Figure 6–22.

FIGURE 6–22

Solution   Because you know the total voltage, you can solve this problem using the voltage-divider formula. First reduce each parallel combination to an equivalent resistance. Since R_1 and R_2 are in parallel between points A and B, combine their values.

$$R_{AB} = \frac{R_1 R_2}{R_1 + R_2} = \frac{(3.3\ \text{k}\Omega)(6.2\ \text{k}\Omega)}{9.5\ \text{k}\Omega} = 2.15\ \text{k}\Omega$$

Since R_4 is in parallel with the series combination of R_5 and R_6 between points C and D, combine these values.

$$R_{CD} = \frac{R_4(R_5 + R_6)}{R_4 + R_5 + R_6} = \frac{(1\ k\Omega)(1.07\ k\Omega)}{2.07\ k\Omega} = 517\ \Omega$$

The equivalent circuit is drawn in Figure 6–23.

FIGURE 6–23

The total circuit resistance is

$$R_T = R_{AB} + R_3 + R_{CD} = 2.15\ k\Omega + 1\ k\Omega + 517\ \Omega = 3.67\ k\Omega$$

Apply the voltage-divider formula to solve for the voltages:

$$V_{AB} = \left(\frac{R_{AB}}{R_T}\right)V_S = \left(\frac{2.15\ k\Omega}{3.67\ k\Omega}\right)8\ V = 4.69\ V$$

$$V_{CD} = \left(\frac{R_{CD}}{R_T}\right)V_S = \left(\frac{517\ k\Omega}{3.67\ k\Omega}\right)8\ V = 1.13\ V$$

$$V_3 = \left(\frac{R_3}{R_T}\right)V_S = \left(\frac{1\ k\Omega}{3.67\ k\Omega}\right)8\ V = 2.18\ V$$

V_{AB} equals the voltage across both R_1 and R_2.

$$V_1 = V_2 = V_{AB} = 4.69\ V$$

V_{CD} is the voltage across R_4 and across the series combination of R_5 and R_6.

$$V_4 = V_{CD} = 1.13\ V$$

Now apply the voltage-divider formula to the series combination of R_5 and R_6 to get V_5 and V_6:

$$V_5 = \left(\frac{R_5}{R_5 + R_6}\right)V_{CD} = \left(\frac{680\ \Omega}{1070\ \Omega}\right)1.13\ V = 718\ mV$$

$$V_6 = \left(\frac{R_6}{R_5 + R_6}\right)V_{CD} = \left(\frac{390\ \Omega}{1070\ \Omega}\right)1.13\ V = 412\ mV$$

Related Exercise   Determine the current and power in each resistor in Figure 6–22.

As you have seen in this section, the analysis of series-parallel circuits can be approached in many ways, depending on what information you need and what circuit values you know. The examples in this section do not represent an exhaustive coverage, but they give you an idea of how to approach series-parallel circuit analysis.

If you know Ohm's law, Kirchhoff's laws, the voltage-divider formula, and the current-divider formula, and if you know how to apply these laws, you can solve most resistive circuit analysis problems. The ability to recognize series and parallel combinations is, of course, essential.

**SECTION 6–2
REVIEW**

1. Find the total resistance between A and B in the circuit of Figure 6–24.
2. Find I_3 in Figure 6–24.
3. Find V_2 in Figure 6–24.
4. Determine R_T and I_T in Figure 6–25.

FIGURE 6–24

FIGURE 6–25

6–3 ■ VOLTAGE DIVIDERS WITH RESISTIVE LOADS

Voltage dividers were introduced in Chapter 4. In this section, you will learn how resistive loads affect the operation of voltage-divider circuits.

After completing this section, you should be able to

■ **Analyze loaded voltage dividers**
   □ Determine the effect of a resistive load on a voltage-divider circuit.
   □ Explain how a voltmeter can alter the value of the voltage being measured

The voltage divider in Figure 6–26(a) produces an output voltage (V_{OUT}) of 5 V because the input is 10 V and the two resistors are of equal value. This voltage is the unloaded output voltage. When a load resistor, R_L, is connected from the output to ground as shown in Figure 6–26(b), the output voltage is reduced by an amount that depends on the value of R_L. The load resistor is in parallel with R_2, reducing the resistance from point A to ground and, as a result, also reducing the voltage across the parallel combination. This is one effect of loading a voltage divider. Another effect of a **load** is that more current is drawn from the source because the total resistance of the circuit is reduced.

The larger R_L is, compared to R_2, the less the output voltage is reduced from its unloaded value, as illustrated in Figure 6–27. When two resistors are connected in parallel and one of the resistors is much greater than the other, the total resistance is close to the value of the smaller resistance.

FIGURE 6–26

A voltage divider with both unloaded and loaded outputs.

(a) Unloaded

(b) Loaded

(a) No load   (b) R_L not significantly greater than R_2 (c) R_L much greater than R_2

FIGURE 6–27

The effect of a load resistor.

EXAMPLE 6–11
(a) Determine the unloaded output voltage of the voltage divider in Figure 6–28.
(b) Find the loaded output voltages of the voltage divider in Figure 6–28 for the following two values of load resistance: $R_L = 10\ \text{k}\Omega$ and $R_L = 100\ \text{k}\Omega$.

FIGURE 6–28

Solution
(a) The unloaded output voltage is

$$V_{\text{OUT(unloaded)}} = \left(\frac{R_2}{R_1 + R_2} \right) V_S = \left(\frac{10\ \text{k}\Omega}{14.7\ \text{k}\Omega} \right) 5\ \text{V} = 3.40\ \text{V}$$

(b) With the 10 kΩ load resistor connected, R_L is in parallel with R_2, which gives 5 kΩ, as shown by the equivalent circuit in Figure 6–29(a). The loaded output voltage is

$$V_{\text{OUT(loaded)}} = \left(\frac{R_2 \parallel R_L}{R_1 + R_2 \parallel R_L} \right) V_S = \left(\frac{5\ \text{k}\Omega}{9.7\ \text{k}\Omega} \right) 5\ \text{V} = 2.58\ \text{V}$$

FIGURE 6–29

(a) $R_L = 10\ \text{k}\Omega$ (b) $R_L = 100\ \text{k}\Omega$

With the 100 kΩ load, the resistance from output to ground is

$$\frac{R_2 R_L}{R_2 + R_L} = \frac{(10 \text{ k}\Omega)(100 \text{ k}\Omega)}{110 \text{ k}\Omega} = 9.1 \text{ k}\Omega$$

The equivalent circuit is shown in Figure 6–29(b). The loaded output voltage is

$$V_{\text{OUT(loaded)}} = \left(\frac{R_2 \parallel R_L}{R_1 + R_2 \parallel R_L}\right) V_S = \left(\frac{9.1 \text{ k}\Omega}{13.8 \text{ k}\Omega}\right) 5 \text{ V} = 3.30 \text{ V}$$

The smaller value of R_L causes a greater reduction in V_{OUT} (3.40 V − 2.58 V = 0.82 V) than does the larger value of R_L (3.40 V − 3.30 V = 0.10 V). This illustrates the loading effect of R_L on the voltage divider.

Related Exercise Determine V_{OUT} in Figure 6–28 for a 1 MΩ load resistance.

Load Current and Bleeder Current

In a loaded voltage-divider circuit, the total current drawn from the source consists of currents through the load resistors and the divider resistors. Figure 6–30 shows a voltage divider with two voltage outputs or two taps. Notice that the total current, I_T, flows through R_1 to point A where the current divides into I_{RL1}, which flows through R_{L1} and into I_2, which flows through R_2. At point B, the current I_2 divides into I_{RL2}, which flows through R_{L2} and into I_3, which flows through R_3. Current I_3 is called the **bleeder current,** which is the current left after the total load current is subtracted from the total current into the circuit:

$$I_{\text{BLEEDER}} = I_T - I_{RL1} - I_{RL2} \tag{6-1}$$

FIGURE 6–30

Currents in a two-tap loaded voltage divider.

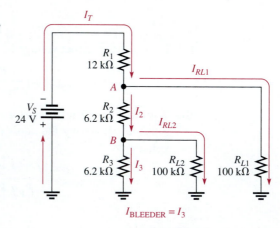

$I_{\text{BLEEDER}} = I_3$

EXAMPLE 6–12

Determine the load currents I_{RL1} and I_{RL2} and the bleeder current I_3 in the two-tap loaded voltage divider in Figure 6–30.

Solution The equivalent resistance from point A to ground is the 100 kΩ load resistor R_{L1} in parallel with the combination of R_2 in series with the parallel combination of R_3 and R_{L2}.

Determine the resistance values first. R_3 in parallel with R_{L2} is designated R_B and is calculated as follows:

$$R_B = \frac{R_3 R_{L2}}{R_3 + R_{L2}} = \frac{(6.2 \text{ k}\Omega)(100 \text{ k}\Omega)}{106.2 \text{ k}\Omega} = 5.84 \text{ k}\Omega$$

R_2 in series with R_B is designated R_{2-B} and is calculated as follows:

$$R_{2\text{-}B} = R_2 + R_B = 6.2 \text{ k}\Omega + 5.84 \text{ k}\Omega = 12.0 \text{ k}\Omega$$

R_{L1} in parallel with R_{2-B} is designated R_A and is calculated as follows:

$$R_A = \frac{R_{L1} R_{2\text{-}B}}{R_{L1} + R_{2\text{-}B}} = \frac{(100 \text{ k}\Omega)(12.0 \text{ k}\Omega)}{112 \text{ k}\Omega} = 10.7 \text{ k}\Omega$$

R_A is the total resistance from point A to ground. The total resistance for the circuit is

$$R_T = R_A + R_1 = 10.7 \text{ k}\Omega + 12.0 \text{ k}\Omega = 22.7 \text{ k}\Omega$$

The voltage across R_{L1} is determined as follows:

$$V_{RL1} = V_A = \left(\frac{R_A}{R_T}\right)V_S = \left(\frac{10.7 \text{ k}\Omega}{22.7 \text{ k}\Omega}\right)24 \text{ V} = 11.3 \text{ V}$$

The load current through R_{L1} is

$$I_{RL1} = \frac{V_{RL1}}{R_{L1}} = \left(\frac{11.3 \text{ V}}{100 \text{ k}\Omega}\right) = 113 \text{ } \mu\text{A}$$

The voltage at point B is determined by the parallel combination of R_{L2} and R_3 (R_B), R_2, and the voltage at point A:

$$V_B = \left(\frac{R_B}{R_2 + R_B}\right)V_A = \left(\frac{5.84 \text{ k}\Omega}{12.0 \text{ k}\Omega}\right)11.3 \text{ V} = 5.50 \text{ V}$$

The load current through R_{L2} is

$$I_{RL2} = \frac{V_{RL2}}{R_{L2}} = \frac{V_B}{R_{L2}} = \frac{5.50 \text{ V}}{100 \text{ k}\Omega} = 55 \text{ } \mu\text{A}$$

The bleeder current is

$$I_3 = \frac{V_B}{R_3} = \frac{5.50 \text{ V}}{6.2 \text{ k}\Omega} = 887 \text{ } \mu\text{A}$$

Related Exercise How can the bleeder current in Figure 6–30 be reduced without affecting the load currents?

Bipolar Voltage Dividers

An example of a voltage divider that produces both positive and negative voltages from a single source is shown in Figure 6–31. Notice that neither the positive nor the negative terminal of the source is connected to ground. The voltages at points A and B are positive with respect to ground, and the voltages at points C and D are negative with respect to ground.

FIGURE 6–31

A bipolar voltage divider. The positive and negative voltages are with respect to ground.

SECTION 6–3 REVIEW

1. A load resistor is connected to an output on a voltage divider. What effect does the load resistor have on the output voltage?
2. A larger-value load resistor will cause the output voltage of a voltage divider to change less than a smaller-value one will (T or F).
3. For the voltage divider in Figure 6–32, determine the unloaded output voltage. Also determine the output voltage with a 10 kΩ load resistor connected from the output to ground.

FIGURE 6–32

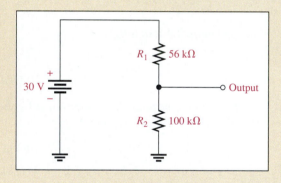

6–4 ■ LOADING EFFECT OF A VOLTMETER

As you have learned, voltmeters must be connected in parallel with a resistor in order to measure the voltage across the resistor. Because of its internal resistance, a voltmeter puts a load on the circuit and will affect, to a certain extent, the voltage that is being measured. Until now, we have ignored the loading effect because the internal resistance of a voltmeter is very high, and normally it has negligible effect on the circuit that is being measured. However, if the internal resistance of the voltmeter is not sufficiently greater than the circuit resistance across which it is connected, the loading effect will cause the measured voltage to be less than its actual value. You should always be aware of this effect.

After completing this section, you should be able to

■ **Determine the loading effect of a voltmeter on a circuit**
  □ Explain why a voltmeter can load a circuit
  □ Discuss the internal resistance of a voltmeter

Why a Voltmeter Can Load a Circuit

When a voltmeter is connected to a circuit as shown, for example, in Figure 6–33(a), its internal resistance appears in parallel with R_3, as shown in part (b). The resistance from

FIGURE 6–33

The loading effect of a voltmeter.

point A to point B is altered by the loading effect of the voltmeter's internal resistance R_M and is equal to $R_3 \parallel R_M$, as indicated in part (c).

If R_M is much greater than R_3, the resistance from A to B changes very little, and the meter indicates the actual voltage. If R_M is not sufficiently greater than R_3, the resistance from A to B is reduced significantly, and the voltage across R_3 is altered by the loading effect of the meter. A good rule of thumb is that *if the meter resistance is at least ten times greater than the resistance across which it is connected, the loading effect can be neglected.*

Internal Resistance of a Voltmeter

Three basic categories of voltmeters are the electromagnetic analog voltmeter, whose internal resistance is determined by its **sensitivity factor;** the electronic analog voltmeter, whose internal resistance is set by an input amplifier and is typically at least 10 MΩ; and the digital voltmeter, whose internal resistance is also typically at least 10 MΩ. The electronic analog voltmeter and the digital voltmeter present fewer loading problems than the electromagnetic type because their internal resistances are much higher.

As mentioned, the internal resistance of an electromagnetic voltmeter is specified by the sensitivity factor. A common sensitivity value is 20,000 Ω/V. The internal resistance depends on the dc voltage range switch setting of the meter and is determined by multiplying the sensitivity by the range setting. For example, if the dc voltage range switch is set at 10 V, the internal resistance is

$$R_M = 20,000 \ \Omega/\text{V} \times 10 \ \text{V} = 200,000 \ \Omega$$

The lower the range setting, the less the internal resistance and the greater the loading effect of the voltmeter on a circuit. Always use the highest possible range setting that will allow an accurate reading from the meter scale. However, you would not use a 60 V setting to measure 1 V because you could not read 1 V accurately on the 60 V meter scale. The following example shows how the range setting affects the loading.

EXAMPLE 6–13   How much does the electromagnetic voltmeter affect the voltage being measured for each range switch setting indicated in Figure 6–34? Assume the meter has a sensitivity of 20,000 Ω/V.

(a)   (b)   (c)

FIGURE 6–34

Solution   The unloaded voltage across R_2 in the voltage-divider circuit is

$$V_{R2} = \left(\frac{R_2}{R_1 + R_2} \right) 15 \text{ V} = \left(\frac{10 \text{ k}\Omega}{28 \text{ k}\Omega} \right) 15 \text{ V} = 5.36 \text{ V}$$

Refer to Figure 6–34(a). The internal resistance of the meter is

$$R_M = 20,000 \text{ }\Omega/\text{V} \times 12 \text{ V} = 240 \text{ k}\Omega$$

The meter's resistance in parallel with R_2 is

$$R_2 \parallel R_M = \left(\frac{R_2 R_M}{R_2 + R_M} \right) = \frac{(10 \text{ k}\Omega)(240 \text{ k}\Omega)}{250 \text{ k}\Omega} = 9.6 \text{ k}\Omega$$

The voltage actually measured by the meter is

$$V_{R2} = \left(\frac{R_2 \parallel R_M}{R_1 + R_2 \parallel R_M} \right) V = \left(\frac{9.6 \text{ k}\Omega}{27.6 \text{ k}\Omega} \right) 15 \text{ V} = 5.22 \text{ V}$$

Refer to Figure 6–34(b). The internal resistance of the meter is

$$R_M = 20,000 \text{ }\Omega/\text{V} \times 60 \text{ V} = 1.2 \text{ M}\Omega$$

and

$$R_2 \parallel R_M = \frac{R_2 R_M}{R_2 + R_M} = \frac{(10 \text{ M}\Omega)(1.2 \text{ M}\Omega)}{1.21 \text{ M}\Omega} = 9.92 \text{ k}\Omega$$

The voltage actually measured is

$$V_{R2} = \left(\frac{R_2 \parallel R_M}{R_1 + R_2 \parallel R_M} \right) V = \left(\frac{9.92 \text{ k}\Omega}{27.9 \text{ k}\Omega} \right) 15 \text{ V} = 5.33 \text{ V}$$

Refer to Figure 6–34(c). The internal resistance of the meter is

$$R_M = 20,000 \text{ }\Omega/\text{V} \times 300 \text{ V} = 6 \text{ M}\Omega$$

and

$$R_2 \parallel R_M = \frac{R_2 R_M}{R_2 + R_M} = \frac{(10 \text{ k}\Omega)(6 \text{ M}\Omega)}{6.01 \text{ M}\Omega} = 9.98 \text{ k}\Omega$$

The voltage actually measured is

$$V_{R2} = \left(\frac{R_2 \| R_M}{R_1 + R_2 \| R_M} \right) V = \left(\frac{9.98 \text{ k}\Omega}{27.98 \text{ k}\Omega} \right) 15 \text{ V} = 5.35 \text{ V}$$

As you can see, the loading decreases as you increase the range setting on the voltmeter because the measured voltage is closer to the unloaded voltage. However, the 12 V setting is the best choice in this case because, although the loading effect is greater, you can read the expected voltage much more accurately than you can on the 60 V range. Of course, using the 300 V range is not a good choice in this case.

Related Exercise Calculate the voltage across R_2 in Figure 6–34 if the range switch is set on 100 V.

SECTION 6–4 REVIEW

1. What is meant when we say a voltmeter loads a circuit?
2. If a voltmeter with a 10 MΩ internal resistance is measuring the voltage across a 1 kΩ resistor, should you normally be concerned about the loading effect?
3. A certain electromagnetic-type voltmeter is set on the 12 V range to measure 2.5 V. If you switch it down to the 3 V range, will there be an increase or a decrease in the measured voltage? What is the advantage of using the 3 V range instead of the 12 V range in this situation?
4. What is the internal resistance of a 30,000 Ω/V voltmeter on its 50 V range setting?

6–5 ■ THE WHEATSTONE BRIDGE

Bridge circuits are widely used in measurement devices and other applications that you will learn later. In this section, you will study the balanced resistive bridge, which can be used to measure unknown resistance values.

After completing this section, you should be able to

■ **Analyze a Wheatstone bridge**
   ☐ Determine when a bridge is balanced
   ☐ Determine an unknown resistance using a balanced bridge
   ☐ Describe how a Wheatstone bridge can be used for temperature measurement

The circuit shown in Figure 6–35(a) is known as a *Wheatstone bridge*. Part (b) is the same circuit electrically, but it is drawn in a different way.

(a)   (b)

FIGURE 6–35
Wheatstone bridge.

A bridge is said to be balanced when the voltage (V_{OUT}) across the output terminals A and B is zero; that is, $V_A = V_B$. In Figure 6–35(b), if V_A equals V_B, then $V_{R1} = V_{R2}$, since the top sides of both R_1 and R_2 are connected to the same point. Also $V_{R3} = V_{R4}$, since the bottom sides of both R_3 and R_4 connect to the same point. The voltage ratios can be written as

$$\frac{V_1}{V_3} = \frac{V_2}{V_4}$$

Substituting by Ohm's law yields

$$\frac{I_1 R_1}{I_1 R_3} = \frac{I_2 R_2}{I_2 R_4}$$

The currents cancel, leaving

$$\frac{R_1}{R_3} = \frac{R_2}{R_4}$$

Solving for R_1,

$$R_1 = R_3 \left(\frac{R_2}{R_4} \right)$$

How can you use this formula to determine an unknown resistance? First, make R_3 a variable resistor and call it R_V. Also, set the ratio R_2/R_4 to a known value. If R_V is adjusted until the bridge is balanced, the product of R_V and the ratio R_2/R_4 is equal to R_1, which is the unknown resistor (R_{UNK}). The formula for unknown resistance is stated as follows:

$$R_{UNK} = R_V \left(\frac{R_2}{R_4} \right) \tag{6–2}$$

The bridge is balanced when the voltage across the output terminals equals zero ($V_A = V_B$). A galvanometer (a meter that measures small currents in either direction and is zero at center scale) is connected between the output terminals. Then R_V is adjusted until the galvanometer shows zero current ($V_A = V_B$), indicating a balanced condition. The setting of R_V multiplied by the ratio R_2/R_4 gives the value of R_{UNK}. Figure 6–36 shows this arrangement. For example, if $R_2/R_4 = \frac{1}{10}$ and $R_V = 680\ \Omega$, then $R_{UNK} = (680\ \Omega)(\frac{1}{10}) = 68\ \Omega$.

FIGURE 6–36
Balanced Wheatstone bridge.

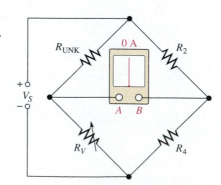

EXAMPLE 6–14   What is R_{UNK} under the balanced bridge conditions shown in Figure 6–37?

FIGURE 6–37

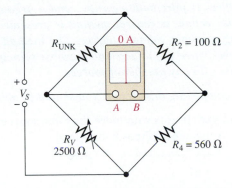

Solution

$$R_{UNK} = R_V \left(\frac{R_2}{R_4} \right) = 2500 \ \Omega \left(\frac{100 \ \Omega}{560 \ \Omega} \right) = 446 \ \Omega$$

Related Exercise   If R_V must be adjusted to 1.8 kΩ in order to balance the bridge, what is R_{UNK}?

A Bridge Application

Bridge circuits are used for many measurements other than that of unknown resistance. One application of the Wheatstone bridge is in accurate temperature measurement. A temperature-sensitive element such as a thermistor is connected in a Wheatstone bridge as shown in Figure 6–38. An amplifier is connected across the output from A to B in order to increase the output voltage from the bridge to a usable value. The bridge is calibrated so that it is balanced at a specified reference temperature. As the temperature changes, the resistance of the sensing element changes proportionately, and the bridge becomes unbalanced. As a result, V_{AB} changes and is amplified (increased) and converted to a form for direct temperature readout on a gauge or a digital-type display.

FIGURE 6–38
A simplified circuit for temperature measurement.

SECTION 6–5 REVIEW

1. Sketch a basic Wheatstone bridge circuit.
2. Under what condition is a bridge balanced?
3. What is the unknown resistance in Figure 6–36 when $R_V = 3.3$ kΩ, $R_2 = 10$ kΩ, and $R_4 = 2.2$ kΩ?

6–6 ■ CIRCUITS WITH MORE THAN ONE VOLTAGE SOURCE

Some circuits require more than one voltage source. For example, certain types of amplifiers require both a positive and a negative voltage source for proper operation. In this section, a general method is presented for evaluating series-parallel circuits having multiple voltage sources. The method is illustrated with two-source circuits. However, once you understand the basic principle, you can easily extend this method to circuits with any number of sources.

After completing this section, you should be able to

■ **Analyze circuits with more than one source**
  ☐ Apply the superposition method

The Superposition Method

The **superposition** method is a way to determine currents and voltages in a circuit that has multiple sources by taking one source at a time. The other sources are replaced by their internal resistances. The ideal voltage source has a zero internal resistance. In this section, all voltage sources will be treated as ideal in order to simplify the coverage.

The steps in applying the superposition method are as follows:

Step 1: Take one voltage source at a time and replace each of the other voltage sources with a short (a short represents zero resistance).

Step 2: Determine the current or voltage that you need just as if there were only that one source in the circuit.

Step 3: Repeat Steps 1 and 2 for each source in turn.

Step 4: To find an actual current or voltage, add or subtract the currents or voltages due to each individual source. If the currents are in the same direction or the voltages of the same polarity, add them. If the currents are in opposite directions or the voltages of opposite polarities, subtract them with the direction of the resulting current or voltage the same as the larger of the original quantities.

An example of the approach to superposition is demonstrated in Figure 6–39 for a series-parallel circuit with two voltage sources. Study the steps in this figure.

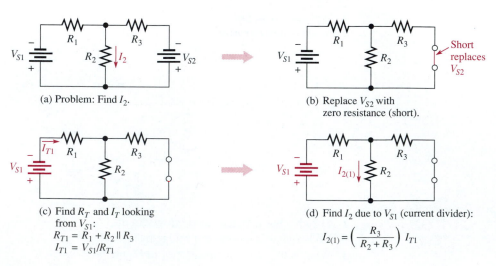

(a) Problem: Find I_2.

(b) Replace V_{S2} with zero resistance (short).

(c) Find R_T and I_T looking from V_{S1}:
$$R_{T1} = R_1 + R_2 \| R_3$$
$$I_{T1} = V_{S1}/R_{T1}$$

(d) Find I_2 due to V_{S1} (current divider):
$$I_{2(1)} = \left(\frac{R_3}{R_2 + R_3}\right) I_{T1}$$

FIGURE 6–39

Demonstration of the superposition method.

(e) Replace V_{S1} with zero resistance (short).

(f) Find R_T and I_T looking from V_{S2}: $R_{T2} = R_3 + R_1 \parallel R_2$
$I_{T2} = V_{S2}/R_{T2}$

(g) Find I_2 due to V_{S2}:

$$I_{2(2)} = \left(\frac{R_1}{R_1 + R_2}\right) I_{T2}$$

(h) Add $I_{2(1)}$ and $I_{2(2)}$ to get the actual I_2 (they are in same direction):
$I_2 = I_{2(1)} + I_{2(2)}$

FIGURE 6–39 *Continued*

EXAMPLE 6–15   Find the total current through R_3 in Figure 6–40.

FIGURE 6–40

Solution

Step 1: Find the current through R_3 due to source V_{S1} by replacing source V_{S2} with a short, as shown in Figure 6–41(a). Looking from V_{S1},

$$R_{T1} = R_1 + \frac{R_2 R_3}{R_2 + R_3} = 1\ \text{k}\Omega + \frac{(1\ \text{k}\Omega)(2.2\ \text{k}\Omega)}{3.2\ \text{k}\Omega} = 1.69\ \text{k}\Omega$$

$$I_{T1} = \frac{V_{S1}}{R_{T1}} = \frac{20\ \text{V}}{1.69\ \text{k}\Omega} = 11.8\ \text{mA}$$

Now apply the current-divider formula to find the current through R_3 due to source V_{S1} as follows:

$$I_{3(1)} = \left(\frac{R_2}{R_2 + R_3}\right) I_{T1} = \left(\frac{1\ \text{k}\Omega}{3.2\ \text{k}\Omega}\right) 11.8\ \text{mA} = 3.69\ \text{mA}$$

Notice that this current flows through R_3 in the direction shown in Figure 6–41(a).

(a)   (b)

FIGURE 6–41

Step 2: Next find I_3 due to source V_{S2} by replacing source V_{S1} with a short, as shown in Figure 6–41(b). Looking from V_{S2},

$$R_{T2} = R_2 + \left(\frac{R_1 R_3}{R_1 + R_3}\right) = 1\ \text{k}\Omega + \frac{(1\ \text{k}\Omega)(2.2\ \text{k}\Omega)}{3.2\ \text{k}\Omega} = 1.69\ \text{k}\Omega$$

$$I_{T2} = \frac{V_{S2}}{R_{T2}} = \frac{15\ \text{V}}{1.69\ \text{k}\Omega} = 8.88\ \text{mA}$$

Now apply the current-divider formula to find the current through R_3 due to source V_{S2} as follows:

$$I_{3(2)} = \left(\frac{R_1}{R_1 + R_3}\right)I_T = \left(\frac{1\ \text{k}\Omega}{3.2\ \text{k}\Omega}\right)8.88\ \text{mA} = 2.78\ \text{mA}$$

Notice that this current flows through R_3 in the direction shown in Figure 6–41(b).

Step 3: Calculation of the total current through R_3 is as follows:

$$I_{3(\text{tot})} = I_{3(1)} - I_{3(2)} = 3.69\ \text{mA} - 2.78\ \text{mA} = 0.91\ \text{mA} = 910\ \mu\text{A}$$

This current flows through R_3 in the same direction as the larger component current, $I_{3(1)}$.

Related Exercise   Determine the voltage at point A in Figure 6–40.

SECTION 6–6 REVIEW	1. State the superposition method. 2. Why is the superposition method useful for analysis of multiple-source linear circuits? 3. Why is a voltage source shorted and a current source opened when the superposition method is applied?

4. Using the superposition method, find the current through R_1 in Figure 6–42.

FIGURE 6–42

5. If, as a result of applying the superposition method, two currents are in opposing directions through a branch of a circuit, in which direction does the net current flow?

6–7 ▪ THEVENIN'S THEOREM

Thevenin's theorem provides a method for simplifying a circuit to a standard equivalent form. In many cases, this theorem can be used to simplify the analysis of complex series-parallel circuits.

After completing this section, you should be able to

▪ **Apply Thevenin's theorem to simplify a circuit for analysis**
  ☐ Describe the form of a Thevenin equivalent circuit
  ☐ Obtain the Thevenin equivalent voltage source
  ☐ Obtain the Thevenin equivalent resistance
  ☐ Explain terminal equivalency in the context of Thevenin's theorem
  ☐ Thevenize a portion of a circuit
  ☐ Thevenize a bridge circuit

The Thevenin equivalent form of any resistive circuit consists of an equivalent voltage source (V_{TH}) and an equivalent resistance (R_{TH}), arranged as shown in Figure 6–43. The values of the equivalent voltage and resistance depend on the values in the original circuit. Any resistive circuit can be simplified regardless of its complexity.

FIGURE 6–43
The general form of a Thevenin equivalent circuit is simply a voltage source in series with a resistance. Any resistive circuit can be reduced to this form.

Thevenin's Equivalent Voltage (V_{TH}) and Equivalent Resistance (R_{TH})

As you have seen, the equivalent voltage, V_{TH}, is one part of the complete Thevenin equivalent circuit. The other part is R_{TH}.

> V_{TH} **is defined to be the open circuit voltage between two specified points in a circuit.**

Any component connected between these two points effectively "sees" V_{TH} in series with R_{TH}. As defined by **Thevenin's theorem,**

> R_{TH} **is the total resistance appearing between two specified points in a given circuit with all sources replaced by their internal resistances.**

Equivalency in Thevenin's Theorem

Although a Thevenin equivalent circuit is not the same as its original circuit, it acts the same in terms of the output voltage and current. For example, as shown in Figure 6–44, place a resistive circuit of any complexity in a box with only the output terminals exposed. Then place the Thevenin equivalent of that circuit in an identical box with, again, only the output terminals exposed. Connect identical load resistors across the output terminals of each box. Next connect a voltmeter and an ammeter to measure the voltage and current for each load as shown in the figure. The measured values will be identical (neglecting tolerance variations), and you will not be able to determine which box contains the original circuit and which contains the Thevenin equivalent. That is, in terms of your observations, both circuits are the same. This condition is sometimes known as **terminal equivalency,** because both circuits look the same from the "viewpoint" of the two output terminals.

FIGURE 6–44

Which box contains the original circuit and which contains the Thevenin equivalent circuit? You cannot tell by observing the meters.

The Thevenin Equivalent of a Circuit

To find the Thevenin equivalent of any circuit, determine the equivalent voltage, V_{TH}, and the equivalent resistance, R_{TH}. For example, in Figure 6–45, the Thevenin equivalent for the circuit between points A and B is found as follows.

In Figure 6–45(a), the voltage across the designated points A and B is the Thevenin equivalent voltage. In this particular circuit, the voltage from A to B is the same as the voltage across R_2 because there is no current through R_3 and, therefore, no voltage drop across it. V_{TH} is expressed as follows for this particular example:

$$V_{TH} = \left(\frac{R_2}{R_1 + R_2}\right)V_S$$

In Figure 6–45(b), the resistance between points A and B with the source replaced by a short (zero internal resistance) is the Thevenin equivalent resistance. In this particular circuit, the resistance from A to B is R_3 in series with the parallel combination of R_1 and R_2. Therefore, R_{TH} is expressed as follows:

$$R_{TH} = R_3 + \frac{R_1 R_2}{R_1 + R_2}$$

The Thevenin equivalent circuit is shown in Figure 6–45(c).

(a) Finding V_{TH}

(b) Finding R_{TH}

(c) Thevenin equivalent circuit

FIGURE 6–45

Example of the simplification of a circuit by Thevenin's theorem.

EXAMPLE 6–16 Find the Thevenin equivalent between the output terminals of the circuit in Figure 6–46.

FIGURE 6–46

Solution V_{TH} equals the voltage across $R_2 + R_3$ as shown in Figure 6–47(a). Use the voltage-divider principle to find V_{TH}.

$$V_{TH} = \left(\frac{R_2 + R_3}{R_1 + R_2 + R_3}\right)V_S = \left(\frac{69\ \Omega}{169\ \Omega}\right)10\ \text{V} = 4.08\ \text{V}$$

(a) The voltage from A to B is V_{TH} and equals $V_{2\text{-}3}$.

(b) Looking from terminals A and B, R_4 appears in series with the combination of R_1 in parallel with $(R_2 + R_3)$.

(c) Thevenin equivalent circuit

FIGURE 6–47

To find R_{TH}, first replace the source with a short to simulate a zero internal resistance. Then R_1 appears in parallel with $R_2 + R_3$, and R_4 is in series with the series-parallel combination of R_1, R_2, and R_3 as indicated in Figure 6–47(b).

$$R_{TH} = R_4 + \frac{R_1(R_2 + R_3)}{R_1 + R_2 + R_3} = 100\ \Omega + \frac{(100\ \Omega)(69\ \Omega)}{169\ \Omega} = 141\ \Omega$$

The resulting Thevenin equivalent circuit is shown in Figure 6–47(c).

Related Exercise Determine V_{TH} and R_{TH} if a 56 Ω resistor is connected in parallel across R_2 and R_3 in Figure 6–46.

Thevenin Equivalency Depends on the Viewpoint

The Thevenin equivalent for any circuit depends on the location of the two points from between which the circuit is "viewed." In Figure 6–46, we viewed the circuit from between the two points labeled A and B. Any given circuit can have more than one Thevenin equivalent, depending on how the viewpoints are designated. For example, if we view the circuit in Figure 6–48 from between points A and C, we obtain a completely different result than if we view it from between points A and B or from between points B and C.

FIGURE 6–48
Thevenin's equivalent depends on viewpoint.

In Figure 6–49(a), when viewed from between points A and C, V_{TH} is the voltage across $R_2 + R_3$ and can be expressed using the voltage-divider formula as

$$V_{TH} = \left(\frac{R_2 + R_3}{R_1 + R_2 + R_3} \right) V_S$$

Also, as shown in Figure 6–49(b), the resistance between points A and C is $R_2 + R_3$ in parallel with R_1 (the source is replaced by a short) and can be expressed as

$$R_{TH} = \frac{R_1(R_2 + R_3)}{R_1 + R_2 + R_3}$$

The resulting Thevenin equivalent circuit is shown in Figure 6–49(c).

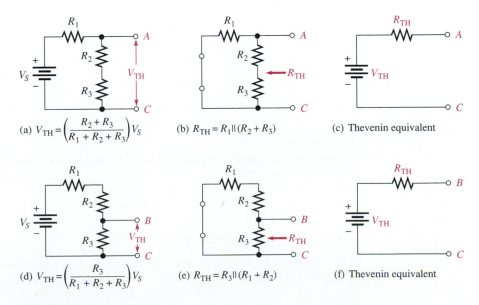

FIGURE 6–49
Example of circuit thevenized from two viewpoints, resulting in two different equivalent circuits. (The V_{TH} and R_{TH} values are different.)

When viewed from between points B and C as indicated in Figure 6–49(d), V_{TH} is the voltage across R_3 and can be expressed as

$$V_{TH} = \left(\frac{R_3}{R_1 + R_2 + R_3} \right) V_S$$

As shown in Figure 6–49(e), the resistance between points B and C is R_3 in parallel with the series combination of R_1 and R_2.

$$R_{TH} = \frac{R_3(R_1 + R_2)}{R_1 + R_2 + R_3}$$

The resulting Thevenin equivalent is shown in Figure 6-49(f).

Thevenizing a Portion of a Circuit

In many cases it helps to thevenize only a portion of a circuit. For example, when we need to know the equivalent circuit as viewed by one particular resistor in the circuit, we remove that resistor and apply Thevenin's theorem to the remaining part of the circuit as viewed from the points between which that resistor was connected. Figure 6–50 illustrates the thevenizing of part of a circuit.

Using this type of approach, we can easily find the voltage and current for a specified resistor for any number of resistor values using only Ohm's law. This method eliminates the necessity of reanalyzing the original circuit for each different resistance value.

(a) Original circuit

(b) Remove R_3 and thevenize

(c) Thevenin equivalent of original circuit with R_3 connected.

FIGURE 6–50

Example of thevenizing a portion of a circuit. In this case, the circuit is thevenized from the viewpoint of R_3.

Thevenizing a Bridge Circuit

The usefulness of Thevenin's theorem is perhaps best illustrated when it is applied to a Wheatstone bridge circuit. For example, when a load resistor is connected to the output terminals of a Wheatstone bridge, as shown in Figure 6–51, the circuit is very difficult to analyze because it is not a straightforward series-parallel arrangement. If you doubt that this analysis is difficult, try to identify which resistors are in parallel and which are in series.

FIGURE 6–51
Wheatstone bridge with a load resistor is not a series-parallel circuit.

Using Thevenin's theorem, we can simplify the bridge circuit to an equivalent circuit viewed from the load resistor as shown step-by-step in Figure 6–52. Study carefully the steps in this figure. Once the equivalent circuit for the bridge is found, the voltage and current for any value of load resistor can easily be determined.

(a) Remove R_L.

(b) Redraw to find V_{TH}.

(c) $V_{TH} = V_A - V_B = \left(\dfrac{R_2}{R_1 + R_2}\right)V_S - \left(\dfrac{R_4}{R_3 + R_4}\right)V_S$

(d) Replace V_S with a short.
Note: The colored lines represent the same electrical point as the colored lines in Part (e).

(e) Redraw to find R_{TH}:
$R_{TH} = R_1 \| R_2 + R_3 \| R_4$

(f) Thevenin's equivalent with R_L reconnected

FIGURE 6–52
Simplifying a Wheatstone bridge with Thevenin's theorem.

EXAMPLE 6–17   Determine the voltage and current for the load resistor, R_L, in the bridge circuit of Figure 6–53.

FIGURE 6–53

Solution

Step 1: Remove R_L.

Step 2: To Thevenize the bridge as viewed from between points A and B, as was shown in Figure 6–52, first determine V_{TH}.

$$V_{TH} = V_A - V_B = \left(\frac{R_2}{R_1 + R_2}\right)V_S - \left(\frac{R_4}{R_3 + R_4}\right)V_S$$

$$= \left(\frac{680\ \Omega}{1010\ \Omega}\right)24\text{ V} - \left(\frac{560\ \Omega}{1240\ \Omega}\right)24\text{ V} = 16.16\text{ V} - 10.84\text{ V} = 5.32\text{ V}$$

Step 3: Determine R_{TH}.

$$R_{TH} = \frac{R_1 R_2}{R_1 + R_2} + \frac{R_3 R_4}{R_3 + R_4}$$

$$= \frac{(330\ \Omega)(680\ \Omega)}{1010\ \Omega} + \frac{(680\ \Omega)(560\ \Omega)}{1240\ \Omega} = 222\ \Omega + 307\ \Omega = 529\ \Omega$$

Step 4: Place V_{TH} and R_{TH} in series to form the Thevenin equivalent circuit.

Step 5: Connect the load resistor from points A to B of the equivalent circuit, and determine the load voltage and current as illustrated in Figure 6–54.

$$V_L = \left(\frac{R_L}{R_L + R_{TH}}\right)V_{TH} = \left(\frac{1\text{ k}\Omega}{1.529\ \Omega}\right)5.32\text{ V} = 3.48\text{ V}$$

$$I_L = \frac{V_L}{R_L} = \frac{3.48\text{ V}}{1\text{ k}\Omega} = 3.48\text{ mA}$$

FIGURE 6–54   Thevenin's equivalent for the Wheatstone bridge

Related Exercise   Calculate I_L for $R_1 = 2.2$ kΩ, $R_2 = 3.3$ kΩ, $R_3 = 3.9$ kΩ, and $R_4 = 2.7$ kΩ in Figure 6–53.

Summary of Thevenin's Theorem

Remember, the Thevenin equivalent circuit is *always* an equivalent voltage source in series with an equivalent resistance regardless of the original circuit that it replaces. The significance of Thevenin's theorem is that the equivalent circuit can replace the original circuit as far as any external load is concerned. Any load resistor connected between the terminals of a Thevenin equivalent circuit will have the same current through it and the same voltage across it as if it were connected to the terminals of the original circuit.

A summary of steps for applying Thevenin's theorem is as follows:

Step 1: Open the two terminals (remove any load) between which you want to find the Thevenin equivalent circuit.

Step 2: Determine the voltage (V_{TH}) across the two open terminals.

Step 3: Determine the resistance (R_{TH}) between the two terminals with all voltage sources shorted and all current sources opened.

Step 4: Connect V_{TH} and R_{TH} in series to produce the complete Thevenin equivalent for the original circuit.

Step 5: Place the load resistor removed in Step 1 across the terminals of the Thevenin equivalent circuit. The load current can now be calculated using only Ohm's law, and it has the same value as the load current in the original circuit.

SECTION 6–7 REVIEW

1. What are the two components of a Thevenin equivalent circuit?
2. Draw the general form of a Thevenin equivalent circuit.
3. How is V_{TH} defined?
4. How is R_{TH} defined?
5. For the original circuit in Figure 6–55, draw the Thevenin equivalent circuit as viewed by R_L.

FIGURE 6–55

6–8 ■ TROUBLESHOOTING

Troubleshooting is the process of identifying and locating a failure or problem in a circuit. Some troubleshooting techniques have already been discussed in relation to both series circuits and parallel circuits. These methods are now extended to the series-parallel networks.

After completing this section, you should be able to

■ **Troubleshoot series-parallel circuits**
  □ Determine the effects of an open circuit
  □ Determine the effects of a short circuit
  □ Locate opens and shorts

Opens and shorts are typical problems that occur in electric circuits. As mentioned in Chapter 4, if a resistor burns out, it will normally produce an open circuit. Bad solder connections, broken wires, and poor contacts can also be causes of open paths. Pieces of foreign material, such as solder splashes, broken insulation on wires, and so on, can often lead to shorts in a circuit. A short is a zero resistance path between two points.

In addition to complete opens or shorts, partial opens or partial shorts can develop in a circuit. A partial open would be a much higher than normal resistance, but not infinitely large. A partial short would be a much lower than normal resistance, but not zero.

The following three examples illustrate troubleshooting series-parallel circuits.

EXAMPLE 6–18

From the indicated voltmeter reading in Figure 6–56, determine if there is a fault. If there is a fault, identify it as either a short or an open.

FIGURE 6–56

Solution First determine what the voltmeter should be indicating. Since R_2 and R_3 are in parallel, their combined resistance is

$$R_{2\text{-}3} = \frac{R_2 R_3}{R_2 + R_3} = \frac{(4.7 \text{ k}\Omega)(10 \text{ k}\Omega)}{14.7 \text{ k}\Omega} = 3.20 \text{ k}\Omega$$

The voltage across the parallel combination is determined by the voltage-divider formula as follows:

$$V_{2\text{-}3} = \left(\frac{R_{2\text{-}3}}{R_1 + R_{2\text{-}3}}\right) V_S = \left(\frac{3.2 \text{ k}\Omega}{18.2 \text{ k}\Omega}\right) 24 \text{ V} = 4.22 \text{ V}$$

This calculation shows that 4.22 V is the voltage reading that you should get on the meter. But the meter reads 9.6 V instead. This value is incorrect, and, because it is higher than it should be, R_2 or R_3 is probably open. Why? Because if either of these two resistors is open, the resistance across which the meter is connected is larger than expected. A higher resistance will drop a higher voltage in this circuit, which is, in effect, a voltage divider.

Start by assuming that R_2 is open. If it is, the voltage across R_3 is

$$V_3 = \left(\frac{R_3}{R_1 + R_3}\right) V_S = \left(\frac{10 \text{ k}\Omega}{25 \text{ k}\Omega}\right) 24 \text{ V} = 9.6 \text{ V}$$

This calculation shows that R_2 is open. Replace R_2 with a new resistor.

Related Exercise What would be the voltmeter reading if R_3 were open in Figure 6–56? If R_1 were open?

EXAMPLE 6–19   Suppose that you measure 24 V with the voltmeter in Figure 6–57. Determine if there is a fault, and, if there is, identify it.

FIGURE 6–57

Solution   There is no voltage drop across R_1 because both sides of the resistor are at +24 V. Either no current is flowing through R_1 from the source, which tells us that R_2 is open in the circuit, or R_1 is shorted.

   If R_1 were open, the meter in Figure 6–57 would not read 24 V. The most logical failure is an open R_2. If it is open, then no current will flow from the source. To verify this, measure across R_2 with the voltmeter as shown in Figure 6-58. If R_2 is open, the meter will indicate 24 V. The right side of R_2 will be at zero volts because no current is flowing through any of the other resistors to cause a voltage drop.

FIGURE 6–58

Related Exercise   What would be the voltage across an open R_5 in Figure 6–57 assuming no other faults?

EXAMPLE 6–20   The two voltmeters in Figure 6–59 indicate the voltages shown. Determine if there are any opens or shorts in the circuit and, if so, where they are located.

FIGURE 6–59

Solution First determine if the voltmeter readings are correct. R_1, R_2, and R_3 act as a voltage divider. The voltage across R_3 (V_A) is calculated as follows:

$$V_A = \left(\frac{R_3}{R_1 + R_2 + R_3} \right) V_S = \left(\frac{3.3 \text{ k}\Omega}{21.6 \text{ k}\Omega} \right) 24 \text{ V} = 3.67 \text{ V}$$

The voltmeter A reading (V_A) is correct.

Now see if the voltmeter B reading (V_B) is correct. $R_6 + R_7$ is in parallel with R_5. The series-parallel combination of R_5, R_6, and R_7 is in series with R_4. The resistance of the R_5, R_6, and R_7 combination is calculated as follows:

$$R_{5\text{-}6\text{-}7} = \frac{(R_6 + R_7)R_5}{R_5 + R_6 + R_7} = \frac{(17.2 \text{ k}\Omega)(10 \text{ k}\Omega)}{27.2 \text{ k}\Omega} = 6.32 \text{ k}\Omega$$

$R_{5\text{-}6\text{-}7}$ and R_4 form a voltage divider, and voltmeter B is measuring the voltage across $R_{5\text{-}6\text{-}7}$. Is it correct? Check as follows:

$$V_B = \left(\frac{R_{5\text{-}6\text{-}7}}{R_4 + R_{5\text{-}6\text{-}7}} \right) V_S = \left(\frac{6.32 \text{ k}\Omega}{11 \text{ k}\Omega} \right) 24 \text{ V} = 13.8 \text{ V}$$

Thus, the actual measured voltage (6.65 V) at this point is incorrect. Some further thought will help to isolate the problem.

R_4 is not open, because if it were, the meter would read 0 V. If there were a short across it, the meter would read 24 V. Since the actual voltage is much less than it should be, $R_{5\text{-}6\text{-}7}$ must be less than the calculated value of 6.32 kΩ. The most likely problem is a short across R_7. If there is a short from the top of R_7 to ground, R_6 is effectively in parallel with R_5. In this case,

$$R_5 \| R_6 = \frac{R_5 R_6}{R_5 + R_6} = \frac{(10 \text{ k}\Omega)(2.2 \text{ k}\Omega)}{12.2 \text{ k}\Omega} = 1.80 \text{ k}\Omega$$

Then V_B is

$$V_B = \left(\frac{1.80 \text{ k}\Omega}{6.5 \text{ k}\Omega} \right) 24 \text{ V} = 6.65 \text{ V}$$

This value for V_B agrees with the voltmeter B reading. So there is a short across R_7. If this were an actual circuit, you would try to find the physical cause of the short.

Related Exercise If R_2 in Figure 6–59 were shorted, what would voltmeter A read? What would voltmeter B read?

SECTION 6–8 REVIEW

1. Name two types of common circuit faults.
2. For the following faults in Figure 6–60, determine what voltage would be measured at point A:
   (a) no faults   (b) R_1 open   (c) short across R_5
   (d) R_3 and R_4 open   (e) R_2 open

FIGURE 6–60

3. In Figure 6–61 one of the resistors in the circuit is open. Based on the meter reading, determine which is the bad resistor.

FIGURE 6–61

6–9 ■ APPLICATION ASSIGNMENT

A voltage divider with three output voltages has been designed and constructed on a PC board. The voltage divider is to be used as part of a portable power supply unit for supplying up to three different reference voltages to measuring instruments in the field. The power supply unit contains a battery pack combined with a voltage regulator that produces a constant +12 V to the voltage-divider circuit board. In this assignment, you will apply your knowledge of loaded voltage dividers, Kirchhoff's laws and Ohm's law to determine the operating parameters of the voltage divider in terms of voltages and currents for all possible load configurations. You will also troubleshoot the circuit for various malfunctions.

Step 1: Draw the Schematic of the Voltage Divider

Draw the schematic and label the resistor values for the circuit board in Figure 6–62.

Step 2: Connect the 12 V Power Supply

Specify how to connect a 12 V power supply to the circuit board so that all resistors are in series and pin 2 has the highest output voltage.

Step 3: Determine the Unloaded Output Voltages

Calculate each of the output voltages with no loads connected. Add these voltage values to a copy of the table in Figure 6–63.

10MΩ Load	$V_{OUT(2)}$	$V_{OUT(3)}$	$V_{OUT(4)}$	% Deviation	$I_{LOAD(2)}$	$I_{LOAD(3)}$	$I_{LOAD(4)}$
None							
Pin 2							
Pin 3							
Pin 4							
Pins 2 and 3				2 3			
Pins 2 and 4				2 4			
Pins 3 and 4				3 4			
Pins 2, 3, and 4				2 3 4			

FIGURE 6–62
Voltage-divider circuit board.

FIGURE 6–63
Table of operating parameters for the power supply voltage divider.

Step 4: Determine the Loaded Output Voltages

The instruments to be connected to the voltage divider each have a 10 MΩ input resistance. This means that when an instrument is connected to a voltage-divider output there is effectively a 10 MΩ resistor from that output to ground (negative side of source). Determine the output voltage across each load for the following load combinations and add these voltage values to a copy of the table in Figure 6–63.

1. A 10 MΩ load connected to pin 2.
2. A 10 MΩ load connected to pin 3.
3. A 10 MΩ load connected to pin 4.
4. 10 MΩ loads connected to pin 2 and pin 3.
5. 10 MΩ loads connected to pin 2 and pin 4.
6. 10 MΩ loads connected to pin 3 and pin 4.
7. 10 MΩ loads connected to pin 2, pin 3, and pin 4.

Step 5: Determine the Percent Deviation of the Output Voltages

Calculate how much each loaded output voltage deviates from its unloaded value for each of the load configurations listed in Step 4 and express as a percentage using the following formula.

$$\text{Percent deviation} = \left(\frac{V_{OUT(unloaded)} - V_{OUT(loaded)}}{V_{OUT(unloaded)}} \right) 100\%$$

Add the values to the table in Figure 6–63.

Step 6: Determine the Load Currents

Calculate the current to each 10 MΩ load for each of the load configurations listed in Step 4. Add these values to the table in Figure 6–63.

Step 7: Specify a Minimum Value for the Fuse

Step 8: Troubleshoot the Circuit Board

The voltage-divider circuit board is connected to a 12 V power supply and to the three instruments to which it provides reference voltages, as shown in Figure 6–64. Voltages at each of the numbered test points are measured with a voltmeter in each of eight different cases. For each case, determine the problem indicated by the voltage measurements.

FIGURE 6–64

The following voltmeter readings are taken at test points 1 through 6 going from left to right. The readings are in volts.

	①	②	③	④	⑤	⑥
Case 1	0	0	0	0	0	0
Case 2	12	0	0	0	0	0
Case 3	12	0	0	0	0	12
Case 4	12	11.6	0	0	0	12
Case 5	12	11.3	10.9	0	0	12
Case 6	12	11	10.3	10	0	12
Case 7	12	5.9	0	0	0	12
Case 8	12	7.8	3.8	0	0	12

SECTION 6–9 REVIEW

1. If the portable unit covered in this section is to supply reference voltages to all three instruments, how many days can a 100 mAh battery be used?
2. Can $\frac{1}{8}$ W resistors be used on the voltage-divider board?
3. If $\frac{1}{8}$ W resistors are used, will an output shorted to ground cause any of the resistors to overheat due to excessive power?

■ **SUMMARY**

- A series-parallel circuit is a combination of both series paths and parallel paths.
- To determine total resistance in a series-parallel circuit, identify the series and parallel relationships, and then apply the formulas for series resistance and parallel resistance from Chapters 4 and 5.
- To find the total current, divide the total voltage by the total resistance.
- To determine branch currents, apply the current-divider formula, Kirchhoff's current law, or Ohm's law. Consider each circuit problem individually to determine the most appropriate method.
- To determine voltage drops across any portion of a series-parallel circuit, use the voltage-divider formula, Kirchhoff's voltage law, or Ohm's law. Consider each circuit problem individually to determine the most appropriate method.
- When a load resistor is connected across a voltage-divider output, the output voltage decreases.
- The load resistor should be large compared to the resistance across which it is connected, in order that the loading effect may be minimized. A *10-times* value is sometimes used as a rule of thumb, but the value depends on the accuracy required for the output voltage.
- To find any current or voltage in a circuit with two or more voltage sources, take the sources one at a time using the superposition method.
- A Wheatstone bridge can be used to measure an unknown resistance.
- A bridge is balanced when the output voltage is zero. The balanced condition produces zero current through a load connected across the output terminals of the bridge.
- Any resistive circuit, no matter how complex, can be replaced by its Thevenin equivalent.
- The Thevenin equivalent circuit is made up of an equivalent resistance (R_{TH}) in series with an equivalent voltage source (V_{TH}).
- Open circuits and short circuits are typical faults.
- Resistors normally open when they burn out.

■ **GLOSSARY**

Bleeder current The current left after the total load current is subtracted from the total current into the circuit.

Load An element (resistor or other component) connected across the output terminals of a circuit that draws current from the circuit.

Sensitivity factor The ohms-per-volt rating of a voltmeter.

Superposition A method for analyzing circuits with two or more sources by examining the effects of each source by itself and then combining the effects.

Terminal equivalency A condition that occurs when two circuits produce the same load voltage and load current when the same value of load resistance is connected to either circuit.

Thevenin's theorem A circuit theorem that provides for reducing any resistive circuit to a single equivalent voltage source in series with an equivalent resistance.

■ **FORMULAS**

(6–1) $I_{\text{BLEEDER}} = I_T - I_{RL1} - I_{RL2}$ Bleeder current

(6–2) $R_{\text{UNK}} = R_V \left(\dfrac{R_2}{R_4} \right)$ Unknown resistance in a Wheatstone bridge

■ SELF-TEST

1. Which of the following statements are true concerning Figure 6–65?
   (a) R_1 and R_2 are in series with R_3, R_4, and R_5
   (b) R_1 and R_2 are in series
   (c) R_3, R_4, and R_5 are in parallel
   (d) The series combination of R_1 and R_2 is in parallel with the series combination of R_3, R_4, and R_5
   (e) both (b) and (d)

FIGURE 6–65

2. The total resistance of Figure 6–65 can be found with which of the following formulas?
   (a) $R_1 + R_2 + R_3 \| R_4 \| R_5$ (b) $R_1 \| R_2 + R_3 \| R_4 \| R_5$
   (c) $(R_1 + R_2) \| (R_3 + R_4 + R_5)$ (d) none of the above

3. If all of the resistors in Figure 6–65 have the same value, when voltage is applied across terminals A and B, the current is
   (a) greatest in R_5 (b) greatest in R_3, R_4, and R_5
   (c) greatest in R_1 and R_2 (d) the same in all the resistors

4. Two 1 kΩ resistors are in series and this series combination is in parallel with a 2.2 kΩ resistor. The voltage across one of the 1 kΩ resistors is 6 V. The voltage across the 2.2 kΩ resistor is
   (a) 6 V (b) 3 V (c) 12 V (d) 13.2 V

5. The parallel combination of a 330 Ω resistor and a 470 Ω resistor is in series with the parallel combination of four 1 kΩ resistors. A 100 V source is connected across the circuit. The resistor with the most current has a value of
   (a) 1 kΩ (b) 330 Ω (c) 470 Ω

6. In the circuit described in Question 5, the resistor(s) with the most voltage has a value of
   (a) 1 kΩ (b) 470 Ω (c) 330 Ω

7. In the circuit described in Question 5, the percentage of the total current through any single 1 kΩ resistor is
   (a) 100% (b) 25% (c) 50% (d) 31.25%

8. The output of a certain voltage divider is 9 V with no load. When a load is connected, the output voltage
   (a) increases (b) decreases (c) remains the same (d) becomes zero

9. A certain voltage divider consists of two 10 kΩ resistors in series. Which of the following load resistors will have the most effect on the output voltage?
   (a) 1 MΩ (b) 20 kΩ (c) 100 kΩ (d) 10 kΩ

10. When a load resistance is connected to the output of a voltage-divider circuit, the current drawn from the source
    (a) decreases (b) increases (c) remains the same (d) is cut off

11. The output voltage of a balanced Wheatstone bridge is
    (a) equal to the source voltage
    (b) equal to zero
    (c) dependent on all of the resistor values in the bridge
    (d) dependent on the value of the unknown resistor

12. The primary method of analyzing a circuit with two or more voltage sources is usually
    (a) Thevenin's theorem (b) Ohm's law
    (c) superposition (d) Kirchhoff's law

13. In a certain two-source circuit, one source acting alone produces 10 mA through a given branch. The other source acting alone produces 8 mA in the opposite direction through the same branch. The total current through the branch is
    (a) 10 mA (b) 8 mA (c) 18 mA (d) 2 mA

14. A Thevenin equivalent circuit consists of
   (a) a voltage source in series with a resistance
   (b) a voltage source in parallel with a resistance
   (c) a current source in parallel with a resistance
   (d) two voltage sources and a resistance

15. You are measuring the voltage at a given point in a circuit that has very high resistance values and the measured voltage is a little lower than it should be. This is possibly because of
   (a) one or more of the resistance values being off
   (b) the loading effect of the voltmeter
   (c) the source voltage is too low
   (d) all the above

■ BASIC PROBLEMS

SECTION 6–1 Identifying Series-Parallel Relationships

1. Identify the series-parallel relationships in Figure 6–66 as seen from the source terminals.

FIGURE 6–66

2. Visualize and sketch the following series-parallel combinations:
   (a) R_1 in series with the parallel combination of R_2 and R_3
   (b) R_1 in parallel with the series combination of R_2 and R_3
   (c) R_1 in parallel with a branch containing R_2 in series with a parallel combination of four other resistors

3. Visualize and sketch the following series-parallel circuits:
   (a) a parallel combination of three branches, each containing two series resistors
   (b) a series combination of three parallel circuits, each containing two parallel resistors

4. In each circuit of Figure 6–67 identify the series and parallel relationships of the resistors viewed from the source.

(a)   (b)

FIGURE 6–67

SECTION 6–2 Analysis of Series-Parallel Circuits

5. A certain circuit is composed of two parallel resistors. The total resistance is 667 Ω. One of the resistors is 1 kΩ. What is the other resistor?

6. For the circuit in Figure 6–68, determine the total resistance between points A and B.

7. Determine the total resistance for each circuit in Figure 6–67.

8. Determine the current through each resistor in Figure 6–66; then calculate each voltage drop.

9. Determine the current through each resistor in both circuits of Figure 6–67; then calculate each voltage drop.

10. In Figure 6–69, find the following:
    (a) total resistance between terminals A and B
    (b) total current drawn from a 6 V source connected from A to B
    (c) current through R_5
    (d) voltage across R_2

FIGURE 6–68

FIGURE 6–69

SECTION 6–3 Voltage Dividers with Resistive Loads

11. A voltage divider consists of two 56 kΩ resistors and a 15 V source. Calculate the unloaded output voltage. What will the output voltage be if a load resistor of 1 MΩ is connected to the output?

12. A 12 V battery output is divided down to obtain two output voltages. Three 3.3 kΩ resistors are used to provide the two outputs with only one output at a time loaded with 10 kΩ. Determine the output voltages in both cases.

13. Which will cause a smaller decrease in output voltage for a given voltage divider, a 10 kΩ load or a 56 kΩ load?

14. In Figure 6–70, determine the current drain on the battery with no load across the output terminals. With a 10 kΩ load, what is the battery current?

FIGURE 6–70

SECTION 6–4 Loading Effect of a Voltmeter

15. On which one of the following voltage range settings will a voltmeter present the minimum load on a circuit?
    (a) 1 V   (b) 10 V   (c) 100 V   (d) 1000 V

16. Determine the internal resistance of a 20,000 Ω/V voltmeter on each of the following range settings.
    (a) 0.5 V   (b) 1 V   (c) 5 V   (d) 50 V   (e) 100 V   (f) 1000 V

17. The voltmeter described in Problem 16 is used to measure the voltage across R_4 in Figure 6–67(a).
    (a) What range should be used?
    (b) How much less is the voltage measured by the meter than the actual voltage?

18. Repeat Problem 17 if the voltmeter is used to measure the voltage across R_4 in the circuit of Figure 6–67(b).

SECTION 6–5   The Wheatstone Bridge

19. A resistor of unknown value is connected to a Wheatstone bridge circuit. The bridge parameters for a balanced condition are set as follows: $R_V = 18$ kΩ and $R_2/R_4 = 0.02$. What is R_{UNK}?

20. A bridge network is shown in Figure 6–71. To what value must R_V be set in order to balance the bridge?

FIGURE 6–71

SECTION 6–6   Circuits with More Than One Voltage Source

21. In Figure 6–72, use the superposition method to find the total current in R_3.

22. In Figure 6–72, what is the total current through R_2?

FIGURE 6–72

SECTION 6–7   Thevenin's Theorem

23. Reduce the circuit in Figure 6–73 to its Thevenin equivalent as viewed from terminals A and B.

24. For each circuit in Figure 6–74, determine the Thevenin equivalent as seen by R_L.

FIGURE 6–73

(a)

(b)

(c)

FIGURE 6–74

SECTION 6–8   Troubleshooting

25. Is the voltmeter reading in Figure 6–75 correct? If not, what is the problem?

26. If R_2 in Figure 6–76 opens, what voltages will be read at points A, B, and C?

27. Check the meter readings in Figure 6–77 and locate any fault that may exist.

FIGURE 6–75

(a) Meter with leads going to circuit board

(b) Circuit board with meter leads and red and black leads from a 12 V dc power supply connected

FIGURE 6–76

FIGURE 6–77

28. Determine the voltage you would expect to measure across each resistor in Figure 6–76 for each of the following faults:

    (a) R_1 open (b) R_3 open (c) R_4 open
    (d) R_5 open (e) point C shorted to ground

29. Determine the voltage you would expect to measure across each resistor in Figure 6–77 for each of the following faults:

    (a) R_1 open (b) R_2 open (c) R_3 open (d) a short across R_4

▪ ADVANCED PROBLEMS

30. In each circuit of Figure 6–78, identify the series and parallel relationships of the resistors viewed from the source.

31. Draw the schematic of the PC board layout in Figure 6–79 showing resistor values and identify the series-parallel relationships. Which resistors, if any, can be removed with no effect on R_T?

32. For the circuit shown in Figure 6–80, calculate the following:

    (a) total resistance across the source (b) total current from the source
    (c) current through the 910 Ω resistor (d) voltage from point A to point B

33. Determine the total resistance and the voltage at points A, B, and C in the circuit of Figure 6–81.

34. Determine the total resistance between terminals A and B of the circuit in Figure 6–82. Also calculate the current in each branch with 10 V between A and B.

35. What is the voltage across each resistor in Figure 6–82?

FIGURE 6–78

(a)

(b)

FIGURE 6–79

FIGURE 6–80

FIGURE 6–81

FIGURE 6–82

36. Determine the voltage, V_{AB}, in Figure 6–83.

37. Find the value of R_2 in Figure 6–84.

38. Determine the total resistance and the voltage at points A, B, and C in the circuit of Figure 6–85.

39. Develop a voltage divider to provide a 6 V output with no load and a minimum of 5.5 V across a 1 kΩ load. The source voltage is 24 V, and the unloaded current is not to exceed 100 mA.

40. Determine the resistance values for a voltage divider that must meet the following specifications: The current under an unloaded condition is not to exceed 5 mA. The source voltage is to be 10 V. A 5 V output and a 2.5 V output are required. Sketch the circuit. Determine the effect on the output voltages if a 1 kΩ load is connected to each output.

41. Using the superposition method, calculate the current in the right-most branch of Figure 6–86.

FIGURE 6–83

FIGURE 6–84

FIGURE 6–85

FIGURE 6–86

42. Find the current through R_L in Figure 6–87.

43. Using Thevenin's theorem, find the voltage across R_4 in Figure 6–88.

44. Determine V_{OUT} for the circuit in Figure 6–89 for the following conditions:
  (a) Switch SW_2 connected to +12 V and the rest to ground
  (b) Switch SW_1 connected to +12 V and the rest to ground

45. Develop a schematic for the double-sided PC board in Figure 6–90 and label the resistor values.

46. Lay out a PC board for the circuit in Figure 6–78(b). The battery is to be connected external to the board.

FIGURE 6–87

FIGURE 6–88

FIGURE 6–89

FIGURE 6–90

Side 1   Side 2

47. The voltage divider in Figure 6–91 has a switched load. Determine the voltage at each tap (V_1, V_2, and V_3) for each position of the switch.

48. Figure 6–92 shows a dc biasing arrangement for a field-effect transistor amplifier. Biasing is a common method for setting up certain dc voltage levels required for proper amplifier operation. Although you may not be familiar with transistor amplifiers at this point, the dc voltages and currents in the circuit can be determined using methods that you already know.
    (a) Find V_G and V_S with respect to ground   **(b)** Determine I_1, I_2, I_D, and I_S
    (c) Find V_{DS} and V_{DG}

49. Look at the voltmeters in Figure 6–93 and determine if there is a fault in the circuit. If there is a fault, identify it.

FIGURE 6–91

FIGURE 6–92

FIGURE 6–93

50. Are the voltmeter readings in Figure 6–94 correct?

FIGURE 6–94

51. There is one fault in Figure 6–95. Based on the voltmeter indications, determine what the fault is.

FIGURE 6–95

52. Look at the voltmeters in Figure 6–96 and determine if there is a fault in the circuit. If there is a fault, identify it.

53. Determine the voltmeter readings in Figure 6–96 if the 4.7 kΩ resistor opens.

FIGURE 6–96

▪ **ANSWERS TO SECTION REVIEWS**

Section 6–1

1. See Figure 6–97.

FIGURE 6–97

2. R_1 and R_2 are in series with the parallel combination of R_3 and R_4.

3. All resistors are in parallel.

4. R_1 and R_2 are in parallel; R_3 and R_4 are in parallel.

5. Yes, the two parallel combinations are in series with each other.

Section 6–2

1. $R_T = R_1 + R_4 + R_2 \| R_3 = 599 \ \Omega$

2. $I_3 = 11.2$ mA

3. $V_2 = I_2 R_2 = 3.7$ V

4. $R_T = 89 \ \Omega; \ I_T = 11.2$ mA

Section 6–3

1. The load resistor decreases the output voltage.

2. True

3. $V_{\text{OUT(unloaded)}} = 19.2$ V, $V_{\text{OUT(loaded)}} = 4.19$ V

Section 6–4

1. A voltmeter loads a circuit because the internal resistance of the meter appears in parallel with the circuit resistance across which it is connected, reducing the resistance between those two points of the circuit and drawing current from the circuit.

2. No, because the meter resistance is much larger than 1 kΩ.

3. There is a decrease in measured voltage because R_M is smaller. You can read the scale more accurately on the 3 V range.

4. $R_M = 1.5$ MΩ

Section 6–5

1. See Figure 6–98.

2. A bridge is balanced when the output voltage is zero.

3. $R_{\text{UNK}} = 15$ kΩ

FIGURE 6–98

Section 6–6

1. The total current in any branch of a multiple-source linear circuit is equal to the algebraic sum of the currents due to the individual sources acting alone, with the other sources replaced by their internal resistances.

2. The superposition method allows each source to be treated independently.

3. A short simulates the internal resistance of an ideal voltage source; an open simulates the internal resistance of an ideal current source.

4. $I_{R1} = 6.67$ mA

5. The net current is in the direction of the larger current.

Section 6–7

1. A Thevenin circuit includes V_{TH} and R_{TH}.

2. See Figure 6–99.

3. V_{TH} is the open circuit voltage between two points in a circuit.

4. R_{TH} is the resistance as viewed from two points in a circuit, with all sources replaced by their internal resistances.

5. See Figure 6–100.

FIGURE 6–99

FIGURE 6–100

Section 6–8

1. Opens and shorts are two common faults.

2. (a) 62.8 V   (b) 62.8 V   (c) 62 V   (d) 100 V   (e) 0 V

3. The 10 kΩ resistor is open.

Section 6–9

1. The battery will last 417 days.

2. Yes, $\frac{1}{8}$ W resistors can be used.

3. No, none of the resistors will overheat.

■ **ANSWERS TO RELATED EXERCISES FOR EXAMPLES**

6–1 The added resistor would be in parallel with the combination of R_4 in series with R_2 and R_3 in parallel.

6–2 The added resistor is in parallel with R_5.

6–3 A to gnd: $R_T = R_3 \parallel (R_1 + R_2) + R_4$
B to gnd: $R_T = R_2 \parallel (R_1 + R_3) + R_4$
C to gnd: $R_T = R_4$

6–4 None

6–5 R_1 and R_4 are effectively removed from the circuit.

6–6 55.1 Ω

6–7 128.3 Ω

6–8 $I_1 = 89.3$ mA; $I_3 = 58.5$ mA; $I_T = 182.3$ mA

6–9 $V_1 = V_2 = 103$ V; $V_3 = 97$ V; $V_4 = 31.6$ V; $V_5 = 65.4$ V

6–10 $I_1 = 1.42$ mA, $P_1 = 6.67$ mW; $I_2 = 756$ μA, $P_2 = 3.55$ mW; $I_3 = 2.18$ mA, $P_3 = 4.75$ mW; $I_4 = 1.13$ mA, $P_4 = 1.28$ mW; $I_5 = 1.06$ mA, $P_5 = 758$ μW; $I_6 = 1.06$ mA, $P_6 = 435$ μW

6–11 3.37 V

6–12 Increase the values of R_1, R_2, and R_3 proportionally.

6–13 5.34 V

6–14 321 Ω

6–15 –2 V

6–16 2.36 V; 124 Ω

6–17 1.17 mA

6–18 5.73 V; 0 V

6–19 9.46 V

6–20 $V_A = 12$ V; $V_B = 13.8$ V

7

MAGNETISM AND ELECTROMAGNETISM

This chapter is somewhat of a departure from the previous six chapters because two new concepts are introduced: magnetism and electromagnetism. The operation of many types of electrical devices is based partially on magnetic or electromagnetic principles.

■ CHAPTER OBJECTIVES

- ☐ Explain the principles of the magnetic field
- ☐ Explain the principles of electromagnetism
- ☐ Describe the principle of operation for several types of electromagnetic devices
- ☐ Explain magnetic hysteresis
- ☐ Discuss the principle of electromagnetic induction
- ☐ Describe some applications of electromagnetic induction

■ APPLICATION ASSIGNMENT

For the application assignment in this chapter, you will apply what you learn in this chapter about relays and other devices to a simple burglar alarm system. You will determine how to connect the components to form a complete system and specify a check-out procedure to make sure the system is working properly. After studying this chapter, you should be able to complete the application assignment in Section 7–7.

7–1 ■ THE MAGNETIC FIELD

A permanent magnet has a magnetic field surrounding it. The magnetic field consists of lines of force that radiate from the north pole to the south pole and back to the north pole through the magnetic material.

After completing this section, you should be able to

■ **Explain the principles of the magnetic field**
- ☐ Define *magnetic flux*
- ☐ Define *magnetic flux density*
- ☐ Discuss how materials are magnetized
- ☐ Explain how a magnetic switch works

A permanent magnet, such as the bar magnet shown in Figure 7–1, has a magnetic field surrounding it. The magnetic field consists of lines of force, or flux lines, that radiate from the north pole (N) to the south pole (S) and back to the north pole through the magnetic material. For clarity, only a few lines of force are shown in the figure. Imagine, however, that many lines surround the magnet in three dimensions. The lines shrink to the smallest possible size and blend together, although they do not touch. This effectively forms a continuous magnetic field surrounding the magnet.

Attraction and Repulsion of Magnetic Poles

When unlike poles of two permanent magnets are placed close together, an attractive force is produced by the magnetic fields, as indicated in Figure 7–2(a). When two like poles are brought close together, they repel each other, as shown in part (b).

(a) Unlike poles attract.

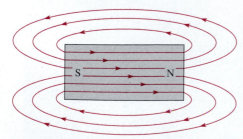

Colored lines represent a few magnetic lines of force in the magnetic field.

(b) Like poles repel.

FIGURE 7–1

Magnetic lines of force around a bar magnet.

FIGURE 7–2

Magnetic attraction and repulsion.

Altering a Magnetic Field

When a nonmagnetic material such as paper, glass, wood, or plastic is placed in a magnetic field, the lines of force are unaltered, as shown in Figure 7–3(a). However, when a magnetic material such as iron is placed in the magnetic field, the lines of force tend to change course and pass through the iron rather than through the surrounding air. They do so because the iron provides a magnetic path that is more easily established than that of air. Figure7–3(b) illustrates this principle.

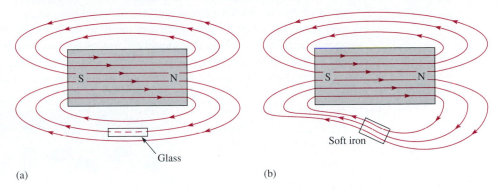

(a)   (b)

FIGURE 7–3
Effect of (a) nonmagnetic and (b) magnetic materials on a magnetic field.

Magnetic Flux (φ)

The group of force lines going from the north pole to the south pole of a magnet is called the **magnetic flux**, symbolized by φ (the lowercase Greek letter phi). The number of lines of force in a magnetic field determines the value of the flux. The more lines of force, the greater the flux and the stronger the magnetic field.

The unit of magnetic flux is the **weber** (Wb). One weber equals 10^8 lines. The weber is a very large unit; thus, in most practical situations, the microweber (μWb) is used. One microweber equals 100 lines of magnetic flux.

Magnetic Flux Density (B)

The **magnetic flux density** is the amount of flux per unit area in the magnetic field. Its symbol is B, and its unit is the **tesla** (T). One tesla equals one weber per square meter (Wb/m²). The following formula expresses the flux density:

$$B = \frac{\phi}{A}$$ (7–1)

where φ is the flux and A is the cross-sectional area of the magnetic field.

EXAMPLE 7–1

Find the flux density in a magnetic field in which the flux in 0.1 m² is 800 μWb.

Solution

$$B = \frac{\phi}{A} = \frac{800\ \mu\text{Wb}}{0.1\ \text{m}^2} = 8000 \times 10^{-6}\ \text{T}$$

Related Exercise Repeat this example for a flux of 1.2 mWb.

How Materials Become Magnetized

Ferromagnetic materials such as iron, nickel, and cobalt become magnetized when placed in the magnetic field of a magnet. We have all seen a permanent magnet pick up paper clips, nails, iron filings, and so on. In these cases, the object becomes magnetized (that is, it actually becomes a magnet itself) under the influence of the permanent magnetic field and becomes attracted to the magnet. When removed from the magnetic field, the object tends to lose its magnetism.

Ferromagnetic materials have minute magnetic domains created within their atomic structure by the orbital motion and spin of electrons. These domains can be viewed as very small bar magnets with north and south poles. When the material is not exposed to an external magnetic field, the magnetic domains are randomly oriented, as shown in Figure 7–4(a). When the material is placed in a magnetic field, the domains align themselves, as shown in part (b). Thus, the object itself effectively becomes a magnet.

FIGURE 7–4

Ferromagnetic domains in (a) an unmagnetized and (b) a magnetized material.

(a) The magnetic domains (N◀ S) are randomly oriented in the unmagnetized material.

(b) The magnetic domains become aligned when the material is magnetized.

An Application

Permanent magnets have almost endless applications, one of which is presented here as an illustration. Figure 7–5 shows a typical magnetically operated, normally closed (NC) switch. When the magnet is near the switch mechanism, the metallic arm is held in its NC position. When the magnet is moved away, the spring pulls the arm up, breaking the contact as shown in Figure 7–6.

(a) Contact is closed.

(b) Contact opens.

FIGURE 7–5

Magnet and switch set (courtesy of Tandy Corp.).

FIGURE 7–6

Operation of a magnetic switch.

FIGURE 7–7
Connection of a typical perimeter alarm system.

Switches of this type are commonly used in perimeter alarm systems to detect entry into a building through windows or doors. As Figure 7–7 shows, several openings can be protected by magnetic switches wired to a common transmitter. When any one of the switches opens, the transmitter is activated and sends a signal to a central receiver and alarm unit.

SECTION 7–1 REVIEW

1. When the north poles of two magnets are placed close together, do they repel or attract each other?
2. What is magnetic flux?
3. What is the flux density when $\phi = 4.5 \ \mu\text{Wb}$ and $A = 5 \times 10^{-3} \ \text{m}^2$?

7–2 ■ ELECTROMAGNETISM

Electromagnetism is the production of a magnetic field by current in a conductor. Many types of useful devices such as tape recorders, electric motors, speakers, solenoids, and relays are based on electromagnetism.

After completing this section, you should be able to

■ **Explain the principles of electromagnetism**
  □ Determine the direction of the magnetic lines of force
  □ Define *permeability*
  □ Define *reluctance*
  □ Define *magnetomotive force*
  □ Describe a basic electromagnet

Current produces a magnetic field, called an **electromagnetic field,** around a conductor, as illustrated in Figure 7–8. The invisible lines of force of the magnetic field form a concentric circular pattern around the conductor and are continuous along its length.

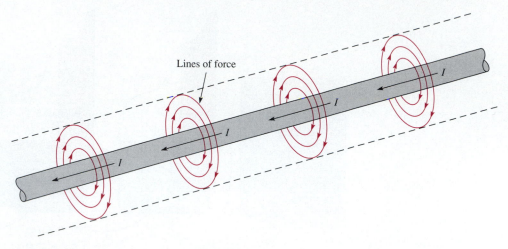

Lines of force

FIGURE 7–8
Magnetic field around a current-carrying conductor.

Iron filings

Compass

(a)   (b)

FIGURE 7–9
Visible effects of an electromagnetic field.

Although the magnetic field cannot be seen, it is capable of producing visible effects. For example, if a current-carrying wire is inserted through a sheet of paper in a perpendicular direction, iron filings placed on the surface of the paper arrange themselves along the magnetic lines of force in concentric rings, as illustrated in Figure 7–9(a). Part (b) of the figure illustrates that the north pole of a compass placed in the electromagnetic field will point in the direction of the lines of force. The field is stronger closer to the conductor and becomes weaker with increasing distance from the conductor.

Direction of the Lines of Force

The direction of the lines of force surrounding the conductor are indicated in Figure 7–10. When the direction of current is right to left, as in part (a), the lines are in a clock-

FIGURE 7–10
Magnetic lines of force around a current-carrying conductor.

(a)   (b)

wise direction. When current is left to right, as in part (b), the lines are in a counterclockwise direction.

Left-Hand Rule

An aid to remembering the direction of the lines of force is illustrated in Figure 7–11. Imagine that you are grasping the conductor with your left hand, with your thumb pointing in the direction of current. Your fingers indicate the direction of the magnetic lines of force.

FIGURE 7–11
Illustration of left-hand rule.

Electromagnetic Properties

Several important properties relating to electromagnetic fields are now discussed.

Permeability (μ)   The ease with which a magnetic field can be established in a given material is measured by the **permeability** of that material. The higher the permeability, the more easily a magnetic field can be established.

The symbol of permeability is μ, and its value varies depending on the type of material. The permeability of a vacuum (μ_0) is $4\pi \times 10^{-7}$ Wb/At·m (weber/ampere-turn·meter) and is used as a reference. Ferromagnetic materials typically have permeabilities hundreds of times larger than that of a vacuum, indicating that a magnetic field can be set up with relative ease in these materials. Ferromagnetic materials include iron, steel, nickel, cobalt, and their alloys.

The relative permeability (μ_r) of a material is the ratio of its absolute permeability (μ) to the permeability of a vacuum (μ_0). Since μ_r is a ratio, it has no units.

$$\mu_r = \frac{\mu}{\mu_0} \qquad (7-2)$$

Reluctance ($\mathcal{R}$)   The opposition to the establishment of a magnetic field in a material is called **reluctance** ($\mathcal{R}$). The value of reluctance is directly proportional to the length (l) of the magnetic path, and inversely proportional to the permeability (μ) and to the cross-sectional area (A) of the material as expressed by the following equation:

$$\mathcal{R} = \frac{l}{\mu A} \qquad (7-3)$$

Reluctance in magnetic circuits in analogous to resistance in electric circuits. The unit of reluctance can be derived using l in meters. A (area) in square meters, and μ in Wb/At·m as follows:

$$\mathcal{R} = \frac{l}{\mu A} = \frac{m}{(Wb/At·m)(m^2)} = \frac{At}{Wb}$$

At/Wb is ampere-turns/weber.

EXAMPLE 7–2

What is the reluctance of a material that has a length of 0.05 m, a cross-sectional area of 0.012 m², and a permeability of 3500×10^{-6} Wb/At·m?

Solution
$$\mathcal{R} = \frac{1}{\mu A} = \frac{0.05 \text{ m}}{(3500 \times 10^{-6} \text{ Wb/At·m})(0.012 \text{ m}^2)} = 1190 \text{ At/Wb}$$

Related Exercise What happens to the reluctance if *l* is doubled and *A* is halved?

Magnetomotive Force (mmf) As you have learned, current in a conductor produces a magnetic field. The force that produces the magnetic field is called the **magnetomotive force** (mmf). The unit of mmf, the **ampere-turn** (At), is established on the basis of the current in a single loop (turn) of wire. The formula for mmf is as follows:

$$F_m = NI \qquad\qquad (7\text{–}4)$$

where F_m is the magnetomotive force, N is the number of turns of wire, and I is the current in amperes.

Figure 7–12 illustrates that a number of turns of wire carrying a current around a magnetic material create a force that sets up flux lines through the magnetic path. The amount of flux depends on the magnitude of the mmf and on the reluctance of the material, as expressed by the following equation:

$$\phi = \frac{F_m}{\mathcal{R}} \qquad\qquad (7\text{–}5)$$

This is known as the *Ohm's law for magnetic circuits* because the flux (ϕ) is analogous to current, the mmf (F_m) is analogous to voltage, and the reluctance ($\mathcal{R}$) is analogous to resistance.

FIGURE 7–12
A basic magnetic circuit.

EXAMPLE 7–3

How much flux is established in the magnetic path of Figure 7–13 if the reluctance of the material is 0.28×10^5 At/Wb?

FIGURE 7–13

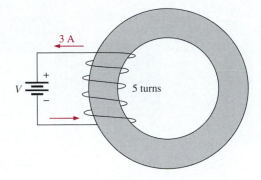

Solution

$$\phi = \frac{F_m}{\mathcal{R}} = \frac{NI}{\mathcal{R}} = \frac{(5 \text{ t})(3 \text{ A})}{0.28 \times 10^5 \text{ At/Wb}} = 5.36 \times 10^{-4} \text{ Wb} = 536 \text{ } \mu\text{Wb}$$

Related Exercise How much flux is established in the magnetic path of Figure 7–13 if the reluctance is 7.5×10^3 At/Wb, the number of turns is 30, and the current is 1.8 A?

EXAMPLE 7–4

There are two amperes of current through a wire with 5 turns.
(a) What is the mmf?
(b) What is the reluctance of the circuit if the flux is 250 μWb?

Solution
(a) $N = 5$ and $I = 2$ A
$F_m = NI = (5 \text{ t})(2 \text{ A}) = 10 \text{ At}$
(b) $\mathcal{R} = \dfrac{F_m}{\phi} = \dfrac{10 \text{ At}}{250 \text{ } \mu\text{Wb}} = 0.04 \times 10^6 \text{ At/Wb} = 4.0 \times 10^4 \text{ At/Wb}$

Related Exercise Rework the example for $I = 850$ mA, $N = 50$, and $\phi = 500$ μWb.

The Electromagnet

An electromagnet is based on the properties that you have just learned. A basic electromagnet is simply a coil of wire wound around a core material that can be easily magnetized.

The shape of the electromagnet can be designed for various applications. For example, Figure 7–14 shows a U-shaped magnetic core. When the coil of wire is connected to a battery and current flows, as shown in part (a), a magnetic field is established as indicated. If the current is reversed, as shown in part (b), the direction of the magnetic field is also reversed. The closer the north and south poles are brought together, the smaller the air gap between them becomes, and the easier it becomes to establish a magnetic field, because the reluctance is lessened.

A good example of one application of an electromagnet is the process of recording on magnetic tape. In this situation, the recording head is an electromagnet with a narrow air gap, as shown in Figure 7–15. Current sets up a magnetic field across the air gap, and

FIGURE 7–14
Reversing the current in the coil causes the electromagnetic field to reverse.

FIGURE 7–15
An electromagnetic recording head records information on magnetic tape by magnetizing the tape as it passes by.

as the recording head passes over the magnetic tape, the tape is permanently magnetized. In digital recording, for example, the tape is magnetized in one direction for a binary 1 and in the other direction for a binary 0, as illustrated in the figure. This magnetization in different directions is accomplished by reversing the coil current in the recording head.

SECTION 7–2 REVIEW	1. Explain the difference between magnetism and electromagnetism. 2. What happens to the magnetic field in an electromagnet when the current through the coil is reversed?

7–3 ▪ ELECTROMAGNETIC DEVICES

In the last section, you learned that the recording head is a type of electromagnetic device. Now, several other common devices are introduced.

After completing this section, you should be able to

▪ **Describe the principle of operation for several types of electromagnetic devices**
   ☐ Discuss how a solenoid works
   ☐ Discuss how a relay works
   ☐ Discuss how a speaker works
   ☐ Discuss the basic analog meter movement

The recording head is one example of an electromagnetic device. Several other examples are now presented.

The Solenoid

The **solenoid** is a type of electromagnetic device that has a movable iron core called a plunger. The movement of this iron core depends on both an electromagnetic field and a mechanical spring force. The basic structure of a solenoid is shown in Figure 7–16. It consists of a cylindrical coil of wire wound around a nonmagnetic hollow form. A stationary iron core is fixed in position at the end of the shaft and a sliding iron core is attached to the stationary core with a spring.

(a) Solenoid

(b) Basic construction

(c) Cutaway view

FIGURE 7–16
Basic solenoid structure.

In the at-rest, or unenergized, state, the plunger is extended as shown in Figure 7–17(a). The solenoid is energized by current through the coil, as shown in part (b). The current sets up an electromagnetic field that magnetizes both iron cores as indicated. The south pole of the stationary core attracts the north pole of the movable core causing it to slide inward, thus retracting the plunger and compressing the spring. As long as the coil current flows, the plunger remains retracted by the attractive force of the magnetic fields. When the current is cut off, the magnetic fields collapse and the force of the compressed spring pushes the plunger back out. The solenoid is used for many applications, such as opening and closing valves, automobile door locks, and so on.

The Relay

Relays differ from solenoids in that the electromagnetic action is used to open or close electrical contacts rather than to provide mechanical movement. Figure 7–18 shows the basic operation of a relay with one normally open (NO) contact and one normally closed (NC) contact (single pole–double throw). When there is no coil current, the armature is

(a) Unenergized — plunger extended

(b) Energized — plunger retracted

FIGURE 7–17
Basic solenoid operation.

(a) Unenergized: continuity from 1 to 2   (b) Energized: continuity from 1 to 3

FIGURE 7–18
Basic structure of a single-pole–double-throw relay.

held against the upper contact by the spring, thus providing continuity from terminal 1 to terminal 2, as shown in Figure 7–18(a). When energized with coil current, the armature is pulled down by the attractive force of the electromagnetic field and makes connection with the lower contact to provide continuity from terminal 1 to terminal 3, as shown in Figure 7–18(b).

A typical relay and its schematic symbol are shown in Figure 7–19.

The Speaker

Permanent-magnet speakers are commonly used in stereos, radios, and TVs, and their operation is based on the principle of electromagnetism. A typical speaker is constructed

FIGURE 7–19
A typical relay.

Contact points

Armature

Spring

Terminals

Electromagnetic coil

Terminals

(a) Structure   (b) Symbol

with a permanent magnet and an electromagnet, as shown in Figure 7–20(a). The cone of the speaker consists of a paper-like diaphragm to which is attached a hollow cylinder with a coil around it, forming an electromagnet. One of the poles of the permanent magnet is positioned within the cylindrical coil. When current flows through the coil in one direction, the interaction of the permanent magnetic field with the electromagnetic field causes the cylinder to move to the right, as indicated in Figure 7–20(b). Current through the coil in the other direction causes the cylinder to move to the left, as shown in Figure 7–20(c).

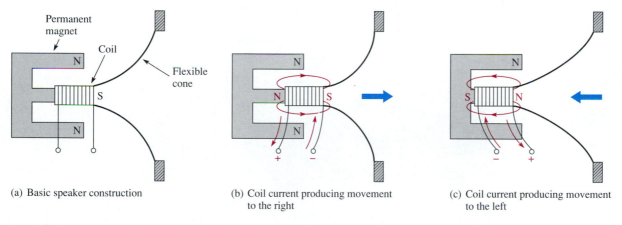

Permanent magnet

Coil

Flexible cone

(a) Basic speaker construction

(b) Coil current producing movement to the right

(c) Coil current producing movement to the left

FIGURE 7–20
Basic speaker operation.

The movement of the coil cylinder causes the flexible diaphragm also to move in or out, depending on the direction of the coil current. The amount of coil current determines the intensity of the magnetic field, which controls the amount that the diaphragm moves.

As shown in Figure 7–21, when an audio signal voltage (voice or music) is applied to the speaker coil, the current varies proportionally in both direction and amount. In response, the diaphragm will vibrate in and out by varying amounts and at varying rates. Vibration in the diaphragm causes the air that is in contact with it to vibrate in the same manner. These air vibrations move through the air as sound waves.

Analog Meter Movement

The d'Arsonval movement is the most common type used in analog multimeters. In this type of meter movement, the pointer is deflected in proportion to the amount of current

(a)   (b)

FIGURE 7–21
The speaker converts audio signal voltages into sound waves.

FIGURE 7–22
The basic d'Arsonval meter movement.

through a coil. Figure 7–22 shows a basic d'Arsonval meter movement. It consists of a coil of wire wound on a bearing-mounted assembly that is placed between the poles of a permanent magnet. A pointer is attached to the moving assembly. With no current through the coil, a spring mechanism keeps the pointer at its left-most (zero) position. When current flows through the coil, electromagnetic forces act on the coil, causing a rotation to the right. The amount of rotation depends on the amount of current.

Figure 7–23 illustrates how the interaction of magnetic fields produces rotation of the coil assembly. The current flows inward at the "cross" and outward at the "dot" in the single winding shown. The inward current produces a counterclockwise electromagnetic field that reinforces the permanent magnetic field below it. The result is an upward force on the left side of the coil as shown. A downward force is developed on the right side of the coil, where the current is outward. These forces produce a clockwise rotation of the coil assembly and are opposed by a spring mechanism. The indicated forces and the spring force are balanced at the value of the current. When current is removed, the spring force returns the pointer to its zero position.

FIGURE 7–23
When the electromagnetic field interacts with the permanent magnetic field, forces are exerted on the rotating coil assembly, causing it to move clockwise and thus deflecting the pointer.

⊕ Current in
⊙ Current out

SECTION 7–3 REVIEW

1. Explain the difference between a solenoid and a relay.
2. What is the movable part of a solenoid called?
3. What is the movable part of a relay called?
4. Upon what basic principle is the d'Arsonval meter movement based?

7–4 ■ MAGNETIC HYSTERESIS

When a magnetizing force is applied to a material, the magnetic flux density in the material changes in a certain way, which we will now examine.

After completing this section, you should be able to

■ **Explain magnetic hysteresis**
  ☐ State the formula for magnetizing force
  ☐ Discuss a hysteresis curve
  ☐ Define *retentivity*

Magnetizing Force (*H*)

The **magnetizing force** in a material is defined to be the magnetomotive force (F_m) per unit length (l) of the material, as expressed by the following equation. The unit of magnetizing force (H) is ampere-turns per meter (At/m).

$$H = \frac{F_m}{l} \tag{7–6}$$

where $F_m = NI$. Note that the magnetizing force depends on the number of turns (N) of the coil of wire, the current (I) through the coil, and the length (l) of the material. It does not depend on the type of material.

Since $\phi = F_m/\mathcal{R}$, as F_m increases, the flux increases. Also, the magnetizing force (H) increases. Recall that the flux density (B) is the flux per unit cross-sectional area ($B = \phi/A$), so B is also proportional to H. The curve showing how these two quantities (B and H) are related is called the *B-H* curve, or the hysteresis curve. The parameters that influence both B and H are illustrated in Figure 7–24.

The Hysteresis Curve and Retentivity

Hysteresis is a characteristic of a magnetic material whereby a change in magnetization lags the application of a magnetizing force. The magnetizing force (H) can be readily in-

FIGURE 7–24
Parameters that determine the magnetizing force (H) and the flux density (B).

creased or decreased by varying the current through the coil of wire, and it can be reversed by reversing the voltage polarity across the coil.

Figure 7–25 illustrates the development of the hysteresis curve. We start by assuming a magnetic core is unmagnetized so that $B = 0$. As the magnetizing force (H) is increased from zero, the flux density (B) increases proportionally as indicated by the curve in Figure 7–25(a). When H reaches a certain value, the value of B begins to level off. As H continues to increase, B reaches a saturation value (B_{sat}) when H reaches a value (H_{sat})

FIGURE 7–25
Development of a magnetic hysteresis curve.

as illustrated in Figure 7–25(b). Once saturation is reached, a further increase in H will not increase B.

Now, if H is decreased to zero, B will fall back along a different path to a residual value (B_R), as shown in Figure 7–25(c). This indicates that the material continues to be magnetized even with the magnetizing force removed ($H = 0$). The ability of a material, once magnetized, to maintain a magnetized state without the presence of a magnetizing force is called **retentivity**. The retentivity of a material is indicated by the ratio of B_R to B_{sat}.

Reversal of the magnetizing force is represented by negative values of H on the curve and is achieved by reversing the current in the coil of wire. An increase in H in the negative direction causes saturation to occur at a value ($-H_{sat}$) where the flux density is at its maximum negative value as indicated in Figure 7–25(d).

When the magnetizing force is removed ($H = 0$), the flux density goes to its negative residual value ($-B_R$) as shown in Figure 7–25(e). From the $-B_R$ value, the flux density follows the curve indicated in part (f) back to its maximum positive value when the magnetizing force equals H_{sat} in the positive direction.

The complete B-H curve is shown in Figure 7–25(g) and is called the *hysteresis curve*. The magnetizing force required to make the flux density zero is called the coercive force, H_C.

Materials with a low retentivity do not retain a magnetic field very well while those with high retentivities exhibit values of B_R very close to the saturation value of B. Depending on the application, retentivity in a magnetic material can be an advantage or a disadvantage. In permanent magnets and memory cores, for example, high retentivity is required. In ac motors, retentivity is undesirable because the residual magnetic field must be overcome each time the current reverses, thus wasting energy.

SECTION 7–4 REVIEW	1. For a given wirewound core, how does an increase in current through the coil affect the flux density? 2. Define *retentivity*.

7–5 ■ ELECTROMAGNETIC INDUCTION

When a conductor is moved through a magnetic field, a voltage is produced across the conductor. This principle is known as electromagnetic induction and the resulting voltage is an induced voltage. Electromagnetic induction is what makes transformers, electrical generators, and many other devices possible.

After completing this section, you should be able to

■ **Discuss the principle of electromagnetic induction**
  □ Explain how voltage is induced in a conductor in a magnetic field
  □ Determine polarity of an induced voltage
  □ Discuss forces on a conductor in a magnetic field
  □ State Faraday's law
  □ State Lenz's law

Relative Motion

When a wire is moved across a magnetic field, there is a relative motion between the wire and the magnetic field. Likewise, when a magnetic field is moved past a stationary wire, there is also relative motion. In either case, there is an induced voltage (v_{ind}) in the wire

(a) Wire moving downward

(b) Magnetic field moving upward

FIGURE 7–26
Relative motion between a wire and a magnetic field.

there is also relative motion. In either case, there is an induced voltage (v_{ind}) in the wire as a result of this motion, as Figure 7–26 indicates. The lowercase v stands for instantaneous voltage.

The amount of the induced voltage depends on the rate at which the wire and the magnetic field move with respect to each other: The faster the relative speed, the greater the induced voltage.

Polarity of the Induced Voltage

If the conductor in Figure 7–26 is moved first one way and then another in the magnetic field, a reversal of the polarity of the induced voltage will be observed. As the wire is moved downward, a voltage is induced with the polarity indicated in Figure 7–27(a). As the wire is moved upward, the polarity is as indicated in part (b) of the figure.

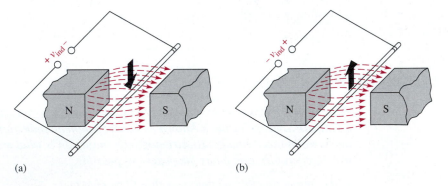

(a)   (b)

FIGURE 7–27
Polarity of induced voltage depends on direction of motion.

Induced Current

When a load resistor is connected to the wire in Figure 7–27, the voltage induced by the relative motion in the magnetic field will cause a current in the load, as shown in Figure 7–28. This current is called the induced current (i_{ind}). The lowercase i stands for instantaneous current.

The action of producing a voltage and a current in a load by moving a conductor across a magnetic field is the basis for electrical generators. Also, the existence of a conductor in a moving magnetic field is fundamental to the concept of inductance in an electrical circuit.

Induced current (i_{ind})

FIGURE 7–28

Induced current in a load as the wire moves through the magnetic field.

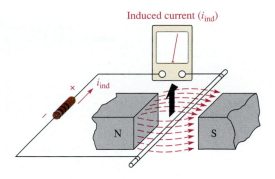

Forces on a Current-Carrying Conductor in a Magnetic Field (Motor Action)

Figure 7–29(a) shows current inward through a wire in a magnetic field. The electromagnetic field set up by the current interacts with the permanent magnetic field; as a result, the permanent lines of force above the wire tend to be deflected down under the wire, because they are opposite in direction to the electromagnetic lines of force. Therefore, the flux density above is reduced, and the magnetic field is weakened. The flux density below the conductor is increased, and the magnetic field is strengthened. An upward force on the conductor results, and the conductor tends to move toward the weaker magnetic field.

Figure 7–29(b) shows the current outward, resulting in a force on the conductor in the downward direction. This force is the basis for electrical motors.

FIGURE 7–29

Forces on a current-carrying conductor in a magnetic field.

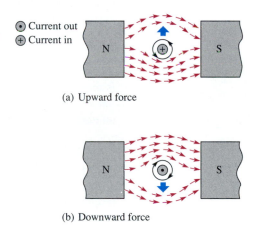

(a) Upward force

(b) Downward force

Faraday's Law

Michael Faraday discovered the principle of **electromagnetic induction** in 1831. He found that moving a magnet through a coil of wire induced a voltage across the coil, and that when a complete path was provided, the induced voltage caused an induced current, as you have seen. Faraday's two observations are as follows:

1. The amount of voltage induced in a coil is directly proportional to the rate of change of the magnetic field with respect to the coil.

2. The amount of voltage induced in a coil is directly proportional to the number of turns of wire in the coil.

Faraday's first observation is demonstrated in Figure 7–30, where a bar magnet is moved through a coil, thus creating a changing magnetic field. In part (a) of the figure, the magnet is moved at a certain rate, and a certain induced voltage is produced as indicated. In part (b), the magnet is moved at a faster rate through the coil, creating a greater induced voltage.

(a) As magnet moves slowly to the right, magnetic field is changing
   with respect to coil, and a voltage is induced.

(b) As magnet moves more rapidly to the right, magnetic field is
   changing more rapidly with respect to coil, and a greater voltage is induced.

FIGURE 7–30

*A demonstration of Faraday's first observation: The amount of
induced voltage is directly proportional to the rate of change of the
magnetic field with respect to the coil.*

Faraday's second observation is demonstrated in Figure 7–31. In part (a), the magnet
is moved through the coil and a voltage is induced as shown. In part (b), the magnet is
moved at the same speed through a coil that has a greater number of turns. The greater
number of turns creates a greater induced voltage.

FIGURE 7–31

*A demonstration of Faraday's
second observation: The
amount of induced voltage is
directly proportional to the
number of turns in the coil.*

(a) Magnet moves through a coil and induces a voltage.

(b) Magnet moves at same rate through a coil with more turns (loops)
   and induces a greater voltage.

Faraday's law is stated as follows:

The voltage induced across a coil of wire equals the number of turns in the coil times the rate of change of the magnetic flux.

Lenz's Law

You have learned that a changing magnetic field induces a voltage in a coil that is directly proportional to the rate of change of the magnetic field and the number of turns in the coil. **Lenz's law** defines the polarity or direction of the induced voltage.

When the current through a coil changes, an induced voltage is created as a result of the changing magnetic field, and the direction of the induced voltage is such that it always opposes the change in current.

SECTION 7–5 REVIEW	1. What is the induced voltage across a stationary conductor in a stationary magnetic field? 2. When the speed at which a conductor is moved through a magnetic field is increased, does the induced voltage increase, decrease, or remain the same? 3. When there is current through a conductor in a magnetic field, what happens?

7–6 ■ APPLICATIONS OF ELECTROMAGNETIC INDUCTION

In this section, two interesting applications of electromagnetic induction are discussed—an automotive crankshaft position sensor and a dc generator. Although there are many varied applications, these two are representative.

After completing this section, you should be able to

■ **Describe some applications of electromagnetic induction**
  □ Explain how a crankshaft position sensor works
  □ Explain how a dc generator works

Automotive Crankshaft Position Sensor

An interesting automotive application is a type of engine sensor that detects the crankshaft position directly using electromagnetic induction. The electronic engine controller in many automobiles uses the position of the crankshaft to set ignition timing and, sometimes, to adjust the fuel control system. Figure 7–32 shows the basic concept. A steel disk is attached to the engine's crankshaft by an extension rod; the protruding tabs on the disk represent specific crankshaft positions.

As illustrated in Figure 7–32, as the disk rotates with the crankshaft, the tabs periodically pass through the air gap of the permanent magnet. Since steel has a much lower reluctance than does air (a magnetic field can be established in steel much more easily than in air), the magnetic flux suddenly increases as a tab comes into the air gap, causing a voltage to be induced across the coil. This process is illustrated in Figure 7–33. The electronic engine control circuit uses the induced voltage as an indicator of the crankshaft position.

FIGURE 7–32

A crankshaft position sensor that produces a voltage when a tab passes through the air gap of the magnet.

(a) There is no changing magnetic field, so there is no induced voltage.

(b) Insertion of the steel tab reduces the reluctance of the air gap, causing the magnetic flux to increase and thus inducing a voltage.

FIGURE 7–33

As the tab passes through the air gap of the magnet, the coil senses a change in the magnetic field, and a voltage is induced.

A DC Generator

Figure 7–34 shows a simplified dc generator consisting of a single loop of wire in a permanent magnetic field. Notice that each end of the loop is connected to a split-ring arrangement. This conductive metal ring is called a *commutator*. As the loop is rotated in the magnetic field, the split commutator ring also rotates. Each half of the split ring rubs against the fixed contacts, called *brushes,* and connects the loop to an external circuit.

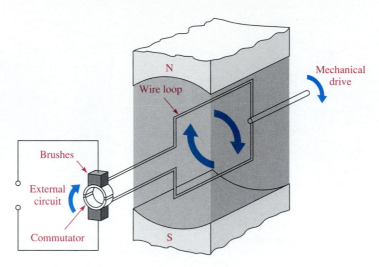

FIGURE 7–34
A basic dc generator.

As the loop rotates through the magnetic field, it "cuts" through the flux lines at varying angles, as illustrated in Figure 7–35. At position *A* in its rotation, the loop of wire is effectively moving parallel with the magnetic field. Therefore, at this instant, the rate at which it is cutting through the magnetic flux lines is zero.

FIGURE 7–35

End view of loop cutting through the magnetic field.

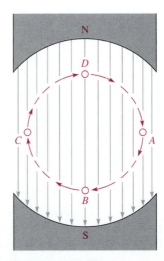

As the loop moves from position *A* to position *B*, it cuts through the flux lines at an increasing rate. At position *B*, it is moving effectively perpendicular to the magnetic field and thus is cutting through a maximum number of lines. As the loop rotates from position *B* to position *C*, the rate at which it cuts the flux lines decreases to minimum (zero) at *C*. From position *C* to position *D*, the rate at which the loop cuts the flux lines increases to a maximum at *D* and then back to a minimum again at *A*.

As you previously learned, when a wire moves through a magnetic field, a voltage is induced, and by Faraday's law, the amount of induced voltage is proportional to the number of loops (turns) in the wire and the rate at which it is moving with respect to the magnetic field. Now you know that the angle at which the wire moves with respect to the magnetic flux lines determines the amount of induced voltage, because the rate at which the wire cuts through the flux lines depends on the angle of motion.

(a) Position *B*: Loop is moving perpendicular to flux lines, and voltage is maximum.

(b) Position *C*: Loop is moving parallel with flux lines, and voltage is zero.

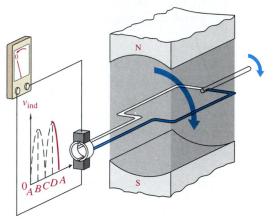

(c) Position *D*: Loop is moving perpendicular to flux lines, and voltage is maximum.

(d) Position *A*: Loop is moving parallel with flux lines, and voltage is zero.

FIGURE 7–36

Operation of a basic dc generator.

Figure 7–36 illustrates how a voltage is induced in the external circuit as the single loop rotates in the magnetic field. Assume that the loop is in its instantaneous horizontal position, so the induced voltage is zero. As the loop continues in its rotation, the induced voltage builds up to a maximum at position *B*, as shown in part (a) of the figure. Then, as the loop continues from *B* to *C*, the voltage decreases to zero at *C*, as shown in part (b).

During the second half of the revolution, shown in Figure 7–36(c) and (d), the brushes switch to opposite commutator sections, so the polarity of the voltage remains the same across the output. Thus, as the loop rotates from position *C* to position *D* and then back to *A*, the voltage increases from zero at *C* to a maximum at *D* and back to zero at *A*.

Figure 7–37 shows how the induced voltage varies as the loop goes through several rotations (three in this case). This voltage is a dc voltage because its polarity does not change. However, the voltage is pulsating between zero and its maximum value.

When more loops are added, the voltages induced across each loop are combined across the output. Since the voltages are offset from each other, they do not reach their maximum or zero values at the same time. A smoother dc voltage results, as shown in Figure 7–38 for two loops. The variations can be further smoothed out by filters to achieve a nearly constant dc voltage. (Filters are discussed later in the book.)

FIGURE 7–37
Induced voltage over three rotations of the loop.

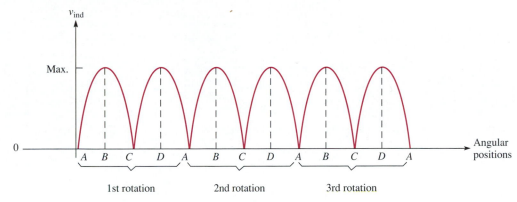

FIGURE 7–38
The induced voltage for a two-loop generator. There is much less variation in the induced voltage.

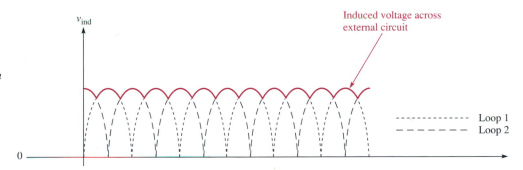

7–7 ■ APPLICATION ASSIGNMENT

The relay is a common type of electromagnetic device that is used in many types of control applications. With a relay, a lower voltage, such as from a battery, can be used to switch a much higher voltage, such as the 110 V from an ac outlet. In this section, you will see how a relay can be used in a basic burglar alarm system.

The schematic in Figure 7–39 shows a basic simplified intrusion alarm system that uses a relay to turn on an audible alarm (siren) and lights. The system operates from a 9 V battery so that even if power to the house is off, the audible alarm will still work.

The magnetic detection switches are normally open (NO) magnetic switches that are parallel connected and located in the windows and doors. The relay is a three-pole, double-throw device that operates with a coil voltage of 9 V dc and draws approximately 50 mA. When an intrusion occurs, one of the switches closes and allows current to flow from the battery through the relay coil, which energizes the relay and causes the three sets of normally open contacts to close. Closure of contact *A* turns on the alarm, which draws 2 A from the battery. Closure of contact *C* turns on a light circuit in the house. Closure of contact *B* latches the relay and keeps it energized even if the intruder closes the door or window through which entry was made. If not for contact *B* in parallel with the detection switches, the alarm and lights would go off as soon as the window or door was shut behind the intruder.

FIGURE 7–39
Simplified burglar alarm system.

The relay contacts are not physically remote in relation to the coil as the schematic indicates. The schematic is drawn this way for functional clarity. The entire relay is housed in the package shown in Figure 7–40. Also shown are the pin diagram and internal schematic for the relay.

FIGURE 7–40
Triple-pole–double-throw relay.

Step 1: Connect the System

Develop a connection diagram and point-to-point wire list for interconnecting the components in Figure 7–41 to create the alarm system shown in the schematic of Figure 7–39. The connection points on the components are indicated by letters.

Step 2: Test the System

Develop a step-by-step procedure to check out the completely wired burglar alarm system.

FIGURE 7–41
Array of burglar alarm components.

■ SUMMARY

- Unlike magnetic poles attract each other, and like poles repel each other.
- Materials that can be magnetized are called *ferromagnetic*.
- When current flows through a conductor, it produces an electromagnetic field around the conductor.
- You can use the left-hand rule to establish the direction of the electromagnetic lines of force around a conductor.
- An electromagnet is basically a coil of wire around a magnetic core.
- When a conductor moves within a magnetic field, or when a magnetic field moves relative to a conductor, a voltage is induced across the conductor.

■ The faster the relative motion between a conductor and a magnetic field, the greater is the induced voltage.

■ Table 7–1 summarizes the magnetic quantities.

TABLE 7–1

SYMBOL	QUANTITY	UNIT
B	Magnetic flux density	Tesla (T)
ϕ	Flux	Weber (Wb)
μ	Permeability	Weber/ampere-turn·meter (Wb/At·m)
$\mathcal{R}$	Reluctance	At/Wb
F_m	Magnetomotive force (mmf)	Ampere-turn (At)
H	Magnetizing force	At/m

■ **GLOSSARY**

Ampere-turn The unit of magnetomotive force (mmf).

Electromagnetic field A formation of a group of magnetic lines of force surrounding a conductor created by electrical current in the conductor.

Electromagnetic induction The phenomenon or process by which a voltage is produced in a conductor when there is relative motion between the conductor and a magnetic or electromagnetic field.

Faraday's law A law stating that the voltage induced across a coil of wire equals the number of turns in the coil times the rate of change of the magnetic flux.

Hysteresis A lagging in the values of resulting magnetization in a magnetic material due to a changing magnetizing force.

Lenz's law A law that states when the current through a coil changes, an induced voltage is created as a result of the changing magnetic field, and the direction of the induced voltage is such that it always opposes the change in current. The current cannot change instantaneously.

Magnetic flux The lines of force between the north and south poles of a permanent magnet or an electromagnet.

Magnetic flux density The number of lines of force per unit area in a magnetic field.

Magnetizing force The amount of mmf per unit length of magnetic material.

Magnetomotive force The force that produces a magnetic field.

Permeability The measure of ease with which a magnetic field can be established in a material.

Relay An electromagnetically controlled mechanical device in which electrical contacts are opened or closed by a magnetizing current.

Reluctance The opposition to the establishment of a magnetic field.

Retentivity The ability of a material, once magnetized, to maintain a magnetized state without the presence of a magnetizing force.

Solenoid An electromagnetically controlled device in which the mechanical movement of a shaft or plunger is activated by a magnetizing current.

Tesla The unit of flux density.

Weber The unit of magnetic flux.

■ **FORMULAS**

(7–1) $B = \dfrac{\phi}{A}$ Magnetic flux density

(7–2) $\mu_r = \dfrac{\mu}{\mu_0}$ Relative permeability

(7–3) $\mathcal{R} = \dfrac{l}{\mu A}$ Reluctance

(7–4) $F_m = NI$ Magnetomotive force

$$(7\text{--}5) \qquad \phi = \frac{F_m}{\mathcal{R}} \qquad \text{Magnetic flux}$$

$$(7\text{--}6) \qquad H = \frac{F_m}{l} \qquad \text{Magnetic flux intensity}$$

■ SELF-TEST

1. When the south poles of two bar magnets are brought close together, there will be
   (a) a force of attraction   (b) a force of repulsion
   (c) an upward force   (d) no force

2. A magnetic field is made up of
   (a) positive and negative charges   (b) magnetic domains
   (c) flux lines   (d) magnetic poles

3. The direction of a magnetic field is from
   (a) north pole to south pole   (b) south pole to north pole
   (c) inside to outside the magnet   (d) from to back

4. Reluctance in a magnetic circuit is analogous to
   (a) voltage in an electric circuit   (b) current in an electric circuit
   (c) power in an electric circuit   (d) resistance in an electric circuit

5. The unit of magnetic flux is the
   (a) tesla   (b) weber   (c) electron-volt   (d) ampere-turn

6. The unit of magnetomotive force is the
   (a) tesla   (b) weber   (c) ampere-turn   (d) electron-volt

7. The unit of flux density is the
   (a) tesla   (b) weber   (c) ampere-turn   (d) electron-volt

8. The electromagnetic activation of a movable shaft is the basis for
   (a) relays   (b) circuit breakers   (c) magnetic switches   (d) solenoids

9. When current flows through a wire placed in a magnetic field,
   (a) the wire will overheat   (b) the wire will become magnetized
   (c) a force is exerted on the wire   (d) the magnetic field will be cancelled

10. A coil of wire is placed in a changing magnetic field. If the number of turns in the coil is increased, the voltage induced across the coil will
    (a) remained unchanged   (b) decrease
    (c) increase   (d) be excessive

11. If a conductor is moved back and forth at a constant rate in a constant magnetic field, the voltage induced in the conductor will
    (a) remain constant   (b) reverse polarity
    (c) be reduced   (d) be increased

12. In the crankshaft position sensor in Figure 7–32, the induced voltage across the coil is caused by
    (a) current in the coil
    (b) rotation of the disk
    (c) a tab passing through the magnetic field
    (d) acceleration of the disk's rotational speed

■ BASIC PROBLEMS

SECTION 7–1   The Magnetic Field

1. The cross-sectional area of a magnetic field is increased, but the flux remains the same. Does the flux density increase or decrease?

2. In a certain magnetic field the cross-sectional area is 0.5 m^2 and the flux is 1500 μWb. What is the flux density?

3. What is the flux in a magnetic material when the flux density is 2500×10^{-6} T and the cross-sectional area is 150 cm^2?

SECTION 7–2   Electromagnetism

4. What happens to the compass needle in Figure 7–9 when the current through the conductor is reversed?

5. What is the relative permeability of a ferromagnetic material whose absolute permeability is 750×10^{-6} Wb/At·m?

6. Determine the reluctance of a material with a length of 0.28 m and a cross-sectional area of 0.08 m^2 if the absolute permeability is 150×10^{-7} Wb/At·m.

7. What is the magnetomotive force in a 50 turn coil of wire with 3 A through it?

SECTION 7–3 Electromagnetic Devices

8. Typically, when a solenoid is activated, is the plunger extended or retracted?

9. **(a)** What force moves the plunger when a solenoid is activated?
   (b) What force causes the plunger to return to its at-rest position?

10. Explain the sequence of events in the circuit of Figure 7–42 starting when switch 1 (SW_1) is closed.

11. What causes the pointer in a d'Arsonval movement to deflect when there is current through the coil?

FIGURE 7–42

SECTION 7–4 Magnetic Hysteresis

12. What is the magnetizing force in Problem 7 if the length of the core is 0.2 m?

13. How can the flux density in Figure 7–43 be changed without altering the physical characteristics of the core?

14. In Figure 7–43; determine the following:
    (a) H **(b)** ϕ **(c)** B

15. Determine from the hysteresis curves in Figure 7–44 which material has the most retentivity.

FIGURE 7–43

FIGURE 7–44

SECTION 7–5 Electromagnetic Induction

16. According to Faraday's law, what happens to the induced voltage across a given coil if the rate of change of magnetic flux doubles?

17. The voltage induced across a certain coil is 100 mV. A 100 Ω resistor is connected to the coil terminals. What is the induced current?

SECTION 7–6 Applications of Electromagnetic Induction

18. In Figure 7–32, why is there no induced voltage when the steel disk is not rotating?

19. Explain the purpose of the commutator and brushes in Figure 7–34.

■ **ADVANCED PROBLEMS**

20. A basic one-loop dc generator is rotated at 60 rps. How many times each second does the dc output voltage peak (reach a maximum)?

21. Assume that another loop, 90° from the first loop, is added to the dc generator in Problem 20. Make a graph of voltage versus time to show how the output voltage appears. Let the maximum voltage be 10 V.

■ **ANSWERS TO SECTION REVIEWS**

Section 7–1

1. The north poles repel.

2. Magnetic flux is the group of lines of force that make up a magnetic field.

3. $B = \phi/A = 900 \times 10^{-6}$ T

Section 7–2

1. Electromagnetism is produced by current through a conductor. An electromagnetic field exists only when current flows. A magnetic field exists independently of current.

2. The direction of the magnetic field also reverses when the current is reversed.

Section 7–3

1. A solenoid provides mechanical movement of a shaft. A relay provides contact closure.

2. The movable part of a solenoid is the plunger.

3. The movable part of a relay is the armature.

4. The d'Arsonval meter movement is based on the interaction of magnetic fields.

Section 7–4

1. In a wirewound core, an increase in current increases the flux density.

2. Retentivity is the ability of a material to remain magnetized after removal of the magnetizing force.

Section 7–5

1. The induced voltage is zero.

2. The induced voltage increases.

3. A force is exerted on the current-carrying conductor in a magnetic field.

Section 7–6

1. The induced voltage is zero.

2. The induced voltage increases.

Section 7–7

1. One or more magnetic detection switches must be closed.

2. Contact A activates alarm, contact B latches the relay, contact C turns on house lights.

■ **ANSWERS TO RELATED EXERCISES FOR EXAMPLES**

7–1	$B = 12000 \times 10^{-6}$ T
7–2	Reluctance quadruples.
7–3	7.2 mWb
7–4	8.5×10^4 At/Wb

PART TWO

AC CIRCUITS

8

INTRODUCTION TO ALTERNATING CURRENT AND VOLTAGE

This chapter provides an introduction to alternating current (ac) circuits. Alternating voltages and currents fluctuate with time and periodically change polarity and direction according to certain patterns called **waveforms.** Particular emphasis is given to the sine wave because of its basic importance in ac circuits. Other types of waveforms are also introduced, including pulse, triangular, and sawtooth. The use of the oscilloscope for displaying and measuring waveforms is discussed.

■ CHAPTER OBJECTIVES

☐ Identify a sinusoidal waveform and measure its characteristics
☐ Describe how sine waves are generated
☐ Determine the voltage and current values of sine waves
☐ Describe angular relationships of sine waves
☐ Mathematically analyze a sinusoidal waveform
☐ Apply the basic circuit laws to ac resistive circuits
☐ Determine total voltages which are composed of both ac and dc components
☐ Use the oscilloscope to measure waveforms
☐ Identify the characteristics of basic nonsinusoidal waveforms

■ APPLICATION ASSIGNMENT

As an electronic technician working for a company that designs and manufactures laboratory instruments, you have been assigned to the testing of a new function generator that produces various types of time-varying voltages with adjustable parameters. The project supervisor has assigned you to measure and record the limits of operation of a prototype model. After studying this chapter, you should be able to complete the application assignment in Section 8–10.

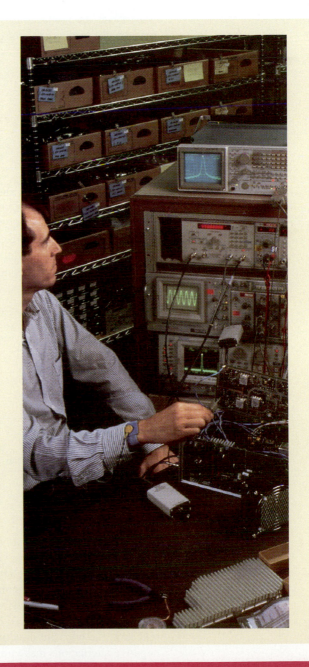

8–1 ■ THE SINE WAVE

The sine wave is a common type of alternating current (ac) and alternating voltage. It is also referred to as a sinusoidal waveform, or, simply, sinusoid. The electrical service provided by the power companies is in the form of sinusoidal voltage and current. In addition, other types of waveforms are composites of many individual sine waves called harmonics.

After completing this section, you should be able to

■ **Identify a sinusoidal waveform and measure its characteristics**
   □ Define and determine the period
   □ Define and determine the frequency
   □ Relate the period and the frequency

Sine waves are produced by two types of sources: rotating electrical machines (ac generators) or electronic oscillator circuits, which are in instruments known as electronic signal generators. Figure 8–1 shows the symbol used to represent either source of sine wave voltage.

FIGURE 8–1
Symbol for a sine wave voltage source.

Figure 8–2 is a graph showing the general shape of a sine wave, which can be either an **alternating current** or voltage. Voltage (or current) is displayed on the vertical axis and time (*t*) is displayed on the horizontal axis. Notice how the voltage (or current) varies with time. Starting at zero, the voltage (or current) increases to a positive maximum (peak), returns to zero, and then increases to a negative maximum (peak) before returning again to zero, thus completing one full cycle.

FIGURE 8–2
Graph of one cycle of sine wave.

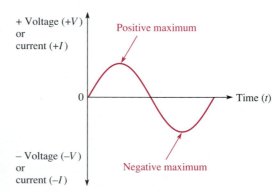

Polarity of a Sine Wave

As you have seen, a sine wave changes polarity at its zero value; that is, it alternates between positive and negative values. When a sine wave voltage is applied to a resistive circuit, as in Figure 8–3, an alternating sine wave current results. When the voltage changes polarity, the current correspondingly changes direction as indicated.

During the positive alternation of the applied voltage V_s, the current is in the direction shown in Figure 8–3(a). During a negative alternation of the applied voltage, the current is in the opposite direction, as shown in Figure 8–3(b). The combined positive and negative alternations make up one **cycle** of a sine wave.

(a) Positive voltage: current direction as shown

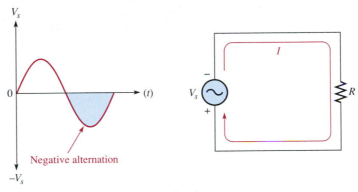

(b) Negative voltage: current reverses direction

FIGURE 8–3
Alternating current and voltage.

Period of a Sine Wave

The time required for a given sine wave to complete one full cycle is called the period (*T*).

Figure 8–4(a) illustrates the **period** of a sine wave. Typically, a sine wave continues to repeat itself in identical cycles, as shown in Figure 8–4(b). Since all cycles of a repetitive sine wave are the same, the period is always a fixed value for a given sine wave. The period of a sine wave can be measured from any point in a given cycle to the corresponding point in the next cycle.

(a)   (b)

FIGURE 8–4
The period of a given sine wave is the same for each cycle.

EXAMPLE 8–1   What is the period of the sine wave in Figure 8–5?

FIGURE 8–5

Solution   As shown in Figure 8–5, it takes four seconds (4 s) to complete each cycle. Therefore, the period is 4 s.

$$T = 4 \text{ s}$$

Related Exercise   What is the period if a given sine wave goes through five cycles in 12 s?

EXAMPLE 8–2   Show three possible ways to measure the period of the sine wave in Figure 8–6. How many cycles are shown?

FIGURE 8–6

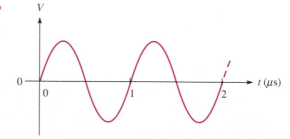

Solution

Method 1:   The period can be measured from one zero crossing to the corresponding zero crossing in the next cycle.

Method 2:   The period can be measured from the positive peak in one cycle to the positive peak in the next cycle.

Method 3:   The period can be measured from the negative peak in one cycle to the negative peak in the next cycle.

These measurements are indicated in Figure 8–7, where two cycles of the sine wave are shown. Keep in mind that you obtain the same value for the period no matter which of these points on the waveform you use.

FIGURE 8–7
Measurement of the period of a sine wave.

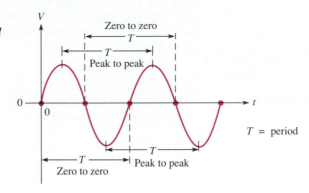

Related Exercise   If a positive peak occurs at 1 ms and the next positive peak occurs at 2.5 ms, what is the period?

Frequency of a Sine Wave

Frequency (f) is the number of cycles that a sine wave completes in 1 s.

The more cycles completed in one second, the higher the frequency. Figure 8–8 shows two sine waves. The sine wave in part (a) completes two full cycles in one second. The one in part (b) completes four cycles in one second. Therefore, the sine wave in part (b) has twice the frequency of the one in part (a).

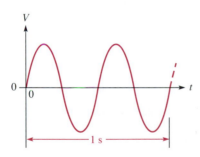

(a) Lower frequency: fewer cycles per second

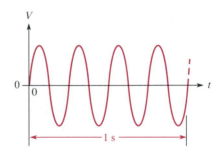

(b) Higher frequency: more cycles per second

FIGURE 8–8
Illustration of the frequency of a sine wave.

Frequency is measured in units of **hertz,** abbreviated Hz. One hertz is equivalent to one cycle per second; 60 Hz is 60 cycles per second; and so on. The symbol for frequency is f.

Relationship of Frequency and Period

The relationship between frequency and period is very important. The formulas for this relationship are as follows:

$$f = \frac{1}{T} \qquad \text{(8–1)}$$

$$T = \frac{1}{f} \qquad \text{(8–2)}$$

There is a reciprocal relationship between f and T. Knowing one, you can calculate the other with the [1/x] key on your calculator. This inverse relationship makes sense because a sine wave with a longer period goes through fewer cycles in one second than one with a shorter period.

EXAMPLE 8–3

Which sine wave in Figure 8–9 has the higher frequency? Determine the period and the frequency of both waveforms.

FIGURE 8–9

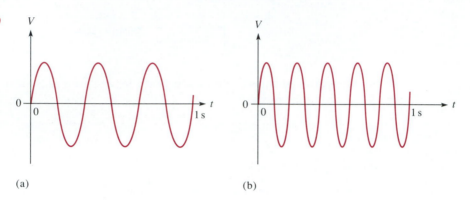

(a)   (b)

Solution   The sine wave in Figure 8–9(b) has the higher frequency because it completes more cycles in 1 s than does the one in part (a).

In Figure 8–9(a), three cycles take 1 s. Therefore, one cycle takes 0.333 s (one-third second), and this is the period.

$$T = 0.333 \text{ s} = 333 \text{ ms}$$

The frequency is

$$f = \frac{1}{T} = \frac{1}{333 \text{ ms}} = 3 \text{ Hz}$$

In Figure 8–9(b), five cycles take 1 s. Therefore, one cycle takes 0.2 s (one-fifth second), and this is the period.

$$T = 0.2 \text{ s} = 200 \text{ ms}$$

The frequency is

$$f = \frac{1}{T} = \frac{1}{200 \text{ ms}} = 5 \text{ Hz}$$

Related Exercise   If the time interval between consecutive negative peaks of a given sine wave is 50 μs, what is the frequency?

EXAMPLE 8–4

The period of a certain sine wave is 10 ms. What is the frequency?

Solution   Use Equation (8–1).

$$f = \frac{1}{T} = \frac{1}{10 \text{ ms}} = \frac{1}{10 \times 10^{-3} \text{ s}} = 100 \text{ Hz}$$

Related Exercise   A certain sine wave goes through four cycles in 20 ms. What is the frequency?

EXAMPLE 8–5   The frequency of a sine wave is 60 Hz. What is the period?

Solution   Use Equation (8–2).

$$T = \frac{1}{f} = \frac{1}{60 \text{ Hz}} = 16.7 \text{ ms}$$

Related Exercise   If $f = 1$ kHz, what is T?

SECTION 8–1 REVIEW

1. Describe one cycle of a sine wave.
2. At what point does a sine wave change polarity?
3. How many maximum points does a sine wave have during one cycle?
4. How is the period of a sine wave measured?
5. Define *frequency*, and state its unit.
6. Determine f when $T = 5$ μs.
7. Determine T when $f = 120$ Hz.

8–2 ■ SINE WAVE VOLTAGE SOURCES

As mentioned in last section, two basic methods of generating sine wave voltages are electromagnetic and electronic. Sine waves are produced electromagnetically by ac generators and electronically by oscillator circuits.

After completing this section, you should be able to

■ **Describe how sine waves are generated**
  □ Discuss the basic operation of an ac generator
  □ Discuss factors that affect frequency in ac generators
  □ Discuss factors that affect voltage in ac generators

The AC Generator

Figure 8–10 is a cutaway view of one type of ac generator. To illustrate the basic operation, Figure 8–11 shows a greatly simplified ac **generator** consisting of a single loop of

FIGURE 8–10
The ac generator, cutaway view (courtesy of General Electric).

FIGURE 8–11
A basic ac generator.

wire in a permanent magnetic field. Notice that each end of the loop is connected to a separate solid conductive ring called a *slip ring*. As the loop rotates in the magnetic field, the slip rings also rotate and rub against the brushes which connect the loop to an external load. Compare this generator to the basic dc generator in Figure 7–34, and note the difference in the ring and brush arrangements.

As you learned in Chapter 7, when a conductor moves through a magnetic field, a voltage is induced. Figure 8–12 illustrates how a sine wave voltage is produced by the basic ac generator as the loop rotates. An oscilloscope is used to display the voltage waveform.

To begin, Figure 8–12(a) shows the loop rotating through the first quarter of a revolution. It goes from an instantaneous horizontal position, where the induced voltage is zero, to an instantaneous vertical position, where the induced voltage is maximum. As shown on the display, this part of the rotation produces the first quarter of the sine wave cycle as the voltage builds up from zero to its positive maximum. Figure 8–12(b) shows the loop completing the first half of a revolution. During this part of the rotation, the voltage decreases from its positive maximum back to zero.

During the second half of the revolution, illustrated in Figures 8–12(c) and 8–12(d), the loop is cutting through the magnetic field in the opposite direction, so the voltage produced has a polarity opposite to that produced during the first half of the revolution. After one complete revolution of the loop, one full cycle of the sine wave voltage has been produced. As the loop continues to rotate, repetitive cycles of the sine wave are generated.

Frequency You have seen that one revolution of the conductor through the magnetic field in the basic ac generator (also called an *alternator*) produces one cycle of induced sinusoidal voltage. It is obvious that the rate at which the conductor is rotated determines the time for completion of one cycle. For example, if the conductor completes 60 revolutions in one second (rps), the period of the resulting sine wave is 1/60 s, corresponding to a frequency of 60 Hz. Thus, the faster the conductor rotates, the higher the resulting frequency of the induced voltage, as illustrated in Figure 8–13.

Another way of achieving a higher frequency is to increase the number of magnetic poles. In the previous discussion, two magnetic poles were used to illustrate the ac generator principle. During one revolution, the conductor passes under a north pole and a south

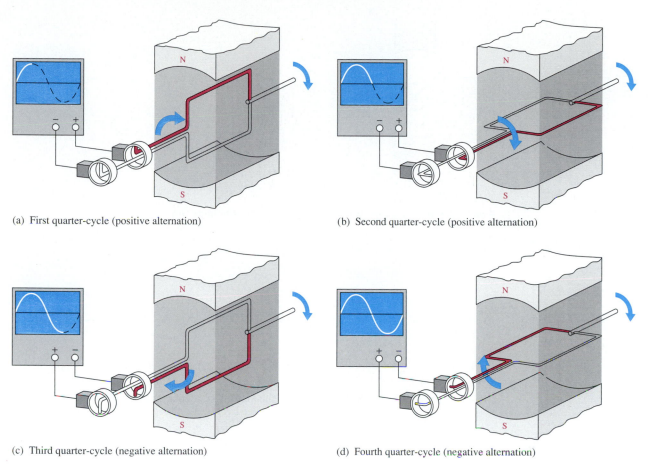

(a) First quarter-cycle (positive alternation)

(b) Second quarter-cycle (positive alternation)

(c) Third quarter-cycle (negative alternation)

(d) Fourth quarter-cycle (negative alternation)

FIGURE 8–12
One revolution of the loop generates one cycle of the sine wave voltage.

FIGURE 8–13
Frequency is directly proportional to the rate of rotation of the generator loop.

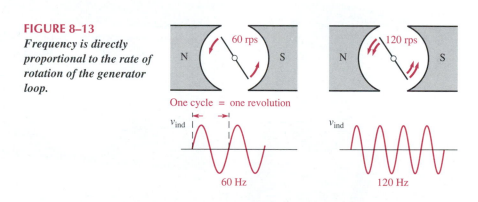

pole, thus producing one cycle of a sine wave. When four magnetic poles are used instead of two, as shown in Figure 8–14, one cycle is generated during one-half a revolution. This doubles the frequency for the same rate of rotation.

An expression for frequency in terms of the number of poles and the number of revolutions per second (rps) is as follows:

$$f = \text{number of pole pairs} \times (\text{rps}) \qquad \text{(8–3)}$$

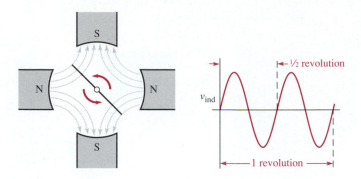

FIGURE 8–14
Four poles achieve a higher frequency than two for the same rps.

EXAMPLE 8–6   A four-pole generator has a rotation speed of 100 rps. Determine the frequency of the output voltage.

Solution f = number of pole pairs × (rps) = 2(100) = 200 Hz

Related Exercise   If the frequency of the output of a four-pole generator is 60 Hz, what is the rps?

(a)

(b)

(c)

FIGURE 8–15
Typical signal generators. Parts (a) and (b) courtesy of B & K Precision. Part (c) courtesy of Hewlett-Packard.

Voltage Magnitude Recall from Chapter 7 that the amount of voltage induced in a conductor depends on the number of turns *(N)* and the rate of change with respect to the magnetic field. Therefore, when the speed of rotation of the conductor is increased, not only does the frequency of the induced voltage increase—so also does the magnitude. Since the frequency value normally is fixed, the most practical method of increasing the amount or magnitude of induced voltage is to increase the number of loops.

Signal Generators

The signal generator is an instrument that electronically produces sine waves for use in testing or controlling electronic circuits and systems. There are a variety of signal generators, ranging from special-purpose instruments that produce only one type of waveform in a limited frequency range, to programmable instruments that produce a wide range of frequencies and a variety of waveforms. All signal generators consist basically of an **oscillator,** which is an electronic circuit that produces sine wave voltages whose amplitude and frequency can be adjusted. Typical signal generators are shown in Figure 8–15.

SECTION 8–2 REVIEW

1. List two types of sine wave sources.
2. How are sine waves generated electromagnetically?
3. Upon what is the frequency of a generator's output voltage dependent?
4. Upon what is the magnitude of a generator's output voltage dependent?

8–3 ▪ VOLTAGE AND CURRENT VALUES OF SINE WAVES

There are several ways to express and measure the value of a sine wave in terms of its voltage or its current magnitude. These are instantaneous, peak, peak-to-peak, rms, and average values.

After completing this section, you should be able to

▪ **Determine the voltage and current values of sine waves**
  ☐ Find the instantaneous value at any point
  ☐ Find the peak value
  ☐ Find the peak-to-peak value
  ☐ Define *rms* and find the rms value
  ☐ Explain why the average value of an alternating sine wave is always zero over a complete cycle
  ☐ Find the half-cycle average value

Instantaneous Value

Figure 8–16 illustrates that at any point in time on a sine wave, the voltage (or current) has an **instantaneous value.** This instantaneous value is different at different points along the curve. Instantaneous values are positive during the positive alternation and negative during the negative alternation. Instantaneous values of voltage and current are symbolized by lowercase v and i, respectively, as shown in part (a). The curve is shown for voltage only, but applies equally for current when v's are replaced with i's. An example is shown in part (b), where the instantaneous voltage is 3.1 V at 1 μs, 7.07 V at 2.5 μs, 10 V at 5 μs, 0 V at 10 μs, −3.1 V at 11 μs, and so on.

(a)

(b)

FIGURE 8–16

Example of instantaneous values of a sine wave voltage.

Peak Value

The **peak value** of a sine wave is the value of voltage (or current) at the positive or the negative maximum (peaks) with respect to zero. Since the peaks are equal in magnitude, a sine wave is characterized by a single peak value, as is illustrated in Figure 8–17. For a given sine wave, the peak value is constant and is represented by V_p or I_p. In this case, the peak value is 8 V.

Peak-to-Peak Value

The **peak-to-peak value** of a sine wave, as illustrated in Figure 8–18, is the voltage (or current) from the positive peak to the negative peak. It is always twice the peak value as expressed in the following equations:

$$V_{pp} = 2V_p \qquad \qquad (8\text{–}4)$$

$$I_{pp} = 2I_p \qquad \qquad (8\text{–}5)$$

Peak-to-peak values are represented by the symbols V_{pp} or I_{pp}. In Figure 8–18, the peak-to-peak value is 16 V.

FIGURE 8–17

Example of the peak value of a sine wave voltage.

FIGURE 8–18

Example of the peak-to-peak value of a sine wave voltage.

RMS Value

The term *rms* stands for **root mean square.** It refers to the mathematical process by which this value is derived. The rms value is also referred to as the **effective value.** Most ac voltmeters display the rms value of a voltage. For example, the 110 V at your wall outlet is an rms value.

The rms value of a sine wave is actually a measure of the heating effect of the sine wave. For example, when a resistor is connected across an ac (sine wave) voltage source, as shown in Figure 8–19(a), a certain amount of heat is generated by the power in the resistor. Figure 8–19(b) shows the same resistor connected across a dc voltage source. In part (a), the value of the ac voltage can be adjusted so that the resistor gives off the same amount of heat as it does when connected to the dc source.

> **The rms value of a sine wave is equal to the dc voltage that produces the same amount of heat in a resistance as does the sinusoidal voltage.**

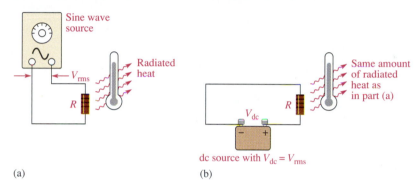

(a)   (b)

FIGURE 8–19

When the same amount of heat is being produced in both cases, the sine wave has an rms value equal to the dc voltage.

The peak value of a sine wave can be converted to the corresponding rms value using the following relationships for either voltage or current:

$$V_{rms} = 0.707V_p \tag{8–6}$$

$$I_{rms} = 0.707I_p \tag{8–7}$$

Using these formulas, you can also determine the peak value knowing the rms value as follows:

$$V_p = \frac{V_{rms}}{0.707}$$

$$V_p = 1.414V_{rms} \tag{8–8}$$

Similarly,

$$I_p = 1.414I_{rms} \tag{8–9}$$

To find the peak-to-peak value, simply double the peak value.

$$V_{pp} = 2.828V_{rms} \qquad \text{(8–10)}$$

and

$$I_{pp} = 2.828I_{rms} \qquad \text{(8–11)}$$

Average Value

The **average value** of a sine wave when taken over one complete cycle is always zero, because the positive values (above the zero crossing) offset the negative values (below the zero crossing).

To be useful for comparison purposes and in determining the average value of a rectified voltage such as found in power supplies, the **average value** of a sine wave is defined over a half-cycle rather than over a full cycle. The average value is expressed in terms of the peak value as follows for both voltage and current sine waves:

$$V_{avg} = 0.637V_p \qquad \text{(8–12)}$$

$$I_{avg} = 0.637 I_p \qquad \text{(8–13)}$$

The half-cycle average value of a sine wave voltage is illustrated in Figure 8–20.

FIGURE 8–20
Half-cycle average value.

EXAMPLE 8–7   Determine V_p, V_{pp}, V_{rms}, and the half-cycle V_{avg} for the sine wave in Figure 8–21.

FIGURE 8–21

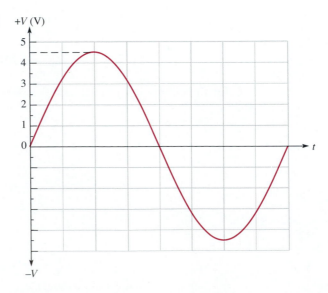

Solution As taken directly from the graph, $V_p = 4.5$ V. From this value calculate the other values:

$$V_{pp} = 2V_p = 2(4.5 \text{ V}) = 9 \text{ V}$$
$$V_{rms} = 0.707V_p = 0.707(4.5 \text{ V}) = 3.18 \text{ V}$$
$$V_{avg} = 0.637V_p = 0.637(4.5 \text{ V}) = 2.87 \text{ V}$$

Related Exercise If $V_p = 25$ V, determine V_{pp}, V_{rms}, and V_{avg} for a sine wave.

SECTION 8–3 REVIEW

1. Determine V_{pp} in each case when
   (a) $V_p = 1$ V   (b) $V_{rms} = 1.414$ V   (c) $V_{avg} = 3$ V
2. Determine V_{rms} in each case when
   (a) $V_p = 2.5$ V   (b) $V_{pp} = 10$ V   (c) $V_{avg} = 1.5$ V
3. Determine the half-cycle V_{avg} in each case when
   (a) $V_p = 10$ V   (b) $V_{rms} = 2.3$ V   (c) $V_{pp} = 60$ V

8–4 ■ ANGULAR MEASUREMENT OF A SINE WAVE

As you have seen, sine waves can be measured along the horizontal axis on a time basis; however, since the time for completion of one full cycle or any portion of a cycle is frequency-dependent, it is often useful to specify points on the sine wave in terms of an angular measurement expressed in degrees or radians. Angular measurement is independent of frequency.

After completing this section, you should be able to

■ **Describe angular relationships of sine waves**
  □ Show how to measure a sine wave in terms of angles
  □ Define *radian*
  □ Convert radians to degrees
  □ Determine the phase of a sine wave

As previously discussed, sine wave voltage can be produced electromagnetically by rotating electromechanical machines. As the rotor of the ac generator goes through a full 360° of rotation, the resulting output is one full cycle of a sine wave voltage. Thus, the angular measurement of a sine wave can be related to the angular rotation of a generator, as shown in Figure 8–22.

Angular Measurement

A **degree** is an angular measurement corresponding to 1/360 of a circle or a complete revolution. A **radian** (rad) is the angular distance along the circumference of a circle equal to the radius of the circle. One radian is equivalent to 57.3°, as illustrated in Figure 8–23. In a 360° revolution, there are 2π radians.

> The Greek letter π (pi) represents the ratio of the circumference of any circle to its diameter and has a constant value of approximately 3.1416.

FIGURE 8–22

FIGURE 8–23
Angular measurement showing relationship of the radian to degrees.

Scientific calculators have a π key so that the actual numerical value does not have to be entered.

Table 8–1 lists several values of degrees and the corresponding radian values. These angular measurements are illustrated in Figure 8–24.

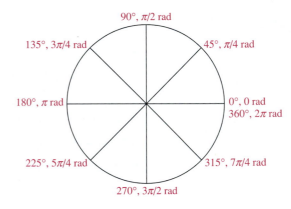

FIGURE 8–24
Angular measurements starting at 0° and going counterclockwise.

TABLE 8–1

DEGREES (°)	RADIANS (rad)
0	0
45	$\pi/4$
90	$\pi/2$
135	$3\pi/4$
180	π
225	$5\pi/4$
270	$3\pi/2$
315	$7\pi/4$
360	2π

Radian/Degree Conversion

Radians can be converted to degrees using Equation (8–14).

$$\text{rad} = \left(\frac{\pi \text{ rad}}{180°} \right) \times \text{degrees} \qquad \textbf{(8–14)}$$

Similarly, degrees can be converted to radians with Equation (8–15).

$$\text{degrees} = \left(\frac{180°}{\pi \text{ rad}} \right) \times \text{rad} \qquad \textbf{(8–15)}$$

EXAMPLE 8–8   **(a)** Convert 60° to radians.   **(b)** Convert π/6 rad to degrees.

Solution

(a) $\text{Rad} = \left(\dfrac{\pi \text{ rad}}{180°}\right)60° = \dfrac{\pi}{3} \text{ rad} = 1.047 \text{ rad}$

(b) $\text{Degrees} = \left(\dfrac{180°}{\pi \text{ rad}}\right)\left(\dfrac{\pi}{6} \text{ rad}\right) = 30°$

Related Exercise   **(a)** Convert 15° to radians.   **(b)** Convert 2π radians to degrees.

Sine Wave Angles

The angular measurement of a sine wave is based on 360° or 2π radians for a complete cycle. A half-cycle is 180° or π radians; a quarter-cycle is 90° or π/2 radians; and so on. Figure 8–25(a) shows angles in degrees over a full cycle of a sine wave; part (b) shows the same points in radians.

(a) Degrees

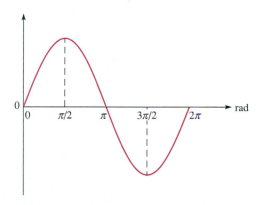

(b) Radians

FIGURE 8–25
Sine wave angles.

Phase of a Sine Wave

The **phase** of a sine wave is an angular measurement that specifies the position of that sine wave relative to a reference. Figure 8–26 shows one cycle of a sine wave to be used

FIGURE 8–26
Phase reference.

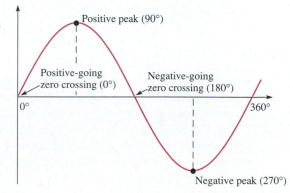

as the reference. Note that the first positive-going crossing of the horizontal axis (zero crossing) is at 0° (0 rad), and the positive peak is at 90° ($\pi/2$ rad). The negative-going zero crossing is at 180° (π rad), and the negative peak is at 270° ($3\pi/2$ rad). The cycle is completed at 360° (2π rad). When the sine wave is shifted left or right with respect to this reference, there is a phase shift.

Figure 8–27 illustrates phase shifts of a sine wave. In part (a), sine wave *B* is shifted to the right by 90° ($\pi/2$ rad). Thus, there is a phase angle of 90° between sine wave *A* and sine wave *B*. In terms of time, the positive peak of sine wave *B* occurs later than the positive peak of sine wave *A*, because time increases to the right along the horizontal axis. In this case, sine wave *B* is said to **lag** sine wave *A* by 90° or $\pi/2$ radians. In other words, sine wave *A* leads sine wave *B* by 90°.

In Figure 8–27(b), sine wave *B* is shown shifted left by 90°. Thus, again there is a phase angle of 90° between sine wave *A* and sine wave *B*. In this case, the positive peak of sine wave *B* occurs earlier in time than that of sine wave *A*; therefore, sine wave *B* is said to **lead** by 90°. In both cases there is a 90° phase angle between the two waveforms.

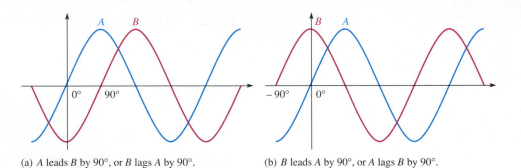

(a) *A* leads *B* by 90°, or *B* lags *A* by 90°.

(b) *B* leads *A* by 90°, or *A* lags *B* by 90°.

FIGURE 8–27
Illustration of a phase shift.

EXAMPLE 8–9

What are the phase angles between the two sine waves in Figure 8–28(a) and 8–28(b)?

FIGURE 8–28

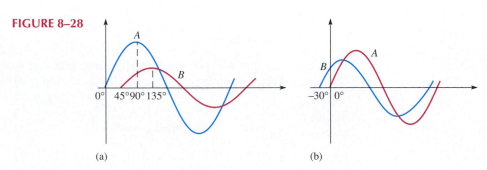

(a)

(b)

Solution In Figure 8–28(a) the zero crossing of sine wave *A* is at 0°, and the corresponding zero crossing of sine wave *B* is at 45°. There is a 45° phase angle between the two waveforms with sine wave *A* leading.

In Figure 8–28(b) the zero crossing of sine wave *B* is at −30°, and the corresponding zero crossing of sine wave *A* is at 0°. There is a 30° phase angle between the two waveforms with sine wave *B* leading.

Related Exercise If the positive-going zero crossing of one sine wave is at 15° relative to the 0° reference and that of the second sine wave is at 23° relative to the 0° reference, what is the phase angle between the sine waves?

SECTION 8–4 REVIEW

1. When the positive-going zero crossing of a sine wave occurs at 0°, at what angle does each of the following points occur?
   (a) positive peak   (b) negative-going zero crossing
   (c) negative peak   (d) end of first complete cycle
2. A half cycle is completed in _____ degrees, or _____ radians.
3. A full cycle is completed in _____ degrees, or _____ radians.
4. Determine the phase angle between the two sine waves in Figure 8–29.

FIGURE 8–29

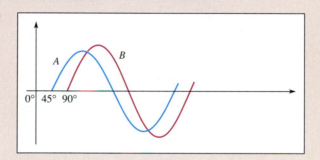

8–5 ▪ THE SINE WAVE FORMULA

A sine wave can be graphically represented by voltage or current values on the vertical axis and by angular measurement (degrees or radians) along the horizontal axis. This graph can be expressed mathematically, as you will see.

After completing this section, you should be able to

▪ **Mathematically analyze a sinusoidal waveform**
   ☐ State the sine wave formula
   ☐ Find instantaneous values using the formula

A generalized graph of one cycle of a sine wave is shown in Figure 8–30. The **amplitude,** *A,* is the maximum value of the voltage or current on the vertical axis, and angular values run along the horizontal axis. The variable *y* is an instantaneous value representing either voltage or current at a given angle, θ.

All sine wave curves follow a specific mathematical formula. The general expression for the sine wave curve in Figure 8–30 is

$$y = A \sin \theta \tag{8–16}$$

This general expression states that any point on the sine wave, represented by an instantaneous value *y,* is equal to the maximum value *A,* times the sine (sin) of the angle θ at that point. For example, if a certain sine wave voltage has a peak value of

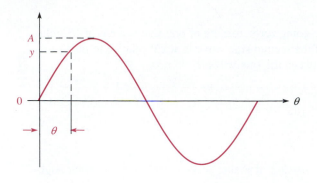

FIGURE 8–30

One cycle of a generalized sine wave showing amplitude and phase.

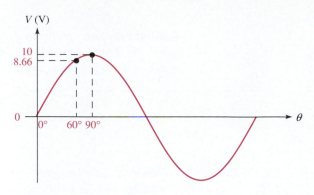

FIGURE 8–31

Illustration of the instantaneous value of a sine wave voltage at θ = 60°.

10 V, the instantaneous voltage at a point 60° along the horizontal axis can be calculated as follows:

$$v = V_p \sin \theta = 10 \sin 60° = 10(0.866) = 8.66 \text{ V}$$

where v has been substituted for y and V_p for A in the general formula. Figure 8–31 shows this instantaneous value on the curve. You can find the sine of any angle on your calculator by first entering the value of the angle and then pressing the [sin] key. Verify that your calculator is in the degree mode, and, if not, use the [DRG] key.

Derivation of the Sine Wave Formula

A rotating arrow called a **phasor** can be used to represent time-varying quantities such as sine waves in terms of their magnitude and angular measurement. One full cycle of a sine wave can be represented by rotation of a phasor through 360°.

Figure 8–32 shows how the phasor "traces out" the sine wave as it goes counterclockwise from 0° to 360°. You can relate this concept to the rotation in an ac generator as mentioned earlier.

Notice in Figure 8–32 that the length of the phasor arrow is equal to the peak value of the sine wave (observe the 90° and the 270° points). The angle of the phasor measured from 0° is the corresponding angular point on the sine wave.

Let's examine a phasor representation at one specific angle. Figure 8–33 shows the voltage phasor at an angular position of 45° and the corresponding point on the sine wave. The instantaneous value, v, of the sine wave at this point is related to both the position and the length of the phasor. The vertical distance from the phasor tip down to the horizontal axis represents the instantaneous value of the sine wave at that point.

FIGURE 8–32

Sine wave represented by a rotating phasor arrow.

FIGURE 8–33
Right triangle derivation of sine wave formula, $v = V_p \sin \theta$.

Notice that when a vertical line is drawn from the phasor tip down to the horizontal axis, a **right triangle** is formed, as shown Figure 8–33. The length of the phasor is the **hypotenuse** of the triangle, and the vertical projection is the opposite side. From trigonometry, *the opposite side of a right triangle is equal to the hypotenuse times the sine of the angle* θ. In this case, the length of the phasor is the peak value of the sine wave voltage, V_p. Thus, the opposite side of the triangle, which is the instantaneous value, can be expressed as

$$v = V_p \sin \theta \qquad \text{(8–17)}$$

This formula also applies to a current sine wave as follows:

$$i = I_p \sin \theta \qquad \text{(8–18)}$$

Expressions for Phase-Shifted Sine Waves

When a sine wave is shifted to the right of the reference (lagging) by a certain angle, ϕ, as illustrated in Figure 8–34(a), the general expression is

$$y = A \sin(\theta - \phi) \qquad \text{(8–19)}$$

where y represents instantaneous voltage or current, and A represents the peak value (amplitude). When a sine wave is shifted to the left of the reference (leading) by a certain angle, ϕ, as shown in Figure 8–34(b), the general expression is

$$y = A \sin(\theta + \phi) \qquad \text{(8–20)}$$

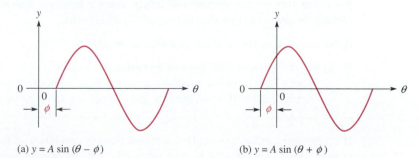

(a) $y = A \sin (\theta - \phi)$

(b) $y = A \sin (\theta + \phi)$

FIGURE 8–34
Shifted sine waves.

EXAMPLE 8–10   Determine the instantaneous value at the 90° reference point on the horizontal axis for each sine wave voltage in Figure 8–35.

FIGURE 8–35

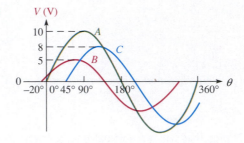

Solution   Sine wave A is the reference. Sine wave B is shifted left 20° with respect to A, so it leads. Sine wave C is shifted right 45° with respect to A, so it lags.

$$v_A = V_p \sin \theta = 10 \sin 90° = 10(1) = 10 \text{ V}$$

$$v_B = V_p \sin(\theta + \phi_B)$$
$$= 5 \sin(90° + 20°) = 5 \sin 110° = 5(0.9397) = 4.7 \text{ V}$$

$$v_C = V_p \sin(\theta - \phi_C)$$
$$= 8 \sin(90° - 45°) = 8 \sin 45° = 8(0.707) = 5.66 \text{ V}$$

Related Exercise   A sine wave has a peak value of 20 V. What is the instantaneous value at +65° from the 0° reference?

SECTION 8–5 REVIEW

1. Determine the sine of the following angles:
   (a) 30°   (b) 60°   (c) 90°
2. Calculate the instantaneous value at 120° for the sine wave in Figure 8–31.
3. Determine the instantaneous value at the 45° point on the reference axis of a sine wave shifted 10° to the left of the zero reference ($V_p = 10$ V).

8–6 ▪ OHM'S LAW AND KIRCHHOFF'S LAWS IN AC CIRCUITS

When time-varying ac voltages such as the sine wave are applied to a circuit, the circuit laws that you studied earlier still apply. Ohm's law and Kirchhoff's laws apply to ac circuits in the same way that they apply to dc circuits.

After completing this section, you should be able to

▪ **Apply the basic circuit laws to ac resistive circuits**
  ☐ Apply Ohm's law to resistive circuits with ac sources
  ☐ Apply Kirchhoff's voltage law and current law to resistive circuits with ac sources

If a sine wave voltage is applied across a resistor as shown in Figure 8–36, a sine wave current flows. The current is zero when the voltage is zero and is maximum when the voltage is maximum. When the voltage changes polarity, the current reverses direction. As a result, the voltage and current are said to be in phase with each other.

FIGURE 8–36

A sine wave voltage produces a sine wave current.

When using Ohm's law in ac circuits, remember that both the voltage and the current must be expressed consistently, that is, both as peak values, both as rms values, both as average values, and so on.

EXAMPLE 8–11   Determine the rms voltage across each resistor and the rms current in Figure 8–37. The source voltage is given as an rms value.

FIGURE 8–37

Solution   The total resistance of the circuit is

$$R_T = R_1 + R_2 = 1 \text{ k}\Omega + 560 \ \Omega = 1.56 \text{ k}\Omega$$

Use Ohm's law to find the rms current.

$$I_{rms} = \frac{V_{s(rms)}}{R_T} = \frac{110 \text{ V}}{1.56 \text{ k}\Omega} = 70.5 \text{ mA}$$

The rms voltage drop across each resistor is

$$V_{1(rms)} = I_{rms}R_1 = (70.5 \text{ mA})(1 \text{ k}\Omega) = 70.5 \text{ V}$$
$$V_{2(rms)} = I_{rms}R_2 = (70.5 \text{ mA})(560 \ \Omega) = 39.5 \text{ V}$$

Related Exercise   Repeat this example for a source voltage of 10 V peak.

Kirchhoff's voltage and current laws apply to ac circuits as well as to dc circuits. Figure 8–38 illustrates Kirchhoff's voltage law in a resistive circuit that has a sine wave voltage source. As you can see, the source voltage is the sum of all the voltage drops across the resistors, just as in a dc circuit.

FIGURE 8–38
Illustration of Kirchhoff's voltage law in an ac circuit.

EXAMPLE 8–12   All values in Figure 8–39 are given in rms.
(a) Find the unknown peak voltage drop in Figure 8–39(a).
(b) Find the total rms current in Figure 8–39(b).

FIGURE 8–39

(a)   (b)

Solution

(a) Use Kirchhoff's voltage law to find V_3.

$$V_s = V_1 + V_2 + V_3$$

$$V_{3(rms)} = V_{s(rms)} - V_{1(rms)} - V_{2(rms)} = 24\text{ V} - 12\text{ V} - 8\text{ V} = 4\text{ V}$$

Convert rms to peak:

$$V_{3(p)} = 1.414 V_{3(rms)} = 1.414(4\text{ V}) = 5.66\text{ V}$$

(b) Use Kirchhoff's current law to find I_T.

$$I_{T(rms)} = I_{1(rms)} + I_{2(rms)} = 10\text{ A} + 3\text{ A} = 13\text{ A}$$

Related Exercise   A series circuit has the following voltage drops: $V_{1(rms)} = 3.50$ V, $V_{2(p)} = 4.25$ V, $V_{3(avg)} = 1.70$ V. Determine the peak-to-peak source voltage.

**SECTION 8–6
REVIEW**

1. A sine wave voltage with a half-cycle average value of 12.5 V is applied to a circuit with a resistance of 330 Ω. What is the peak current in the circuit?
2. The peak voltage drops in a series resistive circuit are 6.2 V, 11.3 V, and 7.8 V. What is the rms value of the source voltage?

8–7 ▪ SUPERIMPOSED DC AND AC VOLTAGES

In many practical circuits, you will find both dc and ac voltages combined. An example of this is in amplifier circuits where ac signal voltages are superimposed on dc operating voltages.

After completing this section, you should be able to

▪ **Determine total voltages which are composed of both ac and dc components**

Figure 8–40 shows a dc source and an ac source in series. These two voltages will add algebraically to produce an ac voltage "riding" on a dc level, as measured across the resistor.

If V_{dc} is greater than the peak value of the sine wave, the combined voltage is a sine wave that never reverses polarity and is therefore nonalternating. That is, we have a sine wave riding on a dc level, as shown in Figure 8–41(a). If V_{dc} is less than the peak value of the sine wave, the sine wave will be negative during a portion of its lower half-cycle, as illustrated in Figure 8–41(b), and is therefore alternating. In either case the sine wave will reach a maximum voltage equal to $V_{dc} + V_p$, and it will reach a minimum voltage equal to $V_{dc} - V_p$.

FIGURE 8–40
Superimposed dc and ac voltages.

(a) $V_{dc} > V_p$. The sine wave never goes negative.

(b) $V_{dc} < V_p$. The sine wave reverses polarity during a portion of its cycle.

FIGURE 8–41
Sine waves with dc levels.

EXAMPLE 8–13   Determine the total voltage across the resistor in each circuit of Figure 8–42.

FIGURE 8–42

(a)   (b)

Solution   In Figure 8–42(a), the maximum voltage across R is

$$V_{max} = V_{dc} + V_p = 12 \text{ V} + 10 \text{ V} = 22 \text{ V}$$

The minimum voltage across R is

$$V_{min} = V_{dc} - V_p = 12 \text{ V} - 10 \text{ V} = 2 \text{ V}$$

Therefore, $V_{R(tot)}$ is a nonalternating sine wave that varies from +22 V to +2 V, as shown in Figure 8–43(a).

In Figure 8–42(b), the maximum voltage across R is

$$V_{max} = V_{dc} + V_p = 6 \text{ V} + 10 \text{ V} = 16 \text{ V}$$

The minimum voltage across R is

$$V_{min} = V_{dc} - V_p = -4 \text{ V}$$

Therefore, $V_{R(tot)}$ is an alternating sine wave that varies from +16 V to –4 V, as shown in Figure 8–43(b).

FIGURE 8–43

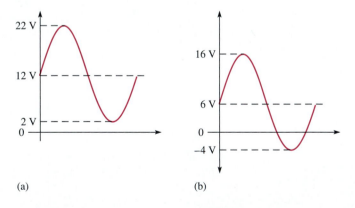

(a)   (b)

Related Exercise   Explain why the waveform in Figure 8–43(a) is nonalternating but the waveform in part (b) is considered to be alternating.

SECTION 8–7 REVIEW

1. What is the maximum positive value of the resulting total voltage when a sine wave with $V_p = 5$ V is added to a dc voltage of +2.5 V?
2. Will the resulting voltage in Question 1 alternate polarity?
3. If the dc voltage in Question 1 is –2.5 V, what is the maximum positive value of the resulting total voltage?

8–8 ■ THE OSCILLOSCOPE

The oscilloscope, or scope for short, is one of the most widely used and versatile test instruments. It displays on a screen the actual shape of voltages that are changing with time so that waveform measurements can be made and one waveform can be compared to another.

After completing this section, you should be able to

■ **Use the oscilloscope to measure waveforms**
   ☐ Describe a basic CRT
   ☐ Identify basic oscilloscope controls
   ☐ Explain how to measure amplitude
   ☐ Explain how to measure period and frequency

Figure 8–44 shows a typical dual-channel oscilloscope. You can think of the **oscilloscope** as essentially a "graphing machine" that graphs out a voltage and shows you how it varies with time. The shape of the sine wave that you have seen throughout this chapter is simply a graph of voltage versus time. Of course, oscilloscopes can display any type of waveform, but we will use the sine wave to illustrate some basic concepts.

FIGURE 8–44
A typical oscilloscope (courtesy of Tektronix, Inc.).

The Oscilloscope Screen

Generally, the screen of an oscilloscope is divided into ten horizontal divisions and eight vertical divisions, as shown in Figure 8–45. Each major division is divided into five small or minor divisions on both the horizontal and the vertical axes. The vertical axis on the scope screen is the voltage scale and the horizontal axis is the time scale. Values of voltage are read vertically and values of time are read horizontally.

Cathode-Ray Tube (CRT)

The oscilloscope is built around the **cathode-ray tube (CRT),** which is the device that displays the waveforms. The screen of the scope is the front of the CRT.

The CRT is a vacuum tube device containing an electron gun that emits a narrow, focused beam of electrons. A phosphorescent coating on the face of the tube forms the screen. The beam is electronically focused and accelerated so that it strikes the screen, causing light to be emitted at the point of impact.

Figure 8–46 shows the basic construction of a CRT. The electron gun assembly contains a heater, a cathode, a control grid, and accelerating and focusing grids. The heater

FIGURE 8–45
The oscilloscope screen.

FIGURE 8–46
Basic construction of a CRT.

carries current that indirectly heats the cathode. The heated cathode emits electrons. The amount of voltage on the control grid determines the flow of electrons and thus the intensity of the beam. The electrons are accelerated by the accelerating grid and are focused by the focusing grid into a narrow beam that converges at the screen. The beam is further accelerated to a high speed after it leaves the electron gun by a high voltage on the anode surfaces of the CRT.

The purpose of the deflection plates in the CRT is to produce a "bending," or deflection, of the electron beam. This deflection allows the position of the point of impact on the screen to be varied. There are two sets of deflection plates: one set for vertical deflection, and the other set for horizontal deflection.

Figure 8–47 illustrates how a waveform (a sine wave in this case) is "plotted out" on the screen. A sawtooth voltage called the *horizontal sweep signal* is internally generated and applied to the horizontal deflection plates. Each cycle of the sawtooth causes the electron beam to sweep across the screen from left to right and then quickly return for another sweep. The external signal that is to be displayed on the screen is applied to the vertical deflection plates. This causes the electron beam to move up or down vertically as it is swept across the screen, tracing out a pattern that follows the variation of the external signal voltage over an interval of time.

FIGURE 8–47
How a waveform is traced out on the scope screen.

Operation of the Oscilloscope

Figure 8–48 shows a typical oscilloscope like the one used in the Application Assignment sections. This is a dual-channel instrument, as most scopes are. *Dual-channel* means that you can display two waveforms from two different inputs on the screen at the same time.

The Volts/Division Control Notice that in Figure 8–48 there are two VOLTS/DIV switches, one for channel 1 (CH1) and one for channel 2 (CH2). The volts/division selector switch sets the number of volts to be represented by each major division on the vertical scale. For example, Figure 8–49(a) shows one cycle of a sine wave on the scope screen with the VOLTS/DIV switch set at 1 V. This means that each of the major vertical divisions is 1 V. The peak of the sine wave is three divisions high, and, since each division is 1 V, assuming a × 1 scope probe is used, the peak value of the sine wave is 3 V (3 divisions × 1 V/division = 3 V).

For the same sine wave, if the VOLTS/DIV setting is changed to 2 V, the sine wave will appear as shown in Figure 8–49(b). Notice that now the peak is 1½ major divisions high. Even though each division is now 2 V, the peak value is still 3 V because 1.5 divisions × 2 V/division = 3 V.

FIGURE 8–48
A typical dual-channel oscilloscope.

(a)   (b)

FIGURE 8–49
Using the VOLTS/DIV control.

So, as you can see, it is relatively easy to measure the peak value by counting the number of major vertical divisions from the zero crossing to the peak of the waveform and multiplying by the VOLTS/DIV setting.

The Seconds/Division Control   The SEC/DIV selector switch sets the number of seconds, milliseconds, or microseconds to be represented by each major division on the horizontal scale. It actually controls how fast the electron beam sweeps horizontally across the screen.

For example, Figure 8–50(a) shows one cycle of a sine wave displayed on the screen. In this case, the entire cycle covers ten horizontal divisions. The SEC/DIV switch is set at 10 μs. This means that each major horizontal division is 10 μs. Since a full cycle of the sine wave covers 10 divisions, the period of the sine wave is 100 μs (10 div × 10 μs/div = 100 μs). From this, $f = 1/100 \ \mu$s = 10 kHz.

For the same sine wave, if the SEC/DIV setting is changed to 20 μs, the sine wave will appear as shown in Figure 8–50(b). Notice that now a full cycle covers only five horizontal divisions and there are two cycles on the screen. Even though each division is now 20 μs, the period is still 100 μs because 5 divisions × 20 μs/division = 100 μs.

(a)   (b)

FIGURE 8–50
Using the SEC/DIV control.

So, as you can see, it is easy to measure the period and then calculate the frequency by counting the number of major divisions covered by one cycle and then multiplying by the SEC/DIV setting. Some scopes automatically measure and display the frequency and other parameters.

Other Oscilloscope Controls

The following descriptions refer to the oscilloscope in Figure 8–48 but apply also to most general-purpose scopes.

Power Switch The power switch turns the power to the scope on and off. A light indicates when the power is on.

Intensity The intensity control knob varies the brightness of the trace on the screen. Caution should be used so that the intensity is not left too high for an extended period of time, especially when the beam forms a motionless dot on the screen. Damage to the screen can result from excessive intensity.

Focus This control focuses the beam so that it converges to a tiny point at the screen. An out-of-focus condition results in a fuzzy trace.

Horizontal Position These control knobs (coarse and fine) adjust the neutral horizontal position of the beam. They are used to reposition horizontally a waveform display for more convenient viewing or measurement.

Vertical Position The two vertical position controls move each trace up or down for easier measurement or observation.

AC-GND-DC Switch This switch, located below the VOLTS/DIV control, allows the input signal to be ac coupled, dc coupled, or grounded. The ac coupling eliminates any dc component on the input signal. The dc coupling permits dc values to be displayed. The ground position allows a 0 V reference to be established on the screen.

Signal Inputs The signals to be displayed are connected into the channel 1 (CH1) and/or channel 2 (CH2) input connectors. These connections are normally done with a special probe that minimizes the loading effect of the scope's input resistance on the circuit being measured. Oscilloscope voltage probes are generally either ×1 (nonattenuating) or ×10 (attenuates by 10). When a ×10 probe is used, the VOLTS/DIV setting must be multiplied by 10. To keep it simple, all applications in this book assume ×1 voltage probes.

Mode Switches These switches provide for displaying either or both channel inputs, inverting channel 2 signal, adding two waveforms, and selecting between alternate and chopped mode of sweep.

Trigger Control The **trigger** controls allow the beam to be triggered from various selected sources. The triggering of the beam causes it to begin its sweep across the screen. It can be triggered from an internally generated signal derived from an input signal, or from the line voltage, or from an externally applied trigger signal. The modes of triggering are auto, normal, single-sweep, and TV. In the auto mode, sweep occurs in the absence of an adequate trigger signal. In the normal mode, a trigger signal must be present for the sweep to occur. The TV mode provides triggering on the TV field or TV line signals. The slope switch allows the triggering to occur on either the positive-going slope or the negative-going slope of the trigger waveform. The level control selects the voltage level on the trigger signal at which the triggering occurs.

Basically, the trigger controls provide for synchronization of the horizontal sweep waveform and the input signal waveform. As a result, the display of the input signal is stable on the screen, rather than appearing to drift across the screen.

EXAMPLE 8–14   Determine the peak value and period of each sine wave in Figure 8–51 from the scope displays and the indicated settings for VOLTS/DIV and SEC/DIV.

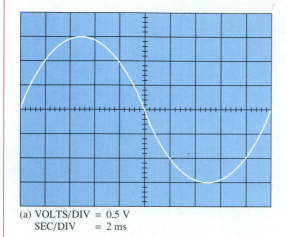

(a) VOLTS/DIV = 0.5 V
    SEC/DIV   = 2 ms

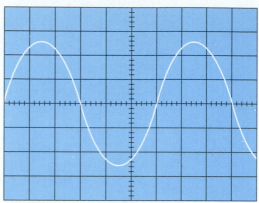

(b) VOLTS/DIV = 50 mV
    SEC/DIV   = 0.1 ms

(c) VOLTS/DIV = 2 V
    SEC/DIV   = 10 μs

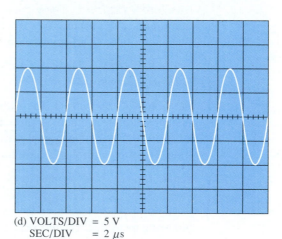

(d) VOLTS/DIV = 5 V
    SEC/DIV   = 2 μs

Sine wave is centered vertically on the screen.

FIGURE 8–51

Solution   Looking at the vertical scale in Figure 8–51(a),

$$V_p = 3 \text{ divisions} \times 0.5 \text{ V/division} = 1.5 \text{ V}$$

From the horizontal scale (one cycle covers ten divisions),

$$T = 10 \text{ divisions} \times 2 \text{ ms/division} = 20 \text{ ms}$$

Looking at the vertical scale in Figure 8–51(b),

$$V_p = 2.5 \text{ divisions} \times 50 \text{ mV/division} = 125 \text{ mV}$$

From the horizontal scale (one cycle covers six divisions),

$$T = 6 \text{ divisions} \times 0.1 \text{ ms/division} = 0.6 \text{ ms} = 600 \text{ μs}$$

Looking at the vertical scale in Figure 8–51(c),

$$V_p = 3.4 \text{ divisions} \times 2 \text{ V/division} = 6.8 \text{ V}$$

From the horizontal scale (one-half cycle covers ten divisions),

$$T = 20 \text{ divisions} \times 10 \text{ } \mu s/\text{division} = 200 \text{ } \mu s$$

Looking at the vertical scale in Figure 8–51(d),

$$V_p = 2 \text{ divisions} \times 5 \text{ V/division} = 10 \text{ V}$$

From the horizontal scale (one cycle covers two divisions),

$$T = 2 \text{ divisions} \times 2 \text{ } \mu s/\text{division} = 4 \text{ } \mu s$$

Related Exercise Determine the rms value and the frequency for each waveform displayed in Figure 8–51.

SECTION 8–8 REVIEW

1. What does *CRT* stand for?
2. On an oscilloscope, voltage is measured (horizontally, vertically) on the screen and time is measured (horizontally, vertically).
3. What can an oscilloscope do that a multimeter cannot?

8–9 ■ NONSINUSOIDAL WAVEFORMS

Sine waves are important in electronics, but they are by no means the only type of ac or time-varying waveform. Two other major types of waveforms, the pulse waveform and the triangular waveform, are discussed next.

After completing this section, you should be able to

■ **Identify the characteristics of basic nonsinusoidal waveforms**
   □ Discuss the properties of a pulse waveform
   □ Define *duty cycle*
   □ Discuss the properties of triangular and sawtooth waveforms
   □ Discuss the harmonic content of a waveform

Pulse Waveforms

Basically, a **pulse** can be described as a very rapid transition (**leading edge**) from one voltage or current level (**baseline**) to another, and then, after an interval of time, a very rapid transition (**trailing edge**) back to the original baseline level. The transitions in level are called *steps*. An ideal pulse consists of two opposite-going steps of equal amplitude. When the leading or trailing edge is positive-going, it is called a **rising edge.** When the leading or trailing edge is negative-going, it is called a **falling edge.**

Figure 8–52(a) shows an ideal positive-going pulse consisting of two equal but opposite instantaneous steps separated by an interval of time called the **pulse width.** Figure 8–52(b) shows an ideal negative-going pulse. The height of the pulse measured from the baseline is its amplitude.

In many applications, analysis is simplified by treating all pulses as ideal (composed of instantaneous steps and perfectly rectangular in shape). Actual pulses, however, are never ideal. All pulses possess certain characteristics that cause them to be different from the ideal.

FIGURE 8–52
Ideal pulses.

In practice, pulses cannot change from one level to another instantaneously. Time is always required for a transition (step), as illustrated in Figure 8–53(a). As you can see, there is an interval of time during which the pulse is rising from its lower value to its higher value. This interval is called the **rise time, t_r**.

> **Rise time is the time required for the pulse to go from 10% of its full amplitude to 90% of its full amplitude.**

The interval of time during which the pulse is falling from its higher value to its lower value is called the **fall time, t_f**.

> **Fall time is the time required for the pulse to go from 90% of its full amplitude to 10% of its full amplitude.**

Pulse width (t_W) also requires a precise definition for the nonideal pulse, because the rising and falling edges are not vertical.

> **Pulse width is the time between the point on the rising edge, where the value is 50% of full amplitude, to the point on the falling edge, where the value is 50% of full amplitude.**

Pulse width is shown in Figure 8–53(b).

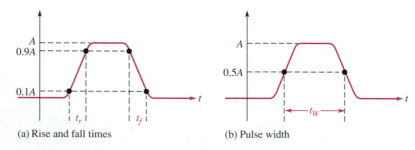

(a) Rise and fall times

(b) Pulse width

FIGURE 8–53
Nonideal pulse.

Repetitive Pulses

Any waveform that repeats itself at fixed intervals is **periodic**. Figure 8–54 shows some examples of periodic pulse waveforms. Notice that in each case, the pulses repeat at regular intervals. The rate at which the pulses repeat is the pulse repetition rate, which is the fundamental frequency of the waveform. The frequency can be expressed in hertz or in pulses per second. The time from one pulse to the corresponding point on the next pulse is the **period (T)**.

FIGURE 8–54
Repetitive pulse waveforms.

(a)

(b)

(c)

An important characteristic of periodic pulse waveforms is the **duty cycle.**

The duty cycle is the ratio of the pulse width (t_W) to the period (T) and is usually expressed as a percentage.

$$\text{Percent duty cycle} = \left(\frac{t_W}{T}\right)100\% \qquad\qquad (8\text{–}21)$$

EXAMPLE 8–15   Determine the period, frequency, and duty cycle for the pulse waveform in Figure 8–55.

FIGURE 8–55

Solution   As indicated in Figure 8–55, the period is

$$T = 10 \; \mu s$$

Use Equations (8–1) and (8–21) to determine the frequency and duty cycle.

$$f = \frac{1}{T} = \frac{1}{10 \; \mu s} = 0.1 \text{ MHz}$$

$$\text{Percent duty cycle} = \left(\frac{t_W}{T}\right)100\% = \left(\frac{1 \; \mu s}{10 \; \mu s}\right)100\% = 10\%$$

Related Exercise   A certain pulse waveform has a frequency of 200 kHz and a pulse width of 0.25 μs. Determine the duty cycle.

Square Waves

A square wave is a pulse waveform with a duty cycle of 50%. Thus, the pulse width is equal to one-half of the period. A square wave is shown in Figure 8–56.

FIGURE 8–56
A square wave has a duty cycle of 50%.

The Average Value of a Pulse Waveform

An alternating square wave that has equal positive and negative amplitudes has a zero average value. If its positive amplitude is greater than its negative amplitude, it has a positive nonzero average value. If its negative amplitude is greater, it has a negative nonzero average value. For periodic pulse waveforms that are not square waves, the duty cycle also affects the average value.

> **The average value (V_{avg}) of a pulse waveform is equal to its baseline value plus the product of its duty cycle and its amplitude.**

The lower level of the waveform is taken as the baseline. The formula is as follows:

$$V_{avg} = \text{baseline} + (\text{duty cycle})(\text{amplitude}) \qquad (8\text{–}22)$$

The following example illustrates the calculation of the average value of pulse waveforms.

EXAMPLE 8–16   Determine the average value of each of the waveforms in Figure 8–57.

FIGURE 8–57

Solution   In Figure 8–57(a), the baseline is at 0 V, the amplitude is 2 V, and the duty cycle is 10%. The average value is

$$V_{avg} = \text{baseline} + (\text{duty cycle})(\text{amplitude})$$
$$= 0 \text{ V} + (0.1)(2 \text{ V}) = 0.2 \text{ V}$$

The waveform in Figure 8–57(b) has a baseline of +1V, an amplitude of 5 V, and a duty cycle of 50%. The average value is

$$V_{avg} = \text{baseline} + (\text{duty cycle})(\text{amplitude})$$
$$= 1\text{ V} + (0.5)(5\text{ V}) = 1\text{ V} + 2.5\text{ V} = 3.5\text{ V}$$

The waveform in Figure 8–57(c) is a square wave with a baseline of −1 V and an amplitude of 2 V. The average value is

$$V_{avg} = \text{baseline} + (\text{duty cycle})(\text{amplitude})$$
$$= -1\text{ V} + (0.5)(2\text{ V}) = -1\text{ V} + 1\text{ V} = 0\text{ V}$$

This is an alternating square wave, and, like an alternating sine wave, it has an average value of zero over a full cycle.

Related Exercise   If the baseline of the waveform in Figure 8–57(a) is shifted to +1 V, what is the average value?

Triangular and Sawtooth Waveforms

Triangular and sawtooth waveforms are formed by voltage or current ramps. A **ramp** is a linear increase or decrease in the voltage or current. Figure 8–58 shows both positive- and negative-going ramps. In part (a) the ramp has a positive slope; in part (b), the ramp has a negative slope. The slope of a voltage ramp is $\pm V/t$ and is expressed in units of V/s. The slope of a current ramp is $\pm I/t$ and is expressed in units of A/s.

(a) Positive ramp

(b) Negative ramp

FIGURE 8–58
Ramps.

EXAMPLE 8–17   What are the slopes of the voltage ramps in Figure 8–59?

FIGURE 8–59

(a)

(b)

Solution In Figure 8–59(a), the voltage increases from 0 V to +10 V in 5 ms. Thus, $V = 10$ V and $t = 5$ ms. The slope is

$$\frac{V}{t} = \frac{10\ V}{5\ ms} = 2\ V/ms = 2\ kV/s$$

In Figure 8–59(b), the voltage decreases from +5 V to 0 V in 100 ms. Thus, $V = -5$ V and $t = 100$ ms. The slope is

$$\frac{V}{t} = \frac{-5\ V}{100\ ms} = -0.05\ V/ms = -50\ V/s$$

Related Exercise A certain voltage ramp has a slope of +12 V/μs. If the ramp starts at zero, what is the voltage at 0.01 ms?

Triangular Waveforms

Figure 8–60 shows that a **triangular waveform** is composed of positive- and negative-going ramps having equal slopes. The period of this waveform is measured from one peak to the next corresponding peak, as illustrated. This particular triangular waveform is alternating and has an average value of zero.

Figure 8–61 depicts a triangular waveform with a nonzero average value. The frequency for triangular waves is determined in the same way as for sine waves, that is, $f = 1/T$.

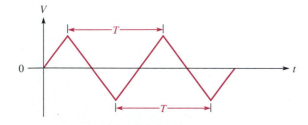

FIGURE 8–60
Alternating triangular waveform.

FIGURE 8–61
Nonalternating triangular waveform.

Sawtooth Waveforms

The **sawtooth wave** is actually a special case of the triangular wave consisting of two ramps, one of much longer duration than the other. Sawtooth waveforms are commonly used in many electronic systems, as you saw in the case of the oscilloscope. Another example is the electron beam that sweeps across the screen of your TV receiver, creating the picture; it also is controlled by sawtooth voltages and currents. One sawtooth wave produces the horizontal beam movement, and the other produces the vertical beam movement. The sawtooth is sometimes called a *sweep*.

Figure 8–62 is an example of a sawtooth wave. Notice that it consists of a positive-going ramp of relatively long duration, followed by a negative-going ramp of relatively short duration.

FIGURE 8–62
Alternating sawtooth waveform.

Harmonics

A repetitive nonsinusoidal waveform is composed of a fundamental frequency and harmonic frequencies. The **fundamental frequency** is the repetition rate of the waveform, and the **harmonics** are higher-frequency sine waves that are multiples of the fundamental.

Odd Harmonics *Odd harmonics* are frequencies that are odd multiples of the fundamental frequency of a waveform. For example, a 1 kHz square wave consists of a fundamental of 1 kHz and odd harmonics of 3 kHz, 5 kHz, 7 kHz, and so on. The 3 kHz frequency in this case is the third harmonic; the 5 kHz frequency is the fifth harmonic; and so on.

Even Harmonics *Even harmonics* are frequencies that are even multiples of the fundamental frequency. For example, if a certain wave has a fundamental of 200 Hz, the second harmonic is 400 Hz, the fourth harmonic is 800 Hz, the sixth harmonic is 1200 Hz, and so on. These are even harmonics.

Composite Waveform Any variation from a pure sine wave produces harmonics. A nonsinusoidal wave is a composite of the fundamental and the harmonics. Some types of waveforms have only odd harmonics, some have only even harmonics, and some contain both. The shape of the wave is determined by its harmonic content. Generally, only the fundamental and the first few harmonics are of significant importance in determining the waveshape.

A square wave is an example of a waveform that consists of a fundamental and only odd harmonics. When the instantaneous values of the fundamental and each odd harmonic are added algebraically at each point, the resulting curve will have the shape of a square wave, as illustrated in Figure 8–63. In part (a), the fundamental and the third harmonic produce a waveshape that begins to resemble a square wave. In part (b), the fundamental, third, and fifth harmonics produce a closer resemblance. When the seventh harmonic is included, as in part (c), the resulting waveshape becomes even more like a square wave. As more harmonics are included, a square wave is approached.

FIGURE 8–63
Odd harmonics combine to produce a square wave.

SECTION 8–9 REVIEW

1. Define the following parameters:
   (a) rise time   (b) fall time   (c) pulse width
2. In a certain repetitive pulse waveform, the pulses are 200 μs wide and occur once every millisecond. What is the frequency of this waveform?
3. Determine the duty cycle, amplitude, and average value of the waveform in Question 2 if the pulses go from −2 V to +6 V.
4. What is the second harmonic of a fundamental frequency of 1 kHz?

8–10 ▪ APPLICATION ASSIGNMENT

You have been assigned to check out a new function generator that produces sine wave, triangular wave, and pulse outputs. For each type of output, you will use the oscilloscope to measure the minimum and maximum frequencies, the minimum and maximum amplitudes, the maximum positive and negative dc offsets, and the minimum and maximum duty cycles for the pulse waveform. You will record your measurements in a logical format.

Step 1: Familiarization with the Function Generator

The function generator is shown in Figure 8–64. Each control is marked by a circled number and a description follows.

FIGURE 8–64
Function generator.

1. *Power on/off switch*   Press this push-button switch to turn the power on. Press again to release the switch and turn the power off.
2. *Function switches*   Press one of the switches to select either a sinusoidal, triangular, or pulse output. Only one switch can be engaged at a time.
3. *Frequency range switches*   These push-button switches are used in conjunction with the frequency adjustment control (4). Press to select the appropriate range in decade increments from 1 Hz to 1 MHz.
4. *Frequency adjustment control*   Turn this dial to set a specific frequency in the range selected by the push button switch (3). For example, if you press the 100 Hz frequency range switch and set the dial to 1.4, the output should have a frequency of 140 Hz (1.4 × 100).
5. *Amplitude control*   This knob adjusts the voltage amplitude of the output. Fully counterclockwise is the minimum and fully clockwise is the maximum. It can be adjusted for any value of amplitude between minimum and maximum.

6. *DC offset control*   This knob adjusts the dc level of the ac output. You can add a positive or a negative dc level to a waveform.

7. *Duty control*   This knob adjusts the duty cycle of the pulse waveform output. The sine wave or triangular wave outputs are not affected by this control.

Step 2: Review the Operation and Controls of the Oscilloscope

Review Section 8–8 on the oscilloscope with particular emphasis on the controls for the scope in Figure 8–48. Note that when the channel 1 coupling switch is set to GND, the screen displays a horizontal trace at the 0 V level. This trace can be set on any of the vertical divisions as a zero-volt reference using the vertical position control. On each screen in the following figures, the zero-volt reference is indicated by "0 V".

Step 3: Measure the Sinusoidal Output

The output of the function generator is connected to the channel 1 (CH1) input of the scope and the sinusoidal output function is selected. The coupling switch on the scope is set to DC.

In Figure 8–65(a), the amplitude and frequency of the function generator are set to the minimum values. Measure and record these values. Express amplitude in both peak and rms. How does the measured frequency compare to the frequency setting on the function generator?

In Figure 8–65(b), the amplitude and frequency are set to the maximum values. Measure and record these values. Express amplitude in both peak and rms. How does the measured frequency compare to the frequency setting on the function generator?

(a)   (b)

FIGURE 8–65

Step 4: Measure the DC Offset

The sinusoidal amplitude and frequency of the function generator is set to an arbitrary value for dc offset measurement. The coupling switch on the scope is set to DC.

In Figure 8–66(a), the dc offset of the function generator is adjusted to the maximum positive value. Measure and record this value.

In Figure 8–66(b), the dc offset is adjusted to the maximum negative value. Measure and record this value.

Step 5: Measure the Triangular Output

The triangular output of the function generator is selected. The coupling switch on the scope is set to AC.

FIGURE 8–66

FIGURE 8–67

FIGURE 8–68

In Figure 8–67(a), the amplitude and frequency of the function generator are set to the minimum values. Measure and record these values.

In Figure 8–67(b), the amplitude and frequency are set to the maximum values. Measure and record these values.

Step 6: Measure the Pulse Output

The pulse output of the function generator is selected. The coupling switch on the scope is set to DC.

In Figure 8–68(a), the amplitude from the base line and frequency of the function generator are set to the minimum values. The duty cycle is adjusted to minimum. Measure and record these values. Express duty cycle as a percentage.

In Figure 8–68(b), the amplitude and frequency are set to the maximum values. The duty cycle is adjusted to maximum. Measure and record these values. Express duty cycle as a percentage.

SECTION 8–10 REVIEW

1. Explain the purpose of each position of the AC-GND-DC switch on each channel of the scope.
2. Generally, what setting of the SEC/DIV control on the scope should be used to obtain the most accurate measurement of frequency?
3. Generally, what setting of the VOLTS/DIV control on the scope should be used to obtain the most accurate measurement of amplitude?

■ SUMMARY

- The sine wave is a time-varying, periodic waveform.
- The sine wave is a common form of ac (alternating current).
- Alternating current changes direction in response to changes in the polarity of the source voltage.
- One cycle of an alternating sine wave consists of a positive alternation and a negative alternation.
- Two common sources of sine waves are the electromagnetic ac generator and the electronic oscillator circuit.
- A full cycle of a sine wave is 360°, or 2π radians. A half-cycle is 180°, or π radian. A quarter-cycle is 90°, or $\pi/2$ radians.
- A sine wave voltage can be generated by a conductor rotating in a magnetic field.
- Phase angle is the difference in degrees (or radians) between two sine waves.
- The angular position of a phasor represents the angle of a sine wave, and the length of the phasor represents the amplitude.
- When applying Ohm's or Kirchhoff's laws in ac circuits, voltages and currents must all be expressed with consistent values.
- Average power in a resistive ac circuit is determined using rms voltage and/or current values.
- A pulse consists of a transition from a baseline level to an amplitude level, followed by a transition back to the baseline level.
- A triangular or sawtooth wave consists of positive-going and negative-going ramps.
- Harmonic frequencies are odd or even multiples of the repetition rate of a nonsinusoidal waveform.
- Conversions of sine wave values are summarized in Table 8–2.

TABLE 8–2

TO CHANGE FROM	TO	MULTIPLY BY
Peak	rms	0.707
Peak	Peak-to-peak	2
Peak	Average	0.637
rms	Peak	1.414
Peak-to-peak	Peak	0.5
Average	Peak	1.57

■ GLOSSARY

Alternating current (ac)   Current that reverses direction in response to a change in source voltage polarity.

Amplitude   The maximum value of a voltage or current.

Average value   The average of a sine wave over one half-cycle. It is 0.637 times the peak value.

Baseline   The normal level of a pulse waveform; the voltage level in the absence of a pulse.

Cathode-ray tube (CRT)   A vacuum tube device containing an electron gun that emits a narrow focused beam of electrons onto a phosphor-coated screen.

Cycle   One repetition of a periodic waveform.

Degree   The unit of angular measure corresponding to 1/360 of a complete revolution.

Duty cycle   A characteristic of a pulse waveform that indicates the percentage of time that a pulse is present during a cycle; the ratio of pulse width to period.

Effective value   A measure of the heating effect of a sine wave; also known as the *rms* (root mean square) *value*.

Falling edge   The negative-going transition of a pulse.

Fall time (t_f)   The time interval required for a pulse to change from 90% to 10% of its full amplitude.

Frequency   A measure of the rate of change of a periodic function; the number of cycles completed in 1 s. The unit of frequency is the hertz.

Fundamental frequency   The repetition rate of a waveform.

Generator   An energy source that produces electrical signals.

Harmonics   The frequencies contained in a composite waveform, which are integer multiples of the repetition frequency (fundamental).

Hertz (Hz)   The unit of frequency. One hertz equals one cycle per second.

Hypotenuse   The longest side of a right triangle.

Instantaneous value   The voltage or current value of a waveform at a given instant in time.

Lag   To be behind; describes a condition of the phase or time relationship of waveforms in which one waveform is behind the other in phase or time.

Lead   To be ahead; describes a condition of the phase or time relationship of waveforms in which one waveform is ahead of the other in phase or time; also, a wire or cable connection to a device or instrument.

Leading edge   The first step or transition of a pulse.

Magnitude   The value of a quantity, such as the number of volts of voltage or the number of amperes of current.

Oscillator   An electronic circuit that produces a time-varying signal without an external input signal using positive feedback.

Oscilloscope   A measurement instrument that displays signal waveforms on a screen.

Peak value   The voltage or current value of a waveform at its maximum positive or negative points.

Peak-to-peak value   The voltage or current value of a waveform measured from its minimum to its maximum points.

Period   The time interval of one complete cycle of a given sine wave or any periodic waveform.

Periodic   Characterized by a repetition at fixed time intervals.

Phase   The relative displacement of a time-varying waveform in terms of its occurrence with respect to a reference.

Phasor   A representation of a sine wave in terms of both magnitude and phase angle.

Pulse   A type of waveform that consists of two equal and opposite steps in voltage or current separated by a time interval.

Pulse width   The time interval between the opposite steps of an ideal pulse. For a nonideal pulse, the time between the 50% points on the leading and trailing edges.

Radian   A unit of angular measurement. There are 2π rads in one complete revolution. One radian equals 57.3°.

Ramp   A type of waveform characterized by a linear increase or decrease in voltage or current.

Right triangle   A triangle with one 90° angle.

Rise time (t_r)   The time interval required for a pulse to change from 10% to 90% of its amplitude.

Rising edge   The positive-going transition of a pulse.

Root mean square (rms)   The value of a sine wave that indicates its heating effect; also known as the *effective value*. It is equal to 0.707 times the peak value.

Sawtooth wave   A type of electrical waveform composed of ramps; a special case of a triangular wave.

Trailing edge   The second step or transition of a pulse.

Triangular wave   A type of electrical waveform that consists of two ramps.

Trigger   The activating mechanism of some electronic devices or instruments.

Waveform   The pattern of variations of a voltage or current showing how the quantity changes with time.

▪ FORMULAS

(8–1)	$f = \dfrac{1}{T}$	Frequency
(8–2)	$T = \dfrac{1}{f}$	Period
(8–3)	$f = \text{number of pole pairs} \times (\text{rps})$	Output frequency of a generator
(8–4)	$V_{pp} = 2V_p$	Peak-to-peak voltage (sine wave)
(8–5)	$I_{pp} = 2I_p$	Peak to peak current (sine wave)
(8–6)	$V_{\text{rms}} = 0.707V_p$	Root-mean-square voltage (sine wave)
(8–7)	$I_{\text{rms}} = 0.707I_p$	Root-mean-square current (sine wave)
(8–8)	$V_p = 1.414V_{\text{rms}}$	Peak voltage (sine wave)
(8–9)	$I_p = 1.414I_{\text{rms}}$	Peak current (sine wave)
(8–10)	$V_{pp} = 2.828V_{\text{rms}}$	Peak-to-peak voltage (sine wave)
(8–11)	$I_{pp} = 2.828I_{\text{rms}}$	Peak-to-peak current (sine wave)
(8–12)	$V_{\text{avg}} = 0.637V_p$	Half-cycle average voltage (sine wave)
(8–13)	$I_{\text{avg}} = 0.637I_p$	Half-cycle average current (sine wave)
(8–14)	$\text{rad} = \left(\dfrac{\pi \text{ rad}}{180°}\right) \times \text{degrees}$	
(8–15)	$\text{degrees} = \left(\dfrac{180°}{\pi \text{ rad}}\right) \times \text{rad}$	
(8–16)	$y = A \sin \theta$	General formula for a sine wave
(8–17)	$v = V_p\sin \theta$	Sine wave voltage
(8–18)	$i = I_p\sin \theta$	Sine wave current
(8–19)	$y = A \sin(\theta - \phi)$	Lagging sine wave
(8–20)	$y = A \sin(\theta + \phi)$	Leading sine wave
(8–21)	$\text{Percent duty cycle} = \left(\dfrac{t_W}{T}\right)100\%$	
(8–22)	$V_{\text{avg}} = \text{baseline} + (\text{duty cycle})(\text{amplitude})$	Average value of pulse waveform

---

▪ SELF-TEST

1. The difference between alternating current (ac) and direct current (dc) is
   (a) ac changes value and dc does not   (b) ac changes direction and dc does not
   (c) both (a) and (b)   (d) neither (a) nor (b)

2. During each cycle, a sine wave reaches a peak value
   (a) one time   (b) two times
   (c) four times   (d) a number of times depending on the frequency

3. A sine wave with a frequency of 12 kHz is changing at a faster rate than a sine wave with a frequency of
   (a) 20 kHz   (b) 15,000 Hz   (c) 10,000 Hz   (d) 1.25 MHz

4. A sine wave with a period of 2 ms is changing at a faster rate than a sine wave with a period of
   (a) 1 ms   (b) 0.0025 s   (c) 1.5 ms   (d) 1000 μs

5. When a sine wave has a frequency of 60 Hz, in 10 s it goes through
   (a) 6 cycles   (b) 10 cycles   (c) 1/16 cycle   (d) 600 cycles

6. If the peak value of a sine wave is 10 V, the peak-to-peak value is
   (a) 20 V   (b) 5 V   (c) 100 V   (d) none of these

7. If the peak value of a sine wave is 20 V, the rms value is
   (a) 14.14 V   (b) 6.37 V   (c) 7.07 V   (d) 0.707 V

8. The average value of a 10 V peak sine wave over one complete cycle is
   (a) 0 V   (b) 6.37 V   (c) 7.07 V   (d) 5 V

9. The average half-cycle value of a sine wave with a 20 V peak is
   (a) 0 V   (b) 6.37 V   (c) 12.74 V   (d) 14.14 V

10. One sine wave has a positive-going zero crossing at 10° and another sine wave has a positive-going zero crossing at 45°. The phase angle between the two waveforms is
    (a) 55°   (b) 35°   (c) 0°   (d) none of these

11. The instantaneous value of a 15 A peak sine wave at a point 32° from its positive-going zero crossing is
    (a) 7.95 A   (b) 7.5 A   (c) 2.13 A   (d) 7.95 V

12. If the rms current through a 10 kΩ resistor is 5 mA, the rms voltage drop across the resistor is
    (a) 70.7 V   (b) 7.07 V   (c) 5 V   (d) 50 V

13. Two series resistors are connected to an ac source. If there is 6.5 V rms across one resistor and 3.2 V rms across the other, the peak source voltage is
    (a) 9.7 V   (b) 9.19 V   (c) 13.72 V   (d) 4.53 V

14. A 10 kHz pulse waveform consists of pulses that are 10 μs wide. Its duty cycle is
    (a) 100%   (b) 10%   (c) 1%   (d) not determinable

15. The duty cycle of a square wave
    (a) varies with the frequency   (b) varies with the pulse width
    (c) both (a) and (b)   (d) is 50%

▪ BASIC PROBLEMS

SECTION 8–1   The Sine Wave

1. Calculate the frequency for each of the following periods:
   (a) 1 s   (b) 0.2 s   (c) 50 ms   (d) 1 ms   (e) 500 μs   (f) 10 μs

2. Calculate the period for each of the following frequencies:
   (a) 1 Hz   (b) 60 Hz   (c) 500 Hz   (d) 1 kHz   (e) 200 kHz   (f) 5 MHz

3. A sine wave goes through 5 cycles in 10 μs. What is its period?

4. A sine wave has a frequency of 50 kHz. How many cycles does it complete in 10 ms?

SECTION 8–2   Sine Wave Voltage Sources

5. The conductive loop on the rotor of a simple two-pole, single-phase generator rotates at a rate of 250 rps. What is the frequency of the induced output voltage?

6. A certain four-pole generator has a speed of rotation of 3600 rpm. What is the frequency of the voltage produced by this generator?

7. At what speed of rotation must a four-pole generator be operated to produce a 400 Hz sine wave voltage?

SECTION 8–3   Voltage and Current Values of Sine Waves

8. A sine wave has a peak value of 12 V. Determine the following voltage values:
   (a) rms   (b) peak-to-peak   (c) half-cycle average

9. A sinusoidal current has an rms value of 5 mA. Determine the following current values:
   (a) peak   (b) half-cycle average   (c) peak-to-peak

10. For the sine wave in Figure 8–69, determine the peak, peak-to-peak, rms, and average values.

FIGURE 8–69

SECTION 8–4   Angular Measurement of a Sine Wave

11. Sine wave A has a positive-going zero crossing at 30°. Sine wave B has a positive-going zero crossing at 45°. Determine the phase angle between the two signals. Which signal leads?

12. One sine wave has a positive peak at 75°, and another has a positive peak at 100°. How much is each sine wave shifted in phase from the 0° reference? What is the phase angle between them?

13. Make a sketch of two sine waves as follows: Sine wave A is the reference, and sine wave B lags A by 90°. Both have equal amplitudes.

14. Convert the following angular values from degrees to radians:
(a) 30° (b) 45° (c) 78° (d) 135° (e) 200° (f) 300°

15. Convert the following angular values from radians to degrees:
(a) $\pi/8$ (b) $\pi/3$ (c) $\pi/2$ (d) $3\pi/5$ (e) $6\pi/5$ (f) 1.8π

SECTION 8–5   The Sine Wave Formula

16. A certain sine wave has a positive-going zero crossing at 0° and an rms value of 20 V. Calculate its instantaneous value at each of the following angles:
(a) 15° (b) 33° (c) 50° (d) 110° (e) 70° (f) 145° (g) 250° (h) 325°

17. For a particular 0° reference sinusoidal current, the peak value is 100 mA. Determine the instantaneous value at each of the following points:
(a) 35° (b) 95° (c) 190° (d) 215° (e) 275° (f) 360°

18. For a 0° reference sine wave with an rms value of 6.37 V, determine its instantaneous value at each of the following points:
(a) $\pi/8$ rad (b) $\pi/4$ rad (c) $\pi/2$ rad (d) $3\pi/4$ rad (e) π rad (f) $3\pi/2$ rad (g) 2π rad

19. Sine wave A lags sine wave B by 30°. Both have peak values of 15 V. Sine wave A is the reference with a positive-going crossing at 0°. Determine the instantaneous value of sine wave B at 30°, 45°, 90°, 180°, 200°, and 300°.

20. Repeat Problem 19 for the case when sine wave A leads sine wave B by 30°.

SECTION 8–6   Ohm's Law and Kirchhoff's Laws in AC Circuits

21. A sinusoidal voltage is applied to the resistive circuit in Figure 8–70. Determine the following:
(a) I_{rms} (b) I_{avg} (c) I_p (d) I_{pp} (e) i at the positive peak

22. Find the half-cycle average values of the voltages across R_1 and R_2 in Figure 8–71. All values shown are rms.

FIGURE 8–70   **FIGURE 8–71**

23. Determine the rms voltage across R_3 in Figure 8–72.

FIGURE 8–72

SECTION 8–7   Superimposed DC and AC Voltages

24. A sine wave with an rms value of 10.6 V is riding on a dc level of 24 V. What are the maximum and minimum values of the resulting waveform?

25. How much dc voltage must be added to a 3 V rms sine wave in order to make the resulting voltage nonalternating (no negative values)?

26. A 6 V peak sine wave is riding on a dc voltage of 8 V. If the dc voltage is lowered to 5 V, how far negative will the sine wave go?

SECTION 8–8   The Oscilloscope

27. Determine the peak value and the period of the sine wave displayed on the scope screen in Figure 8–73.

28. Determine the rms value and the frequency of the sine wave displayed on the scope screen in Figure 8–73.

29. Determine the rms value and the frequency of the sine wave displayed on the scope screen in Figure 8–74.

FIGURE 8–73

FIGURE 8–74

30. Find the amplitude, pulse width, and duty cycle for the pulse waveform displayed on the scope screen in Figure 8–75.

FIGURE 8–75

SECTION 8–9   Nonsinusoidal Waveforms

31. From the graph in Figure 8–76, determine the approximate values of t_r, t_f, t_W, and amplitude.

32. Determine the duty cycle for each pulse waveform in Figure 8–77.

33. Find the average value of each pulse waveform in Figure 8–77.

34. What is the frequency of each waveform in Figure 8–77?

35. What is the frequency of each sawtooth waveform in Figure 8–78?

FIGURE 8–76 V (V)

FIGURE 8–77 V (V)

4 µs, 250 kHz

4 ms, 250 Hz

(a)

(b)

FIGURE 8–78

20 µs, 50 kHz

(a)

100 ms, 10 Hz

(b)

36. A square wave has a period of 40 μs. List the first six odd harmonics.

37. What is the fundamental frequency of the square wave mentioned in Problem 36?

▪ **ADVANCED PROBLEMS**

38. A certain sine wave has a frequency of 2.2 kHz and an rms value of 25 V. Assuming a given cycle begins (zero crossing) at $t = 0$ s, what is the change in voltage from 0.12 ms to 0.2 ms?

39. Figure 8–79 shows a sinusoidal voltage source in series with a dc source. Effectively, the two voltages are superimposed. Sketch the voltage across R_L. Determine the maximum current through R_L and the average voltage across R_L.

40. A nonsinusoidal waveform called a *stairstep* is shown in Figure 8–80. Determine its average value.

FIGURE 8–79   FIGURE 8–80

41. Refer to the oscilloscope in Figure 8–81.
  (a) How many cycles are displayed?
  (b) What is the rms value of the sine wave?
  (c) What is the frequency of the sine wave?

FIGURE 8–81

42. Accurately sketch on a grid representing the scope screen in Figure 8–81 how the sine wave will appear if the volts/division control is moved to the next higher setting.

43. Accurately sketch on a grid representing the scope screen in Figure 8–81 how the sine wave will appear if the seconds/division control is moved to the next lower setting.

44. Based on the instrument settings and an examination of the scope display and the circuit board in Figure 8–82 determine the frequency and peak value of the input signal and output signal. The waveform shown is channel 1. Sketch the channel 2 waveform as it would appear on the scope with the indicated settings.

FIGURE 8–82

45. Examine the circuit board and the oscilloscope display in Figure 8–83 and determine the peak value and the frequency of the unknown input signal.

Unknown input signal

FIGURE 8–83

Section 8–1

1. One cycle goes from the zero crossing through a positive peak, then through zero to a negative peak and back to the zero crossing.

2. A sine wave changes polarity at the zero crossings.

3. A sine wave has two maximum points in one cycle.

4. The period is measured from one zero crossing to the next corresponding zero crossing, or from one peak to the next corresponding peak.

5. Frequency is the number of cycles completed in one second and its unit is the hertz.

6. $f = 1/5 \, \mu s = 200 \, \text{kHz}$ **7.** $T = 1/120 \, \text{Hz} = 8.33 \, \text{ms}$

Section 8–2

1. Two types of sine wave sources are electromagnetic and electronic.

2. Sine waves are generated electromagnetically by rotating a conductor in a magnetic field.

3. Frequency is dependent on the number of poles and the rps.

4. Magnitude is dependent on the number of turns and the rps.

Section 8–3

1. (a) $V_{pp} = 2(1 \, \text{V}) = 2 \, \text{V}$ (b) $V_{pp} = 2(1.414)(1.414 \, \text{V}) = 4 \, \text{V}$ (c) $V_{pp} = 2(1.57)(3 \, \text{V}) = 9.42 \, \text{V}$

2. (a) $V_{\text{rms}} = (0.707)(2.5 \, \text{V}) = 1.77 \, \text{V}$ (b) $V_{\text{rms}} = (0.5)(0.707)(10 \, \text{V}) = 3.54 \, \text{V}$
   (c) $V_{\text{rms}} = (0.707)(1.57)(1.5 \, \text{V}) = 1.66 \, \text{V}$

3. (a) $V_{\text{avg}} = (0.637)(10 \, \text{V}) = 6.37 \, \text{V}$ (b) $V_{\text{avg}} = (0.637)(1.414)(2.3 \, \text{V}) = 2.07 \, \text{V}$
   (c) $V_{\text{avg}} = (0.637)(0.5)(60 \, \text{V}) = 19.1 \, \text{V}$

Section 8–4

1. (a) The positive peak is at 90°.
   (b) The negative-going zero crossing is at 180°.
   (c) The negative peak is at 270°.
   (d) The cycle ends at 360°.
2. A half-cycle is completed in 180° or π radians.
3. A full cycle is completed in 360° or 2π radians.
4. $\theta = 90° - 45° = 45°$

Section 8–5

1. (a) $\sin 30° = 0.5$ (b) $\sin 60° = 0.866$ (c) $\sin 90° = 1$
2. $v = 10 \sin 120° = 8.66$ V   3. $v = 10 \sin(45° + 10°) = 8.19$ V

Section 8–6

1. $I_p = V_p/R = (1.57)(12.5 \text{ V})/330 \ \Omega = 59.5$ mA   2. $V_{s(rms)} = (0.707)(25.3 \text{ V}) = 17.9$ V

Section 8–7

1. $+V_{max} = 5 \text{ V} + 2.5 \text{ V} = 7.5$ V   2. Yes, it will alternate.   3. $+V_{max} = 5 \text{ V} - 2.5 \text{ V} = 2.5$ V

Section 8–8

1. CRT stands for cathode-ray tube.
2. Voltage is measured vertically; time is measured horizontally.
3. An oscilloscope displays the waveform of a voltage. A multimeter provides rms value.

Section 8–9

1. (a) Rise time is the interval from 10% to 90% of the rising edge.
   (b) Fall time is the interval from 90% to 10% of the falling edge.
   (c) Pulse width is the interval from 50% of the leading edge to 50% of the trailing edge.
2. $f = 1/1$ ms $= 1$ kHz
3. D.C. $= (200 \ \mu s/1 \text{ ms})100\% = 20\%$; Amplitude $= 6 \text{ V} - (-2 \text{ V}) = 8$ V;
   $V_{avg} = -2 \text{ V} + (0.2)(8 \text{ V}) = -0.4$ V
4. 2×1 kHz $= 2$ kHz

Section 8–10

1. AC: Couples ac voltage only; GND: Grounds input so trace is at 0 V; DC: Couples ac and dc voltages
2. Lowest SEC/DIV setting that allows frequency to be measured
3. Lowest VOLTS/DIV setting that allows voltage to be measured

---

■ **ANSWERS TO RELATED EXERCISES FOR EXAMPLES**

8–1	2.4 s		8–2	1.5 ms
8–3	20 kHz		8–4	200 Hz
8–5	1 ms		8–6	30 rps
8–7	$V_{pp} = 50$ V; $V_{rms} = 17.7$ V; $V_{avg} = 15.9$ V		8–8	(a) $\pi/12$ (b) 360°
8–9	8°		8–10	18.1 V
8–11	$I_{rms} = 4.53$ mA; $V_{1(rms)} = 4.53$ V; $V_{2(rms)} = 2.54$ V		8–12	23.7 V
8–13	The waveform in part (a) never goes negative. The waveform in part (b) goes negative for a portion of its cycle.		8–14	(a) $V_{rms} = 1.06$ V; $f = 50$ Hz (b) $V_{rms} = 88.4$ mV; $f = 1.67$ kHz (c) $V_{rms} = 4.81$ V; $f = 50$ kHz (d) $V_{rms} = 7.07$ V; $f = 250$ kHz
8–15	5%		8–16	1.2 V
8–17	120 V			

9

CAPACITORS

The capacitor is an electrical device that can store electrical charge, thereby creating an electric field which in turn stores energy; the measure of that energy-storing ability is called *capacitance*. In this chapter, the basic capacitor is introduced and its characteristics are studied. The physical construction and electrical properties of various types of capacitors are discussed.

The basic behavior of capacitors in both dc and ac circuits is studied, and series and parallel combinations are analyzed. Representative applications and methods of testing capacitors also are discussed.

■ CHAPTER OBJECTIVES

☐ Describe the basic structure and characteristics of a capacitor

☐ Discuss various types of capacitors

☐ Analyze series capacitors

☐ Analyze parallel capacitors

☐ Analyze capacitive dc switching circuits

☐ Analyze capacitive ac circuits

☐ Discuss some capacitor applications

☐ Test a capacitor

■ APPLICATION ASSIGNMENT

Capacitors are used in many applications. The application assignment in this chapter focuses on the use of capacitors in amplifier circuits to couple an ac voltage from one point to another while blocking dc voltage. You will determine if voltages measured on an amplifier circuit board with an oscilloscope are correct; and if they are not correct, you will specify the problem. After studying this chapter, you should be able to complete the application assignment in Section 9–9.

9–1 ■ THE BASIC CAPACITOR

In this section, the basic construction and characteristics of capacitors are examined.

After completing this section, you should be able to

■ **Describe the basic structure and characteristics of a capacitor**
  - ☐ Explain how a capacitor stores charge
  - ☐ Define *capacitance* and state its unit
  - ☐ State Coulomb's law
  - ☐ Explain how a capacitor stores energy
  - ☐ Discuss voltage rating and temperature coefficient
  - ☐ Explain capacitor leakage
  - ☐ Specify how the physical characteristics affect the capacitance

Basic Construction

In its simplest form, a **capacitor** is an electrical device constructed of two parallel conductive plates separated by an insulating material called the **dielectric.** Connecting leads are attached to the parallel plates. A basic capacitor is shown in Figure 9–1(a), and the schematic symbol is shown in part (b).

(a) Construction

(b) Symbol

FIGURE 9–1
The basic capacitor.

How a Capacitor Stores Charge

In the neutral state, both plates of a capacitor have an equal number of free electrons, as indicated in Figure 9–2(a). When the capacitor is connected to a voltage source through a resistor, as shown in part (b), electrons (negative charge) are removed from plate *A,* and an equal number are deposited on plate *B*. As plate *A* loses electrons and plate *B* gains electrons, plate *A* becomes positive with respect to plate *B*. During this charging process, electrons flow only through the connecting leads and the source. No electrons flow through the dielectric of the capacitor because it is an insulator. The movement of electrons ceases when the voltage across the capacitor equals the source voltage, as indicated in Figure 9–2(c). If the capacitor is disconnected from the source, it retains the stored charge for a long period of time (the length of time depends on the type of capacitor) and still has the voltage across it, as shown in Figure 9–2(d). Actually, the charged capacitor can be considered as a temporary battery.

Capacitance

The amount of charge per unit of voltage that a capacitor can store is its **capacitance,** designated *C*. That is, capacitance is a measure of a capacitor's ability to store charge.

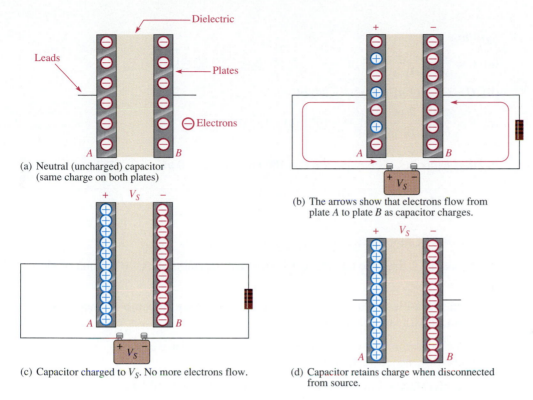

(a) Neutral (uncharged) capacitor (same charge on both plates)

(b) The arrows show that electrons flow from plate A to plate B as capacitor charges.

(c) Capacitor charged to V_S. No more electrons flow.

(d) Capacitor retains charge when disconnected from source.

FIGURE 9–2

Illustration of a capacitor storing charge.

The more charge per unit of voltage that a capacitor can store, the greater its capacitance, as expressed by the following formula:

$$C = \frac{Q}{V} \tag{9–1}$$

where C is capacitance, Q is charge, and V is voltage.

By rearranging Equation (9–1), you can obtain two other forms for calculating Q and V.

$$Q = CV \tag{9–2}$$

$$V = \frac{Q}{C} \tag{9–3}$$

EXAMPLE 9–1

(a) A certain capacitor stores 50 microcoulombs (50 μC) when 10 V are applied across its plates. What is its capacitance?

(b) A 2 microfarad (2 μF) capacitor has 100 V across its plates. How much charge does it store?

(c) Determine the voltage across a 100 picofarad (100 pF) capacitor that is storing 2 microcoulombs (2 μC) of charge.

Solution

(a) $C = \dfrac{Q}{V} = \dfrac{50\ \mu C}{10\ V} = 5\ \mu F$ [5] [0] [EXP] [+/−] [6] [÷] [1] [0] [=]

(b) $Q = CV = (2\ \mu F)(100\ V) = 200\ \mu C$

(c) $V = \dfrac{Q}{C} = \dfrac{2\ \mu C}{100\ pF} = 0.02 \times 10^6 = 20 \times 10^3\ V = 20\ kV$

Related Exercise   Determine V if $C = 1000\ pF$ and $Q = 10\ \mu C$.

The Unit of Capacitance

The **farad (F)** is the basic unit of capacitance, and the **coulomb (C)** is the unit of electrical charge.

> **One farad is the amount of capacitance when one coulomb of charge is stored with one volt across the plates.**

Most capacitors that you will use in electronics work have capacitance values in microfarads (μF) and picofarads (pF).

> **A microfarad is one-millionth of a farad ($1\ \mu F = 1 \times 10^{-6}\ F$).**

> **A picofarad is one-trillionth of a farad ($1\ pF = 1 \times 10^{-12}\ F$).**

Conversions for farads, microfarads, and picofarads are given in Table 9–1.

TABLE 9–1

Conversions for farads, microfarads, and picofarads

TO CONVERT FROM	TO	MOVE THE DECIMAL POINT
Farads	Microfarads	6 places to right ($\times 10^6$)
Farads	Picofarads	12 places to right ($\times 10^{12}$)
Microfarads	Farads	6 places to left ($\times 10^{-6}$)
Microfarads	Picofarads	6 places to right ($\times 10^6$)
Picofarads	Farads	12 places to left ($\times 10^{-12}$)
Picofarads	Microfarads	6 places to left ($\times 10^{-6}$)

EXAMPLE 9–2

Convert the following values to microfarads:
(a) 0.00001 F   **(b)** 0.005 F   **(c)** 1000 pF   **(d)** 200 pF

Solution

(a) $0.00001\ F \times 10^6 = 10\ \mu F$ **(b)** $0.005\ F \times 10^6 = 5000\ \mu F$

(c) $1000\ pF \times 10^{-6} = 0.001\ \mu F$ **(d)** $200\ pF \times 10^{-6} = 0.0002\ \mu F$

Related Exercise   Convert 50,000 pF to microfarads.

EXAMPLE 9–3

Convert the following values to picofarads:
(a) 0.1×10^{-8} F   (b) 0.000025 F   (c) 0.01 μF   (d) 0.005 μF

Solution
(a) 0.1×10^{-8} F $\times 10^{12} = 1000$ pF   (b) 0.000025 F $\times 10^{12} = 25 \times 10^{6}$ pF
(c) 0.01 μF $\times 10^{6} = 10{,}000$ pF   (d) 0.005 μF $\times 10^{6} = 5000$ pF

Related Exercise Convert 100 μF to picofarads.

How a Capacitor Stores Energy

A capacitor stores energy in the form of an electric field that is established by the opposite charges on the two plates. The electric field is represented by lines of force between the positive and negative charges and concentrated within the dielectric, as shown in Figure 9–3.

FIGURE 9–3
The electric field stores energy in a capacitor.

Lines of force

Coulomb's law states

A force exists between two charged bodies that is directly proportional to the product of the two charges and inversely proportional to the square of the distance between the bodies.

This relationship is expressed as

$$F = \frac{kQ_1 Q_2}{d^2} \qquad (9\text{–}4)$$

where F is the force in newtons, Q_1 and Q_2 are the charges in coulombs, d is the distance between the charges in meters, and k is a proportionality constant equal to 9×10^9.

Figure 9–4(a) illustrates the line of force between a positive and a negative charge. Figure 9–4(b) shows that many opposite charges on the plates of a capacitor create many

FIGURE 9–4
Lines of force are created by opposite charges.

Lines of force

(a)   (b)

lines of force, which form an electric field between the plates that stores energy within the dielectric.

The greater the forces between the charges on the plates of a capacitor, the more energy is stored. The amount of energy stored therefore is directly proportional to the capacitance because, from Coulomb's law, the more charge stored, the greater the force.

Also, from the equation $Q = CV$, the amount of charge stored is directly related to the voltage as well as the capacitance. The amount of energy stored is dependent on the square of the voltage across the plates of the capacitor. The formula for the energy stored by a capacitor is as follows:

$$W = \frac{1}{2}CV^2 \tag{9-5}$$

where the energy (W) is in joules when C is in farads and V is in volts.

Voltage Rating

Every capacitor has a limit on the amount of voltage that it can withstand across its plates. The voltage rating specifies the maximum dc voltage that can be applied without risk of damage to the device. If this maximum voltage, commonly called the *breakdown voltage* or *working voltage,* is exceeded, permanent damage to the capacitor can result.

Both the capacitance and the voltage rating must be taken into consideration before a capacitor is used in a circuit application. The choice of capacitance value is based on particular circuit requirements. The voltage rating should always be well above the maximum voltage expected in a particular application.

Dielectric Strength   The breakdown voltage of a capacitor is determined by the **dielectric strength** of the dielectric material used. The dielectric strength is expressed in volts/mil (1 mil = 0.001 in.). Table 9–2 lists typical values for several materials. Exact values vary depending on the specific composition of the material.

TABLE 9–2
Some common dielectric materials and their typical dielectric strengths

MATERIAL	DIELECTRIC STRENGTH (VOLTS/MIL)
Air	80
Oil	375
Ceramic	1000
Paper (paraffined)	1200
Teflon®	1500
Mica	1500
Glass	2000

The dielectric strength can best be explained by an example. Assume that a certain capacitor has a plate separation of 1 mil and that the dielectric material is ceramic. This particular capacitor can withstand a maximum voltage of 1000 V because its dielectric strength is 1000 V/mil. If the maximum voltage is exceeded, the dielectric may break down and conduct current, causing permanent damage to the capacitor. If the ceramic capacitor has a plate separation of 2 mils, its breakdown voltage is 2000 V.

Temperature Coefficient

The **temperature coefficient** indicates the amount and direction of a change in capacitance value with temperature. A positive temperature coefficient means that the capacitance increases with an increase in temperature or decreases with a decrease in temperature. A negative coefficient means that the capacitance decreases with an increase in temperature or increases with a decrease in temperature.

Temperature coefficients typically are specified in parts per million per degree Celsius (ppm/°C). For example, a negative temperature coefficient of 150 ppm/°C for a 1 μF capacitor means that for every degree rise in temperature, the capacitance decreases by 150 pF (there are one million picofarads in one microfarad).

Leakage

No insulating material is perfect. The dielectric of any capacitor will conduct some very small amount of current. Thus, the charge on a capacitor will eventually leak off. Some types of capacitors have higher leakages than others. An equivalent circuit for a nonideal capacitor is shown in Figure 9–5. The parallel resistor represents the extremely high resistance of the dielectric material through which leakage current flows.

FIGURE 9–5
Equivalent circuit for a nonideal capacitor.

Physical Characteristics of Capacitors

The following parameters are important in establishing the capacitance and the voltage rating of a capacitor: plate area, plate separation, and dielectric constant.

Plate Area *Capacitance is directly proportional to the physical size of the plates as determined by the plate area, A.* A larger plate area produces a larger capacitance, and vice versa. Figure 9–6(a) shows that the plate area of a parallel plate capacitor is the area of one of the plates. If the plates are moved in relation to each other, as shown in Figure 9–6(b), the overlapping area determines the effective plate area. This variation in effective plate area is the basis for a certain type of variable capacitor.

Plate Separation *Capacitance is inversely proportional to the distance between the plates.* The plate separation is designated *d*, as shown in Figure 9–7. A greater separation of the plates produces a smaller capacitance, as illustrated in the figure. The breakdown voltage is directly proportional to the plate separation. The further the plates are separated, the greater the breakdown voltage.

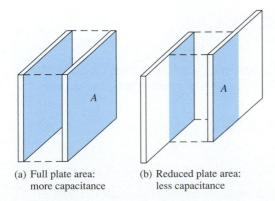

(a) Full plate area:
  more capacitance

(b) Reduced plate area:
  less capacitance

FIGURE 9–6
Capacitance is directly proportional to plate area (A).

(a) More capacitance   (b) Less capacitance

FIGURE 9–7
Capacitance is inversely proportional to the distance d between the plates.

Dielectric Constant   As you know, the insulating material between the plates of a capacitor is called the *dielectric*. Every dielectric material has the ability to concentrate the lines of force of the electric field existing between the oppositely charged plates of a capacitor and thus increase the capacity for energy storage. The measure of a material's ability to establish an electric field is called the **dielectric constant** or *relative permittivity,* symbolized by ϵ_r (the Greek letter epsilon).

Capacitance is directly proportional to the dielectric constant. The dielectric constant of a vacuum is defined as 1, and that of air is very close to 1. These values are used as a reference, and all other materials have values of ϵ_r specified with respect to that of a vacuum or air. For example, a material with $\epsilon_r = 8$ can have a capacitance eight times greater than that of air, with all other factors being equal.

Table 9–3 lists several common dielectric materials and typical dielectric constants for each. Values can vary because they depend on the specific composition of the material.

The dielectric constant (relative permittivity) is dimensionless, because it is a relative measure and is a ratio of the absolute permittivity, ϵ, of a material to the absolute permittivity, ϵ_0, of a vacuum, as expressed by the following formula:

$$\epsilon_r = \frac{\epsilon}{\epsilon_0}$$

(9–6)

The value of ϵ_0 is 8.85×10^{-12} F/m (farads per meter).

TABLE 9–3
Some common dielectric materials and their typical dielectric constants

MATERIAL	TYPICAL ϵ_r VALUES
Air (vacuum)	1.0
Teflon®	2.0
Paper (paraffined)	2.5
Oil	4.0
Mica	5.0
Glass	7.5
Ceramic	1200

Formula for Capacitance in Terms of Physical Parameters

You have seen how capacitance is directly related to plate area A and the dielectric constant ϵ_r, and inversely related to plate separation d. An exact formula for calculating the capacitance in terms of these three quantities is as follows:

$$C = \frac{A\epsilon_r(8.85 \times 10^{-12} \text{ F/m})}{d}$$

(9–7)

where A is in square meters (m²), d is in meters (m), and C is in farads (F). Recall that 8.85×10^{-12} F/m is the absolute permittivity, ϵ_0, of a vacuum. Therefore, the absolute permittivity of a dielectric is

$$\epsilon = \epsilon_r \times \epsilon_0 = \epsilon_r \times (8.85 \times 10^{-12} \text{ F/m})$$

EXAMPLE 9–4

Determine the capacitance of a parallel plate capacitor having a plate area of 0.01 m² and a plate separation of 0.02 m. The dielectric is mica, which has a dielectric constant of 5.0.

Solution Use Equation (9–7).

$$C = \frac{A\epsilon_r(8.85 \times 10^{-12} \text{ F/m})}{d} = \frac{(0.01 \text{ m}^2)(5.0)(8.85 \times 10^{-12} \text{ F/m})}{0.02 \text{ m}} = 22.1 \text{ pF}$$

The calculator sequence is

[.] [0] [1] [×] [5] [×] [8] [.] [8] [5] [EXP] [+/−] [1] [2] [÷] [.] [0] [2] [=]

Related Exercise Determine C where $A = 0.005$ m^2, $d = 0.008$ m, and ceramic is the dielectric.

SECTION 9–1 REVIEW

1. Define *capacitance*.
2. (a) How many microfarads in one farad?
   (b) How many picofarads in one farad?
   (c) How many picofarads in one microfarad?
3. Convert 0.0015 μF to picofarads. To farads.
4. How much energy in joules is stored by a 0.01 μF capacitor with 15 V across its plates?
5. (a) When the plate area of a capacitor is increased, does the capacitance increase or decrease?
   (b) When the distance between the plates is increased, does the capacitance increase or decrease?
6. The plates of a ceramic capacitor are separated by 10 mils. What is the typical breakdown voltage?
7. A ceramic capacitor has a plate area of 0.2 m^2. The thickness of the dielectric is 0.005 m. What is the capacitance?

9–2 ■ TYPES OF CAPACITORS

Capacitors normally are classified according to the type of dielectric material. The most common types of dielectric materials are mica, ceramic, plastic-film, and electrolytic (aluminum oxide and tantalum oxide). In this section, the characteristics and construction of each of these types of capacitors and the variable capacitors are examined.

After completing this section, you should be able to

■ **Discuss various types of capacitors**
   □ Describe the characteristics of mica, ceramic, plastic-film, and electrolytic capacitors
   □ Describe types of variable capacitors
   □ Identify capacitor labeling

Fixed Capacitors

Mica Capacitors Two types of mica capacitors are stacked-foil and silver-mica. The basic construction of the stacked-foil type is shown in Figure 9–8. It consists of alternate layers of metal foil and thin sheets of mica. The metal foil forms the plate, with alternate foil sheets connected together to increase the plate area. More layers are used to increase the plate area, thus increasing the capacitance. The mica/foil stack is encapsulated in an

(a) Stacked layer arrangement

(b) Layers are pressed together
and encapsulated

FIGURE 9–8

Construction of a typical radial-lead mica capacitor.

insulating material such as Bakelite®, as shown in Figure 9–8(b). The silver-mica capacitor is formed in a similar way by stacking mica sheets with silver electrode material screened on them.

Mica capacitors are available with capacitance values ranging from 1 pF to 0.1 μF and voltage ratings from 100 to 2500 V dc and higher. Temperature coefficients from -20 ppm/°C to $+100$ ppm/°C are common. Mica has a typical dielectric constant of 5.

Ceramic Capacitors   Ceramic dielectrics provide very high dielectric constants (1200 is typical). As a result, comparatively high capacitance values can be achieved in a small physical size. Ceramic capacitors are commonly available in a ceramic disk form, as shown in Figure 9–9; in a multilayer radial-lead configuration, as shown in Figure 9–10; or in a leadless ceramic chip, as shown in Figure 9–11, for surface mounting on PC boards.

Ceramic capacitors typically are available in capacitance values ranging from 1 pF to 2.2 μF with voltage ratings up to 6 kV. A typical temperature coefficient for ceramic capacitors is 200,000 ppm/°C.

(a)

Lead wire soldered
to silver electrode

Solder

Ceramic
dielectric

Dipped phenolic coating

Silver electrodes deposited on
top and bottom of ceramic disc

(b)

FIGURE 9–9

A ceramic disk capacitor and its basic construction.

(a)   (b)

FIGURE 9–10
Ceramic capacitors. (a) Typical capacitors with radial leads. (b) Construction view (courtesy of Union Carbide Electronics Corporation).

FIGURE 9–11
Construction view of a typical ceramic chip capacitor used for surface mounting on PC boards.

Plastic-Film Capacitors There are several types of plastic-film capacitors. Polycarbonate, propylene, polyester, polystyrene, polypropylene, and polycarbonate are some of the more common dielectric materials used. Some of these types have capacitance values up to 100 μF.

Figure 9–12 shows a common basic construction used in many plastic-film capacitors. A thin strip of plastic-film dielectric is sandwiched between two thin metal strips

FIGURE 9–12
Basic construction of axial-lead tubular plastic-film dielectric capacitors.

that act as plates. One lead is connected to the inner plate and one to the outer plate as indicated. The strips are then rolled in a spiral configuration and encapsulated in a molded case. Thus, a large plate area can be packaged in a relatively small physical size, thereby achieving large capacitance values. Another method uses metal deposited directly on the film dielectric to form the plates. Figure 9–13 shows a construction view for one type of plastic-film capacitor.

High-purity foil electrodes

Plastic film dielectric

Outer wrap of polyester film

Capacitor section (alternate strips of film dielectric and foil electrodes rolled into cylindrical shape)

Plastic resin end fill bonds with outer wrap and tinned lead

Lead wire soldered to end of section

Solder-coated end assures that all turns of electrode are positively contacted

(a)   (b)

FIGURE 9–13
Plastic-film capacitors. (a) Typical example (courtesy of Siemens Corp.). (b) Construction view of an axial-lead plastic-film capacitor (courtesy of Union Carbide Electronics Corporation).

Electrolytic Capacitors Electrolytic capacitors are polarized so that one plate is positive and the other negative. These capacitors are used for high capacitance values up to over 200,000 μF, but they have relatively low breakdown voltages (350 V is a typical maximum) and high amounts of leakage.

Electrolytic capacitors offer much higher capacitance values than mica or ceramic capacitors, but their voltage ratings are typically lower. Aluminum electrolytics are probably the most commonly used type. While other capacitors use two similar plates, the electrolytic consists of one plate of aluminum foil and another plate made of a conducting electrolyte applied to a material such as plastic film. These two "plates" are separated by a layer of aluminum oxide which forms on the surface of the aluminum plate. Figure 9–14(a) illustrates the basic construction of a typical aluminum electrolytic capacitor with axial leads. Other electrolytics with radial leads are shown in Figure 9–14(b); the symbol for an electrolytic capacitor is shown in part (c).

Tantalum electrolytics can be in either a tubular configuration similar to Figure 9–14 or "tear drop" shape as shown in Figure 9–15 (page 352). In the tear drop configuration, the positive plate is actually a pellet of tantalum powder rather than a sheet of foil. Tantalum oxide forms the dielectric and manganese dioxide forms the negative plate.

Because of the process used for the insulating oxide dielectric, the metallic (aluminum or tantalum) plate is always positive with respect to the electrolyte plate, and,

Solder-coated brass case

Sprayed metal end termination

Plastic insulating cap

Glass-to-metal seal

Wire lead terminal

Aluminum plate

Oxide dielectric

Film

Electrolyte plate

(a) Construction view of an axial-lead electrolytic capacitor

(b) Typical radial-lead electrolytics

(c) Symbol for an electrolytic capacitor

FIGURE 9–14
Electrolytic capacitors.

thus all electrolytic capacitors are polarized. The metal plate (positive lead) is usually indicated by a plus sign or some other obvious marking and must always be connected in a dc circuit where the voltage across the capacitor does not change polarity. Reversal of the polarity of the voltage will usually result in complete destruction of the capacitor.

FIGURE 9–15
Construction view of a typical "tear drop" shaped tantalum electrolytic capacitor.

Variable Capacitors

Variable capacitors are used in a circuit when there is a need to adjust the capacitance value either manually or automatically, for example, in radio or TV tuners. The major types of variable or adjustable capacitors are now introduced. The schematic symbol for a variable capacitor is shown in Figure 9–16.

FIGURE 9–16
Schematic symbol for a variable capacitor.

FIGURE 9–17
A typical variable air capacitor.

Air Capacitor   Variable capacitors with air dielectrics, such as the one shown in Figure 9–17, are sometimes used as tuning capacitors in applications requiring frequency selection. This type of capacitor is constructed of several plates that mesh together. One set of plates can be moved relative to the other, thus changing the effective plate area and the capacitance. The movable plates are linked together mechanically so that they all move when a shaft is rotated.

Trimmers   Adjustable capacitors that normally have screwdriver adjustments and are used for very fine adjustments in a circuit are called **trimmers.** Ceramic or mica is a common dielectric in these types of capacitors, and the capacitance usually is changed by adjusting the plate separation. Figure 9–18 shows some typical devices.

Varactors   The varactor is a semiconductor device that exhibits a capacitance characteristic that is varied by changing the voltage across its terminals. This device usually is covered in detail in a course on electronic devices.

FIGURE 9–18
Trimmer capacitors.

Capacitor Labeling

Capacitor values are indicated on the body of the capacitor either by typographical labels or by color codes. Typographical labels consist of letters and numbers that indicate various parameters such as capacitance, voltage rating, and tolerance.

Some capacitors carry no unit designation for capacitance. In these cases, the units are implied by the value indicated and are recognized by experience. For example, a ceramic capacitor marked .001 or .01 has units of microfarads because picofarad values that small are not available. As another example, a ceramic capacitor labeled 50 or 330 has units of picofarads because microfarad units that large normally are not available in this type. In some cases a 3-digit designation is used. The first two digits are the first two digits of the capacitance value. The third digit is the number of zeros after the second digit. For example, 103 means 10,000 pF. In some instances, the units are labeled as pF or μF; sometimes the microfarad unit is labeled as MF or MFD.

Voltage rating appears on some types of capacitors with WV or WVDC and is omitted on others. When it is omitted, the voltage rating can be determined from information supplied by the manufacturer. The tolerance of the capacitor is usually labeled as a percentage, such as ±10%. The temperature coefficient is indicated by a *parts per million* marking. This type of label consists of a P or N followed by a number. For example, N750 means a negative temperature coefficient of 750 ppm/°C, and P30 means a positive temperature coefficient of 30 ppm/°C. An NP0 designation means that the positive and negative coefficients are zero; thus the capacitance does not change with temperature. Certain types of capacitors are color coded. Refer to Appendix D, for color code information.

SECTION 9–2 REVIEW	**1.** How are capacitors commonly classified? **2.** What is the difference between a fixed and a variable capacitor? **3.** What type of capacitor normally is polarized? **4.** What precautions must be taken when a polarized capacitor is installed in a circuit?

9–3 ■ SERIES CAPACITORS

In this section, you will see why the total capacitance of a series connection of capacitors is less than any of the individual capacitances.

After completing this section, you should be able to

■ **Analyze series capacitors**
   □ Determine total capacitance
   □ Determine capacitor voltages

When capacitors are connected in series, the total capacitance is less than the smallest capacitance value because the effective plate separation increases. The calculation of total series capacitance is analogous to the calculation of total resistance of parallel resistors.

To start, we will use two capacitors in series to show how the total capacitance is determined. Figure 9–19 shows two capacitors, which initially are uncharged, connected in series with a dc voltage source. When the switch is closed, as shown in part (a), current begins to flow.

(a) While charging, $I = Q/t$ is the same at all points.

(b) Both capacitors store the same amount of charge ($Q_T = Q_1 = Q_2$).

FIGURE 9–19

Capacitors in series produce a total capacitance that is less than the smallest value.

Recall that current is the same at all points in a series circuit and that current is defined as the rate of flow of charge ($I = Q/t$). In a certain period of time, a certain amount of charge moves through the circuit. Since current is the same everywhere in the circuit of Figure 9–19(a), the same amount of charge is moved from the negative side of the source to plate A of C_1, and from plate B of C_1 to plate A of C_2, and from plate B of C_2 to the positive side of the source. As a result, of course, the same amount of charge is deposited on the plates of both capacitors in a given period of time, and the total charge (Q_T) moved through the circuit in that period of time equals the charge stored by C_1 and also equals the charge stored by C_2:

$$Q_T = Q_1 = Q_2$$

As the capacitors charge, the voltage across each one increases as indicated.

Figure 9–19(b) shows the capacitors after they have been completely charged and the current has ceased. Both capacitors store an equal amount of charge (Q), and the voltage across each one depends on its capacitance value ($V = Q/C$). By Kirchhoff's voltage law, which applies to capacitive circuits as well as to resistive circuits, the sum of the capacitor voltages equals the source voltage:

$$V_S = V_1 + V_2$$

Using the formula $V = Q/C$, you can substitute into the formula for Kirchhoff's law and get the following relationship (where $Q = Q_T = Q_1 = Q_2$):

$$\frac{Q}{C_T} = \frac{Q}{C_1} + \frac{Q}{C_2}$$

The Q can be factored out of the right side of the equation and canceled with the Q on the left side as follows:

$$\frac{\cancel{Q}}{C_T} = \cancel{Q}\left(\frac{1}{C_1} + \frac{1}{C_2}\right)$$

Thus you can have the following relationship for two capacitors in series:

$$\frac{1}{C_T} = \frac{1}{C_1} + \frac{1}{C_2} \qquad \text{(9–8)}$$

Taking the reciprocal of both sides of Equation (9–8) gives the formula for the total capacitance:

$$C_T = \frac{1}{\left(\dfrac{1}{C_1}\right) + \left(\dfrac{1}{C_2}\right)} \qquad \text{(9–9)}$$

EXAMPLE 9–5   Find C_T in Figure 9–20.

FIGURE 9–20

C_1 — 100 pF, C_2 — 300 pF, V_S

Solution

$$C_T = \frac{1}{\left(\dfrac{1}{C_1}\right) + \left(\dfrac{1}{C_2}\right)} = \frac{1}{\left(\dfrac{1}{100\ \text{pF}}\right) + \left(\dfrac{1}{300\ \text{pF}}\right)} = 75\ \text{pF}$$

The calculator sequence is

$\boxed{1}\ \boxed{0}\ \boxed{0}\ \boxed{\text{EXP}}\ \boxed{+/-}\ \boxed{1}\ \boxed{2}\ \boxed{1/x}\ \boxed{+}\ \boxed{3}\ \boxed{0}\ \boxed{0}\ \boxed{\text{EXP}}\ \boxed{+/-}\ \boxed{1}\ \boxed{2}\ \boxed{1/x}\ \boxed{=}\ \boxed{1/x}$

Related Exercise   Determine C_T if $C_1 = 470$ pF and $C_2 = 680$ pF in Figure 9–20.

General Formula for Total Series Capacitance

Equations (9–8) and (9–9) can be extended to any number of capacitors in series, as shown in Figure 9–21.

FIGURE 9–21
General series circuit with n capacitors.

The expanded formulas for total series capacitance are as follows; the subscript n can be any number.

$$\frac{1}{C_T} = \frac{1}{C_1} + \frac{1}{C_2} + \frac{1}{C_3} + \cdots + \frac{1}{C_n}$$ (9–10)

$$C_T = \frac{1}{\left(\dfrac{1}{C_1}\right) + \left(\dfrac{1}{C_2}\right) + \left(\dfrac{1}{C_3}\right) + \cdots + \left(\dfrac{1}{C_n}\right)}$$ (9–11)

Remember,

The total series capacitance is always less than the smallest capacitance.

EXAMPLE 9–6   Determine the total capacitance in Figure 9–22.

FIGURE 9–22

C_1 C_2 C_3

10 μF   5 μF   8 μF

+ ‖|‖ −

Solution $$\frac{1}{C_T} = \frac{1}{C_1} + \frac{1}{C_2} + \frac{1}{C_3} = \frac{1}{10\ \mu F} + \frac{1}{5\ \mu F} + \frac{1}{8\ \mu F}$$

Taking the reciprocal of both sides yields the following:

$$C_T = \frac{1}{\left(\dfrac{1}{10\ \mu F}\right) + \left(\dfrac{1}{5\ \mu F}\right) + \left(\dfrac{1}{8\ \mu F}\right)} = \frac{1}{0.425}\mu F = 2.35\ \mu F$$

Related Exercise   If a 4.7 μF capacitor is connected in series with the three existing capacitors in Figure 9–22, what is the value of C_T?

Capacitor Voltages

The voltage across each capacitor in a series connection depends on its capacitance value according to the formula $V = Q/C$. *The largest-value capacitor will have the smallest voltage* because of the reciprocal relationship. Likewise, *the smallest capacitance value will have the largest voltage*. The voltage across any individual capacitor in series can be determined using the following formula:

$$V_x = \left(\frac{C_T}{C_x}\right)V_S$$ (9–12)

where C_x is any capacitor in series, such as C_1, C_2, and C_3, and so on, and V_x is the voltage across C_x.

EXAMPLE 9–7

Find the voltage across each capacitor in Figure 9–23.

FIGURE 9–23

C_1 C_2 C_3

0.1 μF   0.5 μF   0.2 μF

V_S
25 V

Solution   Calculate the total capacitance.

$$\frac{1}{C_T} = \frac{1}{C_1} + \frac{1}{C_2} + \frac{1}{C_3} = \frac{1}{0.1\ \mu F} + \frac{1}{0.5\ \mu F} + \frac{1}{0.2\ \mu F}$$

$$C_T = \frac{1}{17}\ \mu F = 0.0588\ \mu F$$

The voltages are as follows:

$$V_1 = \left(\frac{C_T}{C_1}\right)V_S = \left(\frac{0.0588\ \mu F}{0.1\ \mu F}\right)25\ V = 14.7\ V$$

$$V_2 = \left(\frac{C_T}{C_2}\right)V_S = \left(\frac{0.0588\ \mu F}{0.5\ \mu F}\right)25\ V = 2.94\ V$$

$$V_3 = \left(\frac{C_T}{C_3}\right)V_S = \left(\frac{0.0588\ \mu F}{0.2\ \mu F}\right)25\ V = 7.35\ V$$

The calculator sequence for V_1 is

[•] [0] [5] [8] [8] [EXP] [+/−] [6] [÷] [•] [1] [EXP] [+/−] [6] [×] [2] [5] [=]

Related Exercise   A 0.47 μF capacitor is connected in series with the existing capacitors in Figure 9–23. Determine the voltage across the new capacitor.

SECTION 9–3 REVIEW

1. Is the total capacitance of series capacitors less than or greater than the value of the smallest capacitor?
2. The following capacitors are in series: 100 pF, 250 pF, and 500 pF. What is the total capacitance?
3. A 0.01 μF and a 0.015 μF capacitor are in series. Determine the total capacitance.
4. Determine the voltage across the 0.01 μF capacitor in Question 3 if 10 V are connected across the two series capacitors.

9–4 ■ PARALLEL CAPACITORS

In this section, you will see why capacitances add when they are connected in parallel.

After completing this section, you should be able to

■ **Analyze parallel capacitors**
   □ Determine total capacitance

When capacitors are connected in parallel, the total capacitance is the sum of the individual capacitances because the effective plate area increases. The calculation of total parallel capacitance is analogous to the calculation of total series resistance.

Figure 9–24 shows two parallel capacitors connected to a dc voltage source. When the switch is closed, as shown in part (a), current begins to flow. A total amount of charge (Q_T) moves through the circuit in a certain period of time. Part of the total charge is stored by C_1 and part by C_2. The portion of the total charge that is stored by each capacitor depends on its capacitance value according to the relationship $Q = CV$.

Figure 9–24(b) shows the capacitors after they have been completely charged and the current has stopped. Since the voltage across both capacitors is the same, the larger capacitor stores more charge. If the capacitors are equal in value, they store an equal amount of charge. The charge stored by both of the capacitors together equals the total charge that was delivered from the source:

$$Q_T = Q_1 + Q_2$$

Since $Q = CV$, Equation (9–2), you can use substitution to get the following relationship:

$$C_T V_S = C_1 V_S + C_2 V_S$$

Because all the V_S terms are equal, they can be canceled. Therefore, the total capacitance for two capacitors in parallel is

$$C_T = C_1 + C_2 \qquad (9\text{–}13)$$

FIGURE 9–24

Capacitors in parallel produce a total capacitance that is the sum of the individual capacitances.

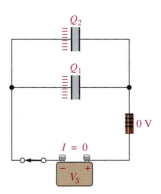

(a) The amount of charge on each capacitor is directly proportional to its capacitance value.

(b) $Q_T = Q_1 + Q_2$

EXAMPLE 9–8

What is the total capacitance in Figure 9–25? What is the voltage across each capacitor?

FIGURE 9–25

Solution The total capacitance is

$$C_T = C_1 + C_2 = 330 \text{ pF} + 220 \text{ pF} = 550 \text{ pF}$$

The voltage across each capacitor is $V_1 = V_2 = 5$ V.

Related Exercise What is C_T if a 100 pF capacitor is connected in parallel with C_1 and C_2 in Figure 9–25?

General Formula for Total Parallel Capacitance

Equation (9–13) can be extended to any number of capacitors in parallel, as shown in Figure 9–26. The expanded formula is as follows; the subscript n can be any number.

$$C_T = C_1 + C_2 + C_3 + \cdots + C_n \tag{9–14}$$

FIGURE 9–26
General parallel circuit with n capacitors.

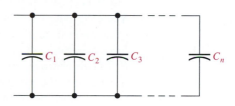

EXAMPLE 9–9

Determine C_T in Figure 9–27.

FIGURE 9–27

Solution

$$C_T = C_1 + C_2 + C_3 + C_4 + C_5 + C_6$$
$$= 0.01 \ \mu\text{F} + 0.02 \ \mu\text{F} + 0.01 \ \mu\text{F} + 0.05 \ \mu\text{F} + 0.02 \ \mu\text{F} + 0.02 \ \mu\text{F} = 0.13 \ \mu\text{F}$$

Related Exercise If three more 0.01 μF capacitors are connected in parallel in Figure 9–27, what is C_T?

SECTION 9–4 REVIEW

1. How is total parallel capacitance determined?
2. In a certain application, you need 0.05 μF. The only values available are 0.01 μF, which are available in large quantities. How can you get the total capacitance that you need?
3. The following capacitors are in parallel: 10 pF, 5 pF, 33 pF, and 50 pF. What is C_T?

9–5 ▪ CAPACITORS IN DC CIRCUITS

A capacitor will charge up when it is connected to a dc voltage source. The buildup of charge across the plates occurs in a predictable manner which is dependent on the capacitance and the resistance in a circuit.

After completing this section, you should be able to

▪ **Analyze capacitive dc switching circuits**
   ☐ Describe the charging and discharging of a capacitor
   ☐ Define *time constant*
   ☐ Relate the time constant to charging and discharging
   ☐ Write equations for the charging and discharging curves
   ☐ Explain why a capacitor blocks constant dc

Charging a Capacitor

A capacitor charges when it is connected to a dc voltage source, as shown in Figure 9–28. The capacitor in part (a) is uncharged; that is, plate A and plate B have equal numbers of free electrons. When the switch is closed, as shown in part (b), the source moves electrons away from plate A through the circuit to plate B as the arrows indicate. As plate A loses electrons and plate B gains electrons, plate A becomes positive with respect to plate B. As this charging process continues, the voltage across the plates builds up rapidly until it is equal to the applied voltage, V_S, but opposite in polarity, as shown in part (c). When the capacitor is fully charged, there is no current.

A capacitor blocks constant dc.

When the charged capacitor is disconnected from the source, as shown in Figure 9–28(d), it remains charged for long periods of time, depending on its leakage resistance, and can cause severe electrical shock. The charge on an electrolytic capacitor generally leaks off more rapidly than in other types of capacitors.

(a) Uncharged

(b) Charging

(c) Fully charged

(d) Retains charge

FIGURE 9–28
Charging a capacitor.

Discharging a Capacitor

When a wire is connected across a charged capacitor, as shown in Figure 9–29, the capacitor will discharge. In this particular case, a very low resistance path (the wire) is

(a) Retains charge   (b) Discharging   (c) Uncharged

FIGURE 9–29
Discharging a capacitor.

connected across the capacitor with a switch. Before the switch is closed, the capacitor is charged to 50 V, as indicated in part (a). When the switch is closed, as shown in part (b), the excess electrons on plate B move through the circuit to plate A (indicated by the arrows); as a result of the current through the low resistance of the wire, the energy stored by the capacitor is dissipated in the wire. The charge is neutralized when the numbers of free electrons on both plates are again equal. At this time, the voltage across the capacitor is zero, and the capacitor is completely discharged, as shown in part (c).

Current and Voltage During Charging and Discharging

Notice in Figures 9–28 and 9–29 that the direction of the current during discharge is opposite to that of the charging current. It is important to understand that *there is no current through the dielectric of the capacitor during charging or discharging because the dielectric is an insulating material.* Current flows from one plate to the other only through the external circuit.

Figure 9–30(a) shows a capacitor connected in series with a resistor and a switch to a dc voltage source. Initially, the switch is open and the capacitor is uncharged with zero volts across its plates. At the instant the switch is closed, the current jumps to its maximum value and the capacitor begins to charge. The current is maximum initially because the capacitor has zero volts across it and, therefore, effectively acts as a short; thus, the current is limited only by the resistance. As time passes and the capacitor charges, the current decreases and the voltage across the capacitor (V_C) increases. The resistor voltage is proportional to the current during this charging period.

After a certain period of time, the capacitor reaches full charge. At this point, the current is zero and the capacitor voltage is equal to the dc source voltage, as shown in Figure 9–30(b). If the switch were opened now, the capacitor would retain its full charge (neglecting any leakage).

In Figure 9–30(c), the voltage source has been removed. When the switch is closed, the capacitor begins to discharge. Initially, the current jumps to a maximum but in a direction opposite to its direction during charging. As time passes, the current and capacitor voltage decrease. The resistor voltage is always proportional to the current. When the capacitor has fully discharged, the current and the capacitor voltage are zero.

Remember the following rules about capacitors in dc circuits:

1. Voltage across a capacitor *cannot* change instantaneously.

2. Current in a capacitive circuit *can* ideally change instantaneously.

3. A fully charged capacitor appears as an *open* to nonchanging current.

4. An uncharged capacitor appears as a *short* to an instantaneous change in current.

Now we will examine in more detail how the voltage and current change with time in a capacitive circuit.

Current jumps to maximum at instant switch is closed; then it decreases.

Voltage is zero at instant switch is closed; then it increases.

Charging current

(a) Charging: Capacitor voltage increases as the current and resistor voltage decrease.

(b) Fully charged: Capacitor voltage equals source voltage. The current is zero.

Current jumps to maximum at instant switch is closed; then it decreases.

Voltage decreases as capacitor discharges.

Discharging current

(c) Discharging: Capacitor voltage, resistor voltage, and the current decrease from initial maximums. Note that the discharge current is opposite to the charge current.

FIGURE 9–30

Current and voltage in a charging and discharging capacitor.

The *RC* Time Constant

In a practical situation, there cannot be capacitance without some resistance in a circuit. It may simply be the small resistance of a wire, or it may be a designed-in resistance. Because of this, the charging and discharging characteristics of a capacitor must always be considered in light of the associated series resistance. The resistance introduces the element of *time* in the charging and discharging of a capacitor.

When a capacitor charges or discharges through a resistance, a certain time is required for the capacitor to charge fully or discharge fully. The voltage across a capacitor cannot change instantaneously because a finite time is required to move charge from one point to another. The rate at which the capacitor charges or discharges is determined by the **time constant** of the circuit.

The time constant of a series *RC* circuit is a time interval that equals the product of the resistance and the capacitance and is expressed in units of seconds when resistance is in ohms and capacitance is in farads.

The time constant is symbolized by τ (Greek letter tau), and the formula is as follows:

$$\tau = RC \tag{9-15}$$

Recall that $I = Q/t$. The current is the amount of charge moved in a given time. When the resistance is increased, the charging current is reduced, thus increasing the charging time of the capacitor. When the capacitance is increased, the amount of charge increases; thus, for the same current, more time is required to charge the capacitor.

EXAMPLE 9–10

A series *RC* circuit has a resistance of 1 MΩ and a capacitance of 5 μF. What is the time constant?

Solution $\tau = RC = (1 \times 10^6 \ \Omega)(5 \times 10^{-6} \ \text{F}) = 5 \ \text{s}$

Related Exercise   A series *RC* circuit has a 270 kΩ resistor and a 3300 pF capacitor. What is the time constant?

During one time-constant interval, the charge on a capacitor changes approximately 63%.

An uncharged capacitor charges to 63% of its fully charged voltage in one time constant. When a capacitor is discharging, its voltage drops to approximately 37% (100% − 63%) of its initial value in one time constant. This change also corresponds to a 63% change.

The Charging and Discharging Curves

A capacitor charges and discharges following a nonlinear curve, as shown in Figure 9–31. In these graphs, the percentage of full charge is shown at each time-constant interval. This type of curve follows a precise mathematical formula and is called an *exponential curve*. The charging curve is an increasing exponential, and the discharging curve is a decreasing exponential. It takes five time constants to approximately reach the final value.

(a) Charging curve with percentages of final value

(b) Discharging curve with percentages of initial value

FIGURE 9–31

Charging and discharging exponential curves for the capacitor voltage in an RC circuit.

A five time-constant interval is accepted as the time to fully charge or discharge a capacitor and is called the *transient time*.

General Formula   The general expressions for either increasing or decreasing exponential curves are given in the following equations for both voltage and current:

$$v = V_F + (V_i - V_F)e^{-t/\tau} \tag{9-16}$$

$$i = I_F + (I_i - I_F)e^{-t/\tau} \tag{9-17}$$

where V_F and I_F are the final values and V_i and I_i are the initial values. The lowercase letters v and i are the instantaneous values of the capacitor voltage and current at time t, and e is the base of natural logarithms with a value of 2.718. The ⌐e^x⌐ key or the ⌐INV⌐ and ⌐ln x⌐ keys on your calculator make it easy to work with these formulas.

Charging from Zero   The formula for the special case in which an increasing exponential voltage curve begins at zero ($V_i = 0$), as shown in Figure 9–31(a), is given in Equation (9–18). It is developed as follows, starting with the general formula in Equation (9–16).

$$v = V_F + (V_i - V_F)e^{-t/\tau}$$
$$= V_F + (0 - V_F)e^{-t/RC}$$
$$= V_F - V_F e^{-t/RC}$$

$$v = V_F(1 - e^{-t/RC}) \tag{9-18}$$

Using Equation (9–18), you can calculate the value of the charging voltage of a capacitor at any instant of time. The same is true for an increasing current.

EXAMPLE 9–11   In Figure 9–32, determine the capacitor voltage 50 μs after the switch is closed if the capacitor initially is uncharged. Sketch the charging curve.

FIGURE 9–32

Solution   The time constant is $RC = (8.2 \text{ k}\Omega)(0.01 \text{ }\mu\text{F}) = 82 \text{ }\mu\text{s}$. The voltage to which the capacitor will fully charge is 50 V (this is V_F). The initial voltage is zero. Notice that 50 μs is less than one time constant; so the capacitor will charge less than 63% of the full voltage in that time.

$$v_C = V_F(1 - e^{-t/RC}) = 50 \text{ V}(1 - e^{-50\mu s/82\mu s})$$
$$= 50 \text{ V}(1 - e^{-0.61}) = 50 \text{ V}(1 - 0.543) = 22.8 \text{ V}$$

Determine the value of $e^{-0.61}$ on the calculator by entering −0.61 and then pressing ⌐e^x⌐ or ⌐INV⌐ ⌐ln x⌐ .

The charging curve for the capacitor is shown in Figure 9–33.

FIGURE 9–33

The calculator sequence is

Related Exercise Determine the capacitor voltage 15 μs after the switch is closed in Figure 9–32.

Discharging to Zero The formula for the special case in which a decreasing exponential voltage curve ends at zero, as shown in Figure 9–31(b), is derived from the general formula as follows:

$$v = V_F + (V_i - V_F)e^{-t/\tau}$$
$$= 0 + (V_i - 0)e^{-t/RC}$$

$$v = V_i e^{-t/RC} \tag{9–19}$$

where V_i is the voltage at the beginning of the discharge. You can use this formula to calculate the discharging voltage at any instant, as Example 9–12 illustrates.

EXAMPLE 9–12

Determine the capacitor voltage in Figure 9–34 at a point in time 6 ms after the switch is closed. Sketch the discharging curve.

FIGURE 9–34

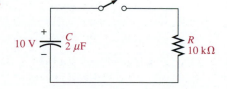

Solution The discharge time constant is

$$RC = (10 \text{ k}\Omega)(2 \text{ }\mu\text{F}) = 20 \text{ ms}$$

The initial capacitor voltage is 10 V. Notice that 6 ms is less than one time constant, so the capacitor will discharge less than 63%. Therefore, it will have a voltage greater than 37% of the initial voltage at 6 ms.

$$v_C = V_i e^{-t/RC} = 10e^{-6ms/20ms}$$
$$= 10e^{-0.3} = 10(0.741) = 7.41 \text{ V}$$

Again, the value of $e^{-0.3}$ can be determined with a calculator.

The discharging curve for the capacitor is shown in Figure 9–35.

FIGURE 9–35

Related Exercise In Figure 9–34, change R to 2.2 kΩ and determine the capacitor voltage 1 ms after the switch is closed.

Graphical Method Using Universal Exponential Curves The universal curves in Figure 9–36 provide a graphic solution of the charge and discharge of capacitors. Example 9–13 illustrates this graphical method.

FIGURE 9–36
Normalized universal exponential curves.

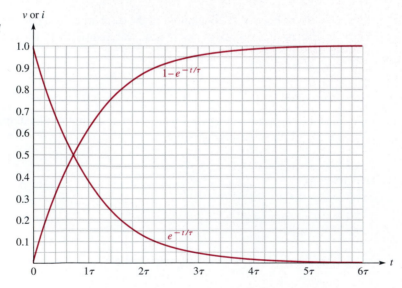

EXAMPLE 9–13 How long will it take the capacitor in Figure 9–37 to charge to 75 V? What is the capacitor voltage 2 ms after the switch is closed? Use the normalized universal curves in Figure 9–36 to determine the answers.

FIGURE 9–37

Solution   The full charge voltage is 100 V, which is the 100% level (1.0) on the graph. 75 V is 75% of the maximum, or 0.75 on the graph. You can see that this value occurs at 1.4 time constants. One time constant is 1 ms. Therefore, the capacitor voltage reaches 75 V at 1.4 ms after the switch is closed.

The capacitor is at approximately 86 V in 2 ms. These graphic solutions are shown in Figure 9–38.

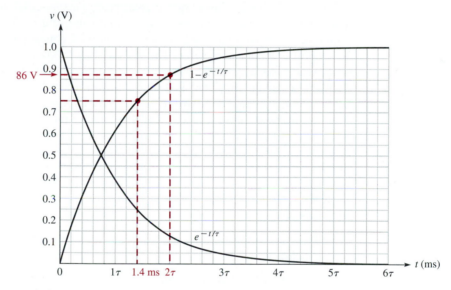

FIGURE 9–38

Related Exercise   Using the normalized universal exponential curves, determine how long it will take the capacitor in Figure 9–37 to charge to 50 V from zero. What is v_C 3 ms after switch closure?

<table>
<tr><td>

SECTION 9–5 REVIEW

</td><td>

1. Determine the time constant when $R = 1.2$ kΩ and $C = 1000$ pF.
2. If the circuit in Question 1 is charged with a 5 V source, how long will it take the initially uncharged capacitor to reach full charge? At full charge, what is the capacitor voltage?
3. A certain circuit has a time constant of 1 ms. If it is charged with a 10 V battery, what will the capacitor voltage be at each of the following intervals: 2 ms, 3 ms, 4 ms, and 5 ms? The capacitor is initially uncharged.
4. A capacitor is charged to 100 V. If it is discharged through a resistor, what is the capacitor voltage at one time constant?

</td></tr>
</table>

9–6 ▪ CAPACITORS IN AC CIRCUITS

As you saw in the last section, a capacitor blocks constant dc. You will learn in this section that a capacitor passes ac but with an amount of opposition that depends on the frequency of the ac.

After completing this section, you should be able to

▪ **Analyze capacitive ac circuits**
  ☐ Define *capacitive reactance*
  ☐ Determine the value of capacitive reactance in a given circuit
  ☐ Explain why a capacitor causes a phase shift between voltage and current
  ☐ Discuss instantaneous, true, and reactive power in a capacitor

Capacitive Reactance, X_C

In Figure 9–39, a capacitor is shown connected to a sinusoidal voltage source. When the source voltage is held at a constant amplitude value and its frequency is increased, the amplitude of the current increases. Also, when the frequency of the source is decreased, the current amplitude decreases. The reason is as follows: When the frequency of the voltage increases, its rate of change also increases. This relationship is illustrated in Figure 9–40, where the frequency is doubled. Now, if the rate at which the voltage is changing increases, the amount of charge moving through the circuit in a given period of time must also increase. More charge in a given period of time means more current. For example, a tenfold increase in frequency means that the capacitor is charging and discharging 10 times as much in a given time interval. Therefore, since the rate of charge movement has increased 10 times, the current must increase by 10 because $I = Q/t$.

An increase in the amount of current for a fixed amount of voltage indicates that opposition to the current has decreased. Therefore, the capacitor offers opposition to current, and that opposition varies *inversely* with frequency.

> **The opposition to sinusoidal current in a capacitor is called capacitive reactance.**

The symbol for **capacitive reactance** is X_C, and its unit is the ohm (Ω).

You have just seen how frequency affects the opposition to current (capacitive reactance) in a capacitor. Now let's see how the capacitance itself affects the reactance. Figure 9–41(a) shows that when a sinusoidal voltage with a fixed amplitude and fixed fre-

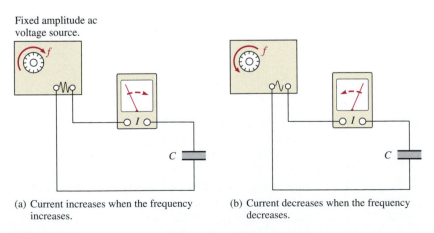

Fixed amplitude ac voltage source.

(a) Current increases when the frequency increases.

(b) Current decreases when the frequency decreases.

FIGURE 9–39

The current in a capacitive circuit varies directly with the frequency of the applied voltage.

FIGURE 9–40

Rate of change of a sine wave increases when frequency increases.

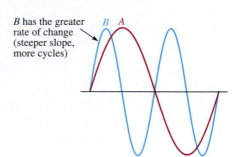

B has the greater rate of change (steeper slope, more cycles)

(a) Less capacitance, less current   (b) More capacitance, more current

FIGURE 9–41

For a fixed voltage and fixed frequency, the current varies directly with the capacitance value.

quency is applied to a 1 μF capacitor, a certain amount of alternating current flows. When the capacitance value is increased to 2 μF, the current increases, as shown in Figure 9–41(b). Thus, when the capacitance is increased, the opposition to current (capacitive reactance) decreases. Therefore, not only is the capacitive reactance *inversely* proportional to frequency, but it is also *inversely* proportional to capacitance. This relationship can be stated as follows:

X_C is proportional to 1/fC.

It can be proven that the constant of proportionality that relates X_C to 1/fC is 1/2π. Therefore, the formula for capacitive reactance (X_C) is

$$X_C = \frac{1}{2\pi fC} \tag{9–20}$$

The 2π term comes from the fact that, as you learned in Chapter 8, a sine wave can be described in terms of rotational motion, and one revolution contains 2π radians.

EXAMPLE 9–14

A sinusoidal voltage is applied to a capacitor, as shown in Figure 9–42. The frequency of the sine wave is 1 kHz. Determine the capacitive reactance.

FIGURE 9–42

Solution
$$X_C = \frac{1}{2\pi fC} = \frac{1}{2\pi(1 \times 10^3 \text{ Hz})(0.005 \times 10^{-6} \text{ F})} = 31.8 \text{ k}\Omega$$

The calculator sequence is

Related Exercise Determine the frequency required to make X_C in Figure 9–42 equal to 10 kΩ.

Ohm's Law in Capacitive AC Circuits

The reactance of a capacitor is analogous to the resistance of a resistor. In fact, both are expressed in ohms. Since both R and X_C are forms of opposition to current, Ohm's law applies to capacitive circuits as well as to resistive circuits and is stated as follows for capacitive reactance:

$$V = IX_C \qquad\qquad (9\text{–}21)$$

When applying Ohm's law in ac circuits, you must express both the current and the voltage in the same way, that is, both in rms, both in peak, and so on.

EXAMPLE 9–15 Determine the rms current in Figure 9–43.

FIGURE 9–43

$V_{rms} = 5 \text{ V}$
$f = 10 \text{ kHz}$

0.005 μF

Solution First, find X_C.

$$X_C = \frac{1}{2\pi fC} = \frac{1}{2\pi(10 \times 10^3 \text{ Hz})(0.005 \times 10^{-6} \text{ F})} = 3.18 \text{ k}\Omega$$

Then, apply Ohm's law.

$$V_{rms} = I_{rms} X_C$$

$$I_{rms} = \frac{V_{rms}}{X_C} = \frac{5 \text{ V}}{3.18 \text{ k}\Omega} = 1.57 \text{ mA}$$

Related Exercise Change the frequency in Figure 9–43 to 25 kHz and determine the rms value of the current.

Current Leads Capacitor Voltage by 90°

A sinusoidal voltage is shown in Figure 9–44. Notice that the rate at which the voltage is changing varies along the sine wave curve, as indicated by the "steepness" of the curve. At the zero crossings, the curve is changing at a faster rate than anywhere else along the curve. At the peaks, the curve has a zero rate of change because it has just reached its maximum and is at the point of changing direction.

The amount of charge stored by a capacitor determines the voltage across it. Therefore, the rate at which the charge is moved ($Q/t = I$) from one plate to the other determines the rate at which the voltage changes. When the current is changing at its maximum rate (at the zero crossings), the voltage is at its maximum value (peak). When the current is changing at its minimum rate (zero at the peaks), the voltage is at its minimum value (zero). This **phase** relationship is illustrated in Figure 9–45. As you can see, the current peaks occur a quarter of a cycle before the voltage peaks. Thus, the current leads the voltage by 90°.

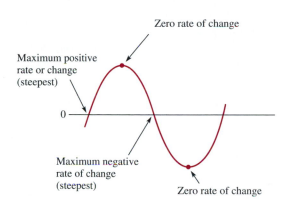

FIGURE 9–44
The rates of change of a sine wave.

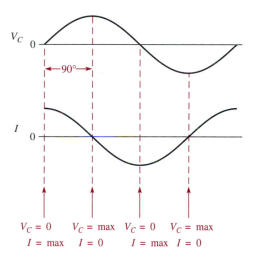

FIGURE 9–45
Current is always leading the capacitor voltage by 90°.

Power in a Capacitor

As discussed earlier in this chapter, a charged capacitor stores energy in the electric field within the dielectric. An ideal capacitor does not dissipate energy; it only stores it. When an ac voltage is applied to a capacitor, energy is stored by the capacitor during a portion of the voltage cycle. Then the stored energy is returned to the source during another portion of the cycle. There is no net energy loss. Figure 9–46 shows the power curve that results from one cycle of capacitor voltage and current.

Instantaneous Power (p) The product of instantaneous voltage, v, and instantaneous current, i, gives instantaneous power, p. At points where v or i is zero, p is also zero. When both v and i are positive, p is also positive. When either v or i is positive and the other negative, p is negative. When both v and i are negative, p is positive. As you can see, the power follows a sinusoidal-type curve. Positive values of power indicate that energy is stored by the capacitor. Negative values of power indicate that energy is returned from the capacitor to the source. Note that the power fluctuates at a frequency twice that of the voltage or current, as energy is alternately stored and returned to the source.

True Power (P_{true}) Ideally, all of the energy stored by a capacitor during the positive portion of the power cycle is returned to the source during the negative portion. No net

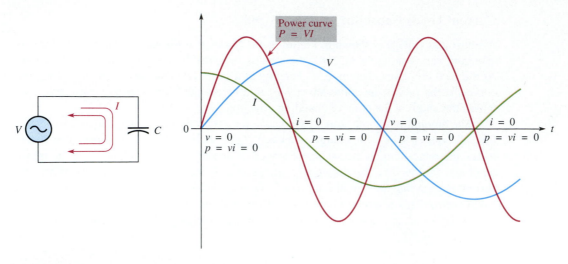

FIGURE 9–46
Power curve for a capacitor.

energy is consumed in the capacitor, so the true power is zero. Actually, because of leakage and foil resistance in a practical capacitor, a small percentage of the total power is dissipated and there is a very small amount of true power, which can normally be neglected.

Reactive Power (P_r)   The rate at which a capacitor stores or returns energy is called its **reactive power**, P_r. The reactive power is a nonzero quantity, because at any instant in time, the capacitor is actually taking energy from the source or returning energy to it. Reactive power does not represent an energy loss. The following formulas apply:

$$P_r = V_{rms}I_{rms} \qquad\qquad (9\text{–}22)$$

$$P_r = \frac{V_{rms}^2}{X_C} \qquad\qquad (9\text{–}23)$$

$$P_r = I_{rms}^2 X_C \qquad\qquad (9\text{–}24)$$

Notice that these equations are of the same form as those introduced in Chapter 3 for true power in a resistor. The voltage and current are expressed as rms values. The unit of reactive power is **volt-ampere reactive** (VAR).

EXAMPLE 9–16   Determine the true power and the reactive power in Figure 9–47.

FIGURE 9–47

V_{rms}
2 V f = 2 kHz   0.01 μF

Solution   The true power P_{true} is always zero for an ideal capacitor. The reactive power is determined by first finding the value for the capacitive reactance.

$$X_C = \frac{1}{2\pi f C} = \frac{1}{2\pi(2 \times 10^3 \text{ Hz})(0.01 \times 10^{-6} \text{ F})} = 7.96 \text{ k}\Omega$$

$$P_r = \frac{V_{rms}^2}{X_C} = \frac{(2 \text{ V})^2}{7.96 \text{ k}\Omega} = 503 \times 10^{-6} \text{ VAR} = 503 \text{ } \mu\text{VAR}$$

Related Exercise   If the frequency is doubled in Figure 9–47, what are the true power and the reactive power?

SECTION 9–6 REVIEW

1. Calculate X_C for $f = 5$ kHz and $C = 50$ pF.
2. At what frequency is the reactance of a 0.1 μF capacitor equal to 2 kΩ?
3. Calculate the rms current in Figure 9–48.

FIGURE 9–48

$V_{rms} = 1$ V
$f = 1$ MHz   0.1 μF

4. State the phase relationship between current and voltage in a capacitor.
5. A 1 μF capacitor is connected to an ac voltage source of 12 V rms. What is the true power?
6. In Question 5, determine reactive power at a frequency of 500 Hz.

9–7 ■ CAPACITOR APPLICATIONS

Capacitors are widely used in electrical and electronic applications. A few typical applications are discussed in this section to illustrate the usefulness of this component.

After completing this section, you should be able to

■ **Discuss some capacitor applications**
  □ Describe a power supply filter
  □ Explain the purposes of coupling and by-pass capacitors
  □ Discuss the basics of capacitors applied to tuned circuits, timing circuits, and computer memories

If you pick up any circuit board, open any power supply, or look inside any piece of electronic equipment, chances are you will find capacitors of one type or another. These components are used for a variety of reasons in both dc and ac applications.

Electrical Storage

One of the most basic applications of a capacitor is as a backup voltage source for low-power circuits such as certain types of semiconductor memories in computers. This particular application requires a very high capacitance value and negligible leakage.

The storage capacitor is connected between the dc power supply input to the circuit and ground. When the circuit is operating from its normal power supply, the capacitor remains fully charged to the dc power supply voltage. If the normal power source is disrupted, effectively removing the power supply from the circuit, the storage capacitor temporarily becomes the power source for the circuit.

The capacitor provides voltage and current to the circuit as long as its charge remains sufficient. As current is drawn by the circuit, charge is removed from the capacitor and the voltage decreases. For this reason, the storage capacitor can only be used as a temporary power source. The length of time that the capacitor can provide sufficient power to the circuit depends on the capacitance and the amount of current drawn by the circuit. The smaller the current and the higher the capacitance, the longer the time.

Power Supply Filtering

A basic dc power supply consists of a circuit known as a **rectifier** followed by a **filter.** The rectifier converts the 110 V, 60 Hz sinusoidal voltage available at a standard outlet to a pulsating dc voltage that can be either a half-wave rectified voltage or a full-wave rectified voltage, depending on the type of rectifier circuit. As shown in Figure 9–49(a), a half-wave rectifier removes each negative half-cycle of the sinusoidal voltage. As shown in Figure 9–49(b), a full-wave rectifier actually reverses the polarity of the negative portion of each cycle. Both half-wave and full-wave rectified voltages are dc because, even though they are changing, they do not alternate polarity.

To be useful for powering electronic circuits, the rectified voltage must be changed to constant dc voltage because all circuits require constant power. The filter eliminates the fluctuations in the rectified voltage and provides a smooth constant-value dc voltage to the load which is the electronic circuit, as indicated in Figure 9–50.

The Capacitor As a Power Supply Filter Capacitors are used as filters in dc power supplies because of their ability to store electrical charge. Figure 9–51(a) shows a dc power supply with a full-wave rectifier and a capacitor filter. The operation can be described from a charging and discharging point-of-view as follows. Assume the capacitor is initially uncharged. When the power supply is first turned on and the first cycle of the rectified voltage occurs, the capacitor will quickly charge through the low resistance of

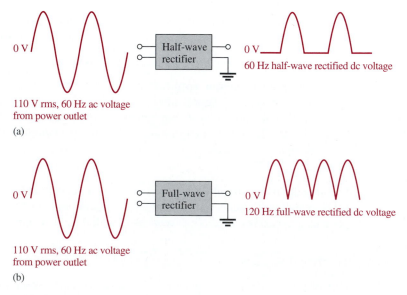

FIGURE 9–49
Half-wave and full-wave rectifier operation.

110 V rms, 60 Hz ac   120 Hz full-wave rectified dc   Constant dc voltage

Full-wave rectifier   Power supply filter   Circuit load

FIGURE 9–50
Basic block diagram and operation of a dc power supply.

the rectifier. The capacitor voltage will follow the rectified voltage curve up to the peak of the rectified voltage. As the rectified voltage passes the peak and begins to decrease, the capacitor will begin to discharge very slowly through the high resistance of the load circuit, as indicated in Figure 9–51(b). The amount of discharge is typically very small and is exaggerated in the figure for purposes of illustration. The next cycle of the rectified voltage will recharge the capacitor back to the peak value by replenishing the small amount of charge lost since the previous peak. This pattern of a small amount of charging and discharging continues as long as the power is on.

A rectifier is designed so that it allows current only in the direction to charge the capacitor. The capacitor will not discharge back through the rectifier but will only discharge a small amount through the relatively high resistance of the load. The small fluctuation in voltage due to the charging and discharging of the capacitor is called the **ripple voltage.**

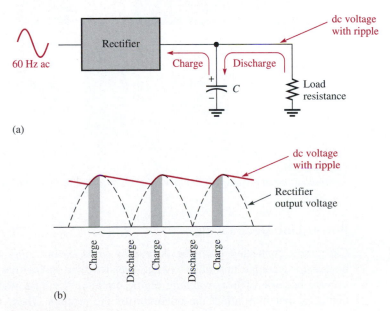

FIGURE 9–51
Basic operation of a power supply filter capacitor.

A good dc power supply has a very small amount of ripple on its dc output. Since the discharge time constant of a power supply filter capacitor depends on its capacitance and the resistance of the load, the higher the capacitance value, the longer the discharge time and, therefore, the smaller the ripple voltage.

DC Blocking and AC Coupling

Capacitors are commonly used to block the constant dc voltage in one part of a circuit from getting to another part. As an example of this, a capacitor is connected between two stages of an amplifier to prevent the dc voltage at the output of stage 1 from affecting the dc voltage at the input of stage 2, as illustrated in Figure 9–52. Assume that, for proper operation, the output of stage 1 has a zero dc voltage and the input to stage 2 has a 3 V dc voltage. The capacitor prevents the 3 V dc at stage 2 from getting to the stage 1 output and affecting its zero value, and vice versa.

If a sinusoidal signal voltage is applied to the input to stage 1, the signal voltage is increased (amplified) and appears on the ouput of stage 1, as shown in Figure 9–52. The amplified signal voltage is then coupled through the capacitor to the input of stage 2 where it is superimposed on the 3 V dc level and then again amplified by stage 2. In order for the signal voltage to be passed through the capacitor without being reduced, the capacitor must be large enough so that its reactance at the frequency of the signal voltage is negligible. In this type of application, the capacitor is known as a *coupling capacitor* which ideally appears as an open to dc and as a short to ac. As the signal frequency is reduced, the capacitive reactance increases and, at some point, the capacitive reactance becomes large enough to cause a significant reduction in ac voltage between stage 1 and stage 2.

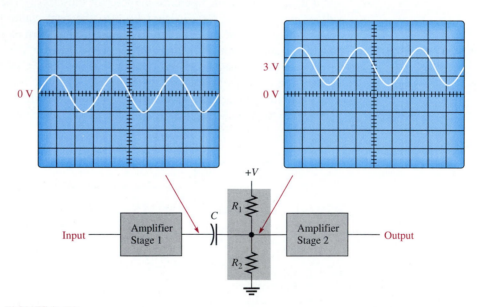

FIGURE 9–52
An application of a capacitor to block dc and couple ac in an amplifier.

Power Line Decoupling

Capacitors connected from the dc supply voltage line to ground are used on circuit boards to decouple unwanted voltage transients or spikes that occur on the dc supply voltage because of fast switching digital circuits. A voltage transient contains high frequencies that may affect the operation of the circuits. These transients are shorted to ground through the very low reactance of the decoupling capacitors. Several decoupling capacitors are often used at various points along the supply voltage line on a circuit board.

Bypassing

Another capacitor application is in bypassing an ac voltage around a resistor in a circuit without affecting the dc voltage across the resistor. In amplifier circuits, for example, dc voltages called *bias voltages* are required at various points. For the amplifier to operate properly, certain bias voltages must remain constant and, therefore, any ac voltages must be removed. A sufficiently large capacitor connected from a bias point to ground provides a low reactance path to ground for ac voltages, leaving the constant dc bias voltage at the given point. This bypass application is illustrated in Figure 9–53. At lower frequencies, the bypass capacitor becomes less effective because of its increased reactance.

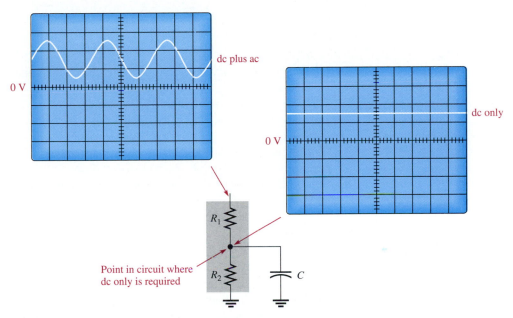

FIGURE 9–53
Example of the operation of a bypass capacitor.

Signal Filters

Capacitors are essential to the operation of a class of circuits called *filters* that are used for selecting one ac signal with a certain specified frequency from a wide range of signals with many different frequencies or for selecting a certain band of frequencies and eliminating all others. A common example of this application is in radio and television receivers where it necessary to select the signal transmitted from a given station and eliminate or filter out the signals transmitted from all the other stations in the area.

When you tune your radio or TV, you are actually changing the capacitance in the tuner circuit (which is a type of filter) so that only the signal from the station or channel you want passes through to the receiver circuitry. Capacitors are used in conjunction with resistors, inductors (covered in the next chapter), and other components in these types of filters. The topic of filters will be covered in detail in later chapters.

The main characteristic of a filter is its frequency selectivity, which is based on the fact that the reactance of a capacitor depends on frequency ($X_C = 1/2\pi fC$).

Timing Circuits

Another important area in which capacitors are used is in timing circuits that generate specified time delays or produce waveforms with specific characteristics. Recall that the time constant of a circuit with resistance and capacitance can be controlled by selecting appropriate values for R and C. The charging time of a capacitor can be used as a basic time delay in various types of circuits. An example is the circuit that controls the turn indicators on your car where the light flashes on and off at regular intervals.

Computer Memories

Dynamic computer memories use capacitors as the basic storage element for binary information, which consists of two digits, 1 and 0. A charged capacitor can represent a stored 1 and a discharged capacitor can represent a stored 0. Patterns of 1s and 0s that make up binary data are stored in a memory that consists of an array of capacitors with associated circuitry. You will study this topic in a computer or digital fundamentals course.

SECTION 9–7 REVIEW	**1.** Explain how half-wave or full-wave rectified dc voltages are smoothed out by a filter capacitor.
	2. Explain the purpose of a coupling capacitor.
	3. How large must a coupling capacitor be?
	4. Explain the purpose of a decoupling capacitor.
	5. Discuss how the relationship of frequency and capacitive reactance is important in frequency-selective circuits such as signal filters.
	6. What characteristics of a capacitor is most important in time-delay applications.

9–8 ■ TESTING CAPACITORS

Capacitors are very reliable devices but their useful life can be extended significantly by operating them well within the voltage rating and at moderate temperatures. In this section, the basic types of failures are discussed and methods for checking for them are introduced.

After completing this section, you should be able to

■ **Test a capacitor**
  ☐ Perform an ohmmeter check
  ☐ Explain what an *LC* meter is

Capacitor failures can be categorized into two areas: catastrophic and degradation. The catastrophic failures are usually a short circuit caused by dielectric breakdown or an open circuit caused by connection failure. Degradation usually results in a gradual decrease in leakage resistance, hence an increase in leakage current or an increase in equivalent series resistance or dielectric absorption.

Ohmmeter Check

When there is a suspected problem, the capacitor can be removed from the circuit and checked with an ohmmeter. First, to be sure that the capacitor is discharged, short its leads, as indicated in Figure 9–54(a). Connect the meter—set on a high ohms range such as R × 1M—to the capacitor, as shown in Figure 9–54(b), and observe the needle. It should initially indicate near zero ohms. Then it should begin to move toward the high-resistance end of the scale as the capacitor charges from the ohmmeter's battery, as shown in Figure 9–54(c). When the capacitor is fully charged, the meter will indicate an extremely high resistance, as shown in Figure 9–54(d).

(a) Discharge by shorting leads

(b) Initially, when the meter is first connected, the pointer jumps immediately to zero.

(c) The needle moves slowly toward infinity as the capacitor charges from the ohmmeter battery.

(d) When the capacitor is fully charged the pointer is at infinity.

FIGURE 9–54
Checking a capacitor with an analog ohmmeter. This check shows a good capacitor.

As mentioned, the capacitor charges from the internal battery of the ohmmeter, and the meter responds to the charging current. The larger the capacitance value, the more slowly the capacitor will charge, as indicated by the needle movement. For small pF values, the meter response may be insufficient to indicate the fast charging action.

If the capacitor is internally shorted, the meter will go to zero and stay there. If it is leaky, the final meter reading will be much less than normal. Most capacitors have a resistance of several hundred megohms. The exception is the electrolytic, which may normally have less than one megohm of leakage resistance. If the capacitor is open, no charging action will be observed, and the meter will indicate an infinite resistance.

Testing for Capacitance Value and Other Parameters with an *LC* Meter

An *LC* meter such as the one shown in Figure 9–55 can be used to check the value of a capacitor. All capacitors change value over a period of time, some more than others. Ceramic capacitors, for example, often exhibit a 10% to 15% change in value during the first year. Electrolytic capacitors are particularly subject to value change due to drying of the electrolyte. In other cases, capacitors may be labeled incorrectly or the wrong value was installed in the circuit. Although a value change represents less than 25% of defective capacitors, a value check should be made to quickly eliminate this as a source of trouble when troubleshooting a circuit.

Typically, values from 1 pF to 200,000 μF can be measured by simply connecting the capacitor, pushing the appropriate button, and reading the value on the display.

Many *LC* meters can also be used to check for leakage current in capacitors. In order to check for leakage, a sufficient voltage must be applied across the capacitor to simulate operating conditions. This is automatically done by the test instrument. Over 40% of all defective capacitors have excessive leakage current and electrolytics are particularly susceptible to this problem.

The problem of dielectric absorption occurs mostly in electrolytic capacitors when they do not completely discharge during use and retain a residual charge. Approximately 25% of defective capacitors have exhibited this condition.

FIGURE 9–55
A typical LC meter (courtesy of SENCORE).

Another defect sometimes found in capacitors is called equivalent series resistance. This problem may be caused by a defective lead to plate contacts, resistive leads, or resistive plates and shows up only under ac conditions. This is the least common capacitor defect and occurs in less than 10% of all defects.

SECTION 9–8 REVIEW

1. How can a capacitor be discharged after removal from the circuit?
2. Describe how the needle of an ohmmeter responds when a good capacitor is checked.
3. List four common capacitor defects.

9–9 ■ APPLICATION ASSIGNMENT

As you learned in this chapter, capacitors are used in certain types of amplifiers for coupling ac signals and blocking dc voltages. In this assignment, an amplifier circuit board contains two coupling capacitors. Your assignment is to check certain voltages on three identical amplifier circuit boards to determine if the capacitors are working properly. A knowledge of amplifier circuits is not necessary for this assignment.

All amplifier circuits contain transistors that require dc voltages to establish proper operating conditions for amplifying ac signals. These dc voltages are referred to as bias voltages. As indicated in Figure 9–56(a), a common type of dc bias circuit used in amplifiers is the voltage divider formed by R_1 and R_2, which sets up the proper dc voltage at the input to the amplifier.

When an ac signal voltage is applied to the amplifier, the input coupling capacitor, C_1, prevents the internal resistance of the ac source from changing the dc bias voltage. Without the capacitor, the internal source resistance would appear in parallel with R_2 and drastically change the value of the dc voltage.

The coupling capacitance is chosen so that its reactance (X_C) at the frequency of the ac signal is very small compared to the bias resistor values. The coupling capacitance

(a) Amplifier schematic

(b) Amplifier board

FIGURE 9–56
Capacitively coupled amplifier.

therefore efficiently couples the ac signal from the source to the input of the amplifier. On the source side of the input coupling capacitor there is only ac but on the amplifier side there is ac plus dc (the signal voltage is riding on the dc bias voltage set by the voltage divider), as indicated in Figure 9–56(a). Capacitor C_2 is the output coupling capacitor which couples the amplified ac signal to another amplifier stage that would be connected to the output.

You will check three amplifier boards like the one in Figure 9–56(b) for the proper input voltages using an oscilloscope. If the voltages are incorrect, you will determine the most likely fault. For all measurements, assume the amplifier has no dc loading effect on the voltage-divider bias circuit.

Step 1: Compare the Printed Circuit Board with the Schematic

Check the printed circuit board in Figure 9–56(b) to make sure it agrees with the amplifier schematic in part (a).

Step 2: Test the Input to Board 1

The oscilloscope probe is connected from channel 1 to the board as shown in Figure 9–57.

The input signal from a sinusoidal voltage source is connected to the board and set to a frequency of 5 kHz with an amplitude of 1 V rms. Determine if the voltage and frequency displayed on the scope are correct. If the scope measurement is incorrect, specify the most likely fault in the circuit.

Step 3: Test the Input to Board 2

The oscilloscope probe is connected from channel 1 to board 2 the same as was shown in Figure 9–57 for board 1.

The input signal from the sinusoidal voltage source is the same as Step 2. Determine if the scope display in Figure 9–58 is correct. If the scope measurement is incorrect, specify the most likely fault in the circuit.

Note: ground reference
has been established at
bottom horizontal line

FIGURE 9–57
Testing board 1.

Note: ground reference
has been established at
bottom horizontal line

FIGURE 9–58
Testing board 2.

Step 4: Test the Input to Board 3

The oscilloscope probe is connected from channel 1 to board 3 the same as was shown in Figure 9–57 for board 1.

The input signal from the sinusoidal voltage source is the same as Step 3. Determine if the scope display in Figure 5–59 is correct. If the scope measurement is incorrect, specify the most likely fault in the circuit.

0 V

Note: ground reference has been established at bottom horizontal line

FIGURE 9–59
Testing board 3.

SECTION 9–9 REVIEW

1. Explain why the input coupling capacitor is necessary when connecting an ac source to the amplifier.
2. Capacitor C_2 in Figure 9–56 is an output coupling capacitor. Generally, what would you expect to measure at point C and at the output of the circuit when an input signal is applied to the amplifier?

■ SUMMARY

- The opposition of a capacitor to ac is inversely proportional to frequency and capacitance.
- A capacitor is composed of two parallel conductive plates separated by an insulating material called the *dielectric.*
- A capacitor stores electrical charge on its plates.
- Energy is stored in a capacitor by the electric field created between the charged plates in the dielectric.
- Capacitance is measured in units of farads (F).
- Capacitance is directly proportional to the plate area and the dielectric constant and inversely proportional to the distance between the plates (the dielectric thickness.)
- The dielectric constant is an indication of the ability of a material to establish an electric field.
- The dielectric strength is one factor that determines the breakdown voltage of a capacitor.
- Capacitors are commonly classified according to the dielectric material. Typical materials are mica, ceramic, plastic, and electrolytic (aluminum oxide and tantalum oxide).
- The total capacitance of series capacitors is less than the smallest capacitance.
- The total capacitance of parallel capacitors is the sum of all the capacitances.

- A capacitor blocks constant dc.
- The time constant determines the charging and discharging time of a capacitor with resistance in series.
- In a circuit with capacitance and resistance, the voltage and current during charging and discharging make a 63% change during each time-constant interval.
- Five time constants are required for a capacitor to fully charge or discharge. This is called the *transient time.*
- The percentage of final charge after each charging time-constant interval is given in Table 9–4.

TABLE 9–4

NUMBER OF TIME CONSTANTS	% OF FINAL CHARGE
1	63
2	86
3	95
4	98
5	99 (considered 100%)

- The percentage of initial charge after each discharging time-constant interval is given in Table 9–5.

TABLE 9–5

NUMBER OF TIME CONSTANTS	% OF INITIAL CHARGE
1	37
2	14
3	5
4	2
5	1 (considered 0)

- Alternating current in a capacitor leads the voltage by 90°.
- A capacitor passes ac to an extent that depends on its reactance.
- Capacitive reactance (X_C) is inversely proportional to the frequency and to the capacitance value.
- Ideally, there is no energy loss in a capacitor and, thus, the true power (watts) is zero. However, in most capacitors there is some small energy loss due to leakage resistance.

■ GLOSSARY

Capacitance   The ability of a capacitor to store electrical charge.

Capacitive reactance   The opposition of a capacitor to sinusoidal current. The unit is the ohm.

Capacitor   An electrical device consisting of two conductive plates separated by an insulating material and possessing the property of capacitance.

Coulomb (C)   The unit of electrical charge.

Coulomb's law   A physical law that states a force exists between two charged bodies that is directly proportional to the product of the two charges and inversely proportional to the square of the distance between them.

Dielectric   The insulating material between the plates of a capacitor.

Dielectric constant   A measure of the ability of a dielectric material to establish an electric field.

Dielectric strength   A measure of the ability of a dielectric material to withstand voltage.

Farad (F)   The unit of capacitance.

Filter   A type of circuit that passes or blocks certain frequencies to the exclusion of all others.

Phase   The relative displacement of a time-varying quantity with respect to a given reference.

Reactive power (P_r)   The rate at which energy is stored and alternately returned to the source by a capacitor. The unit is the VAR.

Rectifier   An electronic circuit that converts ac into pulsing dc; one part of a power supply.

Ripple voltage   The small fluctuation in voltage due to the charging and discharging of a capacitor.

Temperature coefficient   A constant specifying the amount of change in the value of a quantity for a given change in temperature.

Time constant   A fixed interval, equal to the product of the R and C values, that determines the time response of a circuit.

Trimmer   A small variable capacitor.

Volt-ampere reactive (VAR)   The unit of reactive power.

■ FORMULAS

(9–1)	$C = \dfrac{Q}{V}$	Capacitance in terms of charge and voltage
(9–2)	$Q = CV$	Charge in terms of capacitance and voltage
(9–3)	$V = \dfrac{Q}{C}$	Voltage in terms of charge and capacitance
(9–4)	$F = \dfrac{kQ_1Q_2}{d^2}$	Coulomb's law (force between two charged bodies)
(9–5)	$W = \dfrac{1}{2}CV^2$	Energy stored by a capacitor
(9–6)	$\epsilon_r = \dfrac{\epsilon}{\epsilon_0}$	Dielectric constant (relative permittivity)
(9–7)	$C = \dfrac{A\epsilon_r(8.85 \times 10^{-12}\ \text{F/m})}{d}$	Capacitance in terms of physical parameters
(9–8)	$\dfrac{1}{C_T} = \dfrac{1}{C_1} + \dfrac{1}{C_2}$	Reciprocal of total series capacitance (two capacitors)
(9–9)	$C_T = \dfrac{1}{\left(\dfrac{1}{C_1}\right) + \left(\dfrac{1}{C_2}\right)}$	Total series capacitance (two capacitors)
(9–10)	$\dfrac{1}{C_T} = \dfrac{1}{C_1} + \dfrac{1}{C_2} + \dfrac{1}{C_3} + \cdots + \dfrac{1}{C_n}$	Reciprocal of total series capacitance (general)
(9–11)	$C_T = \dfrac{1}{\left(\dfrac{1}{C_1}\right) + \left(\dfrac{1}{C_2}\right) + \left(\dfrac{1}{C_3}\right) + \cdots + \left(\dfrac{1}{C_n}\right)}$	Total series capacitance (general)
(9–12)	$V_x = \left(\dfrac{C_T}{C_x}\right)V_S$	Voltage across series capacitor
(9–13)	$C_T = C_1 + C_2$	Two capacitors in parallel
(9–14)	$C_T = C_1 + C_2 + \cdots + C_n$	n capacitors in parallel
(9–15)	$\tau = RC$	Time constant
(9–16)	$v = V_F + (V_i - V_F)e^{-t/\tau}$	Exponential voltage (general)
(9–17)	$i = I_F + (I_i - I_F)e^{-t/\tau}$	Exponential current (general)
(9–18)	$v = V_F(1 - e^{-t/RC})$	Increasing exponential voltage beginning at zero
(9–19)	$v = V_i e^{-t/RC}$	Decreasing exponential voltage ending at zero
(9–20)	$X_C = \dfrac{1}{2\pi f C}$	Capacitive reactance

$$(9\text{–}21) \qquad V = IX_C \qquad\qquad\qquad \text{Ohm's law for a capacitor}$$

$$(9\text{–}22) \qquad P_r = V_{rms}I_{rms} \qquad\qquad \text{Reactive power in a capacitor}$$

$$(9\text{–}23) \qquad P_r = \frac{V_{rms}^2}{X_C} \qquad\qquad\quad \text{Reactive power in a capacitor}$$

$$(9\text{–}24) \qquad P_r = I_{rms}^2 X_C \qquad\qquad\quad \text{Reactive power in a capacitor}$$

▪ SELF-TEST

1. Which of the the following statement(s) accurately describe(s) a capacitor?
   (a) The plates are conductive.
   (b) The dielectric is an insulator between the plates.
   (c) Constant dc flows through a fully charged capacitor.
   (d) A practical capacitor stores charge indefinitely when disconnected from the source.
   (e) none of the above
   (f) all the above
   (g) only (a) and (b)

2. Which one of the following statements is true?
   (a) There is current through the dielectric of a charging capacitor.
   (b) When a capacitor is connected to a dc voltage source, it will charge to the value of the source.
   (c) An ideal capacitor can be discharged by disconnecting it from the voltage source.

3. A capacitance of 0.01 μF is larger than
   (a) 0.00001 F   (b) 100,000 pF   (c) 1000 pF   (d) all of the above

4. A capacitance of 1000 pF is smaller than
   (a) 0.01 μF   (b) 0.001 μF   (c) 0.00000001 F   (d) both (a) and (c)

5. When the voltage across a capacitor is increased, the stored charge
   (a) increases   (b) decreases   (c) remains constant   (d) fluctuates

6. When the voltage across a capacitor is doubled, the stored charge
   (a) stays the same   (b) is halved   (c) increases by four   (d) doubles

7. The voltage rating of a capacitor is increased by
   (a) increasing the plate separation   (b) decreasing the plate separation
   (c) increasing the plate area   (d) both (b) and (c)

8. The capacitance value is increased by
   (a) decreasing plate area   (b) increasing plate separation
   (c) decreasing plate separation   (d) increasing plate area
   (e) both (a) and (b)   (f) both (c) and (d)

9. A 1 μF, a 2.2 μF, and a 0.05 μF capacitor are connected in series. The total capacitance is less than
   (a) 1 μF   (b) 2.2 μF   (c) 0.05 μF   (d) 0.001 μF

10. Four 0.02 μF capacitors are in parallel. The total capacitance is
    (a) 0.02 μF   (b) 0.08 μF   (c) 0.05 μF   (d) 0.04 μF

11. An uncharged capacitor and a resistor are connected in series with a switch and a 12 V battery. At the instant the switch is closed, the voltage across the capacitor is
    (a) 12 V   (b) 6 V   (c) 24 V   (d) 0 V

12. In Question 11, the voltage across the capacitor when it is fully charged is
    (a) 12 V   (b) 6 V   (c) 24 V   (d) −6 V

13. In Question 11, the capacitor will reach full charge in a time equal to approximately
    (a) RC (b) $5RC$ (c) $12RC$ (d) cannot be predicted

14. A sine wave voltage is applied across a capacitor. When the frequency of the voltage is increased, the current
    (a) increases   (b) decreases   (c) remains constant   (d) ceases

15. A capacitor and a resistor are connected in series to a sine wave generator. The frequency is set so that the capacitive reactance is equal to the resistance and, thus, an equal amount of voltage appears across each component. If the frequency is decreased,
    (a) $V_R > V_C$ (b) $V_C > V_R$ (c) $V_R = V_C$ (d) $V_C < V_R$

16. An ohmmeter is connected across a discharged capacitor and the needle stabilizes at approximately 50 kΩ. The capacitor is
    (a) good   (b) charged   (c) too large   (d) leaky

▪ BASIC PROBLEMS

SECTION 9–1   The Basic Capacitor

1. (a) Find the capacitance when $Q = 50\ \mu C$ and $V = 10$ V.
   (b) Find the charge when $C = 0.001\ \mu F$ and $V = 1$ kV.
   (c) Find the voltage when $Q = 2$ mC and $C = 200\ \mu F$.

2. Convert the following values from microfarads to picofarads:
   (a) $0.1\ \mu F$ (b) $0.0025\ \mu F$ (c) $5\ \mu F$

3. Convert the following values from picofarads to microfarads:
   (a) 1000 pF   (b) 3500 pF   (c) 250 pF

4. Convert the following values from farads to microfarads:
   (a) 0.0000001 F   (b) 0.0022 F   (c) 0.0000000015 F

5. What size capacitor is capable of storing 10 mJ of energy with 100 V across its plates?

6. A mica capacitor has a plate area of 0.04 m^2 and a dielectric thickness of 0.008 m. What is its capacitance?

7. An air capacitor has plates with an area of 0.1 m^2. The plates are separated by 0.01 m. Calculate the capacitance.

8. At ambient temperature (25°C), a certain capacitor is specified to be 1000 pF. It has a negative temperature coefficient of 200 ppm/°C. What is its capacitance at 75°C?

9. A 0.001 μF capacitor has a positive temperature coefficient of 500 ppm/°C. How much change in capacitance will a 25°C increase in temperature cause?

SECTION 9–2   Types of Capacitors

10. In the construction of a stacked-foil mica capacitor, how is the plate area increased?

11. What type of capacitor has the higher dielectric constant, mica or ceramic?

12. Show how to connect an electrolytic capacitor across points A and B in Figure 9–60.

FIGURE 9–60

13. Determine the value of the typographically labeled ceramic disk capacitors in Figure 9–61.

14. Name two types of electrolytic capacitors. How do electrolytics differ from other capacitors?

15. Identify the parts of the ceramic disk capacitor shown in the cutaway view of Figure 9–62 by referring to Figure 9–9(b).

| 0.02 MF | 0.047 | 0.001 MFD | 220 | 104 |
| (a) | (b) | (c) | (d) | (e) |

FIGURE 9–61

FIGURE 9–62

SECTION 9–3 Series Capacitors

16. Five 1000 pF capacitors are in series. What is the total capacitance?

17. Find the total capacitance for each circuit in Figure 9–63.

18. For each circuit in Figure 9–63, determine the voltage across each capacitor.

19. The total charge stored by the series capacitors in Figure 9–64 is 10 μC. Determine the voltage across each of the capacitors.

(a)   (b)   (c)

FIGURE 9–63

FIGURE 9–64

SECTION 9–4 Parallel Capacitors

20. Determine C_T for each circuit in Figure 9–65.

21. Determine C_T for each circuit in Figure 9–66.

22. What is the voltage between points A and B in each circuit in Figure 9–66?

(a)   (b)

FIGURE 9–65

(a)   (b)   (c) $C = 1$ μF for each capacitor

FIGURE 9–66

SECTION 9–5 Capacitors in DC Circuits

23. Determine the time constant for each of the following series *RC* combinations:
   (a) $R = 100\ \Omega$, $C = 1\ \mu F$ **(b)** $R = 10\ M\Omega$, $C = 50\ pF$
   (c) $R = 4.7\ k\Omega$, $C = 0.005\ \mu F$ **(d)** $R = 1.5\ M\Omega$, $C = 0.01\ \mu F$

24. Determine how long it takes the capacitor to reach full charge for each of the following combinations:
   (a) $R = 50\ \Omega$, $C = 50\ \mu F$ **(b)** $R = 3300\ \Omega$, $C = 0.015\ \mu F$
   (c) $R = 22\ k\Omega$, $C = 100\ pF$ **(d)** $R = 5\ M\Omega$, $C = 10\ pF$

25. In the circuit of Figure 9–67, the capacitor initially is uncharged. Determine the capacitor voltage at the following times after the switch is closed:
   (a) 10 μs   **(b)** 20 μs   **(c)** 30 μs   **(d)** 40 μs   **(e)** 50 μs

26. In Figure 9–68, the capacitor is charged to 25 V. Find the capacitor voltage at the following times when the switch is closed:
   (a) 1.5 ms   **(b)** 4.5 ms   **(c)** 6 ms   **(d)** 7.5 ms

FIGURE 9–67   **FIGURE 9–68**

27. Repeat Problem 25 for the following time intervals:
   (a) 2 μs   **(b)** 5 μs   **(c)** 15 μs

28. Repeat Problem 26 for the following time intervals:
   (a) 0.5 ms   **(b)** 1 ms   **(c)** 2 ms

SECTION 9–6 Capacitors in AC Circuits

29. Determine X_C for a 0.047 μF capacitor at each of the following frequencies:
   (a) 10 Hz   **(b)** 250 Hz   **(c)** 5 kHz   **(d)** 100 kHz

30. What is the value of the total capacitive reactance in each circuit in Figure 9–69?

(a)   (b)   (c)

FIGURE 9–69

31. In Figure 9–66, each dc voltage source is replaced by a 10 V rms, 2 kHz ac source. Determine the reactance in each case.

32. In each circuit of Figure 9–69, what frequency is required to produce an X_C of 100 Ω? An X_C of 1 kΩ?

33. A sinusoidal voltage of 20 V rms produces an rms current of 100 mA when connected to a certain capacitor. What is the reactance?

34. A 10 kHz voltage is applied to a 0.0047 μF capacitor, and 1 mA of rms current is measured. What is the rms value of the voltage?

35. Determine the true power and the reactive power in Problem 34.

SECTION 9–7 Capacitor Applications

36. If another capacitor is connected in parallel with the existing capacitor in the power supply filter of Figure 9–51, how is the ripple voltage affected?

37. Ideally, what should the reactance of a bypass capacitor be in order to eliminate a 10 kHz ac voltage at a given point in an amplifier circuit?

SECTION 9–8 Testing Capacitors

38. Assume that you are checking a capacitor with an ohmmeter, and when you connect the leads across the capacitor, the pointer does not move from its left-end scale position. What is the problem?

39. In checking a capacitor with the ohmmeter, you find that the pointer goes all the way to the right end of the scale and stays there. What is the problem?

▪ ADVANCED PROBLEMS

40. Two series capacitors (one 1 μF, the other of unknown value) are charged from a 12 V source. The 1 μF capacitor is charged to 8 V, and the other to 4 V. What is the value of the unknown capacitor?

41. How long does it take C to discharge to 3 V in Figure 9–68?

42. How long does it take C to charge to 8 V in Figure 9–67?

43. Determine the time constant for the circuit in Figure 9–70.

44. In Figure 9–71, the capacitor initially is uncharged. At $t = 10$ μs after the switch is closed, the instantaneous capacitor voltage is 7.2 V. Determine the value of R.

FIGURE 9–70

FIGURE 9–71

45. (a) The capacitor in Figure 9–72 is uncharged when the switch is thrown into position 1. The switch remains in position 1 for 10 ms and then is thrown into position 2, where it remains indefinitely. Sketch the complete waveform for the capacitor voltage.

(b) If the switch is thrown back to position 1 after 5 ms in position 2, and then is left in position 1, how will the waveform appear?

46. Determine the ac voltage across each capacitor and the current in each branch of the circuit in Figure 9–73.

FIGURE 9–72

FIGURE 9–73

47. Find the value of C_1 in Figure 9–74.

48. How much does the voltage across C_5 and C_6 change when the ganged switch is thrown from position 1 to position 2 in Figure 9–75?

FIGURE 9–74

FIGURE 9–75

49. If C_4 in Figure 9–73 is open, determine the voltages that would be measured across the other capacitors.

▪ ANSWERS TO SECTION REVIEWS

Section 9–1

1. Capacitance is the ability (capacity) to store electrical charge.

2. **(a)** There are 1,000,000 microfarads in one farad.
   (b) There are 1×10^{12} picofarads in one farad.
   (c) There are 1,000,000 picofarads in one microfarad.

3. $0.0015 \ \mu F \times 10^6 = 1500 \ pF$; $0.0015 \ \mu F \times 10^{-6} = 0.0000000015 \ F$

4. $W = 1/2 \ CV^2 = 1/2 \ (0.01 \ \mu F)(15 \ V)^2 = 1.125 \ \mu J$

5. **(a)** Capacitance increases when plate area is increased.
   (b) Capacitance decreases when plate separation is increased.

6. $(1000 \ V/mil)(10 \ mils) = 10 \ kV$

7. $C = (0.2 \ m^2)(1200)(8.85 \times 10^{-12} \ F/m)/0.005 \ m = 0.425 \ \mu F$

Section 9–2

1. Capacitors are commonly classified by the dielectric material.

2. A fixed capacitance cannot be changed; a variable can.

3. Electrolytics are polarized.

4. Be sure that the voltage rating is sufficient and connect the positive end to the positive side of the circuit when installing a polarized capacitor.

Section 9–3

1. C_T of series capacitors is less than the smallest value.

2. $C_T = 62.5 \ pF$

3. $C_T = 0.006 \, \mu F$

4. $V = (0.006 \, \mu F/0.01 \, \mu F)10 \text{ V} = 6 \text{ V}$

Section 9–4

1. The individual parallel capacitors are added to get C_T.

2. Use five 0.01 μF capacitors in parallel to get 0.05 μF.

3. $C_T = 98 \text{ pF}$

Section 9–5

1. $\tau = RC = 1.2 \, \mu s$

2. $5\tau = 6 \, \mu s; V_C = 4.97 \text{ V}$

3. $v_C = (0.86)10 \text{ V} = 8.6 \text{ V}; v_C = (0.95)10 \text{ V} = 9.5 \text{ V}; v_C = (0.98)10 \text{ V} = 9.8 \text{ V}; v_C = (0.99)10 \text{ V} = 9.9 \text{ V}$

4. $v_C = (0.37)(100 \text{ V}) = 37 \text{ V}$

Section 9–6

1. $X_C = 1/2 \, \pi \, fC = 637 \text{ k}\Omega$

2. $f = 1/2\pi CX_C = 796 \text{ Hz}$

3. $I_{rms} = 1 \text{ V}/1.59 \, \Omega = 629 \text{ mA}$

4. Current leads voltage by 90°.

5. $P_{true} = 0 \text{ W}$

6. $P_r = (12 \text{ V})^2/318 \, \Omega = 0.453 \text{ VAR}$

Section 9–7

1. Once the capacitor charges to the peak voltage, it discharges very little before the next peak thus smoothing the rectified voltage.

2. A coupling capacitor allows ac to pass from one point to another, but blocks constant dc.

3. A coupling capacitor must be large enough to have a negligible reactance at the frequency that is to be passed without opposition.

4. A decoupling capacitor shorts power line voltage transients to ground.

5. X_C is inversely proportional to frequency and so is the filter's ability to pass ac signals.

6. The time constant is used in delay applications.

Section 9–8

1. Short its leads to discharge.

2. Initially, the needle jumps to zero; then it moves to the high-resistance end of the scale when the capacitor is good.

3. Shorted, open, leakage, and dielectric absorption are possible capacitor faults.

Section 9–9

1. The coupling capacitor prevents the source from affecting the dc voltage but passes the ac input signal.

2. An ac voltage riding on a dc voltage is at point C.

▪ **ANSWERS TO RELATED EXERCISES FOR EXAMPLES**

9–1	10 kV
9–2	0.05 μF
9–3	100,000,000 pF
9–4	6,638 pF
9–5	278 pF
9–6	1.57 μF
9–7	2.78 V
9–8	650 pF
9–9	0.16 μF
9–10	891 μs

9–11	8.36 V
9–12	7.97 V
9–13	0.7 ms; 95 V
9–14	3.18 kHz
9–15	3.93 mA
9–16	0 W; 1.01 mVAR

10

INDUCTORS

Inductance is the property of a coil of wire that opposes a change in current. The basis for inductance is the electromagnetic field that surrounds any conductor when there is current through it. The electrical component designed to have the property of inductance is called an *inductor, coil,* or *choke.* All of these terms refer to the same type of device.

In this chapter, the basic inductor is introduced and its characteristics are studied. Various types of inductors are covered in terms of their physical construction and their electrical properties. The basic behavior of inductors in both dc and ac circuits is studied, and series and parallel combinations are analyzed. A method of testing inductors is discussed.

■ CHAPTER OBJECTIVES

☐ Describe the basic structure and characteristics of an inductor
☐ Discuss various types of inductors
☐ Analyze series inductors
☐ Analyze parallel inductors
☐ Analyze inductive dc switching circuits
☐ Analyze inductive ac circuits
☐ Discuss some inductor applications
☐ Test an inductor

■ APPLICATION ASSIGNMENT

You are working on a piece of defective communications equipment and your supervisor asks you to check out the unmarked coils that have been removed from the system and determine their approximate inductance value. After studying this chapter, you should be able to complete the application assignment in Section 10–9.

10–1 ■ THE BASIC INDUCTOR

In this section, the basic construction and characteristics of inductors are examined.

After completing this section, you should be able to

■ **Describe the basic structure and characteristics of an inductor**
   □ Explain how an inductor stores energy
   □ Define *inductance* and state its unit
   □ State Faraday's law
   □ State Lenz's law
   □ Discuss induced voltage
   □ Specify how the physical characteristics affect inductance
   □ Discuss winding resistance and winding capacitance

When a length of wire is formed into a coil, as shown in Figure 10–1, it becomes a basic **inductor.** Current through the coil produces an electromagnetic field. The magnetic lines of force around each loop (turn) in the **winding** of the coil effectively add to the lines of force around the adjoining loops, forming a strong magnetic field within and around the coil, as shown. The net direction of the total magnetic field creates a north and a south pole.

FIGURE 10–1

A coil of wire forms an inductor. When current flows through it, a three-dimensional electromagnetic field is created, surrounding the coil in all directions.

Self-Inductance

When there is current through an inductor, an electromagnetic field is established. When the current changes, the electromagnetic field also changes. An increase in current expands the field, and a decrease in current reduces it. Therefore, a changing current produces a changing electromagnetic field around the inductor (also known as **coil** or **choke**). In turn, the changing electromagnetic field causes an **induced voltage** across the coil in a direction to oppose the change in current. This property is called *self-inductance,* but is usually referred to as simply **inductance.** Inductance is symbolized by *L*.

> **Inductance is a measure of a coil's ability to establish an induced voltage as a result of a change in its current, and that induced voltage is in a direction to oppose that change in current.**

The Unit Inductance

The **henry,** symbolized by H, is the basic unit of inductance. By definition, the inductance is one henry when current through the coil, changing at the rate of one ampere per second, induces one volt across the coil. In many practical applications, millihenries (mH) and microhenries (μH) are the more common units. A schematic symbol for the inductor is shown in Figure 10–2.

FIGURE 10–2
A symbol for the inductor.

L

Energy Storage

An inductor stores energy in the magnetic field created by the current. The energy stored is expressed as follows:

$$W = \frac{1}{2}LI^2$$

(10–1)

As you can see, the energy stored is proportional to the inductance and the square of the current. When current (I) is in amperes and inductance (L) is in henries, the energy (W) is in joules.

Physical Characteristics of Inductors

The following characteristics are important in establishing the inductance of a coil: the permeability of the core material, the number of turns of wire, the length, and the cross-sectional area.

Core Material As discussed earlier, an inductor is basically a coil of wire. The material around which the coil is formed is called the **core.** Coils are wound on either nonmagnetic or magnetic materials. Examples of nonmagnetic materials are air, wood, copper, plastic, and glass. The permeabilities of these materials are the same as for a vacuum. Examples of magnetic materials are iron, nickel, steel, cobalt, or alloys. These materials have permeabilities that are hundreds or thousands of times greater than that of a vacuum and are classified as *ferromagnetic.* A ferromagnetic core provides a better path for the magnetic lines of force and thus permits a stronger magnetic field.

As you learned in Chapter 7, the permeability (μ) of the core material determines how easily a magnetic field can be established. *The inductance is directly proportional to the permeability of the core material.*

Physical Parameters As indicated in Figure 10–3, the number of turns of wire, the length, and the cross-sectional area of the core are factors in setting the value of inductance. The inductance is inversely proportional to the length of the core and directly proportional to the cross-sectional area. Also, the inductance is directly related to the number of turns squared. This relationship is as follows:

$$L = \frac{N^2 \mu A}{l}$$

(10–2)

where L is the inductance in henries, N is the number of turns, μ is the permeability, A is the cross-sectional area in meters squared, and l is the core length in meters.

FIGURE 10–3
Factors that determine the inductance of a coil.

Length, l

Cross-sectional area, A

Core material

Number of turns, N

EXAMPLE 10–1   Determine the inductance of the coil in Figure 10–4. The permeability of the core is 0.25×10^{-3}.

FIGURE 10–4

Solution

$$L = \frac{N^2 \mu A}{l} = \frac{(5)^2 (0.25 \times 10^{-3})(0.1)}{0.01} = 62.5 \text{ mH}$$

The calculator sequence is

5 x^2 × • 2 5 EXP +/– 3 × • 1 ÷ • 0 1 =

Related Exercise   Determine the inductance of a coil with 12 turns around a core that is 0.05 m long and has a cross-sectional area of 0.15 m². The permeability is 0.25×10^{-3}.

Winding Resistance   When a coil is made of a certain material, for example, insulated copper wire, that wire has a certain resistance per unit of length. When many turns of wire are used to construct a coil, the total resistance may be significant. This inherent resistance is called the *dc resistance* or the *winding resistance* (R_W). Although this resistance is distributed along the length of the wire, as shown in Figure 10–5(a), it is sometimes indicated in a schematic as resistance appearing in series with the inductance of the coil, as shown in Figure 10–5(b). In many applications, the winding resistance can be ignored and the coil considered as an ideal inductor. In other cases, the resistance must be considered.

FIGURE 10–5
Winding resistance of a coil.

(a) The wire has resistance distributed along its length.

(b) Equivalent circuit

Winding Capacitance

When two conductors are placed side by side, there is always some capacitance between them. Thus, when many turns of wire are placed close together in a coil, a certain amount of stray capacitance, called *winding capacitance* (C_W), is a natural side effect. In many applications, this winding capacitance is very small and has no significant effect. In other cases, particularly at high frequencies, it may become quite important.

   The equivalent circuit for an inductor with both its winding resistance (R_W) and its winding capacitance (C_W) is shown in Figure 10–6. The capacitance effectively acts in parallel. The total of the stray capacitances between each loop of the winding is indicated in a schematic as a capacitance appearing in parallel with the coil and its winding resistance, as shown in Figure 10–6(b).

(a) Stray capacitance between each loop appears as a total parallel capacitance.

(b) Equivalent circuit

FIGURE 10–6
Winding capacitance of a coil.

Review of Faraday's law

The principle of Faraday's law was introduced in Chapter 7 and is reviewed here because of its importance in the study of inductors. Faraday found that by moving a magnet through a coil of wire, a voltage was induced across the coil, and that when a complete path was provided, the induced voltage caused an induced current.

> **The amount of induced voltage is directly proportional to the rate of change of the magnetic field with respect to the coil.**

This principle is illustrated in Figure 10–7, where a bar magnet is moved through a coil of wire. An induced voltage is indicated by the voltmeter connected across the coil. The faster the magnet is moved, the greater is the induced voltage.

When a wire is formed into a certain number of loops or turns and is exposed to a changing magnetic field, a voltage is induced across the coil. The induced voltage is proportional to the number of turns of wire in the coil, N, and to the rate at which the magnetic field changes.

FIGURE 10–7
Induced voltage is created by a changing magnetic field.

Lenz's Law

Lenz's law adds to Faraday's law by defining the direction of induced voltage as follows:

> **When the current through a coil changes and an induced voltage is created as a result of the changing magnetic field, the direction of the induced voltage is such that it always opposes the change in current.**

Figure 10–8 illustrates Lenz's law. In part (a), the current is constant and is limited by R_1. There is no induced voltage because the magnetic field is unchanging. In part (b), the switch suddenly is closed, placing R_2 in parallel with R_1 and thus reducing the resistance. Naturally, the current tries to increase and the magnetic field begins to expand, but the induced voltage opposes this attempted increase in current for an instant.

In Figure 10–8(c), the induced voltage gradually decreases, allowing the current to increase. In part (d), the current has reached a constant value as determined by the parallel resistors, and the induced voltage is zero. In part (e), the switch has been sud-

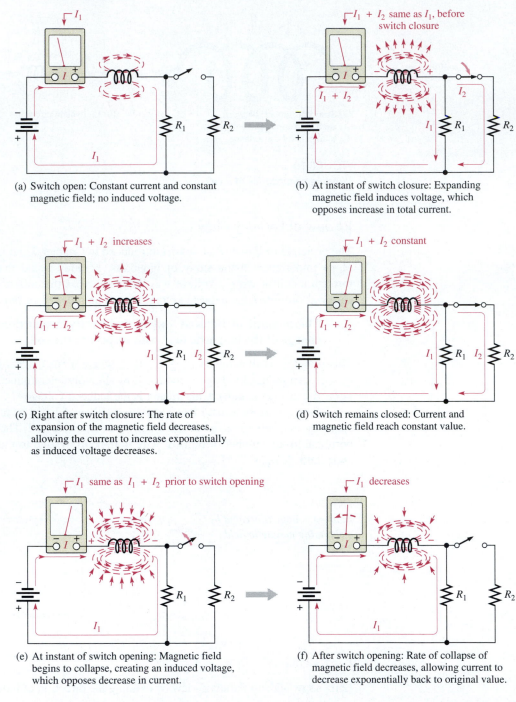

(a) Switch open: Constant current and constant magnetic field; no induced voltage.

(b) At instant of switch closure: Expanding magnetic field induces voltage, which opposes increase in total current.

(c) Right after switch closure: The rate of expansion of the magnetic field decreases, allowing the current to increase exponentially as induced voltage decreases.

(d) Switch remains closed: Current and magnetic field reach constant value.

(e) At instant of switch opening: Magnetic field begins to collapse, creating an induced voltage, which opposes decrease in current.

(f) After switch opening: Rate of collapse of magnetic field decreases, allowing current to decrease exponentially back to original value.

FIGURE 10–8

Demonstration of Lenz's law: When the current tries to change suddenly, the electromagnetic field changes and induces a voltage in a direction that opposes that change in current.

denly opened, and, for an instant, the induced voltage prevents any decrease in current. In part (f), the induced voltage gradually decreases, allowing the current to decrease back to a value determined by R_1. Notice that the induced voltage has a polarity that opposes any current change. The polarity of the induced voltage is opposite that of the battery voltage for an increase in current and aids the battery voltage for a decrease in current.

10–2 ▪ TYPES OF INDUCTORS

Inductors normally are classified according to the type of core material. In this section, the basic types of inductors are examined.

After completing this section, you should be able to

▪ **Discuss various types of inductors**
   ☐ Describe the basic types of fixed inductors
   ☐ Distinguish between fixed and variable inductors

Inductors are made in a variety of shapes and sizes. Basically, they fall into two general categories: fixed and variable. The standard schematic symbols are shown in Figure 10–9.

Both fixed and variable inductors can be classified according to the type of core material. Three common types are the air core, the iron core, and the ferrite core. Each has a unique symbol, as shown in Figure 10–10.

Adjustable (variable) inductors usually have a screw-type adjustment that moves a sliding core in and out, thus changing the inductance. A wide variety of inductors exists, some of which are shown in Figure 10–11.

(a) Fixed   (b) Variable   (a) Air core   (b) Iron core   (c) Ferrite core

FIGURE 10–9
Symbols for (a) fixed and (b) variable inductors.

FIGURE 10–10
Inductor symbols.

(a)   (b)   (c)

FIGURE 10–11
Typical inductors. (a) Fixed molded inductors; (b) variable coils; (c) toroid inductor. Parts (a) and (b) courtesy of Delevan/American Precision. Part (c) courtesy of Dale Electronics.

1. Name two general categories of inductors.
2. Identify the inductor symbols in Figure 10–12.

FIGURE 10–12

(a)   (b)   (c)

10–3 ▪ SERIES INDUCTORS

In this section, you will see that when inductors are connected in series, the total inductance increases.

After completing this section, you should be able to

▪ **Analyze series inductors**
   ☐ Determine total inductance

When inductors are connected in series, as in Figure 10–13, the total inductance, L_T, is the sum of the individual inductances. The formula for L_T is expressed in the following equation for the general case of n inductors in series:

$$L_T = L_1 + L_2 + L_3 + \cdots + L_n$$ (10–3)

Notice that the formula for inductance in series is similar to the formula for resistance in series (Equation (4–1)).

FIGURE 10–13
Inductors in series.

L_1 L_2 L_3 L_n

EXAMPLE 10–2

Determine the total inductance for each of the series connections in Figure 10–14.

1 H   2 H   1.5 H   5 H   5 mH   2 mH   10 mH   1000 μH

(a)   (b)

FIGURE 10–14

Solution   In Figure 10–14(a),

$$L_T = 1\text{ H} + 2\text{ H} + 1.5\text{ H} + 5\text{ H} = 9.5\text{ H}$$

In Figure 10–14(b),

$$L_T = 5\text{ mH} + 2\text{ mH} + 10\text{ mH} + 1\text{ mH} = 18\text{ mH}$$

Note: 1000 μH = 1 mH

Related Exercise   What is the total inductance of ten 50 μH inductors in series?

10–4 ▪ PARALLEL INDUCTORS

In this section, you will see that when inductors are connected in parallel, total inductance is reduced.

After completing this section, you should be able to

▪ **Analyze parallel inductors**
  ☐ Determine total inductance

When inductors are connected in parallel, as in Figure 10–15, the total inductance is less than the smallest inductance. The general formula states that the reciprocal of the total inductance is equal to the sum of the reciprocals of the individual inductances:

$$\frac{1}{L_T} = \frac{1}{L_1} + \frac{1}{L_2} + \frac{1}{L_3} + \cdots + \frac{1}{L_n}$$

(10–4)

Total inductance, L_T, can be found by taking the reciprocals of both sides of Equation (10–4):

$$L_T = \frac{1}{\left(\dfrac{1}{L_1}\right) + \left(\dfrac{1}{L_2}\right) + \left(\dfrac{1}{L_3}\right) + \cdots + \left(\dfrac{1}{L_n}\right)}$$

(10–5)

This formula for total inductance in parallel is similar to that for total parallel resistance or total series capacitance.

FIGURE 10–15
Inductors in parallel.

EXAMPLE 10–3   Determine L_T in Figure 10–16.

FIGURE 10–16

Solution

$$L_T = \cfrac{1}{\cfrac{1}{L_1} + \cfrac{1}{L_2} + \cfrac{1}{L_3}}$$

$$= \cfrac{1}{\left(\cfrac{1}{10\ \text{mH}}\right) + \left(\cfrac{1}{5\ \text{mH}}\right) + \left(\cfrac{1}{2\ \text{mH}}\right)} = \frac{1}{0.8}\ \text{mH} = 1.25\ \text{mH}$$

The calculator sequence is

Related Exercise Determine L_T for the following inductors in parallel: 50 μH, 80 μH, 100 μH, and 150 μH.

SECTION 10–4 REVIEW

1. Compare the total inductance in parallel with the smallest-value individual inductor.
2. The calculation of total parallel inductance is similar to that for parallel resistance (T or F).
3. Determine L_T for each parallel combination:
   (a) 100 mH, 50 mH, and 10 mH
   (b) 40 μH and 60 μH
   (c) Ten 1 H coils

10–5 ▪ INDUCTORS IN DC CIRCUITS

An inductor will energize when it is connected to a dc voltage source. The buildup of current through the inductor occurs in a predictable manner, which is dependent on both the inductance and the resistance in a circuit.

After completing this section, you should be able to

▪ **Analyze inductive dc switching circuits**
   ☐ Describe the energizing and deenergizing of an inductor
   ☐ Define *time constant*
   ☐ Relate the time constant to the energizing and deenergizing of an inductor
   ☐ Describe induced voltage
   ☐ Write the exponential equations for current in an inductor

When there is constant direct current in an inductor, there is no induced voltage. There is, however, a voltage drop due to the winding resistance of the coil. The inductance itself appears as a short to dc. Energy is stored in the magnetic field according to the formula $W = \frac{1}{2}LI^2$. The only energy loss occurs in the winding resistance ($P = I^2 R_W$). This condition is illustrated in Figure 10–17.

The *RL* Time Constant

Because the inductor's basic action is to oppose a change in its current, it follows that current cannot change instantaneously in an inductor. A certain time is required

FIGURE 10–17
Energy storage and loss in an inductor. The only dc voltage drop across the coil is due to the winding resistance.

$P = I^2 R_W$
Energy loss due to winding resistance

Energy stored in magnetic field
$W = \frac{1}{2}LI^2$

for the current to make a change from one value to another. The rate at which the current changes is determined by the **time constant.** The time constant for a series *RL* circuit is

$$\tau = \frac{L}{R} \qquad \qquad (10\text{–}6)$$

where τ is in seconds when L is in henries and R is in ohms.

EXAMPLE 10–4

A series *RL* circuit has a resistance of 1 kΩ and an inductance of 1 mH. What is the time constant?

Solution

$$\tau = \frac{L}{R} = \frac{1 \text{ mH}}{1 \text{ k}\Omega} = \frac{1 \times 10^{-3} \text{ H}}{1 \times 10^{-3} \text{ }\Omega} = 1 \times 10^{-6} \text{ s} = 1 \text{ } \mu\text{s}$$

Related Exercise   Find the time constant for $R = 2.2$ kΩ and $L = 500$ μH.

Energizing Current in an Inductor

In a series *RL* circuit, the current will increase to 63% of its full value in one time-constant interval after the switch is closed. This buildup of current is analogous to the buildup of capacitor voltage during the charging in an *RC* circuit; they both follow an exponential curve and reach the approximate percentages of final value as indicated in Table 10–1 and as illustrated in Figure 10–18.

TABLE 10–1
Percentage of final current after each time-constant interval during current buildup

NUMBER OF TIME CONSTANTS	% OF FINAL VALUE
1	63
2	86
3	95
4	98
5	99 (considered 100%)

The change in current over five time-constant intervals is illustrated in Figure 10–19. When the current reaches its final value at approximately 5τ, it ceases to change. At this

FIGURE 10–18
Energizing current in an inductor.

(a) Initially (*i* = 0)

(b) At 1τ

(c) At 2τ

(d) At 3τ

(e) At 4τ

(f) At 5τ

FIGURE 10–19
Illustration of the exponential buildup of current in an inductor. The current increases another 63% during each time-constant interval. A voltage (v$_L$) is induced in the coil that tends to oppose the increase in current.

time, the inductor acts as a short (except for winding resistance) to the constant current. The final value of the current is

$$I_F = \frac{V_S}{R_W} = \frac{10 \text{ V}}{1 \text{ k}\Omega} = 10 \text{ mA}$$

EXAMPLE 10–5   Calculate the time constant for Figure 10–20. Then determine the current and the time at each time-constant interval, measured from the instant the switch is closed.

FIGURE 10–20

Solution   The time constant is

$$\tau = \frac{L}{R} = \frac{50 \text{ mH}}{100 \text{ }\Omega} = 500 \text{ } \mu s$$

The final current is

$$I_F = \frac{V_S}{R} = \frac{20 \text{ V}}{100 \text{ }\Omega} = 200 \text{ mA}$$

Using the time-constant percentage values from Table 10–1,

At $1\tau = 500$ μs:	$i = 0.63(200 \text{ mA}) = 126 \text{ mA}$
At $2\tau = 1$ ms:	$i = 0.86(200 \text{ mA}) = 172 \text{ mA}$
At $3\tau = 1.5$ ms:	$i = 0.95(200 \text{ mA}) = 190 \text{ mA}$
At $4\tau = 2$ ms:	$i = 0.98(200 \text{ mA}) = 196 \text{ mA}$
At $5\tau = 2.5$ ms:	$i = 0.99(200 \text{ mA}) = 198 \text{ mA} \cong 200 \text{ mA}$

Related Exercise   Repeat the calculations if $R = 680$ Ω and $L = 100$ μH.

Deenergizing Current in an Inductor

Current in an inductor decreases exponentially according to the approximate percentage values shown in Table 10–2 and in Figure 10–21.

Figure 10–22(a) shows a constant current of 1 A (1000 mA) through an inductor. Switch 1 (SW₁) is opened and switch 2 (SW₂) is closed simultaneously, and for an instant

TABLE 10–2
Percentage of initial current after each time-constant interval while current is decreasing

NUMBER OF TIME CONSTANTS	% OF INITIAL VALUE
1	37
2	14
3	5
4	2
5	1 (considered 0)

FIGURE 10–21
Deenergizing current in an inductor.

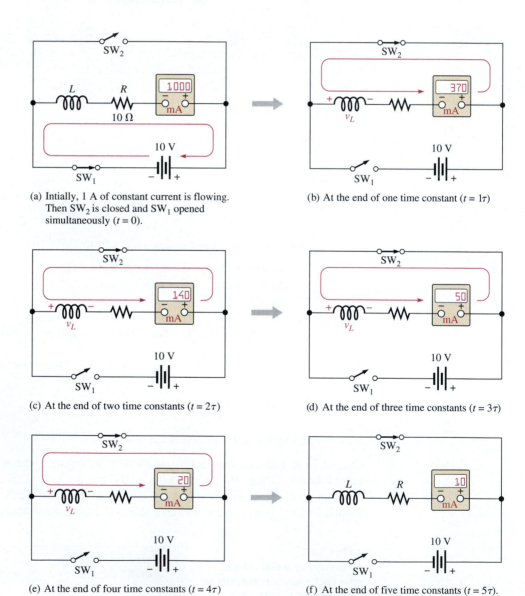

(a) Intially, 1 A of constant current is flowing. Then SW_2 is closed and SW_1 opened simultaneously $(t = 0)$.

(b) At the end of one time constant $(t = 1\tau)$

(c) At the end of two time constants $(t = 2\tau)$

(d) At the end of three time constants $(t = 3\tau)$

(e) At the end of four time constants $(t = 4\tau)$

(f) At the end of five time constants $(t = 5\tau)$. Since only 1% of the current is left, this value is taken as the final zero value.

FIGURE 10–22

Illustration of the exponential decrease of current in an inductor. The current decreases another 63% during each time constant interval.

the induced voltage keeps the 1 A flowing. During the first time-constant interval, the current decreases by 63% down to 370 mA (37% of its initial value), as indicated in Figure 10–22(b). During the second time-constant interval, the current decreases by another 63% to 140 mA (14% of its initial value), as shown in Figure 10–22(c). The continued decrease in the current is illustrated by the remaining parts of Figure 10–22. Part (f) shows that only 1% of the initial current is left at the end of five time constants. Traditionally, this value is accepted as the final value and is approximated as zero current. Notice that until after the five time constants have elapsed, there is an induced voltage across the coil which is trying to keep the current flowing. This voltage follows a decreasing exponential curve, as will be discussed later.

EXAMPLE 10–6

In Figure 10–23, switch 1 (SW$_1$) is opened at the instant that switch 2 (SW$_2$) is closed.
(a) What is the time constant?
(b) What is the initial coil current at the instant of switching?
(c) What is the coil current at 1τ?
Assume steady-state current through the coil prior to switch change.

FIGURE 10–23

Solution

(a) $\tau = \dfrac{L}{R} = \dfrac{200\ \mu\text{H}}{10\ \Omega} = 20\ \mu\text{s}$

(b) Current cannot change instantaneously in an inductor. Therefore, the current at the instant of the switch change is the same as the steady-state current:

$$I = \dfrac{5\ \text{V}}{10\ \Omega} = 500\ \text{mA}$$

(c) At 1τ, the current has decreased to 37% of its initial value:

$$i = 0.37(500\ \text{mA}) = 185\ \text{mA}$$

Related Exercise   Change R to 47 Ω and L to 1 mH in Figure 10–23 and repeat each calculation.

Induced Voltage in the Series *RL* Circuit

As you know, when current changes in an inductor, a voltage is induced. We now examine what happens to the voltages across the resistor and the coil in a series circuit when a change in current occurs.

Look at the circuit in Figure 10–24(a). When the switch is open, there is no current, and the resistor voltage and the coil voltage are both zero. At the instant the switch is closed, as indicated in part (b), V_R is zero and V_L is 10 V. The reason for this change is

(a) Before switched is closed

(b) At instant switch is closed, V_L is equal and opposite to V_S.

(c) During the first five time constants, V_R increases exponentially with current, and V_L decreases exponentially.

(d) After the first five time constants, $V_R = 10$ V and $V_R \approx 0$ V. The current is at a constant maximum value.

FIGURE 10–24

Illustration of how the coil voltage and the resistor voltage change in response to an increase in current.

that the induced voltage across the coil is equal and opposite to the applied voltage, preventing the current from changing instantaneously. Therefore, *at the instant of switch closure, L effectively acts as an open with all the applied voltage across it.*

During the first five time constants, the current is building up exponentially, and the induced coil voltage is decreasing. The resistor voltage increases with the current, as Figure 10–24(c) illustrates.

After five time constants have elapsed, the current has reached its final value, V_S/R. At this time, all of the applied voltage is dropped across the resistor and none across the coil. Thus, *L effectively acts as a short to nonchanging current,* as Figure 10–24(d) illustrates. Keep in mind that the inductor always reacts to a change in current by creating an induced voltage in order to counteract that change.

Now let's examine the case illustrated in Figure 10–25, where the current is switched out, and the inductor discharges through another path. Part (a) shows the steady-state condition, and part (b) illustrates the instant at which the source is removed by opening SW_1 and the discharge path is connected by the closure of SW_2. There was 1 A through L prior to this. Notice that 10 V are induced in L in the direction to aid the 1 A in an effort to keep it from changing. Then, as shown in Figure 10–25(c), the current decays exponentially, and so do V_R and V_L. After 5τ, as shown in Figure 10–25(d), all the energy stored in the magnetic field of L is dissipated, and all values are zero.

(a) Initially, a constant 1 A flows and the voltage is dropped across R.

(b) At the instant that SW_1 is opened and SW_2 closed, 10 V is induced across L.

(c) During the five-time-constant interval, V_R and V_L decrease exponentially with the current.

(d) After the five time constants, V_R, V_L, and I are all zero.

FIGURE 10–25

Illustration of how the coil voltage and the resistor voltage change in response to a decrease in current.

EXAMPLE 10–7

(a) In Figure 10–26(a), what is v_L at the instant the switch (SW) is closed? What is v_L after 5τ?

(b) In Figure 10–26(b), what is v_L at the instant SW_1 opens and SW_2 closes? What is v_L after 5τ?

FIGURE 10–26

(a)

(b)

Solution

(a) At the instant the switch is closed, all the source voltage is across L. Thus, $v_L = 25$ V, with the polarity as shown. After 5τ, L acts as a short, so $v_L = 0$ V.

(b) With SW_1 closed and SW_2 open, the steady-state current is

$$I = \frac{25 \text{ V}}{12 \text{ }\Omega} = 2.08 \text{ A}$$

When the switches are thrown, an induced voltage is created across L sufficient to keep this 2.08 A current for an instant. In this case, it takes an induced voltage

$$v_L = IR_2 = (2.08 \text{ A})(100 \text{ }\Omega) = 208 \text{ V}$$

After 5τ, the inductor voltage is zero.

Related Exercise Repeat part (a) if the voltage source is 9 V. Repeat part (b) if R_1 is 27 Ω with the source voltage at 25 V.

The Exponential Formulas

The formulas for the exponential current and voltage in an RL circuit are similar to those used in the last chapter for the RC circuit, and the universal exponential curves in Figure 9–36 apply to inductors as well as capacitors. The general formulas for RL circuits are stated as follows:

$$v = V_F + (V_i - V_F)e^{-Rt/L} \tag{10–7}$$

$$i = I_F + (I_i - I_F)e^{-Rt/L} \tag{10–8}$$

where V_F and I_F are the final values, V_i and I_i are the initial values, and v and i are the instantaneous values of the inductor voltage or current at time t.

Increasing Current The formula for the special case in which an increasing exponential current curve begins at zero ($I_i = 0$) is

$$i = I_F(1 - e^{-Rt/L}) \tag{10–9}$$

Using Equation (10–9), you can calculate the value of the increasing inductor current at any instant of time. The same is true for voltage.

EXAMPLE 10–8   In Figure 10–27, determine the inductor current 30 μs after the switch is closed.

FIGURE 10–27

Solution   The time constant is

$$\tau = \frac{L}{R} = \frac{100 \text{ mH}}{2.2 \text{ k}\Omega} = 45.5 \ \mu s$$

The final current is

$$I_F = \frac{V_S}{R} = \frac{12 \text{ V}}{2.2 \text{ k}\Omega} = 5.45 \text{ mA}$$

The initial current is zero. Notice that 30 μs is less than one time constant, so the current will reach less than 63% of its final value in that time.

$$i_L = I_F(1 - e^{-Rt/L}) = 5.45 \text{ mA}(1 - e^{-0.66}) = 5.45 \text{ mA}(1 - 0.517) = 2.63 \text{ mA}$$

The calculator sequence is

[3] [0] [EXP] [+/−] [6] [÷] [4] [5] [.] [5] [EXP] [+/−] [6] [=] [+/−] [2nd] [ln x] [+/−]
[+] [1] [=] [×] [5] [.] [4] [5] [EXP] [+/−] [3] [=]

Related Exercise   In Figure 10–27, determine the inductor current 55 μs after the switch is closed.

Decreasing Current   The formula for the special case in which a decreasing exponential current has a final value of zero is as follows:

$$i = I_i e^{-Rt/L} \tag{10–10}$$

This formula can be used to calculate the deenergizing current at any instant, as the following example shows.

EXAMPLE 10–9   Determine the inductor current in Figure 10–28 at a point in time 2 ms after the switches are thrown (SW$_1$ opened and SW$_2$ closed).

FIGURE 10–28

Solution   The deenergizing time constant is

$$\tau = \frac{L}{R} = \frac{200 \text{ mH}}{56 \ \Omega} = 3.57 \text{ ms}$$

The initial current in the inductor is 89.3 mA. Notice that 2 ms is less than one time constant, so the current will show a decrease less than 63%. Therefore, the current will be greater than 37% of its initial value at 2 ms after the switches are thrown.

$$i_L = I_i e^{-Rt/L} = (89.3 \text{ mA})e^{-0.56} = 51.0 \text{ mA}$$

The calculator sequence is

Related Exercise Determine the inductor current in Figure 10–28 at a point in time 6 ms after SW_1 is opened and SW_2 is closed if the source voltage is changed to 10 V.

SECTION 10–5 REVIEW

1. A 15 mH inductor with a winding resistance of 10 Ω has a constant direct current of 10 mA through it. What is the voltage drop across the inductor?
2. A 20 V dc source is connected to a series *RL* circuit with a switch. At the instant of switch closure, what are the values of v_R and v_L?
3. In the same circuit, after a time interval equal to 5τ from switch closure, what are v_R and v_L?
4. In a series *RL* circuit where $R = 1$ kΩ and $L = 500$ μH, what is the time constant? Determine the current 0.25 μs after a switch connects 10 V across the circuit.

10–6 ■ INDUCTORS IN AC CIRCUITS

You will learn in this section that an inductor passes ac but with an amount of opposition that depends on the frequency of the ac.

After completing this section, you should be able to

■ **Analyze inductive ac circuits**
  □ Define *inductive reactance*
  □ Determine the value of inductive reactance in a given circuit
  □ Discuss instantaneous, true, and reactive power in an inductor

Inductive Reactance, X_L

In Figure 10–29, an inductor is shown connected to a sinusoidal voltage source. When the source voltage is held at a constant amplitude value and its frequency is increased, the amplitude of the current decreases. Also, when the frequency of the source is decreased, the current amplitude increases. The reason is as follows: When the frequency of the applied voltage increases, its rate of change also increases, as you already know. Now, if the frequency of the applied voltage is increased, the frequency of the current also increases. According to Faraday's and Lenz's laws, this increase in frequency induces more voltage across the inductor in a direction to oppose the current and cause it to decrease in amplitude. Similarly, a decrease in frequency will cause an increase in current.

A decrease in the amount of current for an increase in frequency with a fixed amount of voltage indicates that opposition to the current has increased. Thus, the inductor offers opposition to current which varies directly with frequency.

(a) Current decreases when the frequency increases.

(b) Current increases when the frequency decreases.

FIGURE 10–29

The current in an inductive circuit varies inversely with the frequency of the applied voltage.

> **The opposition to sinusoidal current in an inductor is called inductive reactance.**

The symbol for **inductive reactance** is X_L, and its unit is the ohm (Ω).

You have just seen how frequency affects the opposition to current (inductive reactance) in an inductor. Now let's see how the inductance, L, affects the reactance. Figure 10–30(a) shows that when a sinusoidal voltage with a fixed amplitude and fixed frequency is applied to a 1 mH inductor, a certain amount of alternating current flows. When the inductance value is increased to 2 mH, the current decreases, as shown in part (b). Thus, when the inductance is increased, the opposition to current (inductive reactance) increases. So not only is the inductive reactance *directly* proportional to frequency, but it is also *directly* proportional to inductance. This relationship can be stated as follows:

> X_L **is proportional to** *fL.*

It can be proven that the constant of proportionality is 2π, so the formula for inductive reactance (X_L) is

$$X_L = 2\pi fL \qquad\qquad (10\text{–}11)$$

As with capacitive reactance, the 2π term comes from the relationship of the sine wave to rotational motion.

(a) Less inductance, more current

(b) More inductance, less current.

FIGURE 10–30

For a fixed voltage and fixed frequency, the current varies inversely with the inductance value.

EXAMPLE 10–10   A sinusoidal voltage is applied to the circuit in Figure 10–31. The frequency is 1 kHz. Determine the inductive reactance.

FIGURE 10–31

Solution   Convert 1 kHz to 1×10^3 Hz and 5 mH to 5×10^{-3} H. Then, the inductive reactance is

$$X_L = 2\pi fL = 2\pi(1 \times 10^3 \text{ Hz})(5 \times 10^{-3} \text{ H}) = 31.4 \text{ } \Omega$$

Related Exercise   What is X_L in Figure 10–31 if the frequency is increased to 3.5 kHz?

Ohm's Law in Inductive AC Circuits

The reactance of an inductor is analogous to the resistance of a resistor. In fact, X_L, just like X_C and R, is expressed in ohms. Since inductive reactance is a form of opposition to current. Ohm's law applies to inductive circuits as well as to resistive circuits and capacitive circuits, and it is stated as follows:

$$V = IX_L \tag{10–12}$$

When applying Ohm's law in ac circuits, you must express both the current and the voltage in the same way, that is, both in rms, both in peak, and so on.

EXAMPLE 10–11   Determine the rms current in Figure 10–32.

FIGURE 10–32

5 V rms
$f = 10$ kHz

L
100 mH

Solution   Convert 10 kHz to 10×10^3 Hz and 100 mH to 100×10^{-3} H. Then calculate X_L:

$$X_L = 2\pi fL = 2\pi(10 \times 10^3 \text{ Hz})(100 \times 10^{-3} \text{ H}) = 6283 \text{ } \Omega$$

Applying Ohm's law,

$$I_{rms} = \frac{V_{rms}}{X_L} = \frac{5 \text{ V}}{6283 \text{ } \Omega} = 796 \text{ } \mu A$$

Related Exercise   Determine the rms current in Figure 10–32 for the following values: $V_{rms} = 12$ V, $f = 4.9$ kHz, and $L = 680$ μH.

Current Lags Inductor Voltage by 90°

As you know, a sinusoidal voltage has a maximum rate of change at its zero crossings and a zero rate of change at the peaks. From Faraday's law (Chapter 7) you know that the amount of voltage induced across a coil is directly proportional to the rate at which the current is changing. Therefore, the coil voltage is maximum at the zero crossings of the current where the rate of change of the current is the greatest. Also, the amount of voltage is zero at the peaks of the current where the rate of change is zero. This phase relationship is illustrated in Figure 10–33. As you can see, the current peaks occur a quarter cycle after the voltage peaks. Thus, the current lags the voltage by 90°.

FIGURE 10–33
Current is always lagging the inductor voltage by 90°.

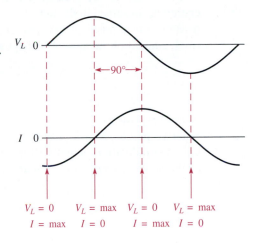

Power in an Inductor

As discussed earlier, an inductor stores energy in its magnetic field when there is current through it. An ideal inductor (assuming no winding resistance) does not dissipate energy; it only stores it. When an ac voltage is applied to an inductor, energy is stored by the inductor during a portion of the cycle; then the stored energy is returned to the source during another portion of the cycle. There is no net energy loss. Figure 10–34 shows the power curve that results from one cycle of inductor current and voltage.

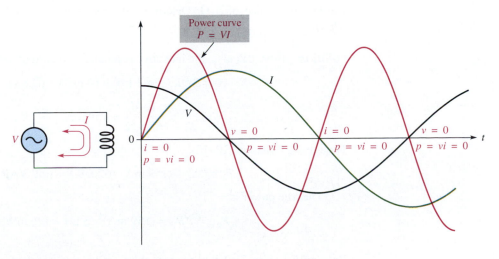

FIGURE 10–34
Power curve for an inductor.

***Instantaneous Power* (*p*)**   The product of instantaneous voltage, *v,* and instantaneous current, *i,* gives instantaneous power, *p*. At points where *v* or *i* is zero, *p* is also zero. When both *v* and *i* are positive, *p* is also positive. When either *v* or *i* is positive and the other negative, *p* is negative. When both *v* and *i* are negative, *p* is positive. As you can see in Figure 10–34, the power follows a sinusoidal-type curve. Positive values of power indicate that energy is stored by the inductor. Negative values of power indicate that energy is returned from the inductor to the source. Note that the power fluctuates at a frequency twice that of the voltage or current, as energy is alternately stored and returned to the source.

***True Power* (P_{true})**   Ideally, all of the energy stored by an inductor during the positive portion of the power cycle is returned to the source during the negative portion. No net energy is consumed in the inductance, so the power is zero. Actually, because of winding resistance in a practical inductor, some power is always dissipated, and there is a very small amount of true power, which can normally be neglected.

$$P_{\text{true}} = I_{\text{rms}}^2 R_W \qquad (10\text{–}13)$$

***Reactive Power* (P_r)**   The rate at which an inductor stores or returns energy is called its **reactive power, P_r**. The reactive power is a nonzero quantity, because at any instant in time, the inductor is actually taking energy from the source or returning energy to it. Reactive power does not represent an energy loss. The following formulas apply:

$$P_r = V_{\text{rms}} I_{\text{rms}} \qquad (10\text{–}14)$$

$$P_r = \frac{V_{\text{rms}}^2}{X_L} \qquad (10\text{–}15)$$

$$P_r = I_{\text{rms}}^2 X_L \qquad (10\text{–}16)$$

The unit of reactive power is volt-ampere reactive (VAR).

EXAMPLE 10–12   A 10 V rms signal with a frequency of 1 kHz is applied to a 10 mH coil with a winding resistance of 5 Ω. Determine the reactive power (P_r) and the true power (P_{true}).

Solution   First, calculate the inductive reactance and current values:

$$X_L = 2\pi f L = 2\pi(1 \text{ kHz})(10 \text{ mH}) = 62.8 \ \Omega$$

$$I = \frac{V_S}{X_L} = \frac{10 \text{ V}}{62.8 \ \Omega} = 159 \text{ mA}$$

Then using Equation (10–16),

$$P_r = I^2 X_L = (159 \text{ mA})^2(62.8 \ \Omega) = 1.59 \text{ VAR}$$

The true power is

$$P_{\text{true}} = I^2 R_W = (159 \text{ mA})^2 (5 \ \Omega) = 126 \text{ mW}$$

Related Exercise   What happens to the reactive power if the frequency increases?

The Quality Factor (Q) of a Coil

The **quality factor** (Q) is the ratio of the reactive power in the inductor to the true power in the winding resistance of the coil or the resistance in series with the coil. It is a ratio of the power in L to the power in R_W. The quality factor is very important in resonant circuits, which are studied in Chapter 14. A formula for Q is developed as follows:

$$Q = \frac{\text{reactive power}}{\text{true power}} = \frac{P_r}{P_{\text{true}}} = \frac{I^2 X_L}{I^2 R_W}$$

In a series circuit, I is the same in L and R; thus, the I^2 terms cancel, leaving

$$Q = \frac{X_L}{R_W} \qquad \qquad (10\text{--}17)$$

When the resistance is just the winding resistance of the coil, the circuit Q and the coil Q are the same. Note that Q is a ratio of like units and, therefore, has no unit itself.

SECTION 10–6 REVIEW

1. State the phase relationship between current and voltage in an inductor.
2. Calculate X_L for $f = 5$ kHz and $L = 100$ mH.
3. At what frequency is the reactance of a 50 μH inductor equal to 800 Ω?
4. Calculate the rms current in Figure 10–35.

FIGURE 10–35

$V_{\text{rms}} = 1$ V
$f = 1$ MHz

L
$10\,\mu$H

5. A 50 mH inductor is connected to a 12 V rms source. What is the true power? What is the reactive power at a frequency of 1 kHz?

10–7 ▪ INDUCTOR APPLICATIONS

Inductors are not as versatile as capacitors and tend to be more limited in their applications due, in part, to size and cost factors. However, there are many practical uses for inductors (coils) such as those discussed in Chapter 7. Recall that relay and solenoid coils, recording and pick-up heads, and sensing elements were introduced as electromagnetic applications of coils. In this section, some additional uses of inductors are presented.

After completing this section, you should be able to

▪ **Discuss some inductor applications**
  ☐ Describe a power supply filter
  ☐ Explain the purpose of an rf choke
  ☐ Discuss the basics of tuned circuits

Power Supply Filter

In Chapter 9, you saw that a capacitor is used to filter the pulsating dc on the output of the rectifier in a power supply. The final output voltage was a dc voltage with a small amount of ripple. In many cases, an inductor is used in the filter, as shown in Figure 10–36(a), to smooth out the ripple voltage. The inductor, placed in series with the load as shown, tends to oppose the current fluctuations caused by the ripple voltage, and thus the voltage developed across the load is more constant, as shown in Figure 10–36(b).

(a)

(b)

FIGURE 10–36
Basic capacitor power supply filter with a series inductor.

RF Choke

Certain types of inductors are used in applications where radio frequencies (rf) must be prevented from getting into parts of a system, such as the power supply or the audio section of a receiver. In these situations, an inductor is used as a series filter and "chokes" off any unwanted rf signals that may be picked up on a line. This filtering action is based on the fact that the reactance of a coil increases with frequency. When the frequency of the current is sufficiently high, the reactance of the coil becomes extremely large and essentially blocks the current. A basic illustration of an inductor used as an rf choke is shown in Figure 10–37.

Tuned Circuits

As mentioned in Chapter 9, inductors are used in conjunction with capacitors to provide frequency selection in communications systems. These tuned circuits allow a narrow

FIGURE 10–37
An inductor used as an rf choke to minimize interfering signals on the power supply line.

band of frequencies to be selected while all other frequencies are rejected. The tuners in your TV and radio receivers are based on this principle and permit you to select one channel or station out of the many that are available.

Frequency selectivity is based on the fact that the reactances of both capacitors and inductors depend on the frequency and on the interaction of these two components when connected in series or parallel. Since the capacitor and the inductor produce opposite phase shifts, their combined opposition to current can be used to obtain a desired response at a selected frequency. Tuned (resonant) *RLC* circuits are covered in detail in Chapter 14.

SECTION 10–7 REVIEW	**1.** Explain how the ripple voltage from a power supply filter can be reduced through use of an inductor. **2.** How does an inductor connected in series act as an rf choke?

10–8 ▪ TESTING INDUCTORS

In this section, two basic types of failures in inductors are discussed and methods for testing inductors are introduced.

After completing this section, you should be able to

▪ **Test an inductor**
  ☐ Perform an ohmmeter check for an open winding
  ☐ Perform an ohmmeter check for shorted windings

The most common failure in an inductor is an open coil. To check for an open, remove the coil from the circuit. If there is an open, an ohmmeter check will indicate infinite resistance, as shown in Figure 10–38(a). If the coil is good, the ohmmeter will show the winding resistance. The value of the winding resistance depends on the wire size and length of the coil. It can be anywhere from one ohm to several hundred ohms. Figure 10–38(b) shows a good reading.

(a) Coil open: Meter reads infinity

(b) Coil good: Meter reads winding resistance

(c) Shorted winding: Meter reads zero or a lower than normal winding resistance depending on the number of turns that are shorted

FIGURE 10–38
Checking a coil by measuring the resistance.

Occasionally, when an inductor is overheated with excessive current, the wire insulation will melt, and some coils will short together. This produces a reduction in the inductance by reducing the effective number of turns and a corresponding reduction in winding resistance, as shown in Figure 10–38(c).

10–9 ■ APPLICATION ASSIGNMENT

Your supervisor has handed you two unmarked coils and wants to know their inductance values. You cannot find an inductance bridge, which is an instrument for measuring inductance directly. After scratching your head for awhile, you decide to use the time-constant characteristic of inductive circuits to determine the unknown inductances. A test setup consisting of a square wave generator and an oscilloscope is used to make the measurements.

The method is to place the coil in series with a resistor of known value and measure the time constant by applying a square wave to the circuit and observing the resulting voltage across the resistor with an oscilloscope. Knowing the time constant and the resistance value, you can calculate the inductance L.

Each time the square wave input voltage goes high, the inductor is energized; and each time the square wave goes back to zero, the inductor is deenergized. The time it takes for the exponential resistor voltage to increase to approximately its final value equals to five time constants. This operation is illustrated in Figure 10–39. To make sure that the winding resistance of the coil can be neglected, it must be measured; and the value of the resistor used in the circuit must be selected to be considerably larger than the winding resistance.

FIGURE 10–39
Circuit for time-constant measurement.

Step 1: Measure the Coil Resistance and Select a Series Resistor

Assume that the winding resistance has been measured with an ohmmeter and found to be 85 Ω. To make the winding resistance negligible, a 10 kΩ series resistor is used in the circuit.

Step 2: Determine the Inductance of Coil 1

Refer to Figure 10–40. To determine the inductance, a 10 V square wave is applied to the breadboarded circuit. The frequency of the square wave is adjusted so that the inductor has time to fully energize during each square wave pulse; the scope is set to view a complete energizing curve as shown. Determine the approximate circuit time constant from the scope display and calculate the inductance of coil 1.

FIGURE 10–40
Testing coil 1.

Step 3: Determine the Inductance of Coil 2

Refer to Figure 10–41. To determine the inductance, a 10 V square wave is applied to the breadboarded circuit. The frequency of the square wave is adjusted so that the inductor has time to fully energize during each square wave pulse; the scope is set to view a complete energizing curve as shown. Determine the approximate circuit time constant from the scope display and calculate the inductance of coil 2. Discuss any difficulty you find with this method.

Step 4: Determine Another Way to Find Unknown Inductance

Determination of the time constant is not the only way that you can find an unknown inductance. Specify a method using a sinusoidal input voltage instead of the square wave.

10 V square wave input

FIGURE 10–41
Testing coil 2.

■ SUMMARY

- Self-inductance is a measure of a coil's ability to establish an induced voltage as a result of a change in its current.
- An inductor opposes a change in its own current.
- Faraday's law states that relative motion between a magnetic field and a coil induces a voltage across the coil.
- Lenz's law states that the polarity of induced voltage is such that the resulting induced current is in a direction that opposes the change in the magnetic field that produced it.
- Energy is stored by an inductor in its magnetic field.
- One henry is the amount of inductance when current, changing at the rate of one ampere per second, induces one volt across the inductor.

- Inductance is directly proportional to the square of the turns, the permeability, and the cross-sectional area of the core. It is inversely proportional to the length of the core.
- The permeability of a core material is an indication of the ability of the material to establish a magnetic field.
- Inductors add in series.
- Total parallel inductance is less than that of the smallest inductor in parallel.
- The time constant for a series RL circuit is the inductance divided by the resistance.
- In an RL circuit, the voltage and current in an energizing or deenergizing inductor make a 63% change during each time-constant interval.
- Energizing and deenergizing current and voltage follow exponential curves.
- Voltage leads current by 90° in an inductor.
- Inductive reactance (X_L) is directly proportional to frequency and inductance.
- The true power in an inductor is zero; that is, there is no energy loss in an ideal inductor, only in its winding resistance.

■ GLOSSARY

Choke   An inductor. The term is more commonly used in connection with inductors used to block or choke off high frequencies.

Coil   A common term for an inductor.

Core   The structure around which the winding of an inductor is formed. The core material influences the electromagnetic characteristics of the inductor.

Henry (H)   The unit of inductance.

Induced voltage   Voltage produced as a result of a changing magnetic field.

Inductance (L)   The property of an inductor that produces an opposition to any change in current.

Inductive reactance (X_L)   The opposition of an inductor to sinusoidal current. The unit is the ohm.

Inductor   An electrical device formed by a wire wound around a core having the property of inductance; also known as a coil or a choke.

Lenz's law   A physical law that states when the current through a coil changes, an induced voltage is created in a direction to oppose the change in current. The current cannot change instantaneously.

Quality factor (Q)   The ratio of reactive power to true power.

Reactive power (P_r)   The rate at which energy is alternately stored and returned to the source by an inductor.

Time constant   A fixed time interval, set by the L and R values, that determines the time response of a circuit.

Winding   The loops of wire in an inductor.

■ FORMULAS

(10–1) $$W = \frac{1}{2}LI^2$$ Energy stored by an inductor

(10–2) $$L = \frac{N^2 \mu A}{l}$$ Inductance in terms of physical parameters

(10–3) $$L_T = L_1 + L_2 + L_3 + \cdots + L_n$$ Series inductance

(10–4) $$\frac{1}{L_T} = \frac{1}{L_1} + \frac{1}{L_2} + \frac{1}{L_3} + \cdots + \frac{1}{L_n}$$ Reciprocal of total parallel inductance

(10–5) $$L_T = \frac{1}{\left(\frac{1}{L_1}\right) + \left(\frac{1}{L_2}\right) + \left(\frac{1}{L_3}\right) + \cdots + \left(\frac{1}{L_n}\right)}$$ Total parallel inductance

(10–6) $$\tau = \frac{L}{R}$$ Time constant

(10–7) $$v = V_F + (V_i - V_F)e^{-Rt/L}$$ Exponential voltage (general)

(10–8) $$i = I_F + (I_i - I_F)e^{-Rt/L}$$ Exponential current (general)

(10–9)	$i = I_F(1 - e^{-Rt/L})$	Increasing exponential current beginning at zero
(10–10)	$i = I_i e^{-Rt/L}$	Decreasing exponential current ending at zero
(10–11)	$X_L = 2\pi fL$	Inductive reactance
(10–12)	$V = IX_L$	Ohm's law
(10–13)	$P_{true} = I_{rms}^2 R_W$	True power
(10–14)	$P_r = V_{rms}I_{rms}$	Reactive power
(10–15)	$P_r = \dfrac{V_{rms}^2}{X_L}$	Reactive power
(10–16)	$P_r = I_{rms}^2 X_L$	Reactive power
(10–17)	$Q = \dfrac{X_L}{R_W}$	Quality factor

▪ SELF-TEST

1. An inductance of 0.05 μH is larger than
   (a) 0.0000005 H   (b) 0.000005 H   (c) 0.000000008 H   (d) 0.00005 mH

2. An inductance of 0.33 mH is smaller than
   (a) 33 μH   (b) 330 μH   (c) 0.05 mH   (d) 0.0005 H

3. When the current through an inductor increases, the amount of energy stored in the electromagnetic field
   (a) decreases   (b) remains constant   (c) increases   (d) doubles

4. When the current through an inductor doubles, the stored energy
   (a) doubles   (b) quadruples   (c) is halved   (d) does not change

5. The winding resistance of a coil can be decreased by
   (a) reducing the number of turns   (b) using a larger wire
   (c) changing the core material   (d) either (a) or (b)

6. The inductance of an iron-core coil increases if
   (a) the number of turns is increased   (b) the iron core is removed
   (c) the length of the core is increased   (d) larger wire is used

7. Four 10 mH inductors are in series. The total inductance is
   (a) 40 mH   (b) 2.5 mH   (c) 40,000 μH   (d) both (a) and (c)

8. A 1 mH, a 3.3 mH, and a 0.1 mH inductor are connected in parallel. The total inductance is
   (a) 4.4 mH   (b) greater than 3.3 mH   (c) less than 0.1 mH   (d) both (a) and (b)

9. An inductor, a resistor, and a switch are connected in series to a 12 V battery. At the instant the switch is closed, the inductor voltage is
   (a) 0 V   (b) 12 V   (c) 6 V   (d) 4 V

10. A sine wave voltage is applied across an inductor. When the frequency of the voltage is increased, the current
    (a) decreases   (b) increases   (c) does not change   (d) momentarily goes to zero

11. An inductor and a resistor are in series with a sinusoidal voltage source. The frequency is set so that the inductive reactance is equal to the resistance. If the frequency is increased, then
    (a) $V_R > V_L$ (b) $V_L < V_R$ (c) $V_L = V_R$ (d) $V_L > V_R$

12. An ohmmeter is connected across an inductor and the pointer indicates an infinite value. The inductor is
    (a) good   (b) open   (c) shorted   (d) resistive

▪ BASIC PROBLEMS

SECTION 10–1   The Basic Inductor

1. Convert the following to millihenries:
   (a) 1 H   (b) 250 μH   (c) 10 μH   (d) 0.0005 H

2. Convert the following to microhenries:
   (a) 300 mH   (b) 0.08 H   (c) 5 mH   (d) 0.00045 mH

3. How many turns are required to produce 30 mH with a coil wound on a cylindrical core having a cross-sectional area of 10×10^{-5} m^2 and a length of 0.05 m? The core has a permeability of 1.2×10^{-6}.

4. A 12 V battery is connected across a coil with a winding resistance of 12 Ω. How much current is there in the coil?

5. How much energy is stored by a 100 mH inductor with a current of 1 A?

6. The current through a 100 mH coil is changing at a rate of 200 mA/s. How much voltage is induced across the coil?

SECTION 10–3 Series Inductors

7. Five inductors are connected in series. The lowest value is 5 μH. If the value of each inductor is twice that of the preceding one, and if the inductors are connected in order of ascending values, what is the total inductance?

8. Suppose that you require a total inductance of 50 mH. You have available a 10 mH coil and a 22 mH coil. How much additional inductance do you need?

SECTION 10–4 Parallel Inductors

9. Determine the total parallel inductance for the following coils in parallel: 75 μH, 50 μH, 25 μH, and 15 μH.

10. You have a 12 mH inductor, and it is your smallest value. You need an inductance of 8 mH. What value can you use in parallel with the 12 mH to obtain 8 mH?

11. Determine the total inductance of each circuit in Figure 10–42.

12. Determine the total inductance of each circuit in Figure 10–43.

FIGURE 10–42

FIGURE 10–43

SECTION 10–5 Inductors in DC Circuits

13. Determine the time constant for each of the following series RL combinations:
    (a) $R = 100$ Ω, $L = 100$ μH   (b) $R = 4.7$ kΩ, $L = 10$ mH   (c) $R = 1.5$ MΩ, $L = 3$ H

14. In a series RL circuit, determine how long it takes the current to build up to its full value for each of the following:
    (a) $R = 56$ Ω, $L = 50$ μH   (b) $R = 3300$ Ω, $L = 15$ mH   (c) $R = 22$ kΩ, $L = 100$ mH

15. In the circuit of Figure 10–44, there is initially no current. Determine the inductor voltage at the following times after the switch is closed:
    (a) 10 μs (b) 20 μs (c) 30 μs (d) 40 μs (e) 50 μs

16. In Figure 10–45, there are 100 mA through the coil. When SW_1 is opened and SW_2 simultaneously closed, find the inductor voltage at the following times:
    (a) initially (b) 1.5 ms (c) 4.5 ms (d) 6 ms

FIGURE 10–44   **FIGURE 10–45**

SECTION 10–6 Inductors in AC Circuits

17. Find the total reactance for each circuit in Figure 10–42 when a voltage with a frequency of 5 kHz is applied across the terminals.

18. Find the total reactance for each circuit in Figure 10–43 when a 400 Hz voltage is applied.

19. Determine the total rms current in Figure 10–46. What are the currents through L_2 and L_3?

$L_{23} = 13.33m$

FIGURE 10–46

$63.33m$

20. What frequency will produce a total rms current of 500 mA in each circuit of Figure 10–43 with an rms input voltage of 10 V?

21. Determine the reactive power in Figure 10–46 neglecting the winding resistance.

$44.77m$

SECTION 10–8 Testing Inductors

22. A certain coil that is supposed to have a 5 Ω winding resistance is measured with an ohmmeter. The meter indicates 2.8 Ω. What is the problem with the coil?

23. Determine the meter indication corresponding to each of the following failures in a coil:
    (a) open (b) shorted (c) partially shorted

■ **ADVANCED PROBLEMS**

24. Determine the time constant for the circuit in Figure 10–47. See Appendix E.

25. Find the inductor current 10 μs after the switch is thrown from position 1 to position 2 in Figure 10–48. For simplicity, assume that the switch makes contact with position 2 at the same instant it breaks contact with position 1.

26. In Figure 10–49, SW_1 is opened and SW_2 is closed at the same instant (t_0). What is the instantaneous voltage across R_2 at t_0?

27. Determine the value of I_{L2} in Figure 10–50.

28. What is the total inductance between points A and B for each switch position in Figure 10–51?

FIGURE 10–47

FIGURE 10–48

FIGURE 10–49

FIGURE 10–50

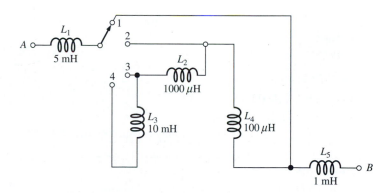

FIGURE 10–51

■ **ANSWERS TO SECTION REVIEWS**

Section 10–1

1. Parameters that determine inductance are turns, permeability, cross-sectional area, and length.

2. **(a)** When N increases, L increases. **(b)** When core length increases, L decreases.
   (c) When area decreases, L decreases. **(d)** For an air core, L decreases.

3. All wire has some resistance, and since inductors are made from turns of wire, there is always resistance in the winding.

4. Adjacent turns act as plates and produce a small capacitance.

Section 10–2

1. Two categories of inductors are fixed and variable.

2. **(a)** air core **(b)** iron core **(c)** variable

Section 10–3

1. Inductances are added in series.
2. $L_T = 2600 \ \mu H$
3. $L_T = 5 \times 100 \ mH = 500 \ mH$

Section 10–4

1. The total parallel inductance is smaller than that of the smallest individual inductor in parallel.
2. True
3. **(a)** $L_T = 7.69 \ mH$ **(b)** $L_T = 24 \ \mu H$ **(c)** $L_T = 100 \ mH$

Section 10–5

1. $V_L = (10 \ mA)(10 \ \Omega) = 100 \ mV$
2. Initially, $v_R = 0 \ V$, $v_L = 20 \ V$
3. After 5τ, $v_R = 20 \ V$, $v_L = 0 \ V$
4. $\tau = 500 \ \mu H/1 \ k\Omega = 500 \ ns$, $i = 3.93 \ mA$

Section 10–6

1. Voltage leads current by $90°$ in an inductor.
2. $X_L = 2 \ \pi f L = 3.14 \ k\Omega$
3. $f = X_L/2\pi L = 2.55 \ MHz$
4. $I_{rms} = 15.9 \ mA$
5. $P_{true} = 0 \ W$, $P_r = 458 \ mVAR$

Section 10–7

1. The inductor tends to level out the ripple because of its opposition to changes in current.
2. The inductive reactance is very high at radio frequencies and blocks these frequencies.

Section 10–8

1. False
2. False

Section 10–9

1. $f_{max} = 125 \ kHz$ ($5\tau = 4 \ ms$)
2. $f_{max} = 3.13 \ kHz$ ($5\tau = 160 \ \mu s$)
3. If $f > f_{max}$, the inductor would not fully energize because $T/2 < 5\tau$.

■ **ANSWERS TO RELATED EXERCISES FOR EXAMPLES**

10–1 108 mH
10–2 500 μH
10–3 20.3 μH
10–4 227 ns
10–5 $I_F = 29.4 \ mA$, $\tau = 147 \ ns$
At 1τ, $i = 18.5 \ mA$
At 2τ, $i = 25.3 \ mA$
At 3τ, $i = 27.9 \ mA$
At 4τ, $i = 28.8 \ mA$
At 5τ, $i = 29.1 \ mA$
10–6 **(a)** 21.3 μs **(b)** 106 mA **(c)** 39.2 mA
10–7 **(a)** 9 V; 0 V **(b)** 92.6 V; 0 V
10–8 3.82 mA
10–9 33.3 mA
10–10 110 Ω
10–11 574 mA
10–12 P_r decreases

11

TRANSFORMERS

In the last chapter, self-inductance was studied. In this chapter, another inductive effect, called *mutual inductance,* is introduced. Mutual inductance occurs when two or more coils are very close together. It is based on the principle of magnetic coupling. A basic transformer is formed by two coils that are magnetically coupled but have no electrical connection between them. The transfer of energy from one coil to the other occurs entirely through a changing magnetic field. There are many applications for transformers, and a basic understanding of their operation is essential.

■ CHAPTER OBJECTIVES

☐ Explain mutual inductance
☐ Describe how a transformer is constructed and how it works
☐ Explain how a step-up transformer works
☐ Explain how a step-down transformer works
☐ Discuss the effect of a resistive load across the secondary winding
☐ Discuss the concept of a reflected load in a transformer
☐ Discuss impedance matching with transformers
☐ Explain how the transformer acts as an isolation device
☐ Describe a practical transformer
☐ Describe several types of transformers
☐ Troubleshoot transformers

■ APPLICATION ASSIGNMENT

Your assignment is to troubleshoot a type of dc power supply that uses a transformer to couple the ac voltage from a standard outlet. By making voltage measurements at various points, you will determine if there is a fault and be able to specify the part of the power supply that is faulty. After studying this chapter, you should be able to complete the application assignment in Section 11–12.

11–1 ■ MUTUAL INDUCTANCE

When two coils are placed close to each other, a changing electromagnetic field produced by the current in one coil will cause an induced voltage in the second coil because of the mutual inductance.

After completing this section, you should be able to

■ **Explain mutual inductance**
  ☐ Discuss magnetic coupling
  ☐ Define *electrical isolation*
  ☐ Define *coefficient of coupling*
  ☐ Identify the factors that affect mutual inductance and state the formula

Recall that the electromagnetic field surrounding a coil expands, collapses, and reverses as the current increases, decreases, and reverses.

When a second coil is placed very close to the first coil so that the changing magnetic flux lines cut through the second coil, the coils are magnetically coupled and a voltage is induced, as indicated in Figure 11–1. When two coils are magnetically coupled, they provide **electrical isolation** because there is no electrical connection between them, only a magnetic link. If the current in the first coil is a sine wave, the voltage induced in the second coil is also a sine wave. The amount of voltage induced in the second coil as a result of the current in the first coil is dependent on the **mutual inductance, L_M,** between the two coils. The mutual inductance is established by the inductance of each coil (L_1 and L_2) and by the amount of coupling (k) between the two coils.

FIGURE 11–1

A voltage is induced in the second coil as a result of the changing current in the first coil producing a changing magnetic field that links the second coil.

Flux lines cutting second coil as the magnetic field expands, collapses, and reverses.

Oscilloscope

Coefficient of Coupling

The **coefficient of coupling, k,** between two coils is the ratio of the flux lines produced by coil 1 that link coil 2 (ϕ_{12}) to the total flux produced by coil 1 (ϕ_1).

$$k = \frac{\phi_{12}}{\phi_1}$$

(11–1)

For example, if half of the total flux produced by coil 1 links coil 2, then $k = 0.5$. A greater value of k means that more voltage is induced in coil 2 for a certain rate of change of current in coil 1. Note that k has no units. Recall that the unit of magnetic lines of force (flux) is the weber, abbreviated Wb.

The coefficient k depends on the physical closeness of the coils and the type of core material on which they are wound. Also, the construction and shape of the cores are factors.

Formula for Mutual Inductance

The three factors that influence mutual inductance, L_M, (k, L_1, and L_2) are shown in Figure 11–2. The formula for mutual inductance is

$$L_M = k\sqrt{L_1 L_2}$$ (11–2)

FIGURE 11–2
The mutual inductance of two coils.

EXAMPLE 11–1
One coil produces a total magnetic flux of 50 μWb, and 20 μWb link coil 2. What is k?

Solution
$$k = \frac{\phi_{12}}{\phi_1} = \frac{20 \ \mu\text{Wb}}{50 \ \mu\text{Wb}} = 0.4$$

Related Exercise   Determine k when $\phi_1 = 500 \ \mu$Wb and $\phi_{12} = 375 \ \mu$Wb.

EXAMPLE 11–2
Two coils are wound on a single core, and the coefficient of coupling is 0.3. The inductance of coil 1 is 10 μH, and the inductance of coil 2 is 15 μH. What is L_M?

Solution $L_M = k\sqrt{L_1 L_2} = 0.3\sqrt{(10 \ \mu\text{H})(15 \ \mu\text{H})} = 3.67 \ \mu\text{H}$

Related Exercise   Determine the mutual inductance when $k = 0.5$, $L_1 = 1$ mH, and $L_2 = 600 \ \mu$H.

SECTION 11–1 REVIEW
1. Define *mutual inductance*.
2. Two 50 mH coils have $k = 0.9$. What is L_M?
3. If k is increased, what happens to the voltage induced in one coil as a result of a current change in the other coil?

11–2 ■ THE BASIC TRANSFORMER

A basic transformer is an electrical device constructed of two coils placed in close proximity to each other so that there is a mutual inductance between the two coils.

After completing this section, you should be able to

■ **Describe how a transformer is constructed and how it works**
   ☐ Identify the parts of a basic transformer
   ☐ Discuss the importance of the core material
   ☐ Define *primary winding* and *secondary winding*
   ☐ Define *turns ratio*
   ☐ Discuss how the direction of windings affect voltage polarities

A schematic of a **transformer** is shown in Figure 11–3(a). One coil is called the **primary winding** and the other coil is called the **secondary winding** as indicated. In standard operation, the source voltage is applied to the primary winding, and a load is connected to the secondary winding, as shown in Figure 11–3(b).

The windings of a transformer are formed around the **core.** The core provides both a physical structure for placement of the windings and a magnetic path so that the magnetic flux lines are concentrated close to the coils. There are three general categories of core material: air, ferrite, and iron. The schematic symbol for each type is shown in Figure 11–4.

Air-core and ferrite-core transformers generally are used for high-frequency applications and consist of windings on an insulating shell which is hollow (air) or constructed of ferrite, such as depicted in Figure 11–5. The wire is typically covered by a varnish-type coating to prevent the windings from shorting together. The amount of **magnetic coupling** between the primary winding and the secondary winding is set by the type of core material and by the relative positions of the windings. In Figure 11–5(a), the windings are loosely coupled because they are separated, and in part (b) they are tightly coupled because they are overlapping. The tighter the coupling, the greater the induced voltage in the secondary for a given current in the primary.

Iron-core transformers generally are used for audio frequency (af) and power applications. These transformers consist of windings on a core constructed from laminated sheets of ferromagnetic material insulated from each other, as shown in Figure 11–6. This construction provides an easy path for the magnetic flux and increases the amount of coupling between the windings. Figures 11–6(a) and 11–6(b) show the basic construction of two major configurations of iron-core transformers. In the core-type construction, shown in part (a), the windings are on separate legs of the laminated core. In the shell-type construction, shown in part (b), both windings are on the same leg. A variety of transformers are shown in Figure 11–7.

FIGURE 11–3
The basic transformer.

Primary winding   Secondary winding

Source   Load

(a) Schematic symbol   (b) Source/load connections

FIGURE 11–4
Schematic symbols based on type of core.

(a) Air core   (b) Ferrite core   (c) Iron core

FIGURE 11–5
Transformers with cylindrical-shaped cores.

Air or ferrite core

(a) Loosely coupled windings

(b) Tightly coupled windings. Cutaway view shows both windings.

Laminated iron core

Primary winding

Secondary winding

(a) Core-type has each winding on a separate leg.

Laminated iron core

(b) Shell-type has both windings on the same leg.

FIGURE 11–6
Iron-core transformer construction with multilayer windings.

FIGURE 11–7
Some common types of transformers (courtesy of Litton Triad-Utrad).

Turns Ratio

An important parameter of a transformer is its **turns ratio.** The turns ratio (n) is the ratio of the number of turns in the secondary winding (N_{sec}) to the number of turns in the primary winding (N_{pri}).

$$n = \frac{N_{sec}}{N_{pri}}$$

(11–3)

In the following sections, you will see how the turns ratio affects the voltages and currents in a transformer. Often, the primary and secondary voltages are specified on a transformer, while the turns ratio is not.

EXAMPLE 11–3

A certain transformer used in a radar system has a primary winding with 100 turns and a secondary winding with 400 turns. What is the turns ratio?

Solution $N_{sec} = 400$ and $N_{pri} = 100$; therefore, the turns ratio is

$$n = \frac{N_{sec}}{N_{pri}} = \frac{400}{100} = 4$$

A turns ratio of 4 can be expressed as 1:4 on a schematic.

Related Exercise A certain transformer has a turns ratio of 10. If $N_{pri} = 50$, what is N_{sec}?

Direction of Windings

Another important transformer parameter is the direction in which the windings are placed around the core. As illustrated in Figure 11–8, the direction of the windings determines the polarity of the secondary voltage with respect to the primary voltage. Phase dots can be used to indicate polarities, as shown in Figure 11–9. If the primary is positive at the dot, the secondary is positive at the dot. If the primary is negative at the dot, the secondary is negative at the dot.

FIGURE 11–8

Relative polarities of the voltages are determined by the direction of the windings.

(a) The primary and secondary voltages are in phase when the windings are in the same effective direction around the magnetic path.

(b) The primary and secondary voltages are 180° out of phase when the windings are in the opposite direction.

Phase dots

(a) Voltages are in phase.

(b) Voltages are out of phase.

FIGURE 11–9
Phase dots indicate corresponding polarities of primary and secondary voltages.

SECTION 11–2 REVIEW

1. Upon what principle is the operation of a transformer based?
2. Define *turns ratio*.
3. Why are the directions of the windings of a transformer important?
4. A certain transformer has a primary winding with 500 turns and a secondary winding with 250 turns. What is the turns ratio?

11–3 ■ THE STEP-UP TRANSFORMER

A step-up transformer has more turns in its secondary winding than in its primary winding and is used to increase ac voltage.

After completing this section, you should be able to

■ **Explain how a step-up transformer works**
  □ State the relationship between primary and secondary voltages and the number of turns in the primary and secondary windings
  □ Identify a step-up transformer by its turns ratio

A transformer in which the secondary voltage is greater than the primary voltage is called a **step-up transformer.** The amount that the voltage is stepped up depends on the turns ratio. For any transformer,

The ratio of secondary voltage (V_{sec}) to primary voltage (V_{pri}) is equal to the ratio of the number of secondary turns (N_{sec}) to the number of primary turns (N_{pri}).

$$\frac{V_{sec}}{V_{pri}} = \frac{N_{sec}}{N_{pri}}$$

(11–4)

From this relationship,

$$V_{sec} = \left(\frac{N_{sec}}{N_{pri}}\right)V_{pri}$$

(11–5)

Equation (11–5) shows that the secondary voltage is equal to the turns ratio times the primary voltage. This condition assumes that all the magnetic flux lines produced in the primary winding pass through the secondary winding.

The turns ratio for a step-up transformer is always greater than 1 since N_{sec} is always greater than N_{pri}.

EXAMPLE 11–4

The transformer in Figure 11–10 has a 200 turn primary winding and a 600 turn secondary winding. What is the voltage across the secondary?

FIGURE 11–10

Solution The secondary voltage is

$$V_{sec} = \left(\frac{N_{sec}}{N_{pri}}\right)V_{pri} = \left(\frac{600}{200}\right)120 \text{ V} = 3(120 \text{ V}) = 360 \text{ V}$$

Note that the turns ratio of 3 is indicated on the schematic as 1:3, meaning that there are three secondary turns for each primary turn.

The calculator sequence for V_{sec} is

Related Exercise The transformer in Figure 11–10 is changed to one with $N_{pri} = 75$ and $N_{sec} = 300$. Determine V_{sec}.

SECTION 11–3 REVIEW

1. What does a step-up transformer do?
2. If the turns ratio is 5, how much greater is the secondary voltage than the primary voltage?
3. When 240 V ac are applied to a transformer with a turns ratio of 10, what is the secondary voltage?

11–4 ■ THE STEP-DOWN TRANSFORMER

A step-down transformer has more turns in its primary winding than in its secondary winding and is used to decrease ac voltage.

After completing this section, you should be able to

■ **Explain how a step-down transformer works**
  □ Identify a step-down transformer by its turns ratio

A transformer in which the secondary voltage is less than the primary voltage is called a **step-down transformer.** The amount by which the voltage is stepped down depends on the turns ratio. Equation (11–5) also applies to a step-down transformer; the turns ratio, however, is always less than 1 since N_{sec} is always less than N_{pri}.

EXAMPLE 11–5

The transformer in Figure 11–11 is part of a laboratory power supply and has 50 turns in the primary and 10 turns in the secondary. What is the secondary voltage?

FIGURE 11–11

Solution   The secondary voltage is

$$V_{\text{sec}} = \left(\frac{N_{\text{sec}}}{N_{\text{pri}}}\right)V_{\text{pri}} = \left(\frac{10}{50}\right)120 \text{ V} = 0.2(120 \text{ V}) = 24 \text{ V}$$

Related Exercise   The transformer in Figure 11–11 is changed to one with $N_{\text{pri}} = 250$ and $N_{\text{sec}} = 120$. Determine the secondary voltage.

SECTION 11–4 REVIEW

1. What does a step-down transformer do?
2. One hundred twenty volts ac are applied to the primary winding of a transformer with a turns ratio of 0.5. What is the secondary voltage?
3. A primary voltage of 120 V ac is reduced to 12 V ac. What is the turns ratio?

11–5 ■ LOADING THE SECONDARY WINDING

When a resistive load is connected to the secondary winding of a transformer, the load current depends on both the primary current and the turns ratio.

After completing this section, you should be able to

■ **Discuss the effect of a resistive load across the secondary winding**
  □ Determine the secondary current when a step-up transformer is loaded
  □ Determine the secondary current when a step-down transformer is loaded
  □ Discuss power in a transformer

When a load resistor (R_L) is connected to the secondary winding, as shown in Figure 11–12, secondary current will flow because of the voltage induced in the secondary coil. It can be shown that the ratio of the primary current (I_{pri}) to the secondary current (I_{sec}) is equal to the turns ratio, as expressed in the following equation:

$$\frac{I_{\text{pri}}}{I_{\text{sec}}} = \frac{N_{\text{sec}}}{N_{\text{pri}}} \tag{11–6}$$

Inverting both sides of the equation gives

$$\frac{I_{\text{sec}}}{I_{\text{pri}}} = \frac{N_{\text{pri}}}{N_{\text{sec}}}$$

Solving for I_{sec},

$$I_{sec} = \left(\frac{N_{pri}}{N_{sec}}\right)I_{pri} \qquad\qquad (11\text{–}7)$$

Note that N_{pri}/N_{sec} is the reciprocal of the turns ratio. Thus, as shown in Figure 11–12(a), for a step-up transformer, in which N_{sec}/N_{pri} is greater than 1, the secondary current is less than the primary current because the reciprocal of the turns ratio is less than 1. For a step-down transformer, shown in Figure 11–12(b), N_{sec}/N_{pri} is less than 1, and I_{sec} is greater than I_{pri} because the reciprocal of the turns ratio is greater than 1.

(a) Step-up transformer: $V_{sec} > V_{pri}$ and $I_{sec} < I_{pri}$

(b) Step-down transformer: $V_{sec} < V_{pri}$ and $I_{sec} > I_{pri}$

FIGURE 11–12
Illustration of voltages and currents in a transformer with a loaded secondary winding.

EXAMPLE 11–6   The transformers in Figures 11–13(a) and 11–13(b) have loaded secondaries. If the primary current is 100 mA in each case, how much current flows through the load?

FIGURE 11–13

Solution   In Figure 11–13(a), the secondary load current is

$$I_L = I_{sec} = \left(\frac{N_{pri}}{N_{sec}}\right)I_{pri} = 0.1(100 \text{ mA}) = 10 \text{ mA}$$

In Figure 11–13(b), the secondary load current is

$$I_L = I_{sec} = \left(\frac{N_{pri}}{N_{sec}}\right)I_{pri} = 2(100 \text{ mA}) = 200 \text{ mA}$$

Related Exercise   What is the secondary current in Figure 11–13(a) if the turns ratio is doubled? What is the secondary current in Figure 11–13(b) if the turns ratio is halved? Assume I_{pri} remains at 100 mA in both cases.

Primary Power Equals Secondary Power

When a load is connected to the secondary winding of a transformer, the power transferred to the load can never be greater than the power in the primary winding. For an ideal transformer, the power in the secondary winding (P_{sec}) equals the power in the primary winding (P_{pri}). When losses are considered, the secondary power is always less.

Power is dependent on voltage and current, and there can be no increase in power in a transformer. Therefore, if the voltage is stepped up, the current is stepped down and vice versa. In an ideal transformer, the secondary power is the same as the primary power regardless of the turns ratio, as shown in the following steps:

The primary power is

$$P_{pri} = V_{pri}I_{pri}$$

The secondary power is

$$P_{sec} = V_{sec}I_{sec}$$

From Equations (11–5) and (11–7),

$$I_{sec} = \left(\frac{N_{pri}}{N_{sec}}\right)I_{pri} \quad \text{and} \quad V_{sec} = \left(\frac{N_{sec}}{N_{pri}}\right)V_{pri}$$

By substitution,

$$P_{sec} = \left(\frac{\cancel{N}_{sec}}{\cancel{N}_{pri}}\right)\left(\frac{\cancel{N}_{pri}}{\cancel{N}_{sec}}\right)V_{pri}I_{pri}$$

Canceling the terms,

$$P_{sec} = V_{pri}I_{pri} = P_{pri}$$

This result is closely approached in practice because of the very high efficiencies of transformers.

SECTION 11–5 REVIEW	1. If the turns ratio of a transformer is 2, is the secondary current greater than or less than the primary current? By how much?
	2. A transformer has 100 primary turns and 25 secondary turns, and I_{pri} is 0.5 A. What is the value of I_{sec}?
	3. In Question 2, how much primary current is necessary to produce a secondary load current of 10 A?

11–6 ■ REFLECTED LOAD

From the viewpoint of the primary circuit, a load connected across the secondary winding of a transformer appears to have a resistance that is not necessarily equal to the actual resistance of the load. The actual load is essentially "reflected" into the primary circuit altered by the turns ratio. This reflected load is what the primary source effectively sees and it determines the amount of primary current.

After completing this section, you should be able to

■ **Discuss the concept of a reflected load in a transformer**
  □ Define *reflected resistance*
  □ Explain how the turns ratio affects the reflected resistance
  □ Calculate reflected resistance

The concept of the **reflected load** is illustrated in Figure 11–14. The load (R_L) in the secondary circuit of a transformer is reflected into the primary circuit by transformer action. The load appears to the source in the primary circuit to be a resistance (R_{pri}) with a value determined by the turns ratio and the actual value of the load resistance. The resistance R_{pri} is called the **reflected resistance.**

The resistance in the primary circuit of Figure 11–14 is $R_{pri} = V_{pri}/I_{pri}$. The resistance in the secondary circuit is $R_L = V_{sec}/I_{sec}$. From Equations (11–4) and (11–6), you know that $V_{sec}/V_{pri} = N_{sec}/N_{pri}$ and $I_{pri}/I_{sec} = N_{sec}/N_{pri}$. Using these relationships, a formula for R_{pri} in terms of R_L is determined as follows:

$$\frac{R_{pri}}{R_L} = \frac{V_{pri}/I_{pri}}{V_{sec}/I_{sec}} = \left(\frac{V_{pri}}{V_{sec}}\right)\left(\frac{I_{sec}}{I_{pri}}\right) = \left(\frac{N_{pri}}{N_{sec}}\right)\left(\frac{N_{pri}}{N_{sec}}\right) = \frac{N_{pri}^2}{N_{sec}^2} = \left(\frac{N_{pri}}{N_{sec}}\right)^2$$

Solving for R_{pri},

$$R_{pri} = \left(\frac{N_{pri}}{N_{sec}}\right)^2 R_L \qquad\qquad (11\text{–}8)$$

Equation (11–8) shows that the resistance reflected into the primary circuit is the square of the reciprocal of the turns ratio times the load resistance.

FIGURE 11–14
Reflected load in a transformer circuit.

EXAMPLE 11–7   Figure 11–15 shows a source that is transformer-coupled to a load resistor of 100 Ω. The transformer has 100 primary turns and 400 secondary turns for a turns ratio of 4. What is the reflected resistance seen by the source?

FIGURE 11–15

Solution The reflected resistance is determined as follows:

$$R_{\text{pri}} = \left(\frac{N_{\text{pri}}}{N_{\text{sec}}}\right)^2 R_L = \left(\frac{100}{400}\right)^2 100\ \Omega = \left(\frac{1}{4}\right)^2 100\ \Omega = \left(\frac{1}{16}\right)100\ \Omega = 6.25\ \Omega$$

The source sees a resistance of 6.25 Ω just as if it were connected directly, as shown in the equivalent circuit of Figure 11–16.

The calculator sequence is

FIGURE 11–16

Resistance "reflected" from secondary

R_{pri}
6.25 Ω

Related Exercise If the turns ratio in Figure 11–15 is 10 and R_L is 600 Ω, what is the reflected resistance?

EXAMPLE 11–8 In Figure 11–15, if a transformer is used having 40 primary turns and 10 secondary turns, what is the reflected resistance? Note that the turns ratio (n) is 0.25.

Solution The reflected resistance is

$$R_{\text{pri}} = \left(\frac{N_{\text{pri}}}{N_{\text{sec}}}\right)^2 R_L = \left(\frac{40}{10}\right)^2 100\ \Omega = (4)^2 100\ \Omega = 1600\ \Omega$$

Related Exercise To achieve a reflected resistance of 800 Ω, what turns ratio is required in Figure 11–15?

Examples 11–7 and 11–8 illustrate that in a step-up transformer (n > 1), the reflected resistance is less than the actual load resistance; and in a step-down transformer (n < 1), the reflected resistance is greater than the load resistance.

SECTION 11–6 REVIEW

1. Define *reflected resistance*.
2. What transformer characteristic determines the reflected resistance?
3. A given transformer has a turns ratio of 10, and the load is 50 Ω. How much resistance is reflected into the primary?
4. What is the turns ratio required to reflect a 4 Ω load resistance into the primary as 400 Ω?

11–7 ■ MATCHING THE LOAD AND SOURCE RESISTANCES

One application of transformers is in the matching of a load resistance to a source resistance in order to achieve maximum transfer of power. This technique is called impedance matching. In audio systems, transformers are often used to get the maximum amount of power from the amplifier to the speaker by proper selection of the turns ratio.

After completing this section, you should be able to

■ **Discuss impedance matching with transformers**
  □ Give a general definition of *impedance*
  □ Discuss the maximum power transfer theorem
  □ Define *impedance matching*
  □ Explain the purpose of impedance matching
  □ Describe a practical application

The term *impedance* will become very familiar to you in the next chapter. Basically, impedance is a general term for the opposition to current, including the effects of both resistance and reactance combined. However, for the time being, we will confine our usage to resistance only.

The concept of **impedance matching** is illustrated in the basic circuit of Figure 11–17. Part (a) shows an ac voltage source with a series resistance representing its internal resistance. Some internal resistance is inherent in all sources due to their internal circuitry or physical makeup. When the source is connected directly to a load, as shown in part (b), often the objective is to transfer as much of the power produced by the source to the load as possible. However, a certain amount of the power produced by the source is lost in its internal resistance, and the remaining power goes to the load.

(a) Voltage source with internal resistence, R_{int}

(b) A portion of the total power is lost in R_{int}

FIGURE 11–17
Power transfer from a nonideal voltage source to a load.

Maximum Power Transfer Theorem

The **maximum power transfer theorem** is important when you need to know the value of the load at which the most power is delivered from the source. The theorem can be stated as follows:

> **When a source is connected to a load, maximum power is delivered to the load when the load resistance is equal to the internal source resistance.**

We will demonstrate this theorem by finding the load power for various values of load resistance and a specific internal source resistance in Example 11–9.

EXAMPLE 11–9   The source in Figure 11–18 has an internal source resistance of 75 Ω. Determine the load power for each of the following values of load resistance:
(a) 25 Ω   (b) 50 Ω   (c) 75 Ω   (d) 100 Ω   (e) 125 Ω

Draw a graph showing the load power versus the load resistance.

FIGURE 11–18

Solution   Use Ohm's law ($I = V/R$) and the power formula ($P = I^2 R$) to find the load power for each value of load resistance.
(a) For $R_L = 25$ Ω,

$$I = \frac{V_s}{R_{int} + R_L} = \frac{10 \text{ V}}{100 \text{ Ω}} = 100 \text{ mA}$$

$$P_L = I^2 R_L = (100 \text{ mA})^2 (25 \text{ Ω}) = 250 \text{ mW}$$

(b) For $R_L = 50$ Ω,

$$I = \frac{V_s}{R_{int} + R_L} = \frac{10 \text{ V}}{125 \text{ Ω}} = 80 \text{ mA}$$

$$P_L = I^2 R_L = (80 \text{ mA})^2 (50 \text{ Ω}) = 320 \text{ mW}$$

(c) For $R_L = 75$ Ω,

$$I = \frac{V_s}{R_{int} + R_L} = \frac{10 \text{ V}}{150 \text{ Ω}} = 66.7 \text{ mA}$$

$$P_L = I^2 R_L = (66.7 \text{ mA})^2 (75 \text{ Ω}) = 334 \text{ mW}$$

(d) For $R_L = 100$ Ω,

$$I = \frac{V_s}{R_{int} + R_L} = \frac{10 \text{ V}}{175 \text{ Ω}} = 57.1 \text{ mA}$$

$$P_L = I^2 R_L = (57.1 \text{ mA})^2 (100 \text{ Ω}) = 326 \text{ mW}$$

(e) For $R_L = 125$ Ω,

$$I = \frac{V_s}{R_{int} + R_L} = \frac{10 \text{ V}}{200 \text{ Ω}} = 50 \text{ mA}$$

$$P_L = I^2 R_L = (50 \text{ mA})^2 (125 \text{ Ω}) = 313 \text{ mW}$$

Notice that the load power is greatest when $R_L = 75$ Ω, which is the same as the internal source resistance. When the load resistance is less than or greater than this value, the load power drops off, as the curve in Figure 11–19 graphically illustrates.
The calculator sequence for $R_L = 25$ Ω is

[1] [0] [÷] [(] [7] [5] [+] [2] [5] [)] [=] [x²] [×] [2] [5] [=]

FIGURE 11–19
Curve showing that the load power is maximum when $R_L = R_{int}$.

Related Exercise What is the maximum power that can be transfered from a 10 V source with an internal source resistance of 30 Ω?

A Practical Application

In most practical situations, the internal source resistance of various types of sources is fixed. Also, in many cases, the resistance of a device that acts as a load is fixed and cannot be altered. If you need to connect a given source to a given load, remember that only by chance will their resistances match. In this situation a transformer comes in handy. You can use the reflected-resistance characteristic of a transformer to make the load resistance appear to have the same value as the source resistance, thereby "fooling" the source into "thinking" that there is a match.

Let's take a practical, everyday situation to illustrate. The typical resistance of the input to a TV receiver is 300 Ω. An antenna must be connected to this input by a lead-in cable in order to receive TV signals. In this situation, the antenna and the lead-in act as the source, and the input resistance of the TV receiver is the load, as illustrated in Figure 11–20.

FIGURE 11–20
An antenna directly coupled to a TV receiver.

(a) The antenna/lead-in is the source; the TV input is the load.

Source–antenna and lead-in   Load–TV receiver
(b) Circuit equivalent of antenna and TV receiver system

It is common for an antenna system to have a characteristic resistance of 75 Ω. Thus, if the 75 Ω source (antenna and lead-in) is connected directly to the 300 Ω TV input, maximum power will not be delivered to the input to the TV, and you will have poor signal reception. The solution is to use a matching transformer connected as indicated in Figure 11–21, in order to match the 300 Ω load resistance to the 75 Ω source resistance.

FIGURE 11–21
Example of a load matched to a source by transformer coupling for maximum power transfer.

To match the resistances, you must select a proper value of turns ratio. You want the 300 Ω load resistance to look like 75 Ω to the source. Use Equation (11–8) to obtain a formula to determine the turns ratio when you know the values for R_L and R_{pri}:

$$R_{pri} = \left(\frac{N_{pri}}{N_{sec}} \right)^2 R_L$$

Transpose terms and divide both sides by R_L:

$$\left(\frac{N_{pri}}{N_{sec}} \right)^2 = \frac{R_{pri}}{R_L}$$

Then take the square root of both sides:

$$\frac{N_{pri}}{N_{sec}} = \sqrt{\frac{R_{pri}}{R_L}}$$

Invert both sides to get the following formula:

$$\frac{N_{sec}}{N_{pri}} = \sqrt{\frac{R_L}{R_{pri}}} \qquad\qquad (11\text{–}9)$$

Finally, solve for this particular turns ratio:

$$n = \frac{N_{sec}}{N_{pri}} = \sqrt{\frac{300\ \Omega}{75\ \Omega}} = \sqrt{4} = 2$$

Therefore, a matching transformer with a turns ratio of 2 must be used in this application.

EXAMPLE 11–10 An amplifier has an 800 Ω internal resistance. In order to provide maximum power to an 8 Ω speaker, what turns ratio must be used in the coupling transformer?

Solution The reflected resistance must equal 800 Ω. Thus, from Equation (11–9), the turns ratio can be determined.

$$n = \frac{N_{sec}}{N_{pri}} = \sqrt{\frac{R_L}{R_{pri}}} = \sqrt{\frac{8\ \Omega}{800\ \Omega}} = \sqrt{0.01} = 0.1$$

There must be ten primary turns for each secondary turn because $n = 0.1$. The diagram and its equivalent reflected circuit are shown in Figure 11–22.

The calculator sequence is

Amplifier equivalent circuit   Speaker/transformer equivalent

FIGURE 11–22

Related Exercise What must be the turns ratio in Figure 11–22 to provide maximum power to two 8 Ω speakers in parallel?

SECTION 11–7 REVIEW
1. What does impedance matching mean?
2. What is the advantage of matching the load resistance to the resistance of a source?
3. A transformer has 100 primary turns and 50 secondary turns. What is the reflected resistance with 100 Ω across the secondary?

11–8 ■ THE TRANSFORMER AS AN ISOLATION DEVICE

Transformers are useful in providing electrical isolation between the primary circuit and the secondary circuit because there is no electrical connection between the two windings. In a transformer, energy is transferred entirely by magnetic coupling.

After completing this section, you should be able to

■ **Explain how the transformer acts as an isolation device**
   ☐ Discuss dc isolation
   ☐ Discuss power line isolation

DC Isolation

If a nonchanging direct current flows through the primary circuit of a transformer, nothing happens in the secondary circuit, as indicated in Figure 11–23(a). The reason is that the primary current must be changing in order to create a changing magnetic field. This will cause voltage to be induced in the secondary circuit, as indicated in Figure 11–23(b). Therefore, the transformer isolates the secondary circuit from any dc voltage in the primary circuit.

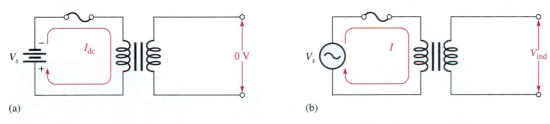

(a)   (b)

FIGURE 11–23
DC isolation and ac coupling.

For example, in some amplifiers, transformers are used to prevent dc voltages at certain points in the circuit, while allowing the signal (ac) voltage to get through. Figure 11–24 is a simplified illustration of transformer coupling in a two-stage audio amplifier.

FIGURE 11–24
Audio amplifier stages with transformer coupling for dc isolation.

Power Line Isolation

Transformers are often used to electrically isolate electronic equipment from the 60 Hz, 110 V ac power line. The reason for using an isolation transformer to couple the 60 Hz ac to an instrument is to prevent a possible shock hazard if the 110 V line is connected to the metal chassis of the equipment. This condition is possible if the line cord plug can be inserted into an outlet either way. Incidentally, to prevent this situation, many plugs have keyed prongs so that they can be plugged in only one way.

Figure 11–25 illustrates how a transformer can prevent the metal chassis from being connected to the 110 V line rather than to neutral (ground), no matter how the cord is plugged into the outlet. When an isolation transformer is used, the secondary circuit is said to be "floating" because it is not referenced to the power line ground. Should a person come in contact with the secondary voltage, there is no complete current path back to ground, and therefore there is no shock hazard. Current must flow through your body in order for you to receive an electrical shock.

(a) Power line direct to instrument

(b) Power line reversed

(c) Power line transformer-coupled to instrument

(d) Power line reversed

FIGURE 11–25
The use of an isolation transformer to prevent a shock hazard.

SECTION 11–8 REVIEW	1. What does the term *electrical isolation* mean? 2. Can a dc voltage be coupled by a transformer?

11–9 ■ NONIDEAL TRANSFORMER CHARACTERISTICS

Up to this point, transformer operation has been discussed from an ideal point of view, and this approach is valid when you are learning new concepts. However, you should be aware of the nonideal characteristics of practical transformers and how they affect performance.

After completing this section, you should be able to

■ **Describe a practical transformer**
  □ List and describe the nonideal characteristics
  □ Explain power rating of a transformer
  □ Define *efficiency* of a transformer

So far, the transformer has been considered as an ideal device. That is, the winding resistance, the winding capacitance, and nonideal core characteristics were all neglected and the transformer was treated as if it had an efficiency of 100%. For studying the basic concepts and in many applications, the ideal model is valid. However, the practical transformer has several nonideal characteristics of which you should be aware.

Winding Resistance

Both the primary and the secondary windings of a practical transformer have winding resistance. (You learned about the winding resistance of inductors in Chapter 10.) The

FIGURE 11–26
Winding resistance in a practical transformer.

winding resistances of a practical transformer are represented as resistors in series with the windings as shown in Figure 11–26.

Winding resistance in a practical transformer results in less voltage across a secondary load. Voltage drops due to the winding resistance effectively subtract from the primary and secondary voltages and result in a load voltage that is less than that predicted by the relationship $V_{sec} = (N_{sec}/N_{pri})V_{pri}$. In most cases the effect is relatively small and can be neglected.

Losses in the Core

There is always some energy loss in the core material of a practical transformer. This loss is seen as a heating of ferrite and iron cores, but it does not occur in air cores. Part of this energy is consumed in the continuous reversal of the magnetic field due to the changing direction of the primary current; this energy loss is called *hysteresis loss*. The rest of the energy loss is caused by eddy currents induced in the core material by the changing magnetic flux. The eddy currents flow in circular patterns in the core resistance, thus causing the energy loss. This loss is greatly reduced by the use of laminated construction of iron cores. The thin layers of ferromagnetic material are insulated from each other to minimize the buildup of eddy currents by confining them to a small area and to keep core losses to a minimum.

Magnetic Flux Leakage

In an ideal transformer, all the magnetic flux produced by the primary current is assumed to pass through the core to the secondary winding and vice versa. In a practical transformer, some of the magnetic flux lines break out of the core and pass through the surrounding air back to the other end of the winding, as illustrated in Figure 11–27 for the magnetic field produced by the primary current. Magnetic flux leakage results in a reduced secondary voltage.

The percentage of magnetic flux that actually reaches the secondary determines the coefficient of coupling of the transformer. For example, if nine out of ten flux lines re-

FIGURE 11–27
Flux leakage in a practical transformer.

Leakage flux

main inside the core, the coefficient of coupling is 0.90 or 90%. Most iron-core transformers have very high coefficients of coupling (greater than 0.99), whereas ferrite-core and air-core devices have lower values.

Winding Capacitance

As you learned in Chapter 10, there is always some stray capacitance between adjacent turns of a winding. These stray capacitances result in an effective capacitance in parallel with each winding of a transformer, as indicated in Figure 11–28.

FIGURE 11–28

Winding capacitance in a practical transformer.

These stray capacitances have very little effect on the transformer's operation at low frequencies, because the reactances (X_C) are very high. However, at higher frequencies, the reactances decrease and begin to produce a bypassing effect across the primary winding and across the secondary load. As a result, less of the total primary current flows through the primary winding, and less of the total secondary current flows through the load. This effect reduces the load voltage as the frequency goes up.

Transformer Power Rating

A transformer is typically rated in volt-amperes (VA), primary/secondary voltage, and operating frequency. For example, a given transformer rating may be specified as 2 kVA, 500/50, 60 Hz. The 2 kVA value is the **apparent power rating.** The 500 and the 50 can be either secondary or primary voltages.

Let's assume, for example, that 50 V is the secondary voltage. In this case the load current is

$$I_L = \frac{P_{\text{sec}}}{V_{\text{sec}}} = \frac{2 \text{ kVA}}{50 \text{ V}} = 40 \text{ A}$$

On the other hand, if 500 V is the secondary voltage, then

$$I_L = \frac{P_{\text{sec}}}{V_{\text{sec}}} = \frac{2 \text{ kVA}}{500 \text{ V}} = 4 \text{ A}$$

These are the maximum currents that the secondary can handle in either case.

The reason that the power rating is in volt-amperes (VA) rather than in watts (true power) is as follows: If the transformer load is purely capacitive or purely inductive, the true power (watts) delivered to the load is ideally zero. However, the current for $V_{\text{sec}} = 500$ V and $X_C = 100 \ \Omega$ at 60 Hz, for example, is 5 A. This current exceeds the maximum of 4 A that the 2 kVA secondary can handle, and even though the true power is zero, the transformer may be damaged. So it is meaningless to specify power in watts for transformers.

Transformer Efficiency

Recall that the secondary power is equal to the primary power in an ideal transformer. Because the nonideal characteristics just discussed result in a power loss in the transformer, the secondary (output) power is always less than the primary (input) power. The **efficiency** (η) of a transformer is a measure of the percentage of the input power that is delivered to the output:

$$\eta = \left(\frac{P_{out}}{P_{in}}\right)100\%$$

(11–10)

Most power transformers have efficiencies in excess of 95%.

EXAMPLE 11–11

A certain type of transformer has a primary current of 5 A and a primary voltage of 4800 V. The secondary current is 90 A and the secondary voltage is 240 V. Determine the efficiency of this transformer.

Solution   The input power is

$$P_{in} = V_{pri}I_{pri} = (4800 \text{ V})(5 \text{ A}) = 24 \text{ kVA}$$

The output power is

$$P_{out} = V_{sec}I_{sec} = (240 \text{ V})(90 \text{ A}) = 21.6 \text{ kVA}$$

The efficiency is

$$\eta = \left(\frac{P_{out}}{P_{in}}\right)100\% = \left(\frac{21.6 \text{ kVA}}{24 \text{ kVA}}\right)100\% = 90\%$$

Related Exercise   A transformer has a primary current of 8 A with a primary voltage of 440 V. The secondary current is 30 A and the secondary voltage is 100 V. What is the efficiency?

SECTION 11–9 REVIEW

1. Explain how a practical transformer differs from the ideal model.
2. The coefficient of coupling of a certain transformer is 0.85. What does this mean?
3. A certain transformer has a rating of 10 kVA. If the secondary voltage is 250 V, how much load current can the transformer handle?

11–10 ■ OTHER TYPES OF TRANSFORMERS

There are several important variations of the basic transformer. They include tapped transformers, multiple-winding transformers, and autotransformers.

After completing this section, you should be able to

■ **Describe several types of transformers**
   □ Describe center-tapped transformers
   □ Describe multiple-winding transformers
   □ Describe autotransformers

Tapped Transformers

A schematic of a transformer with a center-tapped secondary winding is shown in Figure 11–29(a). The **center tap (CT)** is equivalent to two secondary windings with half the total voltage across each.

(a) Center-tapped transformer

(b) Output voltages with respect to the center tap are 180° out of phase with each other and are one-half the magnitude of the secondary voltage.

FIGURE 11–29
Operation of a center-tapped transformer.

The voltages between either end of the secondary winding and the center tap are, at any instant, equal in magnitude but opposite in polarity, as illustrated in Figure 11–29(b). Here, for example, at some instant on the sine wave voltage, the polarity across the entire secondary winding is as shown (top end +, bottom −). At the center tap, the voltage is less positive than the top end but more positive than the bottom end of the secondary winding. Therefore, measured with respect to the center tap, the top end of the secondary winding is positive, and the bottom end is negative. This center-tapped feature is used in some power supply rectifiers in which the ac voltage is converted in dc, as illustrated in Figure 11–30.

Some tapped transformers have taps on the secondary winding at points other than the electrical center. Also, multiple primary and secondary taps are sometimes used in certain applications. Examples of these types of transformers are shown in Figure 11–31.

The rectifier takes these two half-cycles and combines them to get this waveform.

FIGURE 11–30
Application of a center-tapped transformer in ac-to-dc conversion.

(a)　　　　　　　　(b)　　　　　　　　(c)

FIGURE 11–31
Tapped transformers.

One example of a transformer with a multiple-tap primary winding and a center-tapped secondary winding is the utility-pole transformer used by power companies to step down the high voltage from the power line to 110 V/220 V service for residential and commercial customers, as shown in Figure 11–32. The multiple taps on the primary winding are used for minor adjustments in the turns ratio in order to overcome line voltages that are slightly too high or too low.

FIGURE 11–32
Utility-pole transformer in a typical power distribution system.

Multiple-Winding Transformers

Some transformers are designed to operate from either 110 V ac or 220 V ac lines. These transformers usually have two primary windings, each of which is designed for 110 V ac. When the two are connected in series, the transformer can be used for 220 V ac operation, as illustrated in Figure 11–33.

More than one secondary winding can be wound on a common core. Transformers with several secondary windings are often used to achieve several voltages by either stepping up or stepping down the primary voltage. These types are commonly used in power supply applications in which several voltage levels are required for the operation of an electronic instrument.

FIGURE 11–33
Multiple-primary transformer.

(a) Two primary windings

(b) Primary windings in parallel for 110 V ac operation

(c) Primary windings in series for 220 V ac operation

FIGURE 11–34
Multiple-secondary transformer.

A typical schematic of a transformer with multiple secondary windings is shown in Figure 11–34; this transformer has three secondary windings. Sometimes you will find combinations of multiple primary windings, multiple secondary windings, and tapped transformers all in one unit.

EXAMPLE 11–12 The transformer shown in Figure 11–35 has the numbers of turns indicated. One of the secondary windings is also center tapped. If 110 V ac are connected to the primary winding, determine each secondary voltage and the voltages with respect to CT on the middle secondary winding.

FIGURE 11–35

Solution

$$V_{AB} = \left(\frac{5}{100}\right)110 \text{ V} = 5.5 \text{ V}$$

$$V_{CD} = \left(\frac{200}{100}\right)110 \text{ V} = 220 \text{ V}$$

$$V_{(CT)C} = V_{(CT)D} = \frac{220 \text{ V}}{2} = 110 \text{ V}$$

$$V_{EF} = \left(\frac{10}{100}\right)110 \text{ V} = 11 \text{ V}$$

Related Exercise Repeat the calculations for a primary winding with 50 turns.

Autotransformers

In an **autotransformer,** one winding serves as both the primary and the secondary windings. The winding is tapped at the proper points to achieve the desired turns ratio for stepping up or stepping down the voltage.

Autotransformers differ from conventional transformers in that there is no electrical isolation between the primary and the secondary circuits because both are on one winding. Autotransformers normally are smaller and lighter than equivalent conventional transformers because they require a much lower kVA rating for a given load. Many autotransformers provide an adjustable tap using a sliding contact mechanism so that the output voltage can be varied (these are often called *variacs*). Figure 11–36 shows a typical autotransformer and several schematic symbols.

The following example illustrates why an autotransformer has a kVA requirement that is less than the input or output kVA.

(a)   (b) Step-up   (c) Step-down   (d) Variable

FIGURE 11–36
The autotransformer. Part (a) courtesy of Superior Electric Co.

EXAMPLE 11–13 A certain autotransformer is used to change a source voltage of 220 V to a load voltage of 160 V across an 8 Ω load resistance. Determine the input and output power in kVA, and show that the actual kVA requirement is less than this value. Assume that this transformer is ideal.

Solution The circuit is shown in Figure 11–37 with the voltages and currents indicated. The current directions have been assigned arbitrarily for convenience.

FIGURE 11–37

The load current I_3 is determined as follows:

$$I_3 = \frac{V_3}{R_L} = \frac{160 \text{ V}}{8 \text{ } \Omega} = 20 \text{ A}$$

The input power is the total source voltage (V_1) times the total current from the source (I_1):

$$P_{\text{in}} = V_1 I_1$$

The output power is the load voltage V_3 times the load current I_3:

$$P_{\text{out}} = V_3 I_3$$

For an ideal transformer, $P_{\text{in}} = P_{\text{out}}$; thus,

$$V_1 I_1 = V_3 I_3$$

Solving for I_1 yields

$$I_1 = \frac{V_3 I_3}{V_1} = \frac{(160 \text{ V})(20 \text{ A})}{220 \text{ V}} = 14.55 \text{ A}$$

Applying Kirchhoff's current law at the tap junction,

$$I_1 = I_2 + I_3$$

Solving for I_2 yields

$$I_2 = I_1 - I_3 = 14.55 \text{ A} - 20 \text{ A} = -5.45 \text{ A}$$

The minus sign can be dropped because the current directions are arbitrary.

The input and output power are

$$P_{\text{in}} = P_{\text{out}} = V_3 I_3 = (160 \text{ V})(20 \text{ A}) = 3.2 \text{ kVA}$$

The power in winding A is

$$V_2 I_1 = (60 \text{ V})(14.55 \text{ A}) = 873 \text{ VA}$$

The power in winding B is

$$V_3 I_2 = (160 \text{ V})(5.45 \text{ A}) = 872 \text{ VA}$$

Thus, the power rating required for each winding is less than the power that is delivered to the load. The slight difference in the calculated powers in windings A and B is due to rounding.

Related Exercise What happens to the kVA requirement if the load is changed to 4 Ω?

SECTION 11–10 REVIEW

1. A certain transformer has two secondary windings. The turns ratio from the primary winding to the first secondary winding is 10. The turns ratio from the primary winding to the other secondary winding is 0.2. If 220 V ac are applied to the primary winding, what are the secondary voltages?
2. Name one advantage and one disadvantage of an autotransformer over a conventional transformer.

11–11 ▪ TROUBLESHOOTING

The common failures in transformers are opens, shorts, or partial shorts in either the primary or the secondary windings. One cause of such failures is the operation of the device under conditions that exceed its ratings. A few transformer failures and the associated symptoms are covered in this section.

After completing this section, you should be able to

▪ **Troubleshoot transformers**
  □ Find an open primary or secondary winding
  □ Find a shorted or partially shorted primary or secondary winding

Open Primary Winding

When there is an open primary winding, there is no primary current and, therefore, no induced voltage or current in the secondary. This condition is illustrated in Figure 11–38(a), and the method of checking with an ohmmeter is shown in part (b).

(a) Conditions when the primary winding is open

Disconnect source from primary winding.
(b) Checking the primary winding with an ohmmeter

FIGURE 11–38
Open primary winding.

Open Secondary Winding

When there is an open secondary winding, there is no current in the secondary circuit and, as a result, no voltage across the load. Also, an open secondary winding causes the primary current to be very small (only a small magnetizing current flows). In fact, the primary current may be practically zero. This condition is illustrated in Figure 11–39(a), and the ohmmeter check is shown in part (b).

(a) Conditions when the secondary winding is open

(b) Checking the secondary winding with the ohmmeter

FIGURE 11–39
Open secondary winding.

Shorted or Partially Shorted Primary Winding

A completely shorted primary winding will draw excessive current from the source and, unless there is a breaker or a fuse in the circuit, either the source or the transformer or both will burn out. A partial short in the primary winding can cause higher than normal or even excessive primary current.

Shorted or Partially Shorted Secondary Winding

In this case, there is an excessive primary current because of the low reflected resistance due to the short. Often, this excessive current will burn out the primary winding and result in an open. The short-circuit current in the secondary winding causes the load current to be zero (full short) or smaller than normal (partial short), as demonstrated in Figure 11–40(a) and 11–40(b). The ohmmeter check for this condition is shown in part (c).

Normally, when a transformer fails, it is very difficult to repair, and therefore the simplest procedure is to replace it.

FIGURE 11–40
Shorted secondary winding.

(a) Secondary winding completely shorted

FIGURE 11–40
Continued

(b) Secondary winding partially shorted

(c) Checking the secondary winding with the ohmmeter

SECTION 11–11 REVIEW

1. Name two possible failures in a transformer.
2. What is often the cause of transformer failure?

11–12 ▪ APPLICATION ASSIGNMENT

A common application of the transformer is in dc power supplies. The transformer is used to couple the ac line voltage into the power supply circuitry where it is converted to a dc voltage. Your assignment is to troubleshoot four identical transformer-coupled dc power supplies and, based on a series of measurements, determine the fault, if any, in each.

The transformer in the power supply schematic of Figure 11–41 steps the 110 V rms at the ac outlet down to a level that can be converted by the diode bridge rectifier, filtered,

FIGURE 11–41
Basic transformer-coupled dc power supply.

and regulated to obtain a 6 V dc output. The diode rectifier changes the ac to a pulsating full-wave dc voltage that is smoothed by the capacitor filter C_1. The voltage regulator is an integrated circuit that takes the filtered voltage and provides a constant 6 V dc over a range of load values and line voltage variations. Additional filtering is provided by capacitor C_2. You will learn about these circuits in a later course.

Step 1: Familiarization with the Power Supply

You have four identical power supply units to troubleshoot like the one shown in Figure 11–42. The power line to the primary winding of transformer T_1 is fused by F_1. The secondary winding is connected to the circuit board containing the rectifier, filter, and regulator. Measurement points are indicated by the circled numbers.

FIGURE 11–42
Power supply unit (top view).

Step 2: Measure Voltages on Power Supply Unit 1

After plugging the power supply into a standard wall outlet, an autoranging portable multimeter is used to measure the voltages. In an autoranging meter, the appropriate measurement range is automatically selected instead of being manually selected as in a standard multimeter.

Determine from the meter readings in Figure 11–43 whether or not the power supply is operating properly. If it is not, isolate the problem to one of the following: The circuit board containing the rectifier, filter, and regulator, the transformer, the fuse, or the power source. The circled numbers on the meter inputs correspond to the numbered points on the power supply in Figure 11–42.

Step 3: Measure Voltages on Power Supply Units 2, 3, and 4

Determine from the meter readings for units 2, 3, and 4 in Figure 11–44 whether or not each power supply is operating properly. If it is not, isolate the problem to one of the following: the circuit board containing the rectifier, filter, and regulator, the transformer, the fuse, or the power source. Only the meter displays and corresponding measurement points are shown.

FIGURE 11–43
Voltage measurements on power supply unit 1.

Unit 2 measurements

Unit 3 measurements

Unit 4 measurements

FIGURE 11–44
Measurements for power supply units 2, 3, and 4.

SECTION 11–12 REVIEW	
	1. In the case where the transformer was found to be faulty, how can you determine the specific fault (open windings or shorted windings)?
	2. What type of fault in the transformer could cause the fuse to blow?

■ SUMMARY

- A transformer consists of two or more coils that are magnetically coupled on a common core.
- There is mutual inductance between two magnetically coupled coils.
- When current in one coil changes, voltage is induced in the other coil.
- The primary is the winding connected to the source, and the secondary is the winding connected to the load.

- The number of turns in the primary winding and the number of turns in the secondary winding determine the turns ratio.
- The relative polarities of the primary and secondary voltages are determined by the direction of the windings around the core.
- A step-up transformer has a turns ratio greater than 1.
- A step-down transformer has a turns ratio less than 1.
- A transformer cannot increase power.
- In an ideal transformer, the power from the source (input power) is equal to the power delivered to the load (output power).
- If the voltage is stepped up, the current is stepped down, and vice versa.
- A load connected across the secondary winding of a transformer appears to the source as a reflected load having a value dependent on the reciprocal of the turns ratio squared.
- A transformer can match a load resistance to an internal source resistance to achieve maximum power transfer to the load by selection of the proper turns ratio.
- A transformer does not respond to constant dc.
- Energy losses in a practical transformer result from winding resistances, hysteresis loss in the core, eddy currents in the core, and flux leakage.

■ GLOSSARY

Apparent power rating   The method of rating transformers in which the power capability is expressed in volt-amperes (VA).

Autotransformer   A transformer in which the primary and secondary windings are in a single winding.

Center tap (CT)   A connection at the midpoint of the secondary winding of a transformer.

Coefficient of coupling (k)   A constant associated with transformers that is the ratio of secondary flux to primary flux. The ideal value of 1 indicates that all the flux in the primary winding is coupled into the secondary winding.

Core   The physical structure around which the windings are placed.

Efficiency (η)   The ratio of output power to input power expressed as a percentage.

Electrical isolation   The condition that exists when two coils are magnetically linked but have no electrical connection between them.

Impedance matching   A technique used to match a load resistance to an internal source resistance in order to achieve a maximum transfer of power.

Magnetic coupling   The magnetic connection between two coils as a result of the changing magnetic flux lines of one coil cutting through the second coil.

Maximum power transfer theorem   A theorem that states the maximum power is transferred from a source to a load when the load resistance equals the internal source resistance.

Mutual inductance (L_M)   The inductance between two separate coils, such as a transformer.

Primary winding   The input winding of a transformer; also called *primary*.

Reflected load   The load as it appears to the source in the primary circuit.

Reflected resistance   The resistance in the secondary circuit reflected into the primary circuit.

Secondary winding   The output winding of a transformer; also called *secondary*.

Step-down transformer   A transformer in which the secondary voltage is less than the primary voltage.

Step-up transformer   A transformer in which the secondary voltage is greater than the primary voltage.

Transformer   A device formed by two or more windings that are magnetically coupled to each other and providing a transfer of power electromagnetically from one winding to another.

Turns ratio   The ratio of turns in the secondary winding to turns in the primary winding.

■ FORMULAS

$$(11\text{–}1) \qquad k = \frac{\phi_{12}}{\phi_1} \qquad\qquad \text{Coefficient of coupling}$$

$$(11\text{–}2) \qquad L_M = k\sqrt{L_1 L_2} \qquad\qquad \text{Mutual inductance}$$

$$(11\text{–}3) \qquad n = \frac{N_{sec}}{N_{pri}} \qquad\qquad \text{Turns ratio}$$

$$(11\text{–}4) \qquad \frac{V_{sec}}{V_{pri}} = \frac{N_{sec}}{N_{pri}} \qquad\qquad \text{Voltage ratio}$$

$$(11\text{–}5) \qquad V_{sec} = \left(\frac{N_{sec}}{N_{pri}}\right) V_{pri} \qquad\qquad \text{Secondary voltage}$$

$$(11\text{–}6) \qquad \frac{I_{pri}}{I_{sec}} = \frac{N_{sec}}{N_{pri}} \qquad\qquad \text{Current ratio}$$

$$(11\text{–}7) \qquad I_{sec} = \left(\frac{N_{pri}}{N_{sec}}\right) I_{pri} \qquad\qquad \text{Secondary current}$$

$$(11\text{–}8) \qquad R_{pri} = \left(\frac{N_{pri}}{N_{sec}}\right)^2 R_L \qquad\qquad \text{Reflected resistance}$$

$$(11\text{–}9) \qquad \frac{N_{sec}}{N_{pri}} = \sqrt{\frac{R_L}{R_{pri}}} \qquad\qquad \text{Turns ratio for impedance matching}$$

$$(11\text{–}10) \qquad \eta = \left(\frac{P_{out}}{P_{in}}\right) 100\% \qquad\qquad \text{Transformer efficiency}$$

■ **SELF-TEST**

1. A transformer is used for
   (a) dc voltages   (b) ac voltages   (c) both dc and ac

2. The turns ratio of a transformer affects the following:
   (a) primary voltage   (b) dc voltage   (c) secondary voltage   (d) none of these

3. If the windings of a certain transformer with a turns ratio of 1 are in opposite directions around the core, the secondary voltage is
   (a) in phase with the primary voltage   (b) less than the primary voltage
   (c) greater than the primary voltage   (d) out of phase with the primary voltage

4. When the turns ratio of a transformer is 10 and the primary ac voltage is 6 V, the secondary voltage is
   (a) 60 V   (b) 0.6 V   (c) 6 V   (d) 36 V

5. When the turns ratio of a transformer is 0.5 and the primary ac voltage is 100 V, the secondary voltage is
   (a) 200 V   (b) 50 V   (c) 10 V   (d) 100 V

6. A certain transformer has 500 turns in the primary winding and 2500 turns in the secondary winding. The turns ratio is
   (a) 0.2   (b) 2.5   (c) 5   (d) 0.5

7. If 10 W of power are applied to the primary winding of an ideal transformer with a turns ratio of 5, the power delivered to the secondary load is
   (a) 50 W   (b) 0.5 W   (c) 0 W   (d) 10 W

8. In a certain loaded transformer, the secondary voltage is one-third the primary voltage. The secondary current is
   (a) one-third the primary current   (b) three times the primary current
   (c) equal to the primary current   (d) both (b) and (d)

9. When a 1 kΩ load resistor is connected across the secondary winding of a transformer with a turns ratio of 2, the source "sees" a reflected load of
   (a) 250 Ω   (b) 2 kΩ   (c) 4 kΩ   (d) 1 kΩ

10. In Question 9, if the turns ratio is 0.5, the source "sees" a reflected load of
    (a) 1 kΩ   (b) 2 kΩ   (c) 4 kΩ   (d) 500 Ω

11. The turns ratio required to match a 50 Ω source to a 200 Ω load is
    (a) 0.25   (b) 0.5   (c) 4   (d) 2

12. Maximum power is transferred from a source to a load when
    (a) $R_L > R_s$ (b) $R_L < R_s$ (c) $R_L = R_s$ (d) $R_L = nR_s$

13. When a 12 V battery is connected across the primary winding of a transformer with a turns ratio of 4, the secondary voltage is
    (a) 0 V   (b) 12 V   (c) 48 V   (d) 3 V

14. A certain transformer has a turns ratio of 1 and a 0.95 coefficient of coupling. When 1 V ac is applied to the primary, the secondary voltage is
(a) 1 V   (b) 1.95 V   (c) 0.95 V

■ **BASIC PROBLEMS**

SECTION 11–1   Mutual Inductance

1. What is the mutual inductance when $k = 0.75$, $L_1 = 1\ \mu H$, and $L_2 = 4\ \mu H$?

2. Determine the coefficient of coupling when $L_M = 1\ \mu H$, $L_1 = 8\ \mu H$, and $L_2 = 2\ \mu H$.

SECTION 11–2   The Basic Transformer

3. What is the turns ratio of a transformer having 120 primary turns and 360 secondary turns?

4. (a) What is the turns ratio of a transformer having 250 primary turns and 1000 secondary turns?
   (b) What is the turns ratio when the primary winding has 400 turns and the secondary winding has 100 turns?

5. Determine the phase of the secondary voltage with respect to the primary voltage for each transformer in Figure 11–45.

(a)   (b)   (c)

FIGURE 11–45

SECTION 11–3   The Step-Up Transformer

6. If 120 V ac are connected to the primary of a transformer with 100 primary turns and 150 secondary turns, what is the secondary voltage?

7. A certain transformer has 250 turns in its primary winding. In order to double the voltage, how many turns must be in the secondary winding?

8. How many primary volts must be applied to a transformer with a turns ratio of 10 to obtain a secondary voltage of 60 V ac?

9. For each transformer in Figure 11–46, sketch the secondary voltage showing its relationship to the primary voltage. Also indicate the amplitude.

(a)   (b)

FIGURE 11–46

SECTION 11–4   The Step-Down Transformer

10. To step 120 V down to 30 V, what must the turns ratio be?

11. The primary winding of a transformer has 1200 V across it. What is the secondary voltage if the turns ratio is 0.2?

12. How many primary volts must be applied to a transformer with a turns ratio of 0.1 to obtain a secondary voltage of 6 V ac?

SECTION 11–5   Loading the Secondary Winding

13. Determine I_s in Figure 11–47.

14. Determine the following quantities in Figure 11–48:
    (a) secondary voltage   (b) secondary current
    (c) primary current   (d) power in the load

FIGURE 11–47   **FIGURE 11–48**

SECTION 11–6   Reflected Load

15. What is the load resistance as seen by the source in Figure 11–49?

16. What is the resistance reflected into the primary circuit in Figure 11–50?

17. What is the primary current (rms) in Figure 11–50 if the rms source voltage is 115 V?

18. What must the turns ratio be in Figure 11–51 in order to reflect 300 Ω into the primary circuit?

FIGURE 11–49   **FIGURE 11–50**

FIGURE 11–51

SECTION 11–7   Matching the Load and Source Resistances

19. For the circuit in Figure 11–52, find the turns ratio required to deliver maximum power to the 4 Ω speaker.

20. In Figure 11–52, what is the maximum power in watts delivered to the speaker?

21. Determine the value to which R_L must be adjusted in Figure 11–53 for maximum power transfer. The internal source resistance is 50 Ω.

FIGURE 11–52   **FIGURE 11–53**

SECTION 11–8   The Transformer As an Isolation Device

22. What is the voltage across the load in each circuit of Figure 11–54?

23. If the bottom of each secondary winding in Figure 11–54 were grounded, would the values of the load voltages be changed?

24. Determine the unspecified meter readings in Figure 11–55.

FIGURE 11–54

FIGURE 11–55

SECTION 11–9   Nonideal Transformer Characteristics

25. In a certain transformer, the input power to the primary is 100 W. If 5.5 W are lost in the winding resistances, what is the output power to the load, neglecting any other losses?

26. What is the efficiency of the transformer in Problem 25?

27. Determine the coefficient of coupling for a transformer in which 2% of the total flux generated in the primary does not pass through the secondary.

28. A certain transformer is rated at 1 kVA. It operates on 60 Hz, 120 V ac. The secondary voltage is 600 V.
    (a) What is the maximum load current?
    (b) What is the smallest R_L that you can drive?
    (c) What is the largest capacitor that can be connected as a load?

29. What kVA rating is required for a transformer that must handle a maximum load current of 10 A with a secondary voltage of 2.5 kV?

SECTION 11–10   Other Types of Transformers

30. Determine each unknown voltage indicated in Figure 11–56.

31. Using the indicated secondary voltages in Figure 11–57, determine the turns ratio of the primary winding to each tapped section.

32. In Figure 11–58, each primary winding can accommodate 120 V ac. Show how the primaries should be connected for 240 V ac operation. Determine each secondary voltage.

33. Find the secondary voltage for each autotransformer in Figure 11–59.

FIGURE 11–56

FIGURE 11–57

FIGURE 11–58

FIGURE 11–59

SECTION 11–11 Troubleshooting

34. When you apply 120 V ac across the primary winding of a transformer and check the voltage across the secondary winding, you get 0 V. Further investigation shows no primary or secondary current. List the possible faults. What is your next step in isolating the problem?

35. What is likely to happen if the primary winding of a transformer shorts?

36. In checking out a transformer circuit you find that the secondary voltage is less than it should be, although it is not zero. What is the most likely fault?

▪ **ADVANCED PROBLEMS**

37. For the loaded, tapped-secondary transformer in Figure 11–60, determine the following:
   (a) all load voltages and currents
   (b) the resistance looking into the primary

38. A certain transformer is rated at 5 kVA, 2400/120 V, at 60 Hz.
   (a) What is the turns ratio if the 120 V is the secondary voltage?
   (b) What is the current rating of the secondary if 2400 V is the primary voltage?
   (c) What is the current rating of the primary if 2400 V is the primary voltage?

39. Determine the voltage measured by each voltmeter in Figure 11–61. The bench-type meters have one terminal that connects to ground as indicated.

FIGURE 11–60

FIGURE 11–61

(a)

(b)

40. Find the appropriate turns ratio for each switch position in Figure 11–62 in order to transfer the maximum power to each load when the internal source resistance is 10 Ω. Specify the number of turns for the secondary winding if the primary winding has 100 turns.

41. What must the turns ratio be in Figure 11–50 to limit the primary current to 3 mA with a source voltage of 115 V? Assume an ideal transformer and source.

FIGURE 11–62

▪ **ANSWERS TO SECTION REVIEWS**

Section 11–1

1. Mutual inductance is the inductance between two coils, established by the amount of coupling between the coils.

2. $L_M = k\sqrt{L_1 L_2} = 45$ mH

3. Induced voltage increases when k is increased.

Section 11–2

1. The operation of transformers is based on the principle of mutual inductance.

2. Turns ratio is the ratio of secondary turns to primary turns.

3. Directions of the windings determine voltage polarities.

4. $n = N_{sec}/N_{pri} = 0.5$

Section 11–3

1. A step-up transformer increases voltage.

2. The secondary voltage is five times greater.

3. $V_{sec} = (N_{sec}/N_{pri})V_{pri} = 2400$ V

Section 11–4

1. A step-down transformer decreases voltage.

2. $V_{sec} = (N_{sec}/N_{pri})V_{pri} = 60$ V

3. $n = 12$ V$/120$ V $= 0.1$

Section 11–5

1. The secondary current is half the primary current.

2. $I_{sec} = (N_{pri}/N_{sec})I_{pri} = 2$ A

3. $I_{pri} = (N_{sec}/N_{pri})I_{sec} = 2.5$ A

Section 11–6

1. Reflected resistance is the resistance in the secondary circuit as seen from the primary circuit as a function of the turns ratio.

2. The reciprocal of the turns ratio determines reflected resistance.

3. $R_{pri} = (N_{pri}/N_{sec})^2 R_L = 0.5$ Ω

4. $N_{sec}/N_{pri} = \sqrt{R_L/R_{pri}} = 0.1$

Section 11–7

1. Impedance matching is making the load resistance equal the source resistance.

2. Maximum power is delivered to the load when $R_L = R_S$.

3. $R_{pri} = (N_{pri}/V_{sec})^2 R_L = 400$ Ω

Section 11–8

1. Electrical isolation means there is no electrical connection between the primary and secondary circuits.

2. Transformers do not couple dc.

Section 11–9

1. In a practical transformer, energy loss reduces the efficiency. An ideal transformer has an efficiency of 100%.

2. 85% of the magnetic flux generated in the primary winding passes through the secondary winding when the coefficient of coupling is 0.85.

3. $I_L = 10 \text{ kVA}/250 \text{ V} = 40 \text{ A}$

Section 11–10

1. $V_{\text{sec}} = 10(220 \text{ V}) = 2200 \text{ V}$; $V_{\text{sec}} = 0.2(220 \text{ V}) = 44 \text{ V}$

2. Autotransformers are smaller and lighter for same rating. Autotransformers provide no electrical isolation.

Section 11–11

1. Possible failures are open windings and shorted windings.

2. Operating above rated values causes transformer failure.

Section 11–12

1. Use an ohmmeter to check for open windings. Shorted windings are indicated by an incorrect secondary voltage.

2. A short will cause the fuse to blow.

▪ **ANSWERS TO RELATED EXERCISES FOR EXAMPLES**

11–1	0.75
11–2	387 μH
11–3	500 turns
11–4	480 V
11–5	57.6 V
11–6	5 mA; 400 mA
11–7	6 Ω
11–8	0.354
11–9	833 mW
11–10	0.0707 or 14.14:1
11–11	85.2%
11–12	$V_{AB} = 12$ V; $V_{CD} = 480$ V; $V_{(CT)C} = V_{(CT)D} = 240$ V; $V_{EF} = 24$ V
11–13	Increases to 2.13 kVA

12

RC CIRCUITS

An *RC* circuit contains both resistance and capacitance. It is one of the basic types of reactive circuits that you will study. In this chapter, basic series and parallel *RC* circuits and their responses to sine wave voltages are covered. Series-parallel combinations are also examined. Power considerations in *RC* circuits are introduced, and practical aspects of power ratings are discussed. Several basic areas of application are presented to give you an idea of how simple combinations of resistors and capacitors can be applied.

■ CHAPTER OBJECTIVES

☐ Describe the relationship between current and voltage in an *RC* circuit

☐ Determine impedance and phase angle in a series *RC* circuit

☐ Analyze a series *RC* circuit

☐ Determine impedance and phase angle in a parallel *RC* circuit

☐ Analyze a parallel *RC* circuit

☐ Analyze series-parallel *RC* circuits

☐ Determine power in *RC* circuits

☐ Discuss some basic *RC* applications

☐ Troubleshoot *RC* circuits

■ APPLICATION ASSIGNMENT

Your assignment is to run frequency response measurements on a capacitively coupled amplifier used in a certain communications system. You will be concentrating on the *RC* input circuit of the amplifier and how it responds to different frequencies. The results of your measurements will be graphically documented with a frequency response curve. After studying this chapter, you should be able to complete the application assignment in Section 12–10.

12–1 ■ SINUSOIDAL RESPONSE OF *RC* CIRCUITS

When a sinusoidal voltage is applied to an RC circuit, each resulting voltage drop and the current in the circuit are also sinusoidal and have the same frequency as the applied voltage. The capacitance causes a phase shift between the voltage and current that depends on the relative values of the resistance and the capacitive reactance.

After completing this section, you should be able to

■ **Describe the relationship between current and voltage in an *RC* circuit**
   ☐ Discuss voltage and current waveforms
   ☐ Discuss phase shift
   ☐ Describe types of signal generators

As shown in Figure 12–1, the resistor voltage, the capacitor voltage, and the current are all sine waves with the frequency of the source.

Phase shifts are introduced because of the capacitance. As you will learn, the resistor voltage and current are in phase with each other and **lead** the source voltage in phase. The capacitor voltage **lags** the source voltage. The phase angle between the current and the capacitor voltage is always 90°. These generalized phase relationships are indicated in Figure 12–1.

FIGURE 12–1

Illustration of sinusoidal response with general phase relationships of V_R, V_C, and I relative to the source voltage. V_R and I are in phase; V_R leads V_s; V_C lags V_s; and V_R and V_C are 90° out of phase.

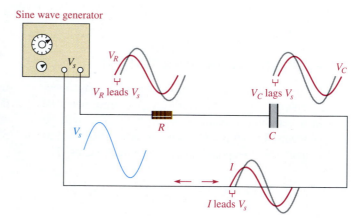

The amplitudes and the phase relationships of the voltages and current depend on the values of the resistance and the capacitive reactance (defined in Section 9–6). When a circuit is purely resistive, the phase angle between the applied (source) voltage and the total current is zero. When a circuit is purely capacitive, the phase angle between the applied voltage and the total current is 90°, with the current leading the voltage. When there is a combination of both resistance and capacitive reactance in a circuit, the phase angle between the applied voltage and the total current is somewhere between zero and 90°, depending on the relative values of the resistance and the reactance.

Signal Generators

When a circuit is hooked up for a laboratory experiment or for troubleshooting, a signal generator similar to those shown in Figure 12–2 is used to provide the source voltage. These instruments, depending on their capability, are classified as sine wave generators which produce only sine waves; sine/square generators, which produce both sine waves and square waves; or function generators, which produce sine waves, pulse waveforms, and triangular (ramp) waveforms.

(a)

(b)

(c)

FIGURE 12–2

Typical signal (function) generators used in circuit testing and troubleshooting. Part (a) courtesy of Hewlett-Packard Company. Part (b) courtesy of Wavetek. Part (c) courtesy of B & K Precision.

SECTION 12–1 REVIEW	1. A 60 Hz sinusoidal voltage is applied to an *RC* circuit. What is the frequency of the capacitor voltage? What is the current? 2. What causes the phase shift between V_s and *I* in a series *RC* circuit? 3. When the resistance in an *RC* circuit is greater than the capacitive reactance, is the phase angle between the applied voltage and the total current closer to 0° or to 90°?

12–2 ▪ IMPEDANCE AND PHASE ANGLE OF A SERIES *RC* CIRCUIT

The impedance of an RC circuit is the total opposition to sinusoidal current and its unit is the ohm. The phase angle is the phase difference between the total current and the source voltage.

After completing this section, you should be able to

▪ **Determine impedance and phase angle in a series *RC* circuit**
  ☐ Define *impedance*
  ☐ Draw an impedance triangle
  ☐ Calculate total impedance magnitude
  ☐ Calculate the phase angle

(a) $Z = R$ (b) $Z = X_C$ (c) Z includes both R and X_C

FIGURE 12–3
Three cases of impedance.

In a purely resistive circuit, the **impedance** is simply equal to the total resistance. In a purely capacitive circuit, the impedance is the total capacitive reactance. The impedance of a series *RC* circuit is determined by both the resistance and the capacitive reactance. These cases are illustrated in Figure 12–3. The magnitude of the impedance is symbolized by *Z*.

The Impedance Triangle

In ac analysis, both *R* and X_C are treated as phasor quantities, as shown in the phasor diagram of Figure 12–4(a), and X_C appearing at a –90° angle with respect to *R*. This relationship comes from the fact that the capacitor voltage in a series *RC* circuit lags the current, and thus the resistor voltage, by 90°.

(a)   (b)   (c)

FIGURE 12–4
Development of the impedance triangle for a series RC circuit.

Since *Z* is the phasor sum of *R* and X_C, its phasor representation is as shown in Figure 12–4(b). A repositioning of the phasors, as shown in Figure 12–4(c), forms a right triangle, called the *impedance triangle*. The length of each phasor represents the magnitude in ohms, and the angle θ (the Greek letter theta) is the phase angle of the *RC* circuit and represents the phase difference between the applied (source) voltage and the current.

From right angle trigonometry (Pythagorean theorem), the magnitude of the impedance can be expressed in terms of the resistance and reactance as

$$Z = \sqrt{R^2 + X_C^2} \tag{12–1}$$

This is the magnitude of *Z*, as shown in Figure 12–5, and is expressed in ohms.
The phase angle, θ, is expressed as

$$\theta = \tan^{-1}\left(\frac{X_C}{R}\right) \tag{12–2}$$

FIGURE 12–5
Impedance of a series RC circuit.

The symbol $\tan^{-1}$ stands for inverse tangent and can be found on some calculators by pressing [2nd] [tan] and on others by pressing [INV] [tan]. We will use the [2nd] [tan] sequence in this book. Another term for inverse tangent is *arctangent* (arctan).

EXAMPLE 12–1

Determine the impedance and the phase angle of the circuit in Figure 12–6.

FIGURE 12–6

Solution   The impedance is

$$Z = \sqrt{R^2 + X_C^2} = \sqrt{(47\ \Omega)^2 + (100\ \Omega)^2} = 110\ \Omega$$

The phase angle is

$$\theta = \tan^{-1}\left(\frac{X_C}{R}\right) = \tan^{-1}\left(\frac{100\ \Omega}{47\ \Omega}\right) = \tan^{-1}(2) = 64.8°$$

The applied voltage lags the current by 64.8°.
   The calculator sequences are as follows:

Z: [4] [7] [x²] [+] [1] [0] [0] [x²] [=] [2nd] [√x]
θ: [1] [0] [0] [÷] [4] [7] [=] [2nd] [tan]

Related Exercise   Find Z and θ for $R = 1\ \text{k}\Omega$ and $X_C = 2.2\ \text{k}\Omega$.

SECTION 12–2 REVIEW

1. Define *impedance*.
2. Does the source voltage lead or lag the current in a series *RC* circuit?
3. What causes the phase angle in the *RC* circuit?
4. A series *RC* circuit has a resistance of 33 kΩ and a capacitive reactance of 50 kΩ. What is the value of the impedance? What is the phase angle?

12–3 ■ ANALYSIS OF SERIES *RC* CIRCUITS

In the previous section, you learned how to express the impedance of a series RC circuit. Now, Ohm's law and Kirchhoff's voltage law are used in the analysis of RC circuits.

After completing this section, you should be able to

■ **Analyze a series *RC* circuit**
  □ Apply Ohm's law and Kirchhoff's voltage law to series *RC* circuits
  □ Determine the phase relationships of the voltages and current
  □ Show how impedance and phase angle vary with frequency

Ohm's Law

The application of Ohm's law to series *RC* circuits involves the use of the quantities of Z, V, and I. The three equivalent forms of Ohm's law are as follows:

$$V = IZ \qquad (12\text{–}3)$$

$$I = \frac{V}{Z} \qquad (12\text{–}4)$$

$$Z = \frac{V}{I} \qquad (12\text{–}5)$$

The following two examples illustrate the use of Ohm's law.

EXAMPLE 12–2

If the current in Figure 12–7 is 0.2 mA, determine the source voltage and the phase angle.

FIGURE 12–7

Solution The capacitive reactance is

$$X_C = \frac{1}{2\pi f C} = \frac{1}{2\pi (1000 \text{ Hz})(0.01 \ \mu\text{F})} = 15.9 \text{ k}\Omega$$

The impedance is

$$Z = \sqrt{R^2 + X_C^2} = \sqrt{(10 \text{ k}\Omega)^2 + (15.9 \text{ k}\Omega)^2} = 18.8 \text{ k}\Omega$$

Applying Ohm's law,

$$V_s = IZ = (0.2 \text{ mA})(18.8 \text{ k}\Omega) = 3.76 \text{ V}$$

The phase angle is

$$\theta = \tan^{-1}\left(\frac{X_C}{R}\right) = \tan^{-1}\left(\frac{15.9 \text{ k}\Omega}{10 \text{ k}\Omega}\right) = 57.8°$$

The source voltage has a magnitude of 3.76 V and lags the current by 57.8°.
The calculator sequence for V_s is

Related Exercise Determine V_s in Figure 12–7 if $f = 2$ kHz and $I = 200 \text{ }\mu$A.

EXAMPLE 12–3   Determine the current in the circuit of Figure 12–8.

FIGURE 12–8

Solution   The capacitive reactance is

$$X_C = \frac{1}{2\pi f C} = \frac{1}{2\pi(1.5 \text{ kHz})(0.02 \text{ }\mu\text{F})} = 5.3 \text{ k}\Omega$$

The impedance is

$$Z = \sqrt{R^2 + X_C^2} = \sqrt{(2.2 \text{ k}\Omega)^2 + (5.3 \text{ k}\Omega)^2} = 5.74 \text{ k}\Omega$$

Applying Ohm's law,

$$I = \frac{V}{Z} = \frac{10 \text{ V}}{5.74 \text{ k}\Omega} = 1.74 \text{ mA}$$

Related Exercise   Determine the phase angle between V_s and I in Figure 12–8.

Relationships of the Current and Voltages in a Series *RC* Circuit

In a series circuit, the current is the same through both the resistor and the capacitor. Thus, the resistor voltage is in phase with the current, and the capacitor voltage lags the current by 90°. Therefore, there is a phase difference of 90° between the resistor voltage, V_R, and the capacitor voltage, V_C, as shown in the waveform diagram of Figure 12–9.

You know from Kirchhoff's voltage law that the sum of the voltage drops must equal the applied voltage. However, since V_R and V_C are not in phase with each other, they must be added as phasor quantities, with V_C lagging V_R by 90°, as shown in Figure

FIGURE 12–9

Phase relation of voltages and current in a series RC circuit.

I leads V_C by 90°.
V_R and *I* are in phase.
Amplitudes are arbitrary.

FIGURE 12–10

Voltage phasor diagram for a series RC circuit.

(a)   (b)

12–10(a). As shown in Figure 12–10(b), V_s is the phasor sum of V_R and V_C, as expressed in the following equation:

$$V_s = \sqrt{V_R^2 + V_C^2} \qquad \textbf{(12–6)}$$

The phase angle between the resistor voltage and the source voltage can be expressed as

$$\theta = \tan^{-1}\left(\frac{V_C}{V_R}\right) \qquad \textbf{(12–7)}$$

Since the resistor voltage and the current are in phase, θ in Equation (12–7) also represents the phase angle between the source voltage and the current and is equivalent to $\tan^{-1}(X_C/R)$.

Figure 12–11 shows a voltage and current phasor diagram representing the waveform diagram of Figure 12–9.

FIGURE 12–11

Voltage and current phasor diagram for the waveforms in Figure 12–9.

EXAMPLE 12–4

Determine the source voltage and the phase angle in Figure 12–12.

FIGURE 12–12

Solution Since V_R and V_C are 90° out of phase, they cannot be added directly. The source voltage is the phasor sum of V_R and V_C:

$$V_s = \sqrt{V_R^2 + V_C^2} = \sqrt{(10 \text{ V})^2 + (15 \text{ V})^2} = 18 \text{ V}$$

The phase angle between the current and source voltage is

$$\theta = \tan^{-1}\left(\frac{V_C}{V_R}\right) = \tan^{-1}\left(\frac{15 \text{ V}}{10 \text{ V}}\right) = 56.3°$$

The calculator sequences are as follows:

V_s: [1] [0] [x^2] [+] [1] [5] [x^2] [=] [2nd] [√x]
θ: [1] [5] [÷] [1] [0] [=] [2nd] [tan]

Related Exercise In a certain series *RC* circuit, $V_s = 10$ V and $V_R = 7$ V. Find V_C.

Variation of Impedance with Frequency

As you know, capacitive reactance varies inversely with frequency. Since $Z = \sqrt{R^2 + X_C^2}$, you can see that when X_C increases, the entire term under the square root sign increases and thus the total impedance also increases; and when X_C decreases, the total impedance also decreases. Therefore, *in an RC circuit, Z is inversely dependent on frequency.*

Figure 12–13 illustrates how the voltages and current in a series *RC* circuit vary as the frequency increases or decreases, with the source voltage held at a constant value. In part (a), as the frequency is increased, X_C decreases; so less voltage is dropped across the capacitor. Also, Z decreases as X_C decreases, causing the current to increase. An increase in the current causes more voltage across R.

In Figure 12–13(b), as the frequency is decreased, X_C increases; so more voltage is dropped across the capacitor. Also, Z increases as X_C increases, causing the current to decrease. A decrease in the current causes less voltage across R.

Changes in Z and X_C can be observed as shown in Figure 12–14. As the frequency increases, the voltage across Z remains constant because V_s is constant. Also, the voltage across C decreases. The increasing current indicates that Z is decreasing. It does so because of the inverse relationship stated in Ohm's law ($Z = V_Z/I$). The increasing current also indicates that X_C is decreasing ($X_C = V_C/I$). The decrease in V_C corresponds to the decrease in X_C.

Variation of the Phase Angle with Frequency

Since X_C is the factor that introduces the phase angle in a series *RC* circuit, a change in X_C produces a change in the phase angle. As the frequency is increased, X_C be-

(a) As frequency is increased, I and V_R increase and V_C decreases.

(b) As frequency is decreased, I and V_R decrease and V_C increases.

FIGURE 12–13

An illustration of how the variation of impedance affects the voltages and current as the source frequency is varied. The source voltage is held at a constant amplitude.

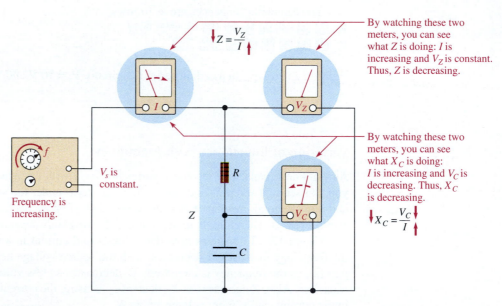

FIGURE 12–14

An illustration of how Z and X_C change with frequency.

comes smaller, and thus the phase angle decreases. As the frequency is decreased, X_C becomes larger, and thus the phase angle increases. This variation is illustrated in Figure 12–15 where a phase meter is connected across the capacitor to measure the angle between V_s and V_R. This angle is the phase angle of the circuit because I is in phase with V_R. By measuring the phase of V_R, you are effectively measuring the phase of I. You can use the phase-meter portion of a vector voltmeter such as the one shown in Figure 12–15(c) to measure the phase of one voltage relative to another. You can also see the variation of phase angle with frequency by using the impedance triangle as shown in Figure 12–16.

(a) As frequency is increased, the phase angle decreases.

(b) As frequency is decreased, the phase angle increases.

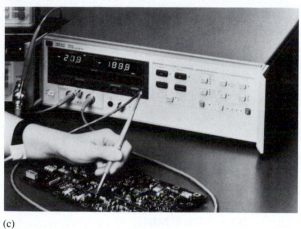

(c)

FIGURE 12–15

The phase meter on a vector voltmeter shows the effect of frequency on the phase angle of a circuit. The phase meter indicates the phase angle between two voltages, V_s and V_R. Since V_R and I are in phase, this angle is the same as the angle between V_s and I. The vector voltmeter also indicates voltage magnitudes (photo courtesy of Hewlett-Packard Co.).

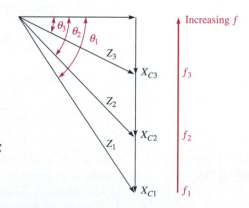

FIGURE 12–16

As the frequency increases, X_C decreases, Z decreases, and θ decreases. Each value of frequency can be visualized as forming a different impedance triangle.

EXAMPLE 12–5   For the series *RC* circuit in Figure 12–17, determine the impedance and phase angle for each of the following values of frequency:
(a) 10 kHz   (b) 20 kHz   (c) 30 kHz

FIGURE 12–17

Solution

(a) For $f = 10$ kHz, the impedance is calculated as follows:

$$X_C = \frac{1}{2\pi f C} = \frac{1}{2\pi(10 \text{ kHz})(0.01 \text{ } \mu\text{F})} = 1.59 \text{ k}\Omega$$

$$Z = \sqrt{R^2 + X_C^2} = \sqrt{(1 \text{ k}\Omega)^2 + (1.59 \text{ k}\Omega)^2} = 1.88 \text{ k}\Omega$$

The phase angle is

$$\theta = \tan^{-1}\left(\frac{X_C}{R}\right) = \tan^{-1}\left(\frac{1.59 \text{ k}\Omega}{1 \text{ k}\Omega}\right) = 57.8°$$

(b) For $f = 20$ kHz,

$$X_C = \frac{1}{2\pi(20 \text{ kHz})(0.01 \text{ } \mu\text{F})} = 796 \text{ } \Omega$$

$$Z = \sqrt{(1 \text{ k}\Omega)^2 + (796 \text{ } \Omega)^2} = 1.28 \text{ k}\Omega$$

$$\theta = \tan^{-1}\left(\frac{796 \text{ } \Omega}{1 \text{ k}\Omega}\right) = 38.5°$$

(c) For $f = 30$ kHz,

$$X_C = \frac{1}{2\pi(30 \text{ kHz})(0.01 \text{ } \mu\text{F})} = 531 \text{ } \Omega$$

$$Z = \sqrt{(1 \text{ k}\Omega)^2 + (531 \text{ } \Omega)^2} = 1.13 \text{ k}\Omega$$

$$\theta = \tan^{-1}\left(\frac{531 \text{ } \Omega}{1 \text{ k}\Omega}\right) = 28.0°$$

Notice that as the frequency increases, X_C, Z, and θ decrease.

Related Exercise Find the total impedance and phase angle in Figure 12–17 for $f = 1$ kHz.

SECTION 12–3 REVIEW

1. In a certain series *RC* circuit, $V_R = 4$ V and $V_C = 6$ V. What is the magnitude of the total voltage?
2. In Question 1, what is the phase angle?
3. What is the phase difference between the capacitor voltage and the resistor voltage in a series *RC* circuit?
4. When the frequency of the applied voltage in a series *RC* circuit is increased, what happens to each of the following?
   (a) the capacitive reactance   **(b)** the impedance   **(c)** the phase angle

12–4 ■ IMPEDANCE AND PHASE ANGLE OF A PARALLEL *RC* CIRCUIT

In this section, you will learn how to determine the impedance and phase angle of a parallel RC circuit. Also conductance (G), capacitive susceptance (B$_C$), and total admittance (Y$_T$) are discussed because of their usefulness in parallel circuit analysis.

After completing this section, you should be able to

■ **Determine impedance and phase angle in a parallel *RC* circuit**
   □ Express total impedance in a product over sum form
   □ Express the phase angle in terms of R and X_C
   □ Determine the conductance, susceptance, and admittance in a parallel *RC* circuit
   □ Convert admittance to impedance

Figure 12–18 shows a basic parallel *RC* circuit.

FIGURE 12–18
Parallel RC circuit.

The expression for the impedance in Equation (12–8) is given in a product over sum form similar to the way two resistors in parallel can be expressed. In this case, the denominator is the phasor sum of R and X_C.

$$Z = \frac{RX_C}{\sqrt{R^2 + X_C^2}} \qquad (12\text{–}8)$$

The phase angle between the applied voltage and the total current can be expressed in terms of R and X_C as shown in Equation (12–9).

$$\theta = \tan^{-1}\left(\frac{R}{X_C}\right) \qquad (12\text{–}9)$$

This formula is derived from an equivalent formula using branch currents, which is introduced in the next section.

EXAMPLE 12–6   For each circuit in Figure 12–19, determine the impedance and the phase angle.

(a)   (b)

FIGURE 12–19

Solution For the circuit in Figure 12–19(a), the impedance and phase angle are

$$Z = \frac{RX_C}{\sqrt{R^2 + X_C^2}} = \frac{(100\ \Omega)(50\ \Omega)}{\sqrt{(100\ \Omega)^2 + (50\ \Omega)^2}} = 44.7\ \Omega$$

$$\theta = \tan^{-1}\left(\frac{R}{X_C}\right) = \tan^{-1}\left(\frac{100\ \Omega}{50\ \Omega}\right) = 63.4°$$

For the circuit in Figure 12–19(b),

$$Z = \frac{(1\ k\Omega)(2\ k\Omega)}{\sqrt{(1\ k\Omega)^2 + (2\ k\Omega)^2}} = 894\ \Omega$$

$$\theta = \tan^{-1}\left(\frac{1\ k\Omega}{2\ k\Omega}\right) = 26.6°$$

The calculator sequences for the circuit in Figure 12–19(a) are as follows:

Z: [1][0][0][×][5][0][÷][(][(][(][1][0][0][x²][+][5][0][x²][)][2nd][√x][)][=]
θ: [1][0][0][÷][5][0][=][2nd][tan]

Related Exercise Determine *Z* in Figure 12–19(a) if the frequency is doubled.

Conductance, Susceptance, and Admittance

In working with parallel circuits, you will often find it easier to use conductance (G), capacitive susceptance (B_C), and admittance (Y) rather than resistance (R), capacitive reactance (X_C) and impedance (Z). **Conductance (G)** is the reciprocal of resistance and is expressed as

$$G = \frac{1}{R} \tag{12–10}$$

Susceptance is the reciprocal of reactance; therefore, **capacitive susceptance (B_C)** is the reciprocal of capacitive reactance and is expressed as

$$B_C = \frac{1}{X_C} \tag{12–11}$$

Admittance (Y) is the reciprocal of impedance and is expressed as

$$Y = \frac{1}{Z} \tag{12–12}$$

The unit of each of these quantities is the siemens (S), which is the reciprocal of the ohm.

In a parallel *RC* circuit, as shown in Figure 12–20, the total admittance is simply the phasor sum of the conductance and the capacitive susceptance:

$$Y_T = \sqrt{G^2 + B_C^2} \tag{12–13}$$

FIGURE 12–20

Admittance is the reciprocal of impedance in a parallel RC circuit.

$$Y = \sqrt{G^2 + B_C^2}$$

EXAMPLE 12–7   Determine the admittance in Figure 12–21, and then convert it to impedance.

FIGURE 12–21

Solution   To determine Y, first calculate the values for G and B_C. Since $R = 330 \ \Omega$, then

$$G = \frac{1}{R} = \frac{1}{330 \ \Omega} = 3.03 \ \text{mS}$$

The capacitive reactance is

$$X_C = \frac{1}{2\pi f C} = \frac{1}{2\pi(1000 \ \text{Hz})(0.2 \ \mu\text{F})} = 796 \ \Omega$$

The capacitive susceptance is

$$B_C = \frac{1}{X_C} = \frac{1}{796 \ \Omega} = 1.26 \ \text{mS}$$

Therefore, the total admittance is

$$Y_T = \sqrt{G^2 + B_C^2} = \sqrt{(3.03 \ \text{mS})^2 + (1.26 \ \text{mS})^2} = 3.28 \ \text{mS}$$

Admittance is converted to impedance as follows:

$$Z = \frac{1}{Y_T} = \frac{1}{3.28 \ \text{mS}} = 305 \ \Omega$$

The calculator sequence for Y is

[3] [•] [0] [3] [EXP] [+/−] [3] [x²] [+] [1] [•] [2] [6] [EXP] [+/−] [3] [x²] [=] [2nd] [√x]

Related Exercise   Calculate the admittance in Figure 12–21 if f is increased to 2.5 kHz.

SECTION 12–4 REVIEW

1. Determine Z if a 1 kΩ resistance is in parallel with a 650 Ω capacitive reactance.
2. Define *conductance, capacitive susceptance,* and *admittance.*
3. If $Z = 100$ Ω, what is the value of Y?
4. In a certain parallel RC circuit, $R = 50$ Ω and $X_C = 75$ Ω. Determine Y.

12–5 ■ ANALYSIS OF PARALLEL *RC* CIRCUITS

In the previous section, you learned how to express the impedance of a parallel RC circuit. Now, Ohm's law and Kirchhoff's current law are used in the analysis of RC circuits. Current and voltage relationships in a parallel RC circuit are examined.

After completing this section, you should be able to

■ *Analyze a parallel RC circuit*
   □ Apply Ohm's law and Kirchhoff's current law to parallel *RC* circuits
   □ Show how impedance and phase angle vary with frequency
   □ Convert from a parallel circuit to an equivalent series circuit

For convenience in the analysis of parallel circuits, the Ohm's law formulas using impedance—Equations (12–3), (12–4), and (12–5)—can be rewritten for admittance using the relation $Y = 1/Z$:

$$V = \frac{I}{Y} \tag{12–14}$$

$$I = VY \tag{12–15}$$

$$Y = \frac{I}{V} \tag{12–16}$$

EXAMPLE 12–8   Determine the total current and the phase angle in Figure 12–22.

FIGURE 12–22

Solution   First determine the total admittance. The capacitive reactance is

$$X_C = \frac{1}{2\pi f C} = \frac{1}{2\pi(1.5 \text{ kHz})(0.02 \text{ } \mu\text{F})} = 5.31 \text{ k}\Omega$$

The capacitive susceptance is

$$B_C = \frac{1}{X_C} = \frac{1}{5.31 \text{ k}\Omega} = 188 \text{ } \mu S$$

The conductance is

$$G = \frac{1}{R} = \frac{1}{2.2 \text{ k}\Omega} = 455 \text{ } \mu S$$

Therefore, the total admittance is

$$Y_T = \sqrt{G^2 + B_C^2} = \sqrt{(455 \text{ } \mu S)^2 + (188 \text{ } \mu S)^2} = 492 \text{ } \mu S$$

Applying Ohm's law, calculate the total current:

$$I_T = VY_T = (10 \text{ V})(492 \text{ } \mu S) = 4.92 \text{ mA}$$

The phase angle is

$$\theta = \tan^{-1}\left(\frac{R}{X_C}\right) = \tan^{-1}\left(\frac{2.2 \text{ k}\Omega}{5.31 \text{ k}\Omega}\right) = 22.5°$$

The total current is 4.92 mA, and it leads the applied voltage by 22.5°.

Related Exercise   What is the total current if the frequency is doubled?

Relationships of the Currents and Voltages in a Parallel *RC* Circuit

Figure 12–23(a) shows all the currents and voltages in a basic parallel *RC* circuit. As you can see, the applied voltage, V_s, appears across both the resistive and the capacitive branches, so V_s, V_R, and V_C are all in phase and of the same magnitude. The total current, I_T, divides at the junction into the two branch currents, I_R and I_C.

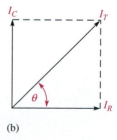

(a)   (b)

FIGURE 12–23
Currents and voltages in a parallel RC circuit. The current directions shown in (a) are
instantaneous and, of course, reverse when the source voltage reverses.

The current through the resistor is in phase with the voltage. The current through the capacitor leads the voltage, and thus the resistive current, by 90°. By Kirchhoff's current law, the total current is the phasor sum of the two branch currents, as shown by the phasor diagram in Figure 12–23(b). The total current is expressed as

$$I_T = \sqrt{I_R^2 + I_C^2} \qquad\qquad (12\text{–}17)$$

The phase angle can be found from the current values as follows:

$$\theta = \tan^{-1}\left(\frac{I_C}{I_R}\right) \qquad\qquad (12\text{--}18)$$

Equation (12–18) is equivalent to the previous expression, $\theta = \tan^{-1}(R/X_C)$ in Equation (12–9).

Figure 12–24 shows a complete current and voltage phasor diagram. Notice that I_C leads I_R by 90° and that I_R is in phase with the voltage ($V_s = V_R = V_C$).

FIGURE 12–24

Current and voltage phasor diagram for a parallel RC circuit (amplitudes are arbitrary).

EXAMPLE 12–9

Determine the value of each current in Figure 12–25, and describe the phase relationship of each with the applied voltage.

FIGURE 12–25

Solution   The resistor current, the capacitor current, and the total current are expressed as follows:

$$I_R = \frac{V_s}{R} = \frac{12\text{ V}}{220\ \Omega} = 54.5\text{ mA}$$

$$I_C = \frac{V_s}{X_C} = \frac{12\text{ V}}{150\ \Omega} = 80\text{ mA}$$

$$I_T = \sqrt{I_R^2 + I_C^2} = \sqrt{(54.5\text{ mA})^2 + (80\text{ mA})^2} = 96.8\text{ mA}$$

The phase angle is

$$\theta = \tan^{-1}\left(\frac{I_C}{I_R}\right) = \tan^{-1}\left(\frac{80\text{ mA}}{54.5\text{ mA}}\right) = 55.7°$$

I_R is in phase with the voltage, I_C leads the voltage by 90°, and I_T leads the voltage by 55.7°.

Related Exercise   In a certain parallel circuit, $I_R = 100$ mA and $I_C = 60$ mA. Determine the total current and the phase angle.

Conversion from Parallel to Series Form

For every parallel *RC* circuit, there is an equivalent series *RC* circuit for any given frequency. Two circuits are equivalent when they both present an equal impedance and phase angle.

To obtain the equivalent series circuit for a given parallel *RC* circuit, first find the impedance and phase angle of the parallel circuit. Then use the values of Z and θ to construct an impedance triangle shown in Figure 12–26. The vertical and horizontal sides of the triangle represent the equivalent series resistance and capacitive reactance as indicated. These values can be found using the following trigonometric relationships:

$$R_{eq} = Z \cos \theta \qquad\qquad (12\text{–}19)$$

$$X_{C(eq)} = Z \sin \theta \qquad\qquad (12\text{–}20)$$

The cosine (cos) and the sine (sin) functions are available on your calculator.

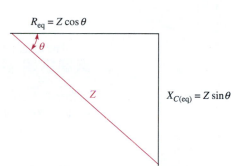

FIGURE 12–26

Impedance triangle for the series equivalent of a parallel RC circuit. Z and θ are the known values for the parallel circuit. R_{eq} and $X_{C(eq)}$ are the series equivalent values.

EXAMPLE 12–10   Convert the parallel circuit in Figure 12–27 to an equivalent series form.

FIGURE 12–27

Solution   First, find the total admittance of the parallel circuit as follows:

$$G = \frac{1}{R} = \frac{1}{18\ \Omega} = 55.6\ \text{mS}$$

$$B_C = \frac{1}{X_C} = \frac{1}{27\ \Omega} = 37.0\ \text{mS}$$

$$Y_T = \sqrt{G^2 + B_C^2} = \sqrt{(55.6\ \text{mS})^2 + (37.0\ \text{mS})^2} = 66.8\ \text{mS}$$

Then, the total impedance is

$$Z_T = \frac{1}{Y_T} = \frac{1}{66.8\ \text{mS}} = 15.0\ \Omega$$

The phase angle is

$$\theta = \tan^{-1}\left(\frac{R}{X_C}\right) = \tan^{-1}\left(\frac{18\ \Omega}{27\ \Omega}\right) = 33.6°$$

The equivalent series values are

$$R_{eq} = Z \cos \theta = 15.0 \cos(33.6°) = 12.5\ \Omega$$

$$X_{C(eq)} = Z \sin \theta = 15.0 \sin(33.6°) = 8.3\ \Omega$$

The series equivalent circuit is shown in Figure 12–28. The value of C can only be determined if the frequency is known.

FIGURE 12–28

The calculator sequence for R_{eq} is

Related Exercise The impedance of a parallel *RC* circuit is 10 kΩ and the phase angle is 26°. Convert to an equivalent series circuit.

Note that a parallel *RC* circuit becomes less reactive when X_C is increased. That is, the circuit phase angle becomes smaller. The reason for this effect is that when X_C is increased relative to R, less current flows through the capacitive branch, and although the in-phase or resistive current does not increase, it becomes a greater percentage of the total current.

SECTION 12–5 REVIEW

1. The admittance of an *RC* circuit is 3.5 mS, and the applied voltage is 6 V. What is the total current?
2. In a certain parallel *RC* circuit, the resistor current is 10 mA, and the capacitor current is 15 mA. Determine the phase angle and the total current.
3. What is the phase angle between the capacitor current and the applied voltage in a parallel *RC* circuit?

12–6 ■ SERIES-PARALLEL ANALYSIS

In this section, the concepts studied in the previous sections are used to analyze circuits with combinations of both series and parallel R and C elements.

After completing this section, you should be able to

■ **Analyze series-parallel *RC* circuits**
  □ Determine total impedance
  □ Calculate currents and voltages
  □ Measure impedance and phase angle

The following example demonstrates the procedures used to analyze a series-parallel circuit.

EXAMPLE 12–11   In the circuit of Figure 12–29, determine the following:
(a) total impedance   (b) total current   (c) phase angle by which I_T leads V_s

FIGURE 12–29

Solution

(a) First, calculate the magnitudes of capacitive reactances:

$$X_{C1} = \frac{1}{2\pi(5 \text{ kHz})(0.1 \ \mu\text{F})} = 318 \ \Omega$$

$$X_{C2} = \frac{1}{2\pi(5 \text{ kHz})(0.05 \ \mu\text{F})} = 637 \ \Omega$$

One approach is to find the series equivalent resistance and capacitive reactance for the parallel portion of the circuit, and then add the resistances to get total resistance and add the reactances to get total reactance. From this, the total impedance can be determined.

Find the impedance of the parallel portion (Z_2) by first finding the admittance:

$$G_2 = \frac{1}{R_2} = \frac{1}{680 \ \Omega} = 1.47 \text{ mS}$$

$$B_{C2} = \frac{1}{X_{C2}} = \frac{1}{637 \ \Omega} = 1.57 \text{ mS}$$

$$Y_2 = \sqrt{G_2^2 + B_{C2}^2} = \sqrt{(1.47 \text{ mS})^2 + (1.57 \text{ mS})^2} = 2.15 \text{ mS}$$

$$Z_2 = \frac{1}{Y_2} = \frac{1}{2.15 \text{ mS}} = 465 \ \Omega$$

The phase angle associated with the parallel portion of the circuit is

$$\theta_p = \tan^{-1}\left(\frac{R_2}{X_{C2}}\right) = \tan^{-1}\left(\frac{680 \ \Omega}{637 \ \Omega}\right) = 46.9°$$

The series equivalent values for the parallel portion are

$$R_{eq} = Z_2 \cos \theta_p = 465 \cos(46.9°) = 318 \ \Omega$$

$$X_{C(eq)} = Z_2 \sin \theta_p = 465 \sin(46.9°) = 340 \ \Omega$$

The total resistance is

$$R_T = R_1 + R_{eq} = 1000 \ \Omega + 318 \ \Omega = 1.32 \ k\Omega$$

The total reactance is

$$X_{C(tot)} = X_{C1} + X_{C(eq)} = 318 \ \Omega + 340 \ \Omega = 658 \ \Omega$$

The total impedance is

$$Z_T = \sqrt{R_T^2 + X_{C(tot)}^2} = \sqrt{(1.32 \ k\Omega)^2 + (658 \ \Omega)^2} = 1.47 \ k\Omega$$

(b) Use Ohm's law to find the total current:

$$I_T = \frac{V_S}{Z_T} = \frac{10 \ V}{1.47 \ k\Omega} = 6.80 \ mA$$

(c) To get the phase angle, view the circuit as a series combination of R_T and $X_{C(tot)}$. The phase angle by which I_T leads V_s is

$$\theta = \tan^{-1}\left(\frac{X_{C(tot)}}{R_T}\right) = \tan^{-1}\left(\frac{658 \ \Omega}{1.32 \ k\Omega}\right) = 26.5°$$

Related Exercise   Determine the voltages across Z_1 and Z_2 in Figure 12–29.

Measurement of Z_T and θ

Now, let's see how the values of Z_T and θ for the circuit in Example 12–11 can be determined by measurement. First, the total impedance is measured as outlined in the following steps and as illustrated in Figure 12–30 (other ways are also possible):

1. Using a sine wave generator, set the source voltage to a known value (10 V) and the frequency to 5 kHz. It is advisable to check the voltage with an ac voltmeter and the frequency with a frequency counter rather than relying on the marked values on the generator controls.

2. Connect an ac ammeter as shown in Figure 12–30, and measure the total current.

3. Calculate the total impedance by using Ohm's law.

FIGURE 12–30
Determining Z_T by measurement of V_s and I_T.

Although you could use a phase meter to measure the phase angle, we will use an oscilloscope in this illustration because it is more commonly available. To measure the phase angle, the source voltage and the total current must be displayed on the screen in the proper time relationship. Two basic types of scope probes are available to measure the quantities with an oscilloscope: the voltage probe and the current probe. Although the current probe is a convenient device, it is often not as readily available as a voltage probe. For this reason, we will confine our phase measurement technique to the use of voltage probes in conjunction with the oscilloscope. A typical oscilloscope voltage probe has two points that are connected to the circuit: the probe tip and the ground lead. Thus, all voltage measurements must be referenced to ground.

Since only voltage probes are to be used, the total current cannot be measured directly. However, for phase measurement, the voltage across R_1 is in phase with the total current and can be used to establish the phase angle of the current.

Before proceeding with the actual phase measurement, there is a problem with displaying V_{R1}. If the scope probe is connected across the resistor, as indicated in Figure 12–31(a), the ground lead of the scope will short point B to ground, thus bypassing the rest of the components and effectively removing them from the circuit electrically, as illustrated in Figure 12–31(b) (assuming that the scope is not isolated from power line ground).

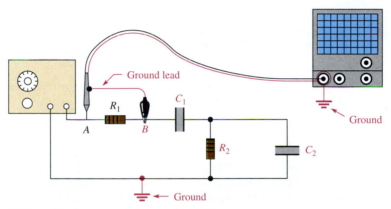

(a) Ground lead on scope probe grounds point B.

(b) The effect of grounding point B is to short out the rest of the circuit.

FIGURE 12–31

Effects of measuring directly across a component when the instrument and the circuit are grounded.

To avoid this problem, you can switch the generator output terminals so that one end of R_1 is connected to the ground terminal, as shown in Figure 12–32(a). Now the scope can be connected across it to display V_{R1}, as indicated in Figure 12–32(b). The other probe is connected across the voltage source to display V_s as indicated. Now channel 1 of

(a) Ground repositioned so that one end of R_1 is grounded.

(b) The scope displays a half-cycle of V_{R1} and V_s.
V_{R1} represents the phase of the total current.

FIGURE 12–32

Repositioning ground so that a direct voltage measurement can be made with respect to ground without shorting out part of the circuit.

the scope has V_{R1} as an input, and channel 2 has V_s. The trigger source switch on the scope should be on internal so that each trace on the screen will be triggered by one of the inputs and the other will then be shown in the proper time relationship to it. Since amplitudes are not important, the volts/div settings are arbitrary. The sec/div settings should be adjusted so that one half-cycle of the waveforms appears on the screen.

Before connecting the probes to the circuit, you must align the two horizontal lines (traces) so that they appear as a single line across the center of the screen. To do so,

FIGURE 12–33

Measurement of the phase angle on the oscilloscope.

ground the probe tips and adjust the vertical position knobs to move the traces toward the center line of the screen until they are superimposed. This procedure ensures that both waveforms have the same zero crossing so that an accurate phase measurement can be made.

The resulting oscilloscope display is shown in Figure 12–33. Since there is 180° in one half-cycle, each of the ten horizontal divisions across the screen represents 18°. Thus, the horizontal distance between the corresponding points of the two waveforms is the phase angle in degrees as indicated.

SECTION 12–6 REVIEW	1. Determine the series equivalent circuit for the series-parallel circuit in Figure 12–29. 2. What is the voltage across R_1 in Figure 12–29?

12–7 ■ POWER IN *RC* CIRCUITS

In a purely resistive ac circuit, all of the energy delivered by the source is dissipated in the form of heat by the resistance. In a purely capacitive ac circuit, all of the energy delivered by the source is stored by the capacitor during a portion of the voltage cycle and then returned to the source during another portion of the cycle so that there is no net energy loss. When there is both resistance and capacitance, some of the energy is alternately stored and returned by the capacitance and some is dissipated by the resistance. The amount of energy loss is determined by the relative values of the resistance and the capacitive reactance.

After completing this section, you should be able to

■ **Determine power in *RC* circuits**
  □ Explain true and reactive power
  □ Draw the power triangle
  □ Define *power factor*
  □ Explain apparent power
  □ Calculate power in an *RC* circuit

It is reasonable to assume that when the resistance is greater than the reactance, more of the total energy delivered by the source is dissipated by the resistance than is stored by the capacitance. Likewise, when the reactance is greater than the resistance, more of the total energy is stored and returned than is dissipated.

The formulas for power in a resistor, sometimes called *true power* (P_{true}), and the power in a capacitor, called *reactive power* (P_r), are restated here. The unit of true power is the watt (W), and the unit of reactive power is the volt-ampere reactive (VAR).

$$P_{\text{true}} = I_T^2 R \tag{12–21}$$

$$P_r = I_T^2 X_C \tag{12–22}$$

The Power Triangle

The generalized impedance phasor diagram is shown in Figure 12–34(a). A phasor relationship for power can also be represented by a similar diagram because the respective

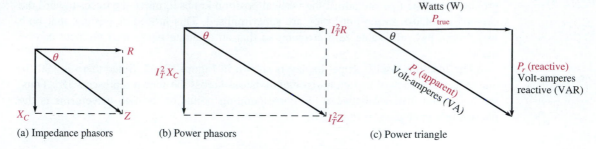

FIGURE 12–34
Development of the power triangle for an RC circuit.

magnitudes of the powers, P_{true} and P_r, differ from R and X_C by a factor of I_T^2, as shown in Figure 12–34(b).

The resultant power phasor, $I_T^2 Z$, represents the **apparent power, P_a**. At any instant in time, P_a is the total power that appears to be transferred between the source and the *RC* circuit. Part of the apparent power is true power, and part of it is reactive power. The unit of apparent power is the volt-ampere (VA). The expression for apparent power is

$$P_a = I_T^2 Z \qquad (12\text{–}23)$$

The power phasor diagram in Figure 12–34(b) can be rearranged in the form of a right triangle as shown in Figure 12–34(c), which is called the *power triangle*. Using the rules of trigonometry, P_{true} can be expressed as

$$P_{\text{true}} = P_a \cos \theta$$

Since P_a equals $I_T^2 Z$ or $V_s I_T$, the equation for the true power can be written as

$$P_{\text{true}} = V_s I_T \cos \theta \qquad (12\text{–}24)$$

where V_s is the applied voltage and I_T is the total current.

For the case of a purely resistive circuit, $\theta = 0°$ and $\cos 0° = 1$, so P_{true} equals $V_s I_T$. For the case of a purely capacitive circuit, $\theta = 90°$ and $\cos 90° = 0$, so P_{true} is zero. As you already know, there is no power dissipated in an ideal capacitor.

Power Factor

The term $\cos \theta$ is called the **power factor** and is stated as follows:

$$PF = \cos \theta \qquad (12\text{–}25)$$

As the phase angle between applied voltage and total current increases, the power factor decreases, indicating an increasingly reactive circuit. The smaller the power factor, the smaller the power dissipation.

The power factor can vary from 0 for a purely reactive circuit to 1 for a purely resistive circuit. In an *RC* circuit, the power factor is referred to as a leading power factor because the current leads the voltage.

EXAMPLE 12–12   Determine the power factor and the true power in the circuit of Figure 12–35.

FIGURE 12–35

Solution   Calculate the capacitive reactance and phase angle as follows:

$$X_C = \frac{1}{2\pi fC} = \frac{1}{2\pi(10 \text{ kHz})(0.005 \ \mu\text{F})} = 3.18 \text{ k}\Omega$$

$$\theta = \tan^{-1}\left(\frac{X_C}{R}\right) = \tan^{-1}\left(\frac{3.18 \text{ k}\Omega}{1 \text{ k}\Omega}\right) = 72.5°$$

The power factor is

$$PF = \cos \theta = \cos(72.5°) = 0.301$$

The impedance is

$$Z = \sqrt{R^2 + X_C^2} = \sqrt{(1 \text{ k}\Omega)^2 + (3.18 \text{ k}\Omega)^2} = 3.33 \text{ k}\Omega$$

Therefore, the current is

$$I = \frac{V_s}{Z} = \frac{15 \text{ V}}{3.33 \text{ k}\Omega} = 4.50 \text{ mA}$$

The true power is

$$P_{\text{true}} = V_s I \cos \theta = (15 \text{ V})(4.50 \text{ mA})(0.301) = 20.3 \text{ mW}$$

Related Exercise   What is the power factor if *f* is reduced by half in Figure 12–35?

The Significance of Apparent Power

As mentioned, apparent power is the power that appears to be transferred between the source and the load, and it consists of two components: a true power component and a reactive power component.

In all electrical and electronic systems, it is the true power that does the work. The reactive power is simply shuttled back and forth between the source and load. Ideally, in terms of performing useful work, all of the power transferred to the load should be true power and none of it reactive power. However, in most practical situations the load has some reactance associated with it, and therefore you must deal with both power components.

In Chapter 11 we discussed the use of apparent power in relation to transformers. For any reactive load, there are two components of the total current: the resistive component and the reactive component. If you consider only the true power (watts) in a load, you are dealing with only a portion of the total current that the load demands from a source. In order to have a realistic picture of the actual current that a load will draw, you must consider apparent power (in VA).

A source such as an ac generator can provide current to a load up to some maximum value. *If the load draws more than this maximum value, the source can be damaged.* Figure 12–36(a) shows a 120 V generator that can deliver a maximum current of 5 A to a load. Assume that the generator is rated at 600 W and is connected to a purely resistive load of 24 Ω (power factor of 1). The ammeter shows that the current is 5 A, and the wattmeter indicates that the power is 600 W. The generator has no problem under these conditions, although it is operating at maximum current and power.

(a) Generator operating at its limits with a resistive load.

(b) Generator is in danger of internal damage due to excess current, even though the wattmeter indicates that the power is below the maximum wattage rating.

FIGURE 12–36
The wattage rating of a source is inappropriate when the load is reactive. The rating should be in VA rather than in watts.

Now, consider what happens if the load is changed to a reactive one with an impedance of 18 Ω and a power factor of 0.6, as indicated in Figure 12–36(b). The current is 120 V/18 Ω = 6.67 A, which exceeds the maximum. Even though the wattmeter reads 480 W, which is less than the power rating of the generator, the excessive current probably will cause damage. This example shows that a true power rating can be deceiving and is inappropriate for ac sources. This particular ac generator should be rated at 600 VA, a rating the manufacturer would use, rather than 600 W.

EXAMPLE 12–13 For the circuit in Figure 12–37, find the true power, the reactive power, and the apparent power. X_C has been determined to be 2 kΩ.

FIGURE 12–37

Solution   First find the total impedance so that the current can be calculated.

$$Z_T = \sqrt{R^2 + X_C^2} = \sqrt{(1 \text{ k}\Omega)^2 + (2 \text{ k}\Omega)^2} = 2.24 \text{ k}\Omega$$

$$I = \frac{V_s}{Z} = \frac{10 \text{ V}}{2.24 \text{ k}\Omega} = 4.46 \text{ mA}$$

The phase angle, θ, is

$$\theta = \tan^{-1}\left(\frac{X_C}{R}\right) = \tan^{-1}\left(\frac{2 \text{ k}\Omega}{1 \text{ k}\Omega}\right) = 63.4°$$

The true power is

$$P_{\text{true}} = V_s I \cos \theta = (10 \text{ V})(4.46 \text{ mA}) \cos(63.4°) = 20 \text{ mW}$$

Note that the same result is realized using the formula $P_{\text{true}} = I^2 R$.
   The reactive power is

$$P_r = I^2 X_C = (4.46 \text{ mA})^2 (2 \text{ k}\Omega) = 39.8 \text{ mVAR}$$

The apparent power is

$$P_a = I^2 Z = (4.46 \text{ mA})^2 (2.24 \text{ k}\Omega) = 44.6 \text{ mVA}$$

The apparent power is also the phasor sum of P_{true} and P_r:

$$P_a = \sqrt{P_{\text{true}}^2 + P_r^2} = 44.6 \text{ mVA}$$

Related Exercise   What is the true power in Figure 12–37 if $X_C = 10 \text{ k}\Omega$?

SECTION 12–7 REVIEW

1. To which component in an *RC* circuit is the power dissipation due?
2. If the phase angle, θ, is 45°, what is the power factor?
3. A certain series *RC* circuit has the following parameter values: $R = 330 \ \Omega$, $X_C = 460 \ \Omega$, and $I = 2$ A. Determine the true power, the reactive power, and the apparent power.

12–8 ▪ BASIC APPLICATIONS

RC circuits are found in a variety of applications, often as part of a more complex circuit. Two major applications, phase shift networks and frequency-selective networks (filters), are covered in this section.

After completing this section, you should be able to

▪ **Discuss some basic *RC* applications**
   ☐ Discuss and analyze the *RC* lag network
   ☐ Discuss and analyze the *RC* lead network
   ☐ Discuss how the *RC* circuit operates as a filter

The *RC* Lag Network

The *RC* lag network is a phase shift circuit in which the output voltage lags the input voltage by a specified angle, ϕ. Phase shift circuits are commonly used in electronic communication systems and in other application areas.

A basic *RC* lag network is shown in Figure 12–38(a). Notice that the output voltage is taken across the capacitor, and the input voltage is the total voltage applied across the circuit. The relationship of the voltages is shown in the phasor diagram in Figure 12–38(b). Keep in mind that the circuit phase angle is the angle between V_{in} and the current. Notice in Figure 12–38(b) that the output voltage, V_C, lags V_{in} by an angle, designated ϕ, that is the difference between 90° and the circuit phase angle θ. The angle ϕ is called the phase lag when the output voltage lags the input voltage and the phase lead when the output voltage leads the input.

Since $\theta = \tan^{-1}(X_C/R)$, the phase lag, ϕ, can be expressed as

$$\phi = 90° - \tan^{-1}\left(\frac{X_C}{R}\right) \tag{12–26}$$

When the input and output waveforms are displayed on an oscilloscope, a relationship similar to that in Figure 12–39 is observed. The exact amount of phase lag between the input and the output depends on the values of the resistance and the capacitive reactance. The magnitude of the output voltage depends on these values also.

(a) A basic *RC* lag network

(b) Phasor voltage diagram showing the phase lag between V_{in} and V_{out}

FIGURE 12–38
The RC lag network ($V_{out} = V_C$).

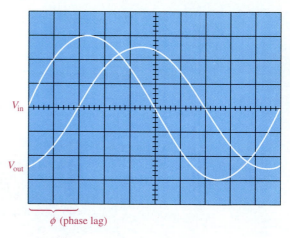

ϕ (phase lag)

FIGURE 12–39
General oscilloscope display of the input and output waveforms of a lag network (V_{out} lags V_{in}).

EXAMPLE 12–14 Determine the amount of phase lag from input to output in the lag network in Figure 12–40.

FIGURE 12–40

Solution   First determine the capacitive reactance.

$$X_C = \frac{1}{2\pi fC} = \frac{1}{2\pi(1 \text{ kHz})(0.1 \ \mu\text{F})} = 1.59 \text{ k}\Omega$$

The phase lag between the output and the input is

$$\phi = 90° - \tan^{-1}\left(\frac{X_C}{R}\right) = 90° - \tan^{-1}\left(\frac{1.59 \text{ k}\Omega}{680 \ \Omega}\right) = 23.2°$$

The output lags the input by 23.2°.

Related Exercise   In a lag network, what happens to the phase lag if the frequency increases?

The phase-lag network can be viewed as a voltage divider with a portion of the input voltage dropped across R and a portion across C. The output voltage can be determined with the following formula:

$$V_{out} = \left(\frac{X_C}{\sqrt{R^2 + X_C^2}}\right)V_{in} \qquad\qquad (12\text{--}27)$$

EXAMPLE 12–15   For the lag network in Figure 12–40 of Example 12–14, determine the output voltage in phasor form when the input voltage has an rms value of 10 V. Sketch the input and output waveforms showing the proper relationships. The values for X_C (1.59 kΩ) and ϕ (23.2°) were found in Example 12–14.

Solution   Use Equation (12–27) to determine the output voltage for the lag network in Figure 12–40.

$$V_{out} = \left(\frac{X_C}{\sqrt{R^2 + X_C^2}}\right)V_{in} = \left(\frac{1.59 \text{ k}\Omega}{\sqrt{(680 \ \Omega)^2 + (1.59 \text{ k}\Omega)^2}}\right)10 \text{ V} = 9.2 \text{ V}$$

The waveforms are shown in Figure 12–41.

FIGURE 12–41

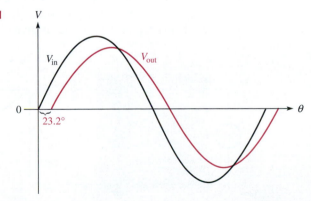

Related Exercise   In a lag network, what happens to the output voltage if the frequency increases?

Effects of Frequency on the Lag Network   Since the circuit phase angle, θ, decreases as frequency increases, the phase lag between the input and the output voltages increases. You can see this relationship by examining Equation (12–26). Also, the magnitude of V_{out} decreases as the frequency increases, because X_C becomes smaller and less of the total input voltage is dropped across the capacitor. Figure 12–42 illustrates the changes in phase lag and output voltage magnitude with frequency.

FIGURE 12–42

Illustration of how the frequency affects the phase lag and the output voltage in an RC lag network, with the amplitude of V_{in} held constant.

(a) When f increases, the phase lag ϕ increases and V_{out} decreases.

(b) When f decreases, ϕ decreases and V_{out} increases.

The *RC* Lead Network

The *RC* lead network is a phase shift circuit in which the output voltage leads the input voltage by a specified angle, ϕ. A basic *RC* lead network is shown in Figure 12–43(a). Notice how it differs from the lag network. Here the output voltage is taken across the resistor. The relationship of the voltages is given in the phasor diagram in Figure 12–43(b).

(a) A basic *RC* lead network

(b) Phasor voltage diagram showing the phase lead between V_{in} and V_{out}

FIGURE 12–43

The RC lead network ($V_{out} = V_R$).

The output voltage, V_R, leads V_{in} by an angle that is the same as the circuit phase angle, because V_R and I are in phase with each other.

When the input and output waveforms are displayed on an oscilloscope, a relationship similar to that in Figure 12–44 is observed. Of course, the exact amount of phase lead and the output voltage magnitude depend on the values of R and X_C. The phase lead, ϕ, is expressed as follows:

$$\phi = \tan^{-1}\left(\frac{X_C}{R}\right)$$

(12–28)

The output voltage is expressed as

$$V_{out} = \left(\frac{R}{\sqrt{R^2 + X_C^2}}\right)V_{in}$$

(12–29)

FIGURE 12–44
General oscilloscope display of the input and output waveforms of a lead network (V_{out} leads V_{in}).

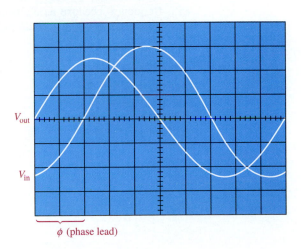

ϕ (phase lead)

EXAMPLE 12–16   Calculate the phase lead and the output voltage for the circuit in Figure 12–45.

FIGURE 12–45

Solution   The phase lead is

$$\phi = \tan^{-1}\left(\frac{X_C}{R}\right) = \tan^{-1}\left(\frac{150\ \Omega}{220\ \Omega}\right) = 34.3°$$

The output leads the input by 34.3°.

Use Equation (12–29) to determine the output voltage:

$$V_{out} = \left(\frac{R}{\sqrt{R^2 + X_C^2}} \right) V_{in} = \left(\frac{220\ \Omega}{\sqrt{(220\ \Omega)^2 + (150\ \Omega)^2}} \right) 10\ V = 8.26\ V$$

The calculator sequence for V_{out} is

| 2 | 2 | 0 | ÷ | (| (| (| 2 | 2 | 0 | x² | + | 1 | 5 | 0 | x² |) | 2nd | √x |) | = | × | 1 | 0 | = |

Related Exercise How does an increase in R affect the phase lead and the output voltage in Figure 12–45?

Effects of Frequency on the Lead Network Since the phase lead is the same as the circuit phase angle θ, it decreases as frequency increases. The output voltage increases with frequency because as X_C becomes smaller, more of the input voltage is dropped across the resistor. Figure 12–46 illustrates this relationship.

FIGURE 12–46
Illustration of how the frequency affects the phase lead and output voltage in an RC lead network with the amplitude of V_{in} held constant.

(a) When f increases, the phase lead ϕ decreases and V_{out} increases.

(b) When f decreases, ϕ increases and V_{out} decreases.

The *RC* Circuit As a Filter

Filters are frequency-selective circuits that permit signals of certain frequencies to pass from the input to the output while blocking all others. That is, all frequencies but the selected ones are filtered out.

Series *RC* circuits exhibit a frequency-selective characteristic and therefore act as simple passive filters. There are two types. The first one that we examine, called a **low-pass filter,** is realized by taking the output across the capacitor, just as in a lag network.

The second type, called a **high-pass filter,** is implemented by taking the output across the resistor, as in a lead network.

Low-Pass Filter   You have seen what happens to the phase and the output voltage in the lag network. In terms of its filtering action, we are interested primarily in how the magnitude of the output voltage varies with frequency.

To illustrate low-pass filter action, Figure 12–47 shows a specific series of measurements in which the frequency starts at zero (dc) and is increased in increments up to 20 kHz. At each value of frequency, the output voltage is measured. As you can see, the capacitive reactance decreases as frequency goes up, thus dropping less voltage across the capacitor while the input voltage is held at a constant 10 V throughout each step. Table 12–1 summarizes the variation of the circuit parameters with frequency.

(a) $f = 0$ Hz, $X_C = \infty$, $V_{OUT} = 10$ V

(b) $f = 1$ kHz, $X_C = 159\ \Omega$, $V_{out} = 8.5$ V

(c) $f = 10$ kHz, $X_C = 15.9\ \Omega$, $V_{out} = 1.57$ V

(d) $f = 20$ kHz, $X_C = 7.96\ \Omega$, $V_{out} = 0.79$ V

FIGURE 12–47
Example of low-pass filter action. As frequency increases, V_{out} decreases.

TABLE 12–1

f	$X_C\ (\Omega)$	$Z_T\ (\Omega)$	I (mA)	V_{out} (V)
0	∞	∞	0	10
1 kHz	159	187.8	53.25	8.5
10 kHz	15.9	101.3	98.7	1.57
20 kHz	7.96	100.3	99.7	0.79

The **frequency response** of the low-pass filter in Figure 12–47 is shown in Figure 12–48, where the measured values are plotted on a graph of V_{out} versus frequency, and a smooth curve is drawn connecting the points. This graph, called a *response curve,* shows that the output voltage is greater at the lower frequencies and decreases as the frequency increases.

FIGURE 12–48

Frequency response curve for the low-pass filter in Figure 12–47.

High-Pass Filter To illustrate high-pass filter action, Figure 12–49 shows a series of specific measurements. Again, the frequency starts at zero (dc) and is increased in increments up to 10 kHz. As you can see, the capacitive reactance decreases as the frequency goes up, thus causing more of the total input voltage to be dropped across the resistor. Table 12–2 summarizes the variation of circuit parameters with frequency.

(a) $f = 0$ Hz, $X_C = \infty$, $V_{OUT} = 0$ V

(b) $f = 100$ Hz, $X_C = 1.59$ kΩ, $V_{out} = 0.63$ V

(c) $f = 1$ kHz, $X_C = 159$ Ω, $V_{out} = 5.32$ V

(d) $f = 10$ kHz, $X_C = 15.9$ Ω, $V_{out} = 9.87$ V

FIGURE 12–49

Example of high-pass filter action. As frequency increases, V_{out} increases.

TABLE 12–2

f	X_C (Ω)	Z_T (Ω)	I (mA)	V_{out} (V)
0	∞	∞	0	0
100 Hz	1590	1593	6.3	0.63
1 kHz	159	187.8	53.2	5.32
10 kHz	15.9	101.3	98.7	9.87

In Figure 12–50, the measured values for the high-pass filter shown in Figure 12–49 have been plotted to produce a response curve for this circuit. As you can see, the output voltage is greater at the higher frequencies and decreases as the frequency is reduced.

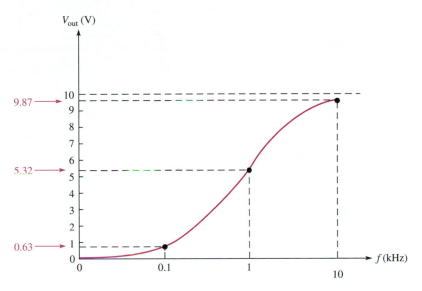

FIGURE 12–50
Frequency response curve for the high-pass filter in Figure 12–49.

The Cutoff Frequency and the Bandwidth of a Filter The frequency at which the capacitive reactance equals the resistance in a low-pass or high-pass RC filter is called the **cutoff frequency** and is designated f_c. This condition is expressed as $1/(2\pi f_c C) = R$. Solving for f_c, we get the following formula:

$$f_c = \frac{1}{2\pi RC}$$

(12–30)

At f_c, the output voltage of the filter is 70.7% of its maximum value. It is standard practice to consider the cutoff frequency as the limit of a filter's performance in terms of passing or rejecting frequencies. For example, in a high-pass filter, all frequencies above f_c are considered to be passed by the filter, and all those below f_c are considered to be rejected. The reverse is true for a low-pass filter.

The range of frequencies that are considered to be passed by a filter is called the **bandwidth.** Figure 12–51 illustrates the bandwidth and the cutoff frequency for a low-pass filter.

FIGURE 12–51

Normalized general response curve of a low-pass filter showing the cutoff frequency and the bandwidth.

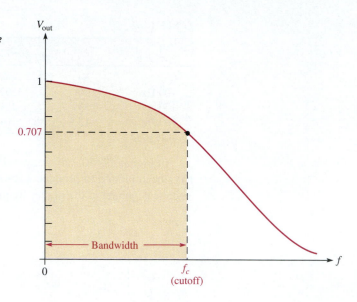

Coupling an AC Signal into a DC Bias Network

Figure 12–52 shows an *RC* network that is used to create a dc voltage level with an ac voltage superimposed on it. This type of circuit is commonly found in amplifiers in which

FIGURE 12–52

Amplifier bias and signal-coupling circuit.

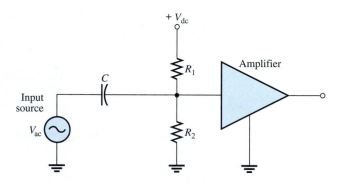

the dc voltage is required to **bias** the amplifier to the proper operating point and the signal voltage to be amplified is coupled through a capacitor and superimposed on the dc level. The capacitor prevents the low internal resistance of the signal source from affecting the dc bias voltage.

In this type of application, a relatively high value of capacitance is selected so that for the frequencies to be amplified, the reactance is very small compared to the resistance of the bias network. When the reactance is very small (ideally zero), there is practically no phase shift or signal voltage dropped across the capacitor. Therefore, all of the signal voltage passes from the source to the input to the amplifier.

Figure 12–53 illustrates the application of the superposition principle to the circuit in Figure 12–52. In part (a), the ac source has been effectively removed from the circuit and replaced with a short to represent its ideal internal resistance. Since *C* is open to dc, the voltage at point *A* is determined by the voltage-divider action of R_1 and R_2 and the dc voltage source.

In Figure 12–53(b), the dc source has been effectively removed from the circuit and replaced with a short to represent its ideal internal resistance. Since *C* appears as a short at the frequency of the ac, the signal voltage is coupled directly to point *A* and appears across the parallel combination of R_1 and R_2. Figure 12–53(c) illustrates that the combined effect of the superposition of the dc and the ac voltages results in the signal voltage "riding" on the dc level.

(a) dc equivalent: ac source replaced by short. C is open to dc. R_1 and R_2 act as dc voltage divider.

(b) ac equivalent: dc source is replaced by short. C is short to ac. All of V_{ac} is coupled to point A.

(c) dc + ac: Voltages are superimposed at point A.

FIGURE 12–53
The superposition of dc and ac voltages in an RC bias and coupling circuit.

SECTION 12–8 REVIEW

1. A certain *RC* lag network consists of a 4.7 kΩ resistor and a 0.022 *μ*F capacitor. Determine the phase lag between input and output at a frequency of 3 kHz.
2. An *RC* lead network has the same component values as the lag network in Question 1. What is the magnitude of the output voltage at 3 kHz when the input is 10 V rms?
3. When an *RC* circuit is used as a low-pass filter, across which component is the output taken?

12–9 ■ TROUBLESHOOTING

In this section, the effects that typical component failures or degradation have on the response of basic RC circuits are considered.

After completing this section, you should be able to

■ **Troubleshoot *RC* circuits**
  □ Find an open resistor
  □ Find an open capacitor
  □ Find a shorted capacitor
  □ Find a leaky capacitor

Effect of an Open Resistor

It is very easy to see how an open resistor affects the operation of a basic series *RC* circuit, as shown in Figure 12–54. Obviously, there is no path for current, so the capacitor voltage remains at zero; thus, the total voltage, V_s, appears across the open resistor.

Effect of an Open Capacitor

When the capacitor is open, there is no current; thus, the resistor voltage remains at zero. The total source voltage is across the open capacitor, as shown in Figure 12–55.

FIGURE 12–54
Effect of an open resistor.

FIGURE 12–55
Effect of an open capacitor.

Effect of a Shorted Capacitor

When a capacitor shorts out, the voltage across it is zero, the current equals V_s/R, and the total voltage appears across the resistor, as shown in Figure 12–56.

FIGURE 12–56
Effect of a shorted capacitor.

Effect of a Leaky Capacitor

When a capacitor exhibits a high leakage current, the leakage resistance effectively appears in parallel with the capacitor, as shown in Figure 12–57(a). When the leakage resistance is comparable in value to the circuit resistance, R, the circuit response is drastically affected. The circuit, looking from the capacitor toward the source, can be Thevenized, as shown in Figure 12–57(b). The Thevenin equivalent resistance is R in parallel with R_{leak} (the source appears as a short), and the Thevenin equivalent voltage is determined by the voltage-divider action of R and R_{leak}.

$$R_{th} = R \| R_{leak}$$

$$R_{th} = \frac{RR_{leak}}{R + R_{leak}} \quad \text{(12–31)}$$

$$V_{th} = \frac{R_{leak}V_{in}}{R + R_{leak}} \quad \text{(12–32)}$$

As you can see, the voltage to which the capacitor will charge is reduced since $V_{th} < V_s$. Also, the circuit time constant is reduced, and the current is increased. The Thevenin equivalent circuit is shown in Figure 12–57(c).

FIGURE 12–57
Effect of a leaky capacitor.

EXAMPLE 12–17   Assume that the capacitor in Figure 12–58 is degraded to a point where its leakage resistance is 10 kΩ. Determine the phase shift from input to output and the output voltage under the degraded condition.

FIGURE 12–58

Solution   The effective circuit resistance is

$$R_{th} = \frac{RR_{leak}}{R + R_{leak}} = \frac{(4.7 \text{ k}\Omega)(10 \text{ k}\Omega)}{14.7 \text{ k}\Omega} = 3.2 \text{ k}\Omega$$

The phase lag is

$$\phi = 90° - \tan^{-1}\left(\frac{X_C}{R_{th}}\right) = 90° - \tan^{-1}\left(\frac{5 \text{ k}\Omega}{3.2 \text{ k}\Omega}\right) = 32.6°$$

To determine the output voltage, first calculate the Thevenin equivalent voltage.

$$V_{th} = \left(\frac{R_{leak}}{R + R_{leak}} \right) V_{in} = \left(\frac{10 \text{ k}\Omega}{14.7 \text{ k}\Omega} \right) 10 \text{ V} = 6.80 \text{ V}$$

$$V_{out} = \left(\frac{X_C}{\sqrt{R_{th}^2 + X_C^2}} \right) V_{th} = \left(\frac{5 \text{ k}\Omega}{\sqrt{(3.2 \text{ k}\Omega)^2 + (5 \text{ k}\Omega)^2}} \right) 6.8 \text{ V} = 5.73 \text{ V}$$

Related Exercise What would the output voltage be if the capacitor were not leaky?

SECTION 12–9 REVIEW

1. Describe the effect of a leaky capacitor on the response of an *RC* circuit.
2. In a series *RC* circuit, if all the applied voltage appears across the capacitor, what is the problem?
3. What faults can cause 0 V across the capacitor in a series *RC* circuit?

12–10 ▪ APPLICATION ASSIGNMENT

In Chapter 9, you worked with the capacitively coupled input to an amplifier with voltage-divider bias. In this assignment, you will check the voltage of a similar amplifier's input circuit to determine how it changes with the frequency of the input signal. If too much voltage is dropped across the capacitor, the overall performance of the amplifier is adversely affected. To do this assignment, you do not have to be familiar with the details of the amplifier circuit but you should review the application assignment in Chapter 9 before proceeding.

As you learned in Chapter 9, the coupling capacitor (C_1) in Figure 12–59 passes the input signal voltage to the input of the amplifier (point *A* to point *B*) without affecting the dc level at point *B* produced by the resistive voltage divider (R_1 and R_2). If the input frequency is high enough so that the reactance of the coupling capacitor is negligibly small, essentially no ac signal voltage is dropped across the capacitor. As the signal frequency is reduced, the capacitive reactance increases and more of the signal voltage is dropped across the capacitor. This lowers the overall voltage gain of the amplifier and thus degrades its performance.

FIGURE 12–59
A capacitively coupled amplifier.

The amount of signal voltage that is coupled from the input source (point A) to the amplifier input (point B) is determined by the values of the capacitor and the dc bias resistors (assuming the amplifier has no loading effect) in Figure 12–59. These components actually form a high-pass RC filter, as shown in Figure 12–60. The voltage-divider bias resistors are effectively in parallel with each other as far as the ac source is concerned. The lower end of R_2 goes to ground and the upper end of R_1 goes to the dc supply voltage as shown in Figure 12–60(a). Since there is no ac voltage at the +18 V dc terminal, the upper end of R_1 is at 0 V ac which is referred to as *ac ground*. The development of the circuit into an effective high-pass RC filter is shown in parts (b) and (c).

(a)

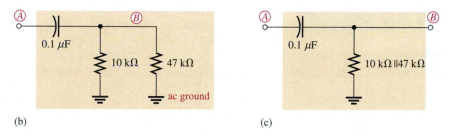

(b)   (c)

FIGURE 12–60
The RC input circuit acts effectively like a high-pass RC filter.

Step 1: Evaluate the Amplifier Input Circuit

Determine the value of the equivalent resistance of the input circuit. Assume the amplifier has no loading effect on the input circuit.

Step 2: Measure the Response at Frequency f_1

Refer to Figure 12–61. The input signal voltage is applied to the amplifier circuit board and displayed on channel 1 of the oscilloscope and channel 2 is connected to a point on the circuit board. Determine to what point on the circuit the channel 2 probe is connected, the frequency, and the voltage that should be displayed.

Step 3: Measure the Response at Frequency f_2

Refer to Figure 12–62 and the circuit board in Figure 12–61. The input signal voltage displayed on channel 1 of the oscilloscope is applied to the amplifier circuit board. Determine the frequency and the voltage that should be displayed on channel 2.

FIGURE 12–61
Measuring the input circuit response at frequency f_1. Circled numbers relate scope inputs to the probes. The channel 1 waveform is shown.

State the difference between the channel 2 waveforms determined in Step 2 and Step 3. Explain the reason for the difference.

Step 4: Measure the Response at Frequency f_3

Refer to Figure 12–63 and the circuit board in Figure 12–61. The input signal voltage displayed on channel 1 of the oscilloscope is applied to the amplifier circuit board. Determine the frequency and the voltage that should be displayed on channel 2.

State the difference between the channel 2 waveforms determined in Step 2 and Step 4. Explain the reason for the difference.

Step 5: Plot a Response Curve for the Amplifier Input Circuit

Determine the frequency at which the signal voltage at point *B* in Figure 12–59 is 70.7% of its maximum value. Plot the response curve using this voltage value and the values at frequencies f_1, f_2, and f_3. How does this curve show that the input circuit acts as a high-pass filter? What can you do to the circuit to lower the frequency at which the voltage is 70.7% of maximum without affecting the dc bias voltage?

FIGURE 12–62
Measuring the input circuit response at frequency f_2. The channel 1 waveform is shown.

FIGURE 12–63
Measuring the input circuit response at frequency f_3. The channel 1 waveform is shown.

SECTION 12–10 REVIEW

1. Explain the effect on the response of the amplifier input circuit of reducing the value of the coupling capacitor.
2. What is the voltage at point *B* in Figure 12–59 if the coupling capacitor opens when the ac input signal is 10 mV rms?
3. What is the voltage at point *B* in Figure 12–59 if resistor R_1 is open when the ac input signal is 10 mV rms?

■ **SUMMARY**

- When a sine wave voltage is applied to an *RC* circuit, the current and all the voltage drops are also sine waves.
- Total current in an *RC* circuit always leads the source voltage.
- The resistor voltage is always in phase with the current.
- The capacitor voltage always lags the current by 90°.
- In an *RC* circuit, the impedance is determined by both the resistance and the capacitive reactance combined.
- Impedance is expressed in units of ohms.
- The circuit phase angle is the angle between the total current and the applied (source) voltage.
- The impedance of a series *RC* circuit varies inversely with frequency.
- The phase angle (θ) of a series *RC* circuit varies inversely with frequency.
- For each parallel *RC* circuit, there is an equivalent series circuit for any given frequency.
- The impedance of a circuit can be determined by measuring the applied voltage and the total current and then applying Ohm's law.
- In an *RC* circuit, part of the power is resistive and part reactive.
- The phasor combination of resistive power (true power) and reactive power is called *apparent power*.
- Apparent power is expressed in volt-amperes (VA).
- The power factor (*PF*) indicates how much of the apparent power is true power.
- A power factor of 1 indicates a purely resistive circuit, and a power factor of 0 indicates a purely reactive circuit.
- In a lag network, the output voltage lags the input voltage in phase.
- In a lead network, the output voltage leads the input voltage.
- A filter passes certain frequencies and rejects others.

■ **GLOSSARY**

Admittance (*Y*)   A measure of the ability of a reactive circuit to permit current; the reciprocal of impedance. The unit is the siemens (S).

Apparent power (*P_a*)   The product of volts and amperes with the unit of volt-amperes (VA).

Bandwidth (*BW*)   The range of frequencies that are considered to be passed by a filter.

Bias   The application of a dc voltage to an electronic device to produce a desired mode of operation.

Capacitive susceptance (*B_C*)   The ability of a capacitor to permit current; the reciprocal of capacitive reactance. The unit is the siemens (S).

Conductance (*G*)   The ability of a circuit to allow current; the reciprocal of resistance. The unit is the siemens (S).

Cutoff frequency (*f_c*)   The frequency at which the output voltage of a filter is 70.7% of the maximum output voltage.

Filter   A type of circuit that passes certain frequencies and rejects all others.

Frequency response   The way the output voltage of a circuit varies as the input frequency is changed.

High-pass filter   A certain type of filter in which higher frequencies are passed and lower frequencies are rejected.

Impedance (*Z*)   The total opposition to sinusoidal current expressed in ohms.

Lag   To be behind; describes a condition of the phase or time relationship of waveforms in which one waveform is behind the other in phase or time.

Lead   To be in front of; describes a condition of the phase or time relationship of waveforms in which one waveform is ahead of the other in phase or time.

Low-pass filter   A certain type of filter in which lower frequencies are passed and higher frequencies are rejected.

Power factor   The relationship between volt-amperes and true power or watts. Volt-amperes multiplied by the power factor equals true power.

■ FORMULAS

(12–1) $Z = \sqrt{R^2 + X_C^2}$ Series RC impedance

(12–2) $\theta = \tan^{-1}\left(\dfrac{X_C}{R}\right)$ Series RC phase angle

(12–3) $V = IZ$ Ohm's law

(12–4) $I = \dfrac{V}{Z}$ Ohm's law

(12–5) $Z = \dfrac{V}{I}$ Ohm's law

(12–6) $V_s = \sqrt{V_R^2 + V_C^2}$ Total voltage in series RC

(12–7) $\theta = \tan^{-1}\left(\dfrac{V_C}{V_R}\right)$ Series RC phase angle

(12–8) $Z = \dfrac{RX_C}{\sqrt{R^2 + X_C^2}}$ Parallel RC impedance

(12–9) $\theta = \tan^{-1}\left(\dfrac{R}{X_C}\right)$ Parallel RC phase angle

(12–10) $G = \dfrac{1}{R}$ Conductance

(12–11) $B_C = \dfrac{1}{X_C}$ Capacitive susceptance

(12–12) $Y = \dfrac{1}{Z}$ Admittance

(12–13) $Y_T = \sqrt{G^2 + B_C^2}$ Total admittance

(12–14) $V = \dfrac{I}{Y}$ Ohm's law

(12–15) $I = VY$ Ohm's law

(12–16) $Y = \dfrac{I}{V}$ Ohm's law

(12–17) $I_T = \sqrt{I_R^2 + I_C^2}$ Total current in parallel RC circuits

(12–18) $\theta = \tan^{-1}\left(\dfrac{I_C}{I_R}\right)$ Parallel RC phase angle

(12–19) $R_{eq} = Z \cos \theta$ Equivalent series resistance

(12–20) $X_{C(eq)} = Z \sin \theta$ Equivalent series reactance

(12–21) $P_{\text{true}} = I_T^2 R$ True power (W)

(12–22) $P_r = I_T^2 X_C$ Reactive power (VAR)

(12–23) $P_a = I_T^2 Z$ Apparent power (VA)

(12–24) $P_{\text{true}} = V_s I_T \cos \theta$ True Power

(12–25) $PF = \cos \theta$ Power factor

(12–26) $\phi = 90° - \tan^{-1}\left(\dfrac{X_C}{R}\right)$ Phase lag

(12–27) $V_{\text{out}} = \left(\dfrac{X_C}{\sqrt{R^2 + X_C^2}}\right)V_{\text{in}}$ Output of lag network

(12–28) $\phi = \tan^{-1}\left(\dfrac{X_C}{R}\right)$ Phase lead

(12–29) $V_{\text{out}} = \left(\dfrac{R}{\sqrt{R^2 + X_C^2}}\right)V_{\text{in}}$ Output of lead network

$$(12\text{--}30) \qquad f_c = \frac{1}{2\pi RC} \qquad\qquad \text{Cutoff frequency of an } RC \text{ filter}$$

$$(12\text{--}31) \qquad R_{th} = \frac{RR_{leak}}{R + R_{leak}} \qquad\qquad \text{Thevenin equivalent resistance}$$

$$(12\text{--}32) \qquad V_{th} = \frac{R_{leak}V_{in}}{R + R_{leak}} \qquad\qquad \text{Thevenin equivalent voltage}$$

■ SELF-TEST

1. In a series *RC* circuit, the voltage across the resistance is
   (a) in phase with the source voltage (b) lagging the source voltage by 90°
   (c) in phase with the current (d) lagging the current by 90°

2. In a series *RC* circuit, the voltage across the capacitor is
   (a) in phase with the source voltage (b) lagging the resistor voltage by 90°
   (c) in phase with the current (d) lagging the source voltage by 90°

3. When the frequency of the voltage applied to a series *RC* circuit is increased, the impedance
   (a) increases (b) decreases (c) remains the same (d) doubles

4. When the frequency of the voltage applied to a series *RC* circuit is decreased, the phase angle
   (a) increases (b) decreases (c) remains the same (d) becomes erratic

5. In a series *RC* circuit when the frequency and the resistance are doubled, the impedance
   (a) doubles (b) is halved (c) is quadrupled (d) cannot be determined without values

6. In a series *RC* circuit, 10 V rms is measured across the resistor and 10 V rms is also measured across the capacitor. The rms source voltage is
   (a) 20 V (b) 14.14 V (c) 28.28 V (d) 10 V

7. The voltages in Problem 6 are measured at a certain frequency. To make the resistor voltage greater than the capacitor voltage, the frequency
   (a) must be increased (b) must be decreased (c) is held constant (d) has no effect

8. When $R = X_C$, the phase angle is
   (a) 0° (b) +90° (c) –90° (d) 45°

9. To decrease the phase angle below 45°, the following condition must exist:
   (a) $R = X_C$ (b) $R < X_C$ (c) $R > X_C$ (d) $R = 10X_C$

10. When the frequency of the source voltage is increased, the impedance of a parallel *RC* circuit
    (a) increases (b) decreases (c) does not change

11. In a parallel *RC* circuit, there is 1 A rms through the resistive branch and 1 A rms through the capacitive branch. The total rms current is
    (a) 1 A (b) 2 A (c) 2.28 A (d) 1.414 A

12. A power factor of 1 indicates that the circuit phase angle is
    (a) 90° (b) 45° (c) 180° (d) 0°

13. For a certain load, the true power is 100 W and the reactive power is 100 VAR. The apparent power is
    (a) 200 VA (b) 100 VA (c) 141.4 VA (d) 141.4 W

14. Energy sources are normally rated in
    (a) watts (b) volt-amperes (c) volt-amperes reactive (d) none of these

15. If the bandwidth of a certain low-pass filter is 1 kHz, the cutoff frequency is
    (a) 0 Hz (b) 500 Hz (c) 2 kHz (d) 1000 Hz

■ BASIC PROBLEMS

SECTION 12–1 Sinusoidal Response of *RC* Circuits

1. A 8 kHz sinusoidal voltage is applied to a series *RC* circuit. What is the frequency of the voltage across the resistor? The capacitor?

2. What is the waveshape of the current in the circuit of Problem 1?

SECTION 12–2 Impedance and Phase Angle of a Series *RC* Circuit

3. Find the impedance of each circuit in Figure 12–64.

4. Determine the impedance and the phase angle in each circuit in Figure 12–65.

FIGURE 12–64

FIGURE 12–65

5. For the circuit of Figure 12–66, determine the impedance for each of the following frequencies:
(a) 100 Hz (b) 500 Hz (c) 1 kHz (d) 2.5 kHz

6. Repeat Problem 5 for $C = 0.005 \ \mu F$.

FIGURE 12–66

SECTION 12–3 Analysis of Series *RC* Circuits

7. Calculate the total current in each circuit of Figure 12–64.

8. Repeat Problem 7 for the circuits in Figure 12–65.

9. For the circuit in Figure 12–67, draw the phasor diagram showing all voltages and the total current. Indicate the phase angles.

10. For the circuit in Figure 12–68, determine the following:
(a) Z (b) I (c) V_R (d) V_C

FIGURE 12–67

FIGURE 12–68

FIGURE 12–69

11. To what value must the rheostat be set in Figure 12–69 to make the total current 10 mA? What is the resulting phase angle?

SECTION 12–4 Impedance and Phase Angle of a Parallel *RC* Circuit

12. Determine the impedance for the circuit in Figure 12–70.
13. Determine the impedance and the phase angle in Figure 12–71.
14. Repeat Problem 13 for the following frequencies:
   (a) 1.5 kHz **(b)** 3 kHz **(c)** 5 kHz **(d)** 10 kHz
15. Determine the impedance and phase angle in Figure 12–72.

FIGURE 12–70

FIGURE 12–71

FIGURE 12–72

SECTION 12–5 Analysis of Parallel *RC* Circuits

16. For the circuit in Figure 12–73, find all the currents and voltages.
17. For the parallel circuit in Figure 12–74, find each branch current and the total current. What is the phase angle between the applied voltage and the total current?

FIGURE 12–73

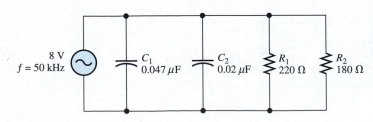

FIGURE 12–74

18. For the circuit in Figure 12–75, determine the following:
    (a) Z (b) I_R (c) I_C (d) I_T (e) θ
19. Repeat Problem 18 for $R = 5 \text{ k}\Omega$, $C = 0.05 \ \mu\text{F}$, and $f = 500$ Hz.
20. Convert the circuit in Figure 12–76 to an equivalent series form.

FIGURE 12–75   **FIGURE 12–76**

SECTION 12–6   Series-Parallel Analysis

21. Determine the voltages across each element in Figure 12–77. Sketch the voltage phasor diagram.
22. Is the circuit in Figure 12–77 predominantly resistive or predominantly capacitive?
23. Find the current through each branch and the total current in Figure 12–77. Sketch the current phasor diagram.
24. For the circuit in Figure 12–78, determine the following:
    (a) I_T (b) θ (c) V_{R1} (d) V_{R2} (e) V_{R3} (f) V_C

FIGURE 12–77   **FIGURE 12–78**

SECTION 12–7   Power in *RC* Circuits

25. In a certain series *RC* circuit, the true power is 2 W, and the reactive power is 3.5 VAR. Determine the apparent power.
26. In Figure 12–68, what is the true power and the reactive power?
27. What is the power factor for the circuit of Figure 12–76?
28. Determine P_{true}, P_r, P_a, and *PF* for the circuit in Figure 12–78. Sketch the power triangle.

SECTION 12–8   Basic Applications

29. For the lag network in Figure 12–79, determine the phase lag between the input voltage and the output voltage for each of the following frequencies:
    (a) 1 Hz   (b) 100 Hz   (c) 1 kHz   (d) 10 kHz
30. The lag network in Figure 12–79 also acts as a low-pass filter. Draw a response curve for this circuit by plotting the output voltage versus frequency for 0 Hz to 10 kHz in 1 kHz increments.
31. Repeat Problem 29 for the lead network in Figure 12–80.
32. Plot the frequency response curve for the lead network in Figure 12–80 for a frequency range of 0 Hz to 10 kHz in 1 kHz increments.
33. Draw the voltage phasor diagram for each circuit in Figures 12–79 and 12–80 for a frequency of 5 kHz with $V_{\text{in}} = 1$ V rms.

FIGURE 12–79

FIGURE 12–80

34. The rms value of the signal voltage out of amplifier *A* in Figures 12–81 is 50 mV. If the input resistance to amplifier *B* is 10 kΩ, how much of the signal is lost due to the coupling capacitor (*C_c*) when the frequency is 3 kHz?

35. Determine the cutoff frequency for each circuit in Figures 12–79 and 12–80.

36. Determine the bandwidth of the circuit in Figure 12–79.

FIGURE 12–81

SECTION 12–9   Troubleshooting

37. Assume that the capacitor in Figure 12–82 is excessively leaky. Show how this degradation affects the output voltage and phase angle, assuming that the leakage resistance is 5 kΩ and the frequency is 10 Hz.

FIGURE 12–82

38. Each of the capacitors in Figure 12–83 has developed a leakage resistance of 2 kΩ. Determine the output voltages under this condition for each circuit.

39. Determine the output voltage for the circuit in Figure 12–83(a) for each of the following failure modes, and compare it to the correct output:
    (a) *R_1* open   (b) *R_2* open   (c) *C* open   (d) *C* shorted

40. Determine the output voltage for the circuit in Figure 12–83(b) for each of the following failure modes, and compare it to the correct output.
    (a) *C* open   (b) *C* shorted   (c) *R_1* open   (d) *R_2* open   (e) *R_3* open

(a)

(b)

FIGURE 12–83

■ ADVANCED PROBLEMS

41. A single 220 V, 60 Hz source drives two loads. Load A has an impedance of 50 Ω and a power factor of 0.85. Load B has an impedance of 72 Ω and a power factor of 0.95.
 (a) How much current does each load draw?
 (b) What is the reactive power in each load?
 (c) What is the true power in each load?
 (d) What is the apparent power in each load?
 (e) Which load has more voltage drop along the lines connecting it to the source?

42. What value of coupling capacitor is required in Figure 12–84 so that the signal voltage at the input of amplifier 2 is at least 70.7% of the signal voltage at the output of amplifier 1 when the frequency is 20 Hz? Disregard the input resistance of the amplifier.

FIGURE 12–84

43. Determine the value to which R_1 must be adjusted to get a phase angle of 30° between the source voltage and the total current in Figure 12–85.

44. Sketch the voltage and current phasor diagram for Figure 12–86.

FIGURE 12–85

FIGURE 12–86

45. A certain load dissipates 1.5 kW of power with an impedance of 12 Ω and a power factor of 0.75. What is its reactive power? What is its apparent power?

46. Determine the series element or elements that are in the block of Figure 12–87 to meet the following overall circuit requirements:
 (a) $P_{true} = 400$ W   **(b)** leading power factor (I_T leads V_s)

47. Determine the value of C_2 in Figure 12–88 when $V_A = V_B$.

FIGURE 12–87

FIGURE 12–88

FIGURE 12–89

(a) Oscilloscope display

From signal generator 10 V peak

Scope probe and ground clip

0.1 μF

(b) Circuit with leads connected

48. Draw the schematic for the circuit in Figure 12–89 and determine if the waveform on the scope is correct. If there is a fault in the circuit, identify it.

■ **ANSWERS TO SECTION REVIEWS**

Section 12–1

1. V_C is 60 Hz, I is 60 Hz.

2. Capacitive reactance

3. θ is closer to $0°$ when $R > X_C$.

Section 12–2

1. Impedance is the opposition to sinusoidal current.

2. V_s lags I.

3. The capacitive reactance produces the phase angle.

4. $Z = \sqrt{R^2 + X_C^2} = 59.9 \ \Omega$; $\theta = \tan^{-1}(X_C/R) = 56.6°$

Section 12–3

1. $V_T = \sqrt{V_R^2 + V_C^2} = 7.2 \ \text{V}$

2. $\theta = \tan^{-1}(V_C/V_R) = 56.3°$

3. $\theta = 90°$

4. **(a)** X_C decreases when f increases.   **(b)** Z decreases when f increases.

 (c) θ decreases when f increases.

Section 12–4

1. $Z = RX_C/\sqrt{R^2 + X_C^2} = 545 \ \Omega$

2. *Conductance* is the reciprocal of resistance; *capacitive susceptance* is the reciprocal of capacitive reactance; and *admittance* is the reciprocal of impedance.

3. $Y = 1/Z = 10 \ \text{mS}$

4. $Y = \sqrt{G^2 + B_C^2} = 24 \ \text{mS}$

Section 12–5

1. $I_T = V_s Y = 21$ mA
2. $\theta = \tan^{-1}(I_C/I_R) = 56.3°$; $I_T = \sqrt{I_R^2 + I_C^2} = 18$ mA
3. $\theta = 90°$

Section 12–6

1. See Figure 12–90.
2. $V_1 = I_T R_1 = 6.80$ V

FIGURE 12–90

Section 12–7

1. Power is in the resistance.
2. $PF = \cos 45° = 0.707$
3. $P_{\text{true}} = I_T^2 R = 1.32$ kW; $P_r = I_T^2 X_C = 1.84$ kVAR; $P_a = I_T^2 Z = 2.26$ kVA

Section 12–8

1. $\phi = 90° - \tan^{-1}(X_C/R) = 62.8°$
2. $V_{\text{out}} = (X_C / \sqrt{R^2 + X_C^2})V_{\text{in}} = 8.9$ V rms
3. Output is across the capacitor.

Section 12–9

1. The leakage resistance acts in parallel with C, which alters the circuit time constant.
2. The capacitor is open.
3. A shorted capacitor or an open resistor can cause 0 V across the capacitor.

Section 12–10

1. A lower value coupling capacitor will increase the frequency at which a significant drop in voltage occurs.
2. $V_B = 3.16$ V dc
3. $V_B = 10$ mV rms

▪ **ANSWERS TO RELATED EXERCISES FOR EXAMPLES**

12–1	2.42 kΩ; 65.6°	**12–2**	2.56 V
12–3	67.5°	**12–4**	7.14 V
12–5	15.9 kΩ; 86.4°	**12–6**	24.3 Ω
12–7	4.36 mS	**12–8**	5.91 mA
12–9	117 mA; 31°	**12–10**	$R_{\text{eq}} = 8.99$ kΩ; $X_{C(\text{eq})} = 4.38$ kΩ
12–11	$V_1 = 7.14$ V; $V_2 = 3.16$ V	**12–12**	0.155
12–13	990 μW	**12–14**	Phase lag ϕ increases.
12–15	Output voltage decreases.		
12–16	Phase lead ϕ decreases; output voltage increases.		
12–17	7.29 V		

13

RL CIRCUITS

An *RL* circuit contains both resistance and inductance. In this chapter, basic series and parallel *RL* circuits and their responses to sine wave voltages are covered. In addition, series-parallel combinations are examined. Power considerations in *RL* circuits are introduced, and practical aspects of the power factor are discussed. A method of improving the power factor is presented, and basic areas of application are covered. As you study this chapter, note both the differences and the similarities in the response of *RL* circuits compared to *RC* circuits.

■ CHAPTER OBJECTIVES

☐ Describe the relationship between current and voltage in an *RL* circuit

☐ Determine impedance and phase angle in a series *RL* circuit

☐ Analyze a series *RL* circuit

☐ Determine impedance and phase angle in a parallel *RL* circuit

☐ Analyze a parallel *RL* circuit

☐ Analyze series-parallel *RL* circuits

☐ Determine power in *RL* circuits

☐ Discuss some basic *RL* applications

☐ Troubleshoot *RL* circuits

■ APPLICATION ASSIGNMENT

Your assignment is to identify the types of *RL* circuits contained in two sealed modules that have been removed from a communications system. You will use your knowledge of *RL* circuits and basic measurements to determine the circuit arrangement and component values. After studying this chapter, you should be able to complete the application assignment in Section 13–10.

13–1 ■ SINUSOIDAL RESPONSE OF *RL* CIRCUITS

As with the RC circuit, all currents and voltages in an RL circuit are sinusoidal when the input voltage is sinusoidal. The inductance causes a phase shift between the voltage and the current that depends on the relative values of the resistance and the inductive reactance.

After completing this section, you should be able to

■ **Describe the relationship between current and voltage in an *RL* circuit**
    □ Discuss voltage and current waveforms
    □ Discuss phase shift

As you will learn, the resistor voltage and the current lag the source voltage. The inductor voltage leads the source voltage. The phase angle between the current and the inductor voltage is always 90°. These generalized phase relationships are indicated in Figure 13–1. Notice that they are opposite from those of the *RC* circuit, as discussed in Chapter 12.

FIGURE 13–1

Illustration of sinusoidal response with general phase relationships of V_R, V_L, and I relative to V_s. V_R and I are in phase, V_R lags V_s, and V_L leads V_s. V_R and V_L are 90° out of phase with each other.

The amplitudes and the phase relationships of the voltages and current depend on the values of the resistance and the inductive reactance (defined in Section 10–6). When a circuit is purely inductive, the phase angle between the applied voltage and the total current is 90°, with the current lagging the voltage. When there is a combination of both resistance and inductive reactance in a circuit, the phase angle is somewhere between zero and 90°, depending on the relative values of R and X_L.

SECTION 13–1 REVIEW

1. A 1 kHz sinusoidal voltage is applied to an *RL* circuit. What is the frequency of the resulting current?
2. When the resistance in an *RL* circuit is greater than the inductive reactance, is the phase angle between the applied voltage and the total current closer to 0° or to 90°?

13–2 ■ IMPEDANCE AND PHASE ANGLE OF SERIES *RL* CIRCUITS

The impedance of an RL circuit is the total opposition to sinusoidal current and its unit is the ohm. The phase angle is the phase difference between the total current and the source voltage.

After completing this section, you should be able to

■ **Determine impedance and phase angle in a series *RL* circuit**
  □ Draw the impedance triangle
  □ Calculate impedance magnitude
  □ Calculate the phase angle

The impedance of a series *RL* circuit is determined by the resistance and the inductive reactance, as indicated in Figure 13–2.

FIGURE 13–2
Impedance of a series RL circuit.

The Impedance Triangle

In ac analysis, both R and X_L are treated as phasor quantities, as shown in the phasor diagram of Figure 13–3(a), with X_L appearing at a +90° angle with respect to R. This relationship comes from the fact that the inductor voltage leads the current, and thus the resistor voltage, by 90°. Since Z is the phasor sum of R and X_L, its phasor representation is as shown in Figure 13–3(b). A repositioning of the phasors, as shown in part (c), forms a right triangle. This formation, as you have learned, is called the *impedance triangle*. The length of each phasor represents the magnitude of the quantity, and θ is the phase angle between the applied voltage and the current in the *RL* circuit.

(a)　　　　　　　　(b)　　　　　　　　(c)

FIGURE 13–3
Development of the impedance triangle for a series RL circuit.

The impedance, Z, of the series RL circuit can be expressed in terms of the resistance and reactance as

$$Z = \sqrt{R^2 + X_L^2} \tag{13–1}$$

where Z is expressed in ohms.

The phase angle, θ, is expressed as

$$\theta = \tan^{-1}\left(\frac{X_L}{R}\right) \tag{13–2}$$

EXAMPLE 13–1   Determine the impedance and phase angle of the circuit in Figure 13–4.

FIGURE 13–4

Solution   The impedance is

$$Z = \sqrt{R^2 + X_L^2} = \sqrt{(56\ \Omega)^2 + (100\ \Omega)^2} = 115\ \Omega$$

The phase angle is

$$\theta = \tan^{-1}\left(\frac{X_L}{R}\right) = \tan^{-1}\left(\frac{100\ \Omega}{56\ \Omega}\right) = 60.8°$$

The applied voltage leads the current by 60°.

The calculator sequences are as follows:

Z: [5] [6] [x²] [+] [1] [0] [0] [x²] [=] [2nd] [√x]
θ: [1] [0] [0] [÷] [5] [6] [=] [2nd] [tan]

Related Exercise   In a series RL circuit, $R = 1.8\ k\Omega$ and $X_L = 950\ \Omega$. Determine the impedance and phase angle.

SECTION 13–2 REVIEW

1. Does the source voltage lead or lag the current in a series RL circuit?
2. What causes the phase angle in the RL circuit?
3. How does the response of an RL circuit differ from that of an RC circuit?
4. A series RL circuit has a resistance of 33 kΩ and an inductive reactance of 50 kΩ. Determine Z and θ.

13–3 ■ ANALYSIS OF SERIES *RL* CIRCUITS

In the previous section, you learned how to express the impedance of a series RL circuit. Now, Ohm's law and Kirchhoff's voltage law are used in the analysis of RL circuits.

After completing this section, you should be able to

■ **Analyze a series *RL* circuit**
   □ Apply Ohm's law and Kirchhoff's voltage law to series *RL* circuits
   □ Determine the phase relationships of the voltages and current
   □ Show how impedance and phase angle vary with frequency

Ohm's Law

The application of Ohm's law to series *RL* circuits involves the use of the quantities of Z, V, and I. The three equivalent forms of Ohm's law were stated in Chapter 12 for *RC* circuits. They apply also to *RL* circuits and are restated here for convenience:

$$V = IZ \qquad\qquad I = \frac{V}{Z} \qquad\qquad Z = \frac{V}{I}$$

The following example illustrates the use of Ohm's law.

EXAMPLE 13–2   The current in Figure 13–5 is 200 μA. Determine the source voltage.

FIGURE 13–5

Solution   From Equation (10–11), the inductive reactance is

$$X_L = 2\pi fL = 2\pi(10 \text{ kHz})(100 \text{ mH}) = 6.28 \text{ k}\Omega$$

The impedance is

$$Z = \sqrt{R^2 + X_L^2} = \sqrt{(10 \text{ k}\Omega)^2 + (6.28 \text{ k}\Omega)^2} = 11.8 \text{ k}\Omega$$

Applying Ohm's law,

$$V_s = IZ = (200 \text{ }\mu\text{A})(11.8 \text{ k}\Omega) = 2.36 \text{ V}$$

Related Exercise   If the source voltage in Figure 13–5 were 5 V, what would be the current?

Relationships of the Current and Voltages in a Series *RL* Circuit

In a series *RL* circuit, the current is the same through both the resistor and the **inductor**. Thus, the resistor voltage is in phase with the current, and the inductor voltage

FIGURE 13–6
Phase relation of current and voltages in a series RL circuit.

leads the current by 90°. Therefore, there is a phase difference of 90° between the resistor voltage, V_R, and the inductor voltage, V_L, as shown in the waveform diagram of Figure 13–6.

From Kirchhoff's voltage law, the sum of the voltage drops must equal the applied voltage. However, since V_R and V_L are not in phase with each other, they must be added as phasor quantities with V_L leading V_R by 90°, as shown in Figure 13–7(a). As shown in part (b), V_s is the phasor sum of V_R and V_L. This equation can be expressed as

$$V_s = \sqrt{V_R^2 + V_L^2} \tag{13–3}$$

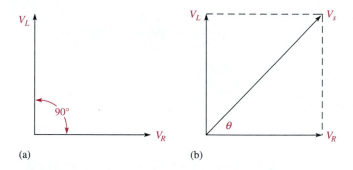

(a)   (b)

FIGURE 13–7
Voltage phasor diagram for a series RL circuit.

The phase angle between the resistor voltage and the source voltage can be expressed as

$$\theta = \tan^{-1}\left(\frac{V_L}{V_R}\right) \tag{13–4}$$

Since the resistor voltage and the current are in phase, θ in Equation (13–4) also represents the phase angle between the source voltage and the current and is equivalent to $\tan^{-1}(X_L/R)$.

Figure 13–8 shows a voltage and current phasor diagram that represents the waveform diagram of Figure 13–6.

FIGURE 13–8
Voltage and current phasor diagram for the waveforms in Figure 13–6.

EXAMPLE 13–3   Determine the source voltage and the phase angle in Figure 13–9.

FIGURE 13–9

Solution   Since V_R and V_L are 90° out of phase, they cannot be added directly and must be added as phasor quantities. The source voltage is

$$V_s = \sqrt{V_R^2 + V_L^2} = \sqrt{(50 \text{ V})^2 + (35 \text{ V})^2} = 61 \text{ V}$$

The phase angle between the current and the source voltage is

$$\theta = \tan^{-1}\left(\frac{V_L}{V_R}\right) = \tan^{-1}\left(\frac{35 \text{ V}}{50 \text{ V}}\right) = 35°$$

Related Exercise   With the information given, can you determine the current in Figure 13–9?

Variation of Impedance with Frequency

As you know, inductive reactance varies directly with frequency. When X_L increases, the total impedance also increases; and when X_L decreases, the total impedance decreases. Thus, *in an RL circuit, Z is directly dependent on frequency.*

Figure 13–10 illustrates how the voltages and current in a series *RL* circuit vary as the frequency increases or decreases, with the source voltage held at a constant value. In

(a) As frequency is increased, I and V_R decrease and V_L increases.

(b) As frequency is decreased, I and V_R increase and V_L decreases.

FIGURE 13–10

An illustration of how the variation of impedance affects the voltages and current as the source frequency is varied. The source voltage is held at a constant amplitude.

part (a), as the frequency is increased, X_L increases; so more of the total voltage is dropped across the inductor. Also, Z increases as X_L increases, causing the current to decrease. A decrease in current causes less voltage across R.

In Figure 13–10(b), as the frequency is decreased, X_L decreases; so less voltage is dropped across the inductor. Also, Z decreases as X_L decreases, causing the current to increase. An increase in current causes more voltage across R.

Changes in Z and X_L can be observed as shown in Figure 13–11. As the frequency increases, the voltage across Z remains constant because V_s is constant, but the voltage across L increases. The decreasing current indicates that Z is increasing. It does so because of the inverse relationship stated in Ohm's law ($Z = V_Z/I$). The decreasing current also indicates that X_L is increasing. The increase in V_L corresponds to the increase in X_L.

By watching thses two meters, you can see what Z is doing: I is decreasing, V_Z is constant and thus Z is increasing.

$$\uparrow Z = \frac{V_Z}{I} \downarrow$$

By watching thses two meters, you can see what X_L is doing: I is decreasing, V_L is increasing and thus X_L is increasing.

$$\uparrow X_L = \frac{V_L \uparrow}{I \downarrow}$$

V_s is constant

Frequency is increasing

FIGURE 13–11

Observing changes in Z and X_L with frequency by watching the meters and recalling Ohm's law.

Variation of the Phase Angle with Frequency

Since X_L is the factor that introduces the phase angle in a series *RL* circuit, a change in X_L produces a change in the phase angle. As the frequency is increased, X_L becomes greater, and thus the phase angle increases. As the frequency is decreased, X_L becomes smaller, and thus the phase angle decreases. This relationship is illustrated in Figure 13–12 where a phase meter is connected to V_s and to V_R. This connection gives the angle between V_s and V_R, which is the phase angle of the circuit because *I* is in phase with V_R. By measuring the phase of V_R, you are effectively measuring the phase of *I*.

The variations of phase angle with frequency are illustrated with the impedance triangle as shown in Figure 13–13.

(a) As frequency is increased, the phase angle increases.

(b) As frequency is decreased, the phase angle decreases.

FIGURE 13–12

A phase meter indicates that the phase angle between V_s and V_R varies directly with frequency. V_R is the same phase as I.

FIGURE 13–13

As the frequency increases, the phase angle θ increases.

EXAMPLE 13–4   For the series *RL* circuit in Figure 13–14, determine the impedance and the phase angle for each of the following frequencies:
(a) 10 kHz   **(b)** 20 kHz   **(c)** 30 kHz

FIGURE 13–14

Solution

(a) For $f = 10$ kHz, the impedance is calculated as follows:

$$X_L = 2\pi fL = 2\pi(10 \text{ kHz})(20 \text{ mH}) = 1.26 \text{ k}\Omega$$

$$Z = \sqrt{R^2 + X_L^2} = \sqrt{(1 \text{ k}\Omega)^2 + (1.26 \text{ k}\Omega)^2} = 1.61 \text{ k}\Omega$$

The phase angle is

$$\theta = \tan^{-1}\left(\frac{X_L}{R}\right) = \tan^{-1}\left(\frac{1.26 \text{ k}\Omega}{1 \text{ k}\Omega}\right) = 51.6°$$

(b) For $f = 20$ kHz,

$$X_L = 2\pi(20 \text{ kHz})(20 \text{ mH}) = 2.51 \text{ k}\Omega$$

$$Z = \sqrt{(1 \text{ k}\Omega)^2 + (2.51 \text{ k}\Omega)^2} = 2.70 \text{ k}\Omega$$

$$\theta = \tan^{-1}\left(\frac{2.51 \text{ k}\Omega}{1 \text{ k}\Omega}\right) = 68.3°$$

(c) For $f = 30$ kHz,

$$X_L = 2\pi(30 \text{ kHz})(20 \text{ mH}) = 3.77 \text{ k}\Omega$$

$$Z = \sqrt{(1 \text{ k}\Omega)^2 + (3.77 \text{ k}\Omega)^2} = 3.90 \text{ k}\Omega$$

$$\theta = \tan^{-1}\left(\frac{3.77 \text{ k}\Omega}{1 \text{ k}\Omega}\right) = 75.1°$$

Notice that as the frequency increases, X_L, Z, and θ also increase.

Related Exercise   Determine Z and θ in Figure 13–14 if $f = 100$ kHz.

SECTION 13–3 REVIEW

1. In a certain series *RL* circuit, $V_R = 2$ V and $V_L = 3$ V. What is the magnitude of the total voltage?
2. In Question 1, what is the phase angle?
3. When the frequency of the applied voltage in a series *RL* circuit is increased, what happens to the inductive reactance? To the impedance? To the phase angle?

13–4 ■ IMPEDANCE AND PHASE ANGLE OF PARALLEL *RL* CIRCUITS

In this section, you will learn how to determine the impedance and phase angle of a parallel RL circuit. Also, inductive susceptance and admittance of a parallel RL circuit are introduced.

After completing this section, you should be able to

■ **Determine impedance and phase angle in a parallel *RL* circuit**
  ☐ Express total impedance in a product over sum form
  ☐ Express the phase angle in terms of R and X_L
  ☐ Determine susceptance and admittance in a parallel *RL* circuit
  ☐ Convert admittance to impedance

A basic parallel *RL* circuit is shown in Figure 13–15. The expression for the impedance is

$$Z = \frac{RX_L}{\sqrt{R^2 + X_L^2}}$$

(13–5)

The phase angle between the applied voltage and the total current is

$$\theta = \tan^{-1}\left(\frac{R}{X_L}\right)$$

(13–6)

FIGURE 13–15
Parallel RL circuit.

EXAMPLE 13–5   For each circuit in Figure 13–16, determine the impedance and the phase angle.

(a)   (b)

FIGURE 13–16

Solution   For the circuit in Figure 13–16(a), the impedance and phase angle are

$$Z = \frac{RX_L}{\sqrt{R^2 + X_L^2}} = \frac{(100\ \Omega)(50\ \Omega)}{\sqrt{(100\ \Omega)^2 + (50\ \Omega)^2}} = 44.7\ \Omega$$

$$\theta = \tan^{-1}\left(\frac{R}{X_L}\right) = \tan^{-1}\left(\frac{100\ \Omega}{50\ \Omega}\right) = 63.4°$$

For the circuit in Figure 13–16(b),

$$Z = \frac{(1 \text{ k}\Omega)(2 \text{ k}\Omega)}{\sqrt{(1 \text{ k}\Omega)^2 + (2 \text{ k}\Omega)^2}} = 894 \ \Omega$$

$$\theta = \tan^{-1}\left(\frac{1 \text{ k}\Omega}{2 \text{ k}\Omega}\right) = 26.6°$$

The voltage leads the current, as opposed to the parallel *RC* case where the voltage lags the current.

The calculator sequences for the circuit in Figure 13–16(a) are as follows:

Z: [1] [0] [0] [×] [5] [0] [÷] [(] [(] [(] [1] [0] [0] [x²] [+] [5] [0] [x²] [)] [2nd] [√x] [)] [=]

θ: [1] [0] [0] [÷] [5] [0] [=] [2nd] [tan]

Related Exercise In a parallel circuit $R = 10 \text{ k}\Omega$ and $X_L = 14 \text{ k}\Omega$. Find Z and θ.

Conductance, Susceptance, and Admittance

As you know from Section 12–4, conductance (*G*) is the reciprocal of resistance, susceptance (*B*) is the reciprocal of reactance, and admittance (*Y*) is the reciprocal of impedance.

For parallel *RL* circuits, conductance (*G*) is expressed as

$$G = \frac{1}{R} \tag{13–7}$$

Inductive susceptance (B_L) is expressed as

$$B_L = \frac{1}{X_L} \tag{13–8}$$

Admittance (*Y*) is expressed as

$$Y = \frac{1}{Z} \tag{13–9}$$

As with the *RC* circuit, the unit for *G*, B_L, and *Y* is the siemens (S).

In the basic parallel *RL* circuit shown in Figure 13–17, the total admittance is the phasor sum of the conductance and the inductive susceptance and is expressed as follows:

$$Y_T = \sqrt{G^2 + B_L^2} \tag{13–10}$$

FIGURE 13–17

Admittance is the reciprocal of impedance in a parallel RL circuit.

EXAMPLE 13–6   Determine the total admittance in Figure 13–18; then convert it to impedance.

FIGURE 13–18

Solution   To determine Y, first calculate the values for G and B_L. Since $R = 330 \, \Omega$,

$$G = \frac{1}{R} = \frac{1}{330 \, \Omega} = 3.03 \text{ mS}$$

The inductive reactance is

$$X_L = 2\pi fL = 2\pi(1000 \text{ Hz})(100 \text{ mH}) = 628 \, \Omega$$

The inductive susceptance is

$$B_L = \frac{1}{X_L} = 1.59 \text{ mS}$$

Therefore, the total admittance is

$$Y_T = \sqrt{G^2 + B_L^2} = \sqrt{(3.03 \text{ mS})^2 + (1.59 \text{ mS})^2} = 3.42 \text{ mS}$$

Converting to impedance,

$$Z = \frac{1}{Y_T} = \frac{1}{3.42 \text{ mS}} = 292 \, \Omega$$

Related Exercise   What is the admittance of the circuit in Figure 13–18 if f is increased to 2 kHz?

SECTION 13–4 REVIEW

1. If $Y = 50$ mS, what is the value of Z?
2. In a certain parallel *RL* circuit, $R = 47 \, \Omega$ and $X_L = 75 \, \Omega$. Determine the admittance.
3. In the circuit of Question 2, does the total current lead or lag the applied voltage and by what phase angle?

13–5 ■ ANALYSIS OF PARALLEL *RL* CIRCUITS

In the previous section, you learned how to express the impedance of a parallel RL circuit. Now, Ohm's law and Kirchhoff's current law are used in the analysis of RL circuits. Current and voltage relationships in a parallel RL circuit are examined.

After completing this section, you should be able to

■ **Analyze a parallel *RL* circuit**
  ☐ Apply Ohm's law and Kirchhoff's current law to parallel *RL* circuits
  ☐ Determine total current and phase angle

The following example applies Ohm's law to the analysis of a parallel *RL* circuit.

EXAMPLE 13–7   Determine the total current and the phase angle in the circuit of Figure 13–19.

FIGURE 13–19

Solution   First determine the total admittance. The inductive reactance is

$$X_L = 2\pi fL = 2\pi(1.5 \text{ kHz})(150 \text{ mH}) = 1.41 \text{ k}\Omega$$

The inductive susceptance is

$$B_L = \frac{1}{X_L} = \frac{1}{1.41 \text{ k}\Omega} = 709 \text{ }\mu S$$

The conductance is

$$G = \frac{1}{R} = \frac{1}{2.2 \text{ k}\Omega} = 455 \text{ }\mu S$$

Therefore, the total admittance is

$$Y_T = \sqrt{G^2 + B_L^2} = \sqrt{(455 \text{ }\mu S)^2 + (709 \text{ }\mu S)^2} = 842 \text{ }\mu S$$

Applying Ohm's law to get the total current,

$$I_T = VY_T = (10 \text{ V})(842 \text{ }\mu S) = 8.42 \text{ mA}$$

The phase angle is

$$\theta = \tan^{-1}\left(\frac{R}{X_L}\right) = \tan^{-1}\left(\frac{2.2 \text{ k}\Omega}{1.41 \text{ k}\Omega}\right) = 57.3°$$

The total current lags the source voltage by 57.3°.

Related Exercise   Determine the total current and the phase angle if the frequency is reduced to 800 Hz in Figure 13–19.

Relationships of the Currents and Voltages in a Parallel *RL* Circuit

Figure 13–20(a) shows all the currents and voltages in a basic parallel *RL* circuit. As you can see, the applied voltage, V_s, appears across both the resistive and the inductive branches, so V_s, V_R, and V_L are all in phase and of the same magnitude. The total current, I_T, divides at the junction into the two branch currents, I_R and I_L.

The current through the resistor is in phase with the voltage. The current through the inductor lags the voltage and the resistor current by 90°. By Kirchhoff's current law, the

FIGURE 13–20

Currents and voltages in a parallel RL circuit. The current directions shown are instantaneous and, of course, reverse when the source voltage reverses.

(a)   (b)

total current is the phasor sum of the two branch currents, as shown by the phasor diagram in Figure 13–20(b). The total current is expressed as

$$I_T = \sqrt{I_R^2 + I_L^2} \qquad\qquad (13\text{–}11)$$

The phase angle between the resistor current and the total current is

$$\theta = \tan^{-1}\left(\frac{I_L}{I_R}\right) \qquad\qquad (13\text{–}12)$$

Figure 13–21 shows a complete current and voltage phasor diagram.

FIGURE 13–21

Current and voltage phasor diagram for a parallel RL circuit (amplitudes are arbitrary).

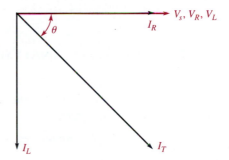

EXAMPLE 13–8   Determine the value of each current in Figure 13–22, and describe the phase relationship of each with the applied voltage.

FIGURE 13–22

Solution   The resistor current, the inductor current, and the total current are calculated as follows:

$$I_R = \frac{V_s}{R} = \frac{12\text{ V}}{220\ \Omega} = 54.5\text{ mA}$$

$$I_L = \frac{V_s}{X_L} = \frac{12\text{ V}}{150\ \Omega} = 80\text{ mA}$$

$$I_T = \sqrt{I_R^2 + I_L^2} = \sqrt{(54.5\text{ mA})^2 + (80\text{ mA})^2} = 96.8\text{ mA}$$

The phase angle is

$$\theta = \tan^{-1}\left(\frac{R}{X_L}\right) = \tan^{-1}\left(\frac{220 \ \Omega}{150 \ \Omega}\right) = 55.7°$$

The resistor current is 54.5 mA and is in phase with the applied voltage. The inductor current is 80 mA and lags the applied voltage by 90°. The total current is 96.8 mA and lags the voltage by 55.7°.

Related Exercise Find I_T and θ if $X_L = 300 \ \Omega$ in Figure 13–22.

SECTION 13–5 REVIEW

1. The admittance of a parallel *RL* circuit is 4 mS, and the applied voltage is 8 V. What is the total current?
2. In a certain parallel *RL* circuit, the resistor current is 12 mA, and the inductor current is 20 mA. Determine the phase angle and the total current.
3. What is the phase angle between the inductor current and the applied voltage in a parallel *RL* circuit?

13–6 ▪ SERIES-PARALLEL ANALYSIS

In this section, the concepts studied in the previous sections are used to analyze circuits with combinations of both series and parallel R and L elements. Examples are used to illustrate.

After completing this section, you should be able to

▪ **Analyze series-parallel *RL* circuits**
   ☐ Determine total impedance
   ☐ Calculate currents and voltages

The following two examples illustrate the basic procedures used to analyze a series-parallel circuit.

EXAMPLE 13–9

In the circuit of Figure 13–23, determine the values of the following:
(a) Z_T (b) I_T (c) θ

FIGURE 13–23

Solution

(a) First, calculate the inductive reactances:

$$X_{L1} = 2\pi f L_1 = 2\pi(5 \text{ kHz})(250 \text{ mH}) = 7.85 \text{ k}\Omega$$

$$X_{L2} = 2\pi f L_2 = 2\pi(5 \text{ kHz})(100 \text{ mH}) = 3.14 \text{ k}\Omega$$

One approach is to find the series equivalent resistance and inductive reactance for the parallel portion of the circuit. Then you can add the resistances to get the total resistance, add the reactances to get the total reactance, and then determine the total impedance.

Determine the impedance of the parallel portion (Z_2) as follows:

$$G_2 = \frac{1}{R_2} = \frac{1}{3.3 \text{ k}\Omega} = 303 \text{ } \mu S$$

$$B_{L2} = \frac{1}{X_{L2}} = \frac{1}{3.14 \text{ k}\Omega} = 318 \text{ } \mu S$$

$$Y_2 = \sqrt{G_2^2 + B_L^2} = \sqrt{(303 \text{ } \mu S)^2 + (318 \text{ } \mu S)^2} = 439 \text{ } \mu S$$

Then

$$Z_2 = \frac{1}{Y_2} = \frac{1}{439 \text{ } \mu S} = 2.28 \text{ k}\Omega$$

The phase angle associated with the parallel portion of the circuit is

$$\theta_p = \tan^{-1}\left(\frac{R_2}{X_{L2}}\right) = \tan^{-1}\left(\frac{3.3 \text{ k}\Omega}{3.14 \text{ k}\Omega}\right) = 46.4°$$

The series equivalent values for the parallel portion are

$$R_{eq} = Z_2\cos\theta_p = (2.28 \text{ k}\Omega) \cos(46.4°) = 1.57 \text{ k}\Omega$$

$$X_{L(eq)} = Z_2\sin\theta_p = (2.28 \text{ k}\Omega) \sin(46.4°) = 1.65 \text{ k}\Omega$$

The total circuit resistance is

$$R_T = R_1 + R_{eq} = 4.7 \text{ k}\Omega + 1.57 \text{ k}\Omega = 6.27 \text{ k}\Omega$$

The total circuit reactance is

$$X_{L(tot)} = X_{L1} + X_{L(eq)} = 7.85 \text{ k}\Omega + 1.65 \text{ k}\Omega = 9.50 \text{ k}\Omega$$

The total circuit impedance is

$$Z_T = \sqrt{R_T^2 + X_{L(tot)}^2} = \sqrt{(6.27 \text{ k}\Omega)^2 = (9.50 \text{ k}\Omega)^2} = 11.4 \text{ k}\Omega$$

(b) Use Ohm's law to find the total current:

$$I_T = \frac{V_s}{Z_T} = \frac{10 \text{ V}}{11.4 \text{ k}\Omega} = 877 \text{ } \mu A$$

(c) To find the phase angle, view the circuit as a series combination of R_T and $X_{L(tot)}$. The phase angle by which I_T lags V_s is

$$\theta = \tan^{-1}\left(\frac{X_{L(tot)}}{R_T}\right) = \tan^{-1}\left(\frac{9.5 \text{ k}\Omega}{6.27 \text{ k}\Omega}\right) = 56.6°$$

Related Exercise **(a)** Determine the voltage across the series part of the circuit in Figure 13–23. **(b)** Determine the voltage across the parallel part of the circuit.

EXAMPLE 13–10   Determine the voltage across each element in Figure 13–24. Sketch a voltage phasor diagram.

FIGURE 13–24

Solution   First calculate X_{L1} and X_{L2}:

$$X_{L1} = 2\pi fL_1 = 2\pi(2 \text{ MHz})(50 \text{ }\mu\text{H}) = 628 \text{ }\Omega$$

$$X_{L2} = 2\pi fL_2 = 2\pi(2 \text{ MHz})(100 \text{ }\mu\text{H}) = 1.26 \text{ k}\Omega$$

Now, determine the impedance of each branch:

$$Z_1 = \sqrt{R_1^2 + X_{L1}^2} = \sqrt{(330 \text{ }\Omega)^2 + (628 \text{ }\Omega)^2} = 709 \text{ }\Omega$$

$$Z_2 = \sqrt{R_2^2 + X_{L2}^2} = \sqrt{(1 \text{ k}\Omega)^2 + (1.26 \text{ k}\Omega)^2} = 1.61 \text{ k}\Omega$$

Calculate each branch current:

$$I_1 = \frac{V_s}{Z_1} = \frac{10 \text{ V}}{709 \text{ }\Omega} = 14.1 \text{ mA}$$

$$I_2 = \frac{V_s}{Z_2} = \frac{10 \text{ V}}{1.61 \text{ k}\Omega} = 6.21 \text{ mA}$$

Now, use Ohm's law to find the voltage across each element:

$$V_{R1} = I_1 R_1 = (14.1 \text{ mA})(330 \text{ }\Omega) = 4.65 \text{ V}$$

$$V_{L1} = I_1 X_{L1} = (14.1 \text{ mA})(628 \text{ }\Omega) = 8.85 \text{ V}$$

$$V_{R2} = I_2 R_2 = (6.21 \text{ mA})(1 \text{ k}\Omega) = 6.21 \text{ V}$$

$$V_{L2} = I_2 X_{L2} = (6.21 \text{ mA})(1.26 \text{ k}\Omega) = 7.82 \text{ V}$$

Now determine the angles associated with each parallel branch:

$$\theta_1 = \tan^{-1}\left(\frac{X_{L1}}{R_1}\right) = \tan^{-1}\left(\frac{628 \text{ }\Omega}{330 \text{ }\Omega}\right) = 62.3°$$

$$\theta_2 = \tan^{-1}\left(\frac{X_{L2}}{R_2}\right) = \tan^{-1}\left(\frac{1.26 \text{ k}\Omega}{1 \text{ k}\Omega}\right) = 51.6°$$

Thus, I_1 lags V_s by 62.3°, and I_2 lags V_s by 51.6°, as indicated in Figure 13–25(a), where the negative signs indicate lagging angles.

The phase relationships of the voltages are determined as follows:

☐ V_{R1} is in phase with I_1 and therefore lags V_s by 62.3°.

☐ V_{L1} leads I_1 by 90°, so its angle is 90° − 62.3° = 27.7°.

☐ V_{R2} is in phase with I_2 and therefore lags V_s by 51.6°.

☐ V_{L2} leads I_2 by 90°, so its angle is 90° − 51.6° = 38.4°.

These phase relationships are shown in Figure 13–25(b).

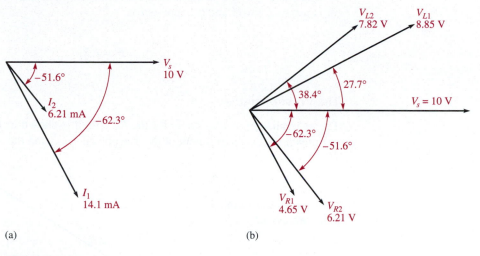

FIGURE 13–25

Related Exercise What effect does an increase in frequency have on the total current in Figure 13–24?

SECTION 13–6 REVIEW	1. Determine the total current for the circuit in Figure 13–24.
	2. What is the total impedance of the circuit in Figure 13–24?

13–7 ■ POWER IN *RL* CIRCUITS

In a purely resistive ac circuit, all of the energy delivered by the source is dissipated in the form of heat by the resistance. In a purely inductive ac circuit, all of the energy delivered by the source is stored by the inductor in its magnetic field during a portion of the voltage cycle and then returned to the source during another portion of the cycle so that there is no net energy loss. When there is both resistance and inductance, some of the energy is alternately stored and returned by the inductance and some is dissipated by the resistance. The amount of energy loss is determined by the relative values of the resistance and the inductive reactance.

After completing this section, you should be able to

■ **Determine power in *RL* circuits**
  □ Explain true and reactive power
  □ Draw the power triangle
  □ Define *power factor*
  □ Explain power factor correction

When the resistance is greater than the inductive reactance, more of the total energy delivered by the source is dissipated by the resistance than is stored and returned by the inductor. When the reactance is greater than the resistance, more of the total energy is stored and returned than is dissipated.

As you know, the power in a resistor is called the true power. The power in an inductor is reactive power and is expressed as

$$P_r = I^2 X_L \qquad \text{(13–13)}$$

The Power Triangle

The generalized power triangle for the *RL* circuit is shown in Figure 13–26. The **apparent power, P_a,** is the resultant of the true power, P_{true}, and the reactive power, P_r.

FIGURE 13–26
Generalized power triangle for an RL circuit.

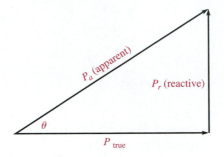

Recall that the power factor equals the cosine of θ ($PF = \cos\theta$). As the phase angle between the applied voltage and the total current increases, the power factor decreases, indicating an increasingly reactive circuit. The smaller the power factor, the smaller the true power is compared to the reactive power. The power factor of inductive loads is called a *lagging power factor* because the current lags the source voltage.

EXAMPLE 13–11 Determine the power factor, the true power, the reactive power, and the apparent power in Figure 13–27.

FIGURE 13–27

Solution The impedance of the circuit is

$$Z = \sqrt{R^2 + X_L^2} = \sqrt{(1\ \text{k}\Omega)^2 + (2\ \text{k}\Omega)^2} = 2.24\ \text{k}\Omega$$

The current is

$$I = \frac{V_s}{Z} = \frac{10\ \text{V}}{2.24\ \text{k}\Omega} = 4.46\ \text{mA}$$

The phase angle is

$$\theta = \tan^{-1}\left(\frac{X_L}{R}\right) = \tan^{-1}\left(\frac{2\ \text{k}\Omega}{1\ \text{k}\Omega}\right) = 63.4°$$

Therefore, the power factor is

$$PF = \cos \theta = \cos(63.4°) = 0.448$$

The true power is

$$P_{\text{true}} = V_s I \cos \theta = (10 \text{ V})(4.46 \text{ mA})(0.448) = 20 \text{ mW}$$

The reactive power is

$$P_r = I^2 X_L = (4.46 \text{ mA})^2 (2 \text{ k}\Omega) = 39.8 \text{ mVAR}$$

The apparent power is

$$P_a = I^2 Z = (4.46 \text{ mA})^2 (2.24 \text{ k}\Omega) = 44.6 \text{ mVA}$$

Related Exercise If the frequency in Figure 13–27 is increased, what happens to P_{true}, P_r, and P_a?

Significance of the Power Factor

As you learned in Chapter 12, the **power factor** (*PF*) is very important in determining how much useful power (true power) is transferred to a load. The highest power factor is 1, which indicates that all of the current to a load is in phase with the voltage (resistive). When the power factor is 0, all of the current to a load is 90° out of phase with the voltage (reactive).

Generally, a power factor as close to 1 as possible is desirable, because then most of the power transferred from the source to the load is useful or true power. True power goes only one way—from source to load—and performs work on the load in terms of energy utilization. Reactive power simply goes back and forth between the source and the load with no net work being done. Energy must be used in order for work to be done.

Many practical loads have inductance as a result of their particular function, and it is essential for their proper operation. Examples are transformers, electric motors, and speakers, to name a few. Therefore, inductive (and capacitive) loads are a fact of life and we must live with them.

To see the effect of the power factor on system requirements, refer to Figure 13–28 which shows a representation of a typical inductive load consisting effectively of induc-

(a) A lower power factor means more total current for a given power dissipation (watts). A larger source is required to deliver the true power (watts).

(b) A higher power factor means less total current for a given power dissipation. A smaller source can deliver the same true power (watts).

FIGURE 13–28

Illustration of the effect of the power factor on system requirements such as source rating (VA) and conductor size.

tance and resistance in parallel. Part (a) shows a load with a relatively low power factor (0.75), and part (b) shows a load with a relatively high power factor (0.95). Both loads dissipate equal amounts of power as indicated by the wattmeters. Thus, an equal amount of work is done on both loads.

Although both loads are equivalent in terms of the amount of work done (true power), the low-power factor load in Figure 13–28(a) draws more current from the source than does the high-power factor load in Figure 13–28(b), as indicated by the ammeters. Therefore, the source in part (a) must have a higher VA rating than the one in part (b). Also, the lines connecting the source to the load must be a larger wire gage than those in part (b), a condition that becomes significant when very long transmission lines are required, such as in power distribution.

Figure 13–28 has demonstrated that a higher power factor is an advantage in delivering power more efficiently to a load.

Power Factor Correction

The power factor of an inductive load can be increased by the addition of a capacitor in parallel, as shown in Figure 13–29. The capacitor compensates for the phase lag of the total current by creating a capacitive component of current that is 180° out of phase with the inductive component. This has a canceling effect and reduces the phase angle, thus increasing the power factor as well as the total current, as illustrated in the figure.

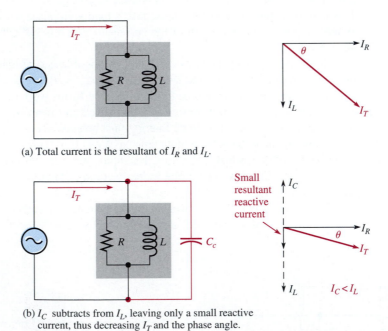

(a) Total current is the resultant of I_R and I_L.

(b) I_C subtracts from I_L, leaving only a small reactive current, thus decreasing I_T and the phase angle.

FIGURE 13–29

Example of how the power factor can be increased by the addition of a compensating capacitor.

SECTION 13–7 REVIEW

1. To which component in an *RL* circuit is the power dissipation due?
2. Calculate the power factor when $\theta = 50°$.
3. A certain series *RL* circuit consists of a 470 Ω resistor and an inductive reactance of 620 Ω at the operating frequency. Determine P_{true}, P_r, and P_a when $I = 100$ mA.

13–8 ■ BASIC APPLICATIONS

Two basic applications, phase shift networks and frequency-selective networks (filters), are covered in this section.

After completing this section, you should be able to

■ **Discuss some basic *RL* applications**
   □ Discuss and analyze the *RL* lag network
   □ Discuss and analyze the *RL* lead network
   □ Describe how the *RL* circuit operates as a filter

The *RL* Lag Network

The *RL* lag network is a phase shift circuit in which the output voltage lags the input voltage by a specified angle, ϕ. A basic *RL* lag network is shown in Figure 13–30(a). Notice that the output voltage is taken across the resistor and the input voltage is the total voltage applied across the circuit. The relationship of the voltages is shown in the phasor diagram in Figure 13–30(b), and a waveform diagram is shown in Figure 13–30(c). Notice that the output voltage, V_R, lags V_{in} by an angle, designated ϕ, that is the same as the circuit phase angle. The angles are equal, of course, because V_R and I are in phase with each other.

The phase lag, ϕ, can be expressed as

$$\phi = \tan^{-1}\left(\frac{X_L}{R}\right) \quad\quad (13\text{–}14)$$

(a) A basic *RL* lag network

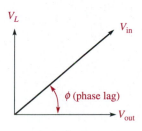

(b) Phasor voltage diagram showing phase lag between V_{in} and V_{out}

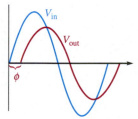

(c) Input and output waveforms

FIGURE 13–30
The RL lag network ($V_{out} = V_R$).

EXAMPLE 13–12   Calculate the phase lag for each circuit in Figure 13–31.

FIGURE 13–31

(a)

(b)

Solution For the circuit in Figure 13–31(a),

$$\phi = \tan^{-1}\left(\frac{X_L}{R}\right) = \tan^{-1}\left(\frac{5 \text{ k}\Omega}{15 \text{ k}\Omega}\right) = 18.4°$$

The output lags the input by 18.4°.

For the circuit in Figure 13–31(b), first determine the inductive reactance:

$$X_L = 2\pi fL = 2\pi(1 \text{ kHz})(1 \text{ mH}) = 6.28 \text{ }\Omega$$

The phase lag is

$$\phi = \tan^{-1}\left(\frac{X_L}{R}\right) = \tan^{-1}\left(\frac{6.28 \text{ }\Omega}{10 \text{ }\Omega}\right) = 32.1°$$

The output lags the input by 32.1°.

Related Exercise In a certain lag network, $R = 5.6 \text{ k}\Omega$ and $X_L = 3.5 \text{ k}\Omega$. Determine the phase lag between input and output.

The phase-lag network can be considered as a voltage divider with a portion of the input voltage dropped across L and a portion across R. The output voltage can be determined with the following formula:

$$V_{out} = \left(\frac{R}{\sqrt{R^2 + X_L^2}}\right)V_{in} \qquad \textbf{(13–15)}$$

EXAMPLE 13–13   The input voltage in Figure 13–31(b) of Example 13–12 has an rms value of 10 V. Determine the output voltage for the lag network shown in Figure 13–31(b). Sketch the waveform relationships for the input and output voltages. The phase lag (32.1°) and X_L (6.28 Ω) were found in Example 13–12.

Solution Use Equation (13–15) to determine the output voltage for the lag network in Figure 13–31(b).

$$V_{out} = \left(\frac{R}{\sqrt{R^2 + X_L^2}}\right)V_{in} = \left(\frac{10 \text{ }\Omega}{11.8 \text{ }\Omega}\right)10 \text{ V} = 8.47 \text{ V rms}$$

The waveforms are shown in Figure 13–32.

FIGURE 13–32

The calculator sequence is

[1] [0] [÷] [(] [(] [1] [0] [x^2] [+] [6] [.] [2] [8] [x^2] [)] [2nd] [$\sqrt{x}$] [)] [=] [×] [1] [0] [=]

Related Exercise In a lag network, $R = 4.7$ kΩ and $X_L = 6$ kΩ. If the rms input voltage is 20 V, what is the output voltage?

Effects of Frequency on the Lag Network Since the circuit phase angle and the phase lag are the same, an increase in frequency causes an increase in phase lag. Also, an increase in frequency causes a decrease in the magnitude of the output voltage because X_L becomes greater and more of the total voltage is dropped across the inductor and less across the resistor. Figure 13–33 illustrates the changes in phase lag and output voltage magnitude with frequency.

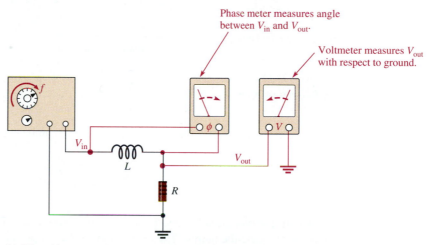

Phase meter measures angle between V_{in} and V_{out}.

Voltmeter measures V_{out} with respect to ground.

(a) When f increases, the phase lag ϕ increases and V_{out} decreases.

(b) When f decreases, ϕ decreases and V_{out} increases.

FIGURE 13–33

Illustration of how the frequency affects the phase lag and the output voltage in an RL lag network with the amplitude of V_{in} held constant.

The *RL* Lead Network

The *RL* lead network is a phase shift circuit in which the output voltage leads the input voltage by a specified angle, ϕ. A basic *RL* lead network is shown in Figure 13–34(a). Notice how this network differs from the lag network. Here the output voltage is taken across the inductor rather than across the resistor. The relationship of the voltages is shown in the phasor diagram of Figure 13–34(b) and in the waveform plot of Figure 13–34(c). Notice that the output voltage, V_L, leads V_{in} by an angle (phase lead) that is the difference between 90° and the circuit phase angle θ.

(a) A basic *RL* lead network

(b) Phasor voltage diagram showing V_{out} leading V_{in}

(c) Input and output waveforms

FIGURE 13–34
The RL lead network ($V_{out} = V_L$).

Since $\theta = \tan^{-1}(X_L/R)$, the amount of phase lead can be expressed as follows:

$$\phi = 90° - \tan^{-1}\left(\frac{X_L}{R}\right) \qquad (13\text{–}16)$$

Again, the phase-lead network can be considered as a voltage divider with the voltage across L being the output. The expression for the output voltage is

$$V_{out} = \left(\frac{X_L}{\sqrt{R^2 + X_L^2}}\right)V_{in} \qquad (13\text{–}17)$$

EXAMPLE 13–14 Determine the amount of phase lead and output voltage in the lead network in Figure 13–35.

FIGURE 13–35

Solution First determine the inductive reactance:

$$X_L = 2\pi fL = 2\pi(1\text{ kHz})(50\text{ mH}) = 314\ \Omega$$

The phase lead is

$$\phi = 90° - \tan^{-1}\left(\frac{X_L}{R}\right) = 90° - \tan^{-1}\left(\frac{314\ \Omega}{680\ \Omega}\right) = 65.2°$$

The output leads the input by 65.2°.
    The output voltage is

$$V_{out} = \left(\frac{X_L}{\sqrt{R^2 + X_L^2}}\right)V_{in} = \left(\frac{314\ \Omega}{\sqrt{(680\ \Omega)^2 + (314\ \Omega)^2}}\right)5\ V = 2.1\ V$$

Related Exercise   In a certain lead network, $R = 2.2\ k\Omega$ and $X_L = 1\ k\Omega$. What is the phase lead?

Effects of Frequency on the Lead Network   Since the circuit phase angle, θ, increases as frequency increases, the phase lead between the input and the output voltages decreases. You can see this relationship by examining Equation (13–16). Also, the amplitude of the output voltage increases as the frequency increases because X_L becomes greater and more of the total input voltage is dropped across the inductor. Figure 13–36 illustrates the changes in the phase lead and output voltage with frequency.

FIGURE 13–36
Illustration of how the frequency affects the phase lead and output voltage in an RL lead network with the amplitude of V_{in} held constant.

(a) When f increases, the phase lead ϕ decreases and V_{out} increases.

(b) When f decreases, ϕ increases and V_{out} decreases.

The *RL* Circuit As a Filter

Like *RC* circuits, series *RL* circuits also exhibit a frequency-selective characteristic and therefore act as basic filters.

Low-Pass Filter You have seen what happens to the phase and the output voltage in the lag network. In terms of its filtering action, the change in the amplitude of the output voltage with frequency is the primary consideration.

To illustrate low-pass filter action, Figure 13–37 shows a specific series of measurements in which the frequency starts at zero (dc) and is increased in increments up to 20 kHz. At each value of frequency, the output voltage is measured. As you can see, the inductive reactance increases as frequency goes up, thus causing less voltage to be dropped across the resistor while the input voltage is held at a constant 10 V throughout each step. The response curve for these particular values would appear the same as the response curve in Figure 12–48 for the *RC* low-pass filter.

(a) $f = 0$ Hz (dc), $X_L = 0$, $V_{OUT} = 10$ V

(b) $f = 1$ kHz, $X_L = 62.8$ Ω, $V_{out} = 8.5$ V

(c) $f = 10$ kHz, $X_L = 628$ Ω, $V_{out} = 1.57$ V

(d) $f = 20$ kHz, $X_L = 1256$ Ω, $V_{out} = 0.79$ V

FIGURE 13–37
Example of low-pass filter action. Winding resistance has been neglected.

High-Pass Filter To illustrate *RL* high-pass filter action, Figure 13–38 shows a series of specific measurements. The frequency starts at zero (dc) and is increased in increments up to 10 kHz. As you can see, the inductive reactance increases as the frequency goes up, thus causing more voltage to be dropped across the inductor. Again, when the values are plotted, the response curve is the same as the one for the *RC* high-pass filter as shown in Figure 12–50.

The Cutoff Frequency of an RL Filter The frequency at which the inductive reactance equals the resistance in a low-pass or high-pass *RL* filter is called the *cutoff frequency* and is designated f_c. This condition is expressed as $2\pi f_c L = R$. Solving for f_c, you get the following formula:

$$f_c = \frac{R}{2\pi L}$$

(13–18)

(a) $f = 0$ Hz (dc), $X_L = 0$, $V_{OUT} = 0$ V

(b) $f = 100$ Hz, $X_L = 6.28$ Ω, $V_{out} = 0.63$ V

(c) $f = 1$ kHz, $X_L = 62.8$ Ω, $V_{out} = 5.32$ V

(d) $f = 10$ kHz, $X_L = 628$ Ω, $V_{out} = 9.88$ V

FIGURE 13–38
Example of high-pass filter action. Winding resistance has been neglected.

As with the *RC* filter, the output voltage is 70.7% of its maximum value at f_c. In a high-pass filter, all frequencies above f_c are considered to be passed by the filter, and all those below f_c are considered to be rejected. The reverse, of course, is true for a low-pass filter. Filter bandwidth, defined in Chapter 12, applies to both *RC* and *RL* filter circuits.

SECTION 13–8 REVIEW

1. A certain *RL* lead network consists of a 3.3 kΩ resistor and a 15 mH inductor. Determine the phase lead between input and output at a frequency of 5 kHz.
2. An *RL* lag network has the same component values as the lead network in Question 1. What is the magnitude of the output voltage at 5 kHz when the input is 10 V rms?
3. When an *RL* circuit is used as a low-pass filter, across which component is the output taken?

13–9 ▪ TROUBLESHOOTING

In this section, the effects that typical component failures have on the response of basic RL circuits are considered.

After completing this section, you should be able to

▪ **Troubleshoot *RL* circuits**
  ☐ Find an open inductor
  ☐ Find an open resistor
  ☐ Find an open in a parallel circuit
  ☐ Find an inductor with shorted windings

Effect of an Open Inductor

The most common failure mode for inductors occurs when the winding opens as a result of excessive current or a mechanical contact failure. It is easy to see how an open coil affects the operation of a basic series *RL* circuit, as shown in Figure 13–39. Obviously, there is no current path—so the resistor voltage is zero, and the total applied voltage appears across the inductor.

FIGURE 13–39
Effect of an open coil.

Effect of an Open Resistor

When the resistor is open, there is no current, and the inductor voltage is zero. The total input voltage is across the open resistor, as shown in Figure 13–40.

FIGURE 13–40
Effect of an open resistor.

Open Components in Parallel Circuits

In a parallel *RL* circuit, an open resistor or inductor will cause the total current to decrease, because the total impedance will increase. Obviously, the branch with the open component will have zero current. Figure 13–41 illustrates these conditions.

(a) Before failure   (b) After *L* opens

FIGURE 13–41
Effect of an open component in a parallel circuit with V_s constant.

Effect of an Inductor with Shorted Windings

It is possible for some of the windings of coils to short together as a result of damaged insulation. This failure mode is much less likely than the open coil. Shorted windings result in a reduction in inductance because the inductance of a coil is proportional to the square of the number of turns.

SECTION 13–9 REVIEW

1. Describe the effect of an inductor with shorted windings on the response of a series *RL* circuit.
2. In the circuit of Figure 13–42, indicate whether I_T, V_{R1} and V_{R2} increase or decrease as a result of *L* opening.

FIGURE 13–42

13–10 ▪ APPLICATION ASSIGNMENT

You are given two sealed modules that have been removed from a communications system that is being modified. Each module has three terminals and are labeled as RL filters, but no specifications are given. Your assignment is to test the modules to determine the type of filters and the component values.

The sealed modules have three terminals labeled IN, GND, and OUT as shown in Figure 13–43. You will apply your knowledge of series *RL* circuits and some basic measurements to determine the internal circuit configuration and the component values.

FIGURE 13–43
Ohmmeter measurements of module 1.

Step 1: Ohmmeter Measurements of Module 1

Determine the arrangement of the two components and the values of the resistor and winding resistance for module 1 indicated by the meter readings in Figure 13–43.

Step 2: AC Measurements of Module 1

Determine the inductance value for module 1 indicated by the test setup in Figure 13–44.

Step 3: Ohmmeter Measurements of Module 2

Determine the arrangement of the two components and the values of the resistor and the winding resistance for module 2 indicated by the meter readings in Figure 13–45.

Step 4: AC Measurements of Module 2

Determine the inductance value for module 2 indicated by the test setup in Figure 13–46.

FIGURE 13–44

AC measurements for module 1.

FIGURE 13–45
Ohmmeter measurements of module 2.

Top waveform is CH1.

FIGURE 13–46
AC measurements for module 2.

■ **SUMMARY**

- When a sine wave voltage is applied to an *RL* circuit, the current and all the voltage drops are also sine waves.
- Total current in an *RL* circuit always lags the source voltage.
- The resistor voltage is always in phase with the current.
- The inductor voltage always leads the current by 90°.
- In an *RL* circuit, the impedance is determined by both the resistance and the inductive reactance combined.
- Impedance is expressed in units of ohms.
- The impedance of an *RL* circuit varies directly with frequency.
- The phase angle (θ) of a series *RL* circuit varies directly with frequency.
- You can determine the impedance of a circuit by measuring the applied voltage and the total current and then applying Ohm's law.
- In an *RL* circuit, part of the power is resistive and part reactive.
- The phasor combination of resistive power (true power) and reactive power is called *apparent power*.
- The power factor indicates how much of the apparent power is true power.
- A power factor of 1 indicates a purely resistive circuit, and a power factor of 0 indicates a purely reactive circuit.
- In a lag network, the output voltage lags the input voltage in phase.
- In a lead network, the output voltage leads the input voltage in phase.
- A filter passes certain frequencies and rejects others.

■ **GLOSSARY**

Admittance (*Y*)   A measure of the ability of a reactive circuit to permit current; the reciprocal of impedance. The unit is siemens (S).

Apparent power (*P_a*)   The product of volts and amperes, with the unit of volt-amperes (VA).

Inductive susceptance (*B_L*)   The reciprocal of inductive reactance. The unit is siemens (S).

Inductor   An electrical device formed by a coil of wire around a core material having the property of inductance; also known as a choke or coil.

Power factor   The relationship between volt-amperes and watts. Volt-amperes multiplied by the power factor equals watts.

■ **FORMULAS**

(13–1) $Z = \sqrt{R^2 + X_L^2}$ Series *RL* impedance

(13–2) $\theta = \tan^{-1}\left(\dfrac{X_L}{R}\right)$ Series *RL* phase angle

(13–3) $V_s = \sqrt{V_R^2 + V_L^2}$ Total voltage in a series *RL* circuit

(13–4) $\theta = \tan^{-1}\left(\dfrac{V_L}{V_R}\right)$ Series *RL* phase angle

(13–5) $Z = \dfrac{RX_L}{\sqrt{R^2 + X_L^2}}$ Parallel *RL* impedance

(13–6) $\theta = \tan^{-1}\left(\dfrac{R}{X_L}\right)$ Parallel *RL* phase angle

$$(13-7) \qquad G = \frac{1}{R} \qquad\qquad\qquad \text{Conductance}$$

$$(13-8) \qquad B_L = \frac{1}{X_L} \qquad\qquad\qquad \text{Inductive susceptance}$$

$$(13-9) \qquad Y = \frac{1}{Z} \qquad\qquad\qquad \text{Admittance}$$

$$(13-10) \qquad Y_T = \sqrt{G^2 + B_L^2} \qquad\qquad \text{Total admittance}$$

$$(13-11) \qquad I_T = \sqrt{I_R^2 + I_L^2} \qquad\qquad \text{Total current in parallel } RL \text{ circuit}$$

$$(13-12) \qquad \theta = \tan^{-1}\!\left(\frac{I_L}{I_R}\right) \qquad\qquad \text{Parallel } RL \text{ phase angle}$$

$$(13-13) \qquad P_r = I^2 X_L \qquad\qquad\qquad \text{Reactive power}$$

$$(13-14) \qquad \phi = \tan^{-1}\!\left(\frac{X_L}{R}\right) \qquad\qquad \text{Phase lag}$$

$$(13-15) \qquad V_{\text{out}} = \left(\frac{R}{\sqrt{R^2 + X_L^2}}\right) V_{\text{in}} \qquad \text{Output of lag network}$$

$$(13-16) \qquad \phi = 90° - \tan^{-1}\!\left(\frac{X_L}{R}\right) \qquad \text{Phase lead}$$

$$(13-17) \qquad V_{\text{out}} = \left(\frac{X_L}{\sqrt{R^2 + X_L^2}}\right) V_{\text{in}} \qquad \text{Output of lead network}$$

$$(13-18) \qquad f_c = \frac{R}{2\pi L} \qquad\qquad\qquad \text{Cutoff frequency of } RL \text{ filter}$$

■ **SELF-TEST**

1. In a series RL circuit, the resistor voltage
   (a) leads the applied voltage   (b) lags the applied voltage
   (c) is in phase with the applied voltage   (d) is in phase with the current
   (e) both (a) and (d)   (f) both (b) and (d)

2. When the frequency of the voltage applied to a series RL circuit is increased, the impedance
   (a) decreases   (b) increases   (c) does not change

3. When the frequency of the voltage applied to a series RL circuit is decreased, the phase angle
   (a) decreases   (b) increases   (c) does not change

4. If the frequency is doubled and the resistance is halved, the impedance of a series RL circuit
   (a) doubles   (b) halves   (c) remains constant   (d) cannot be determined without values

5. To reduce the current in a series RL circuit, the frequency should be
   (a) increased   (b) decreased   (c) constant

6. In a series RL circuit, 10 V rms is measured across the resistor, and 10 V rms is measured across the inductor. The peak value of the source voltage is
   (a) 14.14 V   (b) 28.28 V   (c) 10 V   (d) 20 V

7. The voltages in Question 6 are measured at a certain frequency. To make the resistor voltage greater than the inductor voltage, the frequency is
   (a) increased   (b) decreased   (c) doubled   (d) not a factor

8. When the resistor voltage in a series RL circuit becomes greater than the inductor voltage, the phase angle
   (a) increases   (b) decreases   (c) is not affected

9. When the frequency is increased, the impedance of a parallel RL circuit
   (a) increases   (b) decreases   (c) remains constant

10. In a parallel RL circuit, there are 2 A rms in the resistive branch and 2 A rms in the inductive branch. The total rms current is
    (a) 4 A   (b) 5.656 A   (c) 2 A   (d) 2.828 A

11. You are observing two voltage waveforms on an oscilloscope. The time base (sec/div) of the scope is adjusted so that one half-cycle of the waveforms covers the ten horizontal divisions. The positive-going zero crossing of one waveform is at the leftmost division, and the positive-

going zero crossing of the other is three divisions to the right. The phase angle between these two waveforms is

(a) 18° (b) 36° (c) 54° (d) 180°

12. Which of the following power factors results in less energy loss in an *RL* circuit?

(a) 1 (b) 0.9 (c) 0.5 (d) 0.1

13. If a load is purely inductive and the reactive power is 10 VAR, the apparent power is

(a) 0 VA (b) 10 VA (c) 14.14 VA (d) 3.16 VA

14. For a certain load, the true power is 10 W and the reactive power is 10 VAR. The apparent power is

(a) 5 VA (b) 20 VA (c) 14.14 VA (d) 100 VA

15. The cutoff frequency of a certain low-pass *RL* filter is 20 kHz. The filter's bandwidth is

(a) 20 kHz (b) 40 kHz (c) 0 kHz (d) unknown

▪ BASIC PROBLEMS

SECTION 13–1   Sinusoidal Response of *RL* Circuits

1. A 15 kHz sinusoidal voltage is applied to a series *RL* circuit. Determine the frequency of I, V_R, and V_L.

2. What are the waveshapes of I, V_R, and V_L in Problem 1?

SECTION 13–2   Impedance and Phase Angle of Series *RL* Circuits

3. Find the impedance of each circuit in Figure 13–47.

4. Determine the impedance and phase angle in each circuit in Figure 13–48.

5. In Figure 13–49, determine the impedance at each of the following frequencies:

(a) 100 Hz (b) 500 Hz (c) 1 kHz (d) 2 kHz

FIGURE 13–47

FIGURE 13–48

FIGURE 13–49

6. Determine the values of R and X_L in a series *RL* circuit for the following values of impedance and phase angle:

(a) $Z = 20\ \Omega$, $\theta = 45°$ (b) $Z = 500\ \Omega$, $\theta = 35°$

(c) $Z = 2.5\ k\Omega$, $\theta = 72.5°$ (d) $Z = 998\ \Omega$, $\theta = 45°$

SECTION 13–3   Analysis of Series *RL* Circuits

7. Find the current for each circuit of Figure 13–47.

8. Calculate the total current in each circuit of Figure 13–48.

9. Determine θ for the circuit in Figure 13–50.

10. If the inductance in Figure 13–50 is doubled, does θ increase or decrease, and by how many degrees?

11. Sketch the waveforms for V_s, V_R, and V_L in Figure 13–50. Show the proper phase relationships.

12. For the circuit in Figure 13–51, find V_R and V_L for each of the following frequencies:
  (a) 60 Hz **(b)** 200 Hz **(c)** 500 Hz **(d)** 1 kHz

FIGURE 13–50   **FIGURE 13–51**

SECTION 13–4   Impedance and Phase Angle of Parallel *RL* Circuits

13. What is the impedance for the circuit in Figure 13–52?

14. Repeat Problem 13 for the following frequencies:
  (a) 1.5 kHz **(b)** 3 kHz **(c)** 5 kHz **(d)** 10 kHz

15. At what frequency does X_L equal R in Figure 13–52?

FIGURE 13–52

SECTION 13–5   Analysis of Parallel *RL* Circuits

16. Find the total current and each branch current in Figure 13–53.

17. Determine the following quantities in Figure 13–54:
  (a) Z **(b)** I_R **(c)** I_L **(d)** I_T **(e)** θ

18. Convert the circuit in Figure 13–55 to an equivalent series form.

FIGURE 13–53   **FIGURE 13–54**

FIGURE 13–55

SECTION 13–6 Series-Parallel Analysis

19. Determine the voltage across each element in Figure 13–56.

20. Is the circuit in Figure 13–56 predominantly resistive or predominantly inductive?

21. Find the current in each branch and the total current in Figure 13–56.

FIGURE 13–56

SECTION 13–7 Power in *RL* Circuits

22. In a certain *RL* circuit, the true power is 100 mW, and the reactive power is 340 mVAR. What is the apparent power?

23. Determine the true power and the reactive power in Figure 13–50.

24. What is the power factor in Figure 13–53?

25. Determine P_{true}, P_r, P_a, and *PF* for the circuit in Figure 13–56. Sketch the power triangle.

SECTION 13–8 Basic Applications

26. For the lag network in Figure 13–57, determine the phase lag of the output voltage with respect to the input for the following frequencies:
  (a) 1 Hz **(b)** 100 Hz **(c)** 1 kHz **(d)** 10 kHz

27. Plot the response curve for the circuit in Figure 13–57. Show the output voltage versus frequency in 1 kHz increments from 0 Hz to 5 kHz.

28. Repeat Problem 26 for the lead network in Figure 13–58.

29. Using the same procedure as in Problem 27, plot the response curve for Figure 13–58.

30. Sketch the voltage phasor diagram for each circuit in Figures 13–57 and 13–58 for a frequency of 8 kHz.

FIGURE 13–57

FIGURE 13–58

SECTION 13–9 Troubleshooting

31. Determine the voltage across each element in Figure 13–59 if L_1 were open.

FIGURE 13–59

32. Determine the output voltage in Figure 13–60 for each of the following failure modes:
   (a) L_1 open   **(b)** L_2 open   **(c)** R_1 open   **(d)** a short across R_2

FIGURE 13–60

▪ **ADVANCED PROBLEMS**

33. Determine the voltage across the inductors in Figure 13–61.

34. Is the circuit in Figure 13–61 predominantly resistive or predominantly inductive?

35. Find the total current in Figure 13–61.

36. For the circuit in Figure 13–62, determine the following:
   (a) Z_T **(b)** I_T **(c)** θ **(d)** V_L **(e)** V_{R3}

FIGURE 13–61

FIGURE 13–62

37. For the circuit in Figure 13–63, determine the following:
   (a) I_{R1} **(b)** I_{L1} **(c)** I_{L2} **(d)** I_{R2}

38. Determine the phase shift and attenuation (ratio of V_{out} to V_{in}) from the input to the output for the network in Figure 13–64.

39. Determine the attenuation from the input to the output for the circuit in Figure 13–65.

FIGURE 13–63

FIGURE 13–64

FIGURE 13–65

40. Design an ideal inductive switching circuit that will provide a momentary voltage of 2.5 kV from a 12 V dc source when a switch is thrown instantaneously from one position to another. The drain on the source must not exceed 1 A.

41. Draw the schematic for the circuit in Figure 13–66 and determine if the waveforms on the scope are correct. If there is a fault in the circuit, identify it.

(a) Oscilloscope display

(b) Circuit with leads connected

FIGURE 13–66

■ **ANSWERS TO SECTION REVIEWS**

Section 13–1

1. The current is at 1 kHz.

2. θ is closer to 0° when $R > X_L$.

Section 13–2

1. V_s leads I.

2. The inductive reactance produces the phase angle.

3. RL has an opposite phase angle compared to RC.

4. $Z = \sqrt{R^2 + X_L^2} = 59.9 \text{ k}\Omega;\ \theta = \tan^{-1}\left(\dfrac{X_L}{R}\right) = 56.6°$

Section 13–3

1. $V_T = \sqrt{V_R^2 + V_L^2} = 3.61 \text{ V}$

2. $\theta = \tan^{-1}\left(\dfrac{X_L}{R}\right) = 56.3°$

3. When f increases, X_L increases, Z increases, and θ increases.

Section 13–4

1. $Z = \dfrac{1}{Y} = 20 \ \Omega$

2. $Y = \sqrt{G^2 + B_L^2} = 25 \ \text{mS}$

3. I_T lags V_s by $32.1°$.

Section 13–5

1. $I_T = V_s Y = 32 \ \text{mA}$

2. $\theta = \tan^{-1}(I_L/I_R) = 59.0°$; $I_T = \sqrt{I_R^2 + I_L^2} = 23.3 \ \text{mA}$

3. $\theta = 90°$

Section 13–6

1. $I_T = \sqrt{I_1^2 + I_2^2} = 20.2 \ \text{mA}$

2. $Z = V_s/I_T = 495 \ \Omega$

Section 13–7

1. Power is in the resistor.

2. $PF = \cos 50° = 0.643$

3. $P_{\text{true}} = I^2 R = 4.7 \ \text{W}$; $P_r = I^2 X_L = 6.2 \ \text{VAR}$; $P_a = \sqrt{P_{\text{true}}^2 + P_r^2} = 7.78 \ \text{VA}$

Section 13–8

1. $\phi = 90° - \tan^{-1}(X_L/R) = 81.9°$

2. $V_{\text{out}} = \left(R/\sqrt{R^2 + X_L^2} \right) V_{\text{in}} = 1.41 \ \text{V}$

3. Output is across the resistor.

Section 13–9

1. Shorted windings reduce L and thereby reduce X_L at any given frequency.

2. I_T decreases, V_{R1} decreases, and V_{R2} increases when L opens.

Section 13–10

1. 0 V on output

2. Output same as input

▪ **ANSWERS TO RELATED EXERCISES FOR EXAMPLES**

13–1	2.04 kΩ; 27.8°
13–2	423 μA
13–3	No
13–4	12.6 kΩ; 85.5°
13–5	8.14 kΩ; 35.5°
13–6	3.13 mS
13–7	14.0 mA
13–8	67.6 mA; 36.3°
13–9	(a) 8.04 V (b) 2.00 V
13–10	Current decreases.
13–11	$P_{\text{true}}, P_r,$ and P_a decrease.
13–12	32°
13–13	12.3 V rms
13–14	65.6°

14

RLC CIRCUITS AND RESONANCE

In this chapter, you will learn about frequency response of *RLC* circuits—that is, circuits with combinations of resistance, capacitance, and inductance. Both series and parallel *RLC* circuits, including the concepts of series and parallel resonance will be discussed.

Because it is the basis for frequency selectivity, resonance in electrical circuits is very important to the operation of many types of electronic systems, particularly in the area of communications. For example, the ability of a radio or television receiver to select a certain frequency transmitted by a certain station and to eliminate frequencies from other stations is based on the principle of resonance.

The operation of both band-pass and band-stop filters is based on resonance of circuits containing inductance and capacitance, and these filters are discussed in this chapter. Also, certain system applications are presented.

■ CHAPTER OBJECTIVES

☐ Determine the impedance and phase angle of a series *RLC* circuit
☐ Analyze series *RLC* circuits
☐ Analyze a circuit for series resonance
☐ Analyze series resonant filters
☐ Analyze parallel *RLC* circuits
☐ Analyze a circuit for parallel resonance
☐ Analyze the operation of parallel resonant filters
☐ Discuss some system applications of resonant circuits

■ APPLICATION ASSIGNMENT

Your application assignment in this chapter requires that you plot a frequency response curve for a resonant filter with unknown characteristics. From the frequency response measurements, you will identify the type of filter and determine the resonant frequency and the bandwidth. After studying this chapter, you should be able to complete the application assignment in Section 14–9.

14–1 ▪ IMPEDANCE AND PHASE ANGLE OF SERIES *RLC* CIRCUITS

A series RLC circuit contains both inductance and capacitance. Since inductive reactance and capacitive reactance have opposite effects on the circuit phase angle, the total reactance is less than either individual reactance.

After completing this section, you should be able to

▪ **Determine the impedance and phase angle of a series *RLC* circuit**
   ☐ Calculate total reactance
   ☐ Determine whether a circuit is predominately inductive or capacitive

A series *RLC* circuit is shown in Figure 14–1. It contains resistance, inductance, and capacitance.

FIGURE 14–1
Series RLC circuit.

As you know, inductive reactance (X_L) causes the total current to lag the applied voltage. Capacitive reactance (X_C) has the opposite effect: It causes the current to lead the voltage. Thus, X_L and X_C tend to offset each other. When they are equal, they cancel, and the total reactance is zero. In any case, the total reactance in the series circuit is

$$X_T = |X_L - X_C| \tag{14–1}$$

The term $|X_L - X_C|$ means the absolute value of the difference of the two reactances. That is, the sign of the result is considered positive no matter which reactance is greater. For example, $3 - 7 = -4$, but the absolute value is

$$|3 - 7| = 4$$

When $X_L > X_C$, the circuit is predominantly inductive, and when $X_C > X_L$, the circuit is predominantly capacitive.

The total impedance for the series *RLC* circuit is given by

$$Z_T = \sqrt{R^2 + (X_L - X_C)^2} \tag{14–2}$$

and the phase angle between V_s and I is

$$\theta = \tan^{-1}\left(\frac{X_T}{R}\right) \tag{14–3}$$

EXAMPLE 14–1   Determine the total impedance and the phase angle in Figure 14–2.

FIGURE 14–2

Solution First find X_C and X_L:

$$X_C = \frac{1}{2\pi fC} = \frac{1}{2\pi(100 \text{ Hz})(500 \ \mu\text{F})} = 3.18 \ \Omega$$

$$X_L = 2\pi fL = 2\pi(100 \text{ Hz})(10 \text{ mH}) = 6.28 \ \Omega$$

In this case, X_L is greater than X_C, and thus the circuit is more inductive than capacitive. The magnitude of the total reactance is

$$X_T = |X_L - X_C| = |6.28 \ \Omega - 3.18 \ \Omega| = 3.10 \ \Omega \qquad \text{(inductive)}$$

The total circuit impedance is

$$Z_T = \sqrt{R^2 + X_T^2} = \sqrt{(5.6 \ \Omega)^2 + (3.10 \ \Omega)^2} = 6.40 \ \Omega$$

The phase angle (between I and V_s) is

$$\theta = \tan^{-1}\left(\frac{X_T}{R}\right) = \tan^{-1}\left(\frac{3.10 \ \Omega}{5.6 \ \Omega}\right) = 29.0° \qquad \text{(current lagging } V_s\text{)}$$

The calculator sequences are as follows:

Z: [2] [×] [2nd] [π] [×] [1] [0] [0] [×] [1] [0] [EXP] [+/−] [3] [=] [STO] [2] [×]
[2nd] [π] [×] [1] [0] [0] [×] [5] [0] [0] [EXP] [+/−] [6] [=] [1/x] [+/−] [SUM] [RCL]
[x²] [+] [5] [•] [6] [x²] [=] [2nd] [√x]

θ: [EXP] [÷] [5] [•] [6] [=] [2nd] [tan]

Related Exercise Increase the frequency in Figure 14–2 to 200 Hz and determine Z and θ.

As you have seen, when the inductive reactance is greater than the capacitive reactance, the circuit appears inductive, and the current lags the applied voltage. When the capacitive reactance is greater, the circuit appears capacitive, and the current leads the applied voltage.

SECTION 14–1 REVIEW

1. Explain how you can determine if a series *RLC* circuit is inductive or capacitive.
2. In a given series *RLC* circuit, X_C is 150 Ω and X_L is 80 Ω. What is the total reactance in ohms? Is it inductive or capacitive?
3. Determine the impedance for the circuit in Question 2 when $R = 45 \ \Omega$. What is the phase angle? Is the current leading or lagging the applied voltage?

14–2 ▪ ANALYSIS OF SERIES *RLC* CIRCUITS

Recall that capacitive reactance varies inversely with frequency and that inductive reactance varies directly with frequency. In this section, the combined effects of the reactances as a function of frequency are examined.

After completing this section, you should be able to

▪ **Analyze series *RLC* circuits**
  ☐ Determine current in a series *RLC* circuit
  ☐ Determine the voltages in a series *RLC* circuit
  ☐ Determine the phase angle

As shown in Figure 14–3, for a typical series *RLC* circuit the total reactance behaves as follows: Starting at a very low frequency, X_C is high, X_L is low, and the circuit is predominantly capacitive. As the frequency is increased, X_C decreases and X_L increases until a value is reached where $X_C = X_L$ and the two reactances cancel, making the circuit purely resistive. This condition is **series resonance** and will be studied in Section 14–3. As the frequency is increased further, X_L becomes greater than X_C, and the circuit is predominantly inductive. Example 14–2 illustrates how the impedance and phase angle change as the source frequency is varied.

Note that the graph of X_L is a straight line and the graph of X_C is curved in Figure 14–3. The reason is as follows: The general equation for a straight line is $y = mx + b$, where *m* is the slope of the line and *b* is the y-axis intercept point. The formula $X_L = 2\pi fL$ fits this general straight-line formula, where $y = X_L$ (a variable), $m = 2\pi L$ (a constant), $x = f$ (a variable), and $b = 0$ as follows: $X_L = 2\pi Lf + 0$.

The X_C curve is called a *hyperbola,* and the general equation of a hyperbola is $xy = k$. The equation for capacitive reactance, $X_C = 1/2\pi fC$, can be rearranged as $X_C f = 1/2\pi C$ where $x = X_C$ (a variable), $y = f$ (a variable), and $k = 1/2\pi C$ (a constant).

FIGURE 14–3
How X_C and X_L vary with frequency.

EXAMPLE 14–2

For each of the following frequencies of the source voltage, find the impedance and the phase angle for the circuit in Figure 14–4. Note how the impedance and the phase angle change with frequency.
(a) $f = 1$ kHz **(b)** $f = 3.5$ kHz **(c)** $f = 5$ kHz

FIGURE 14–4

Solution
(a) At $f = 1$ kHz,

$$X_C = \frac{1}{2\pi fC} = \frac{1}{2\pi(1\text{ kHz})(0.02\ \mu\text{F})} = 7.96\text{ k}\Omega$$

$$X_L = 2\pi fL = 2\pi(1\text{ kHz})(100\text{ mH}) = 628\ \Omega$$

The circuit is highly capacitive because X_C is much larger than X_L. The impedance is

$$Z = \sqrt{R^2 + (X_L - X_C)^2} = \sqrt{(3.3 \text{ k}\Omega)^2 + (628 \ \Omega - 7.96 \text{ k}\Omega)^2} = 8.04 \text{ k}\Omega$$

The phase angle is

$$\theta = \tan^{-1}\left(\frac{X_T}{R}\right) = \tan^{-1}\left(\frac{|X_L - X_C|}{R}\right) = \tan^{-1}\left(\frac{7.33 \text{ k}\Omega}{3.3 \text{ k}\Omega}\right) = 65.8°$$

I leads V_s by 65.8°.

(b) At $f = 3.5$ kHz,

$$X_C = \frac{1}{2\pi(3.5 \text{ kHz})(0.02 \ \mu\text{F})} = 2.27 \text{ k}\Omega$$

$$X_L = 2\pi(3.5 \text{ kHz})(100 \text{ mH}) = 2.20 \text{ k}\Omega$$

The circuit is very close to being purely resistive but is still slightly capacitive because X_C is slightly larger than X_L. The impedance and phase angle are

$$Z = \sqrt{(3.3 \text{ k}\Omega)^2 + (2.20 \text{ k}\Omega - 2.27 \text{ k}\Omega)^2} = 3.30 \text{ k}\Omega$$

$$\theta = \tan^{-1}\left(\frac{X_T}{R}\right) = \tan^{-1}\left(\frac{70 \ \Omega}{3.3 \text{ k}\Omega}\right) = 1.22°$$

I leads V_s by 1.22°.

(c) At $f = 5$ kHz,

$$X_C = \frac{1}{2\pi(5 \text{ kHz})(0.02 \ \mu\text{F})} = 1.59 \text{ k}\Omega$$

$$X_L = 2\pi(5 \text{ kHz})(100 \text{ mH}) = 3.14 \text{ k}\Omega$$

The circuit is now predominantly inductive because $X_L > X_C$. The impedance and phase angle are

$$Z = \sqrt{(3.3 \text{ k}\Omega)^2 + (3.14 \text{ k}\Omega - 1.59 \text{ k}\Omega)^2} = 3.65 \text{ k}\Omega$$

$$\theta = \tan^{-1}\left(\frac{X_T}{R}\right) = \tan^{-1}\left(\frac{1.55 \text{ k}\Omega}{3.3 \text{ k}\Omega}\right) = 25.2°$$

I lags V_s by 25.2°.

Notice how the circuit changed from capacitive to inductive as the frequency increased. The phase condition changed from the leading current to the lagging current. It is interesting to note that both the impedance and the phase angle decreased to a minimum and then began increasing again as the frequency went up.

Related Exercise   Determine Z for $f = 7$ kHz in Figure 14–4.

In a series *RLC* circuit, the capacitor voltage and the inductor voltage are always 180° out of phase with each other. For this reason, V_C and V_L subtract from each other, and thus the voltage across *L* and *C* combined is always less than the larger individual voltage across either element, as illustrated in Figure 14–5 and in the waveform diagram of Figure 14–6.

FIGURE 14–5
The voltage across the series combination of C and L is always less than the larger individual voltage.

FIGURE 14–6
Inductor voltage and capacitor voltage effectively subtract.

In the next example, Ohm's law is used to find the voltages in a series *RLC* circuit.

EXAMPLE 14–3
Find the voltages across each element in Figure 14–7, and draw a complete voltage phasor diagram. Also find the voltage across *L* and *C* combined.

FIGURE 14–7

Solution   First find the total impedance:

$$Z = \sqrt{R^2 + (X_L - X_C)^2} = \sqrt{(75 \ \Omega)^2 + (25 \ \Omega - 60 \ \Omega)^2}$$
$$= \sqrt{(75 \ \Omega)^2 + (35 \ \Omega)^2} = 82.8 \ \Omega$$

Apply Ohm's law to find the current:

$$I = \frac{V_s}{Z} = \frac{10 \ \text{V}}{82.8 \ \Omega} = 121 \ \text{mA}$$

Now apply Ohm's law to find the voltages across *R*, *L*, and *C*:

$$V_R = IR = (121 \ \text{mA})(75 \ \Omega) = 9.08 \ \text{V}$$
$$V_L = IX_L = (121 \ \text{mA})(25 \ \Omega) = 3.03 \ \text{V}$$
$$V_C = IX_C = (121 \ \text{mA})(60 \ \Omega) = 7.26 \ \text{V}$$

The voltage across L and C combined is

$$V_{LC} = V_C - V_L = 7.26 \text{ V} - 3.03 \text{ V} = 4.23 \text{ V}$$

The circuit phase angle is

$$\theta = \tan^{-1}\left(\frac{X_T}{R}\right) = \tan^{-1}\left(\frac{|X_L - X_C|}{R}\right) = \tan^{-1}\left(\frac{35 \text{ }\Omega}{75 \text{ }\Omega}\right) = 25°$$

Since the circuit is capacitive ($X_C > X_L$), the current leads the source voltage by 25°.

The phasor diagram is shown in Figure 14–8. Notice that V_L is leading V_R by 90°, and V_C is lagging V_R by 90°. Also, there is a 180° phase difference between V_L and V_C. If the current phasor were shown, it would be at the same angle as V_R.

FIGURE 14–8

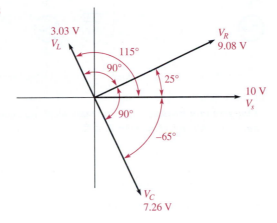

Related Exercise   What will happen to the current if the frequency in Figure 14–7 is increased?

SECTION 14–2 REVIEW

1. The following voltages occur in a certain series *RLC* circuit: $V_R = 24$ V, $V_L = 15$ V, $V_C = 45$ V. Determine the source voltage.
2. When $R = 10 \text{ }\Omega$, $X_C = 18 \text{ }\Omega$, and $X_L = 12 \text{ }\Omega$, does the current lead or lag the applied voltage? Why?
3. Determine the total reactance in Question 2.

14–3 ▪ SERIES RESONANCE

In a series RLC circuit, series resonance occurs when $X_L = X_C$. The frequency at which resonance occurs is called the resonant frequency and is designated f_r.

After completing this section, you should be able to

▪ **Analyze a circuit for series resonance**
   □ Define *series resonance*
   □ Determine the impedance at resonance
   □ Explain why the reactances cancel at resonance
   □ Determine the series resonant frequency
   □ Calculate the current, voltages, and phase angle at resonance

Figure 14–9 illustrates the series resonant condition. Since $X_L = X_C$, the reactances effectively cancel, and the impedance is purely resistive. These resonant conditions are stated in the following equations:

$$X_L = X_C \tag{14-4}$$

$$Z_r = R \tag{14-5}$$

FIGURE 14–9
At the resonant frequency (f_r), the reactances are equal in magnitude and effectively cancel, leaving Z = R.

EXAMPLE 14–4   For the series *RLC* circuit in Figure 14–10, determine X_C and Z at resonance.

FIGURE 14–10

Solution X_L equals X_C at resonance; therefore,

$$X_L = X_C = 50 \ \Omega$$

Since the reactances cancel each other, the impedance equals the resistance.

$$Z_r = R = 100 \ \Omega$$

Related Exercise   Just below the resonant frequency, is the circuit more inductive or more capacitive?

X_L and X_C Effectively Cancel at Resonance

At the series **resonant frequency,** the voltages across C and L are equal in magnitude because the reactances are equal and because the same current flows through both since they are in series ($IX_C = IX_L$). Also, V_L and V_C are always 180° out of phase with each other.

During any given cycle, the polarities of the voltages across C and L are opposite, as shown in Figures 14–11(a) and 14–11(b). The equal and opposite voltages across C and L cancel, leaving zero volts from point A to point B as shown in the figure. Since there is no voltage drop from A to B but there is still current, the total reactance must be zero, as indicated in Figure 14–11(c). Also, the voltage phasor diagram in part (d) shows that V_C and V_L are equal in magnitude and 180° out of phase with each other.

(a)

(b)

(c)

(d)

FIGURE 14–11

At the resonant frequency, f_r, the voltages across C and L are equal in magnitude. Since they are 180° out of phase with each other, they cancel, leaving 0 V across the LC combination. The section of the circuit from A to B effectively looks like a short at resonance (neglecting winding resistance).

The Series Resonant Frequency

For a given series RLC circuit, resonance occurs at only one specific frequency. A formula for this resonant frequency is developed as follows:

$$X_L = X_C$$

Substitute the reactance formulas:

$$2\pi f_r L = \frac{1}{2\pi f_r C}$$

Solve for the resonant frequency (f_r):

$$(2\pi f_r L)(2\pi f_r C) = 1$$
$$4\pi^2 f_r^2 LC = 1$$
$$f_r^2 = \frac{1}{4\pi^2 LC}$$

Take the square root of both sides. The formula for resonant frequency is

$$f_r = \frac{1}{2\pi\sqrt{LC}}$$

(14–6)

EXAMPLE 14–5   Find the series resonant frequency for the circuit in Figure 14–12.

FIGURE 14–12

R
10 Ω

C
50 pF

L
5 mH

V_s

Solution   The resonant frequency is

$$f_r = \frac{1}{2\pi\sqrt{LC}} = \frac{1}{2\pi\sqrt{(5 \text{ mH})(50 \text{ pF})}} = 318 \text{ kHz}$$

The calculator sequence is

5 EXP +/− 3 × 5 0 EXP +/− 1 2 = 2nd √x × 2 × 2nd π = 1/x

Related Exercise   If $C = 0.01 \ \mu F$ in Figure 14–12, what is the resonant frequency?

Voltages and Current in a Series *RLC* Circuit

It is interesting to see how the current and the voltages in a series *RLC* circuit vary as the frequency is increased from below the resonant frequency, through resonance, and then above resonance. Figure 14–13 illustrates the general response of a circuit in terms of changes in the current and in the voltage drops.

Below the Resonant Frequency   Figure 14–13(a) indicates the response as the source frequency is increased from zero toward f_r. At $f = 0$ Hz (dc), the capacitor appears open and blocks the current. Thus, there is no voltage across R or L, and the entire source voltage appears across C. The impedance of the circuit is infinitely large at 0 Hz because X_C is infinite (C is open). As the frequency begins to increase, X_C decreases and X_L increases, causing the total reactance, $X_C - X_L$, to decrease. As a result, the impedance decreases and the current increases. As the current increases, V_R also increases, and both V_C and V_L increase. The voltage across C and L combined decreases from its maximum value of V_s because as V_C and V_L approach the same value, their difference becomes less.

At the Resonant Frequency   When the frequency reaches its resonant value, f_r, as shown in Figure 14–13(b), V_C and V_L can each be much larger than the source voltage. V_C and V_L cancel leaving 0 V across the combination, because they are equal in magnitude but opposite in phase. At this point the total impedance is equal to R and is at its minimum value because the total reactance is zero. Thus, the current is at its maximum value, V_s/R, and V_R is at its maximum value which is equal to the source voltage.

Above the Resonant Frequency   As the frequency is increased above resonance, as indicated in Figure 14–13(c), X_L continues to increase and X_C continues to decrease, causing the total reactance, $X_L - X_C$, to increase. As a result, there is an increase in impedance and a decrease in current. As the current decreases, V_R also decreases and V_C and V_L decrease. As V_C and V_L decrease, their difference becomes greater, so V_{CL} increases. As the frequency becomes very high, the current approaches zero, both V_R and V_C approach zero, and V_L approaches V_s.

(a) Below resonance ($X_C > X_L$)

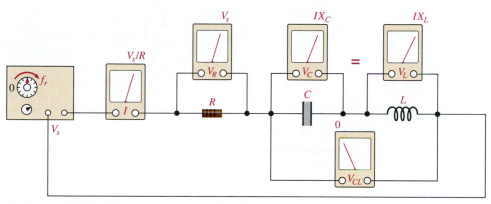

(b) At resonance ($X_C = X_L$)

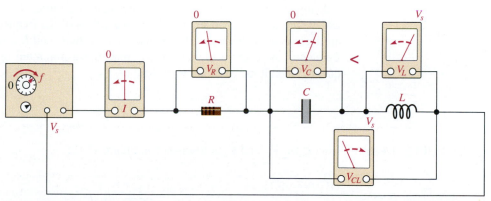

(c) Above resonance ($X_L > X_C$)

FIGURE 14–13

An illustration of how the voltages and the current respond in a series RLC circuit as the frequency is increased from below to above its resonant value. The source voltage is held at a constant amplitude.

Figure 14–14 (a) and (b) illustrate the responses of voltage and current to increasing frequency. As frequency is increased, the current increases below resonance, reaches a peak at the resonant frequency, and then decreases above resonance. The resistor voltage responds in the same way as the current.

The general slopes of the V_C and V_L curves are indicated in Figure 14–14 (c) and (d). The voltages are maximum at resonance but drop off above and below f_r. The voltages

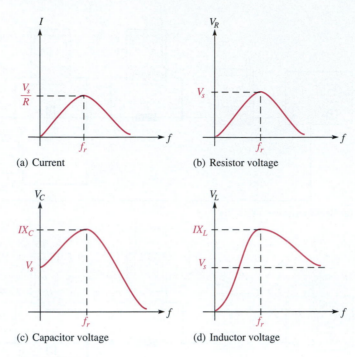

(a) Current

(b) Resistor voltage

(c) Capacitor voltage

(d) Inductor voltage

FIGURE 14–14
Current and voltages as functions of frequency in a series RLC circuit.

across L and C at resonance are exactly equal in magnitude but 180° out of phase, so they cancel. Thus, the total voltage across both L and C is zero, and $V_R = V_s$ at resonance. Individually, V_L and V_C can be much greater than the source voltage. Keep in mind that V_L and V_C are always opposite in polarity regardless of the frequency, but only at resonance are their magnitudes equal. The voltage across the C and L combination decreases as the frequency increases below resonance, reaching a minimum of zero at the resonant frequency; then it increases above resonance.

EXAMPLE 14–6

Find I, V_R, V_L, and V_C at resonance in Figure 14–15.

FIGURE 14–15

Solution   At resonance, I is maximum and equal to V_s/R:

$$I = \frac{V_s}{R} = \frac{50 \text{ V}}{22 \text{ }\Omega} = 2.27 \text{ A}$$

Applying Ohm's law, obtain the following voltages:

$$V_R = IR = (2.27 \text{ A})(22 \text{ }\Omega) = 50 \text{ V}$$
$$V_L = IX_L = (2.27 \text{ A})(100 \text{ }\Omega) = 227 \text{ V}$$
$$V_C = IX_C = (2.27 \text{ A})(100 \text{ }\Omega) = 227 \text{ V}$$

Notice that all of the source voltage is dropped across the resistor. Also, of course, V_L and V_C are equal in magnitude but opposite in phase. This causes the voltage to cancel, making the total reactive voltage zero.

Related Exercise   What is the current at resonance in Figure 14–15 if $X_L = X_C = 1 \text{ k}\Omega$?

The Impedance of a Series *RLC* Circuit

Figure 14–16 shows a general graph of impedance versus frequency superimposed on the curves for X_C and X_L. At zero frequency, both X_C and Z are infinitely large and X_L is zero because the capacitor looks like an open at 0 Hz and the inductor looks like a short. As the frequency increases, X_C decreases and X_L increases. Since X_C is larger than X_L at frequencies below f_r, Z decreases along with X_C. At f_r, $X_C = X_L$ and $Z = R$. At frequencies above f_r, X_L becomes increasingly larger than X_C, causing Z to increase.

FIGURE 14–16
Series RLC impedance as a function of frequency.

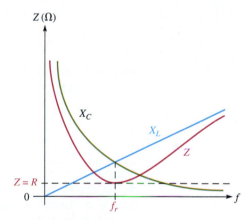

EXAMPLE 14–7   For the circuit in Figure 14–17, determine the impedance at the following frequencies:
(a) f_r (b) 1000 Hz below f_r (c) 1000 Hz above f_r

FIGURE 14–17

Solution
(a) At f_r, the impedance is equal to R:

$$Z = R = 10 \ \Omega$$

To determine the impedance above and below f_r, first calculate the resonant frequency as follows:

$$f_r = \frac{1}{2\pi\sqrt{LC}} = \frac{1}{2\pi\sqrt{(100 \text{ mH})(0.01 \ \mu\text{F})}} = 5.03 \text{ kHz}$$

(b) At 1000 Hz below f_r, the frequency and reactances are as follows:

$$f = f_r - 1 \text{ kHz} = 5.03 \text{ kHz} - 1 \text{ kHz} = 4.03 \text{ kHz}$$

$$X_C = \frac{1}{2\pi fC} = \frac{1}{2\pi(4.03 \text{ kHz})(0.01 \text{ } \mu F)} = 3.95 \text{ k}\Omega$$

$$X_L = 2\pi fL = 2\pi(4.03 \text{ kHz})(100 \text{ mH}) = 2.53 \text{ k}\Omega$$

Therefore, the impedance at $f_r - 1$ kHz is

$$Z = \sqrt{R^2 + (X_L - X_C)^2} = \sqrt{(10 \text{ } \Omega)^2 + (2.53 \text{ k}\Omega - 3.95 \text{ k}\Omega)^2} = 1.42 \text{ k}\Omega$$

(c) At 1000 Hz above f_r,

$$f = 5.03 \text{ kHz} + 1 \text{ kHz} = 6.03 \text{ kHz}$$

$$X_C = \frac{1}{2\pi(6.03 \text{ kHz})(0.01 \text{ } \mu F)} = 2.64 \text{ k}\Omega$$

$$X_L = 2\pi(6.03 \text{ kHz})(100 \text{ mH}) = 3.79 \text{ k}\Omega$$

Therefore, the impedance at $f_r + 1$ kHz is

$$Z = \sqrt{(10 \text{ } \Omega)^2 + (3.79 \text{ k}\Omega - 2.64 \text{ k}\Omega)^2} = 1.15 \text{ k}\Omega$$

X_L is greater than X_C, so Z is inductive.

Related Exercise   What happens to the impedance if f is decreased below 4.03 kHz? Above 6.03 kHz?

The Phase Angle of a Series *RLC* Circuit

At frequencies below resonance, $X_C > X_L$, and the current leads the source voltage, as indicated in Figure 14–18(a). The phase angle decreases as the frequency approaches the resonant value and is 0° at resonance, as indicated in part (b). At frequencies above reso-

FIGURE 14–18
The phase angle as a function of frequency in a series RLC circuit.

(a) Below f_r, I leads V_s.

(b) At f_r, I is in phase with V_s.

(c) Above f_r, I lags V_s.

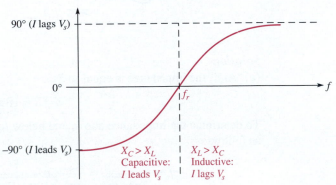

(d) Phase angle versus frequency.

nance, $X_L > X_C$, and the current lags the source voltage, as indicated in part (c). As the frequency goes higher, the phase angle approaches 90°. A plot of phase angle versus frequency is shown in Figure 14–18(d).

SECTION 14–3 REVIEW	**1.** What is the condition for series resonance? **2.** Why is the current maximum at the resonant frequency? **3.** Calculate the resonant frequency for $C = 1000$ pF and $L = 1000$ μH. **4.** In Question 3, is the circuit inductive or capacitive at 50 kHz?

14–4 ■ SERIES RESONANT FILTERS

A common use of series RLC circuits is in filter applications. In this section, you will learn the basic configurations for band-pass and band-stop filters and several important filter characteristics.

After completing this section, you should be able to

■ **Analyze series resonant filters**
   ☐ Identify a basic series resonant band-pass filter
   ☐ Define and determine *bandwidth*
   ☐ Define the *half-power frequency*
   ☐ Discuss dB measurement
   ☐ Define *selectivity*
   ☐ Discuss filter quality factor (Q)
   ☐ Identify a series resonant band-stop filter

The Band-Pass Filter

A basic series resonant **band-pass filter** is shown in Figure 14–19. Notice that the series *LC* portion is placed between the input and the output and that the output is taken across the resistor.

FIGURE 14–19

A basic series resonant band-pass filter.

A band-pass filter allows signals at the resonant frequency and at frequencies within a certain band (or range) extending below and above the resonant value to pass from input to output without a significant reduction in amplitude. Signals at frequencies lying outside this specified band (called the **pass band**) are reduced in amplitude to below a certain level and are considered to be rejected by the filter.

The filtering action is the result of the impedance characteristic of the filter. As you learned in Section 14–3, the impedance is minimum at resonance and has increasingly higher values below and above the resonant frequency. At very low frequencies, the impedance is very high and tends to block the current. As the frequency increases, the impedance drops, allowing more current and thus more voltage across the output resistor. At the resonant frequency, the impedance is very low and equal to the winding resistance of the coil. At this point a maximum current is flowing and the resulting output voltage is maximum. As the frequency goes above resonance, the impedance again increases, causing the current and the resulting output voltage to drop. Figure 14–20 illustrates the general frequency response of a series resonant band-pass filter.

(a) As the frequency increases to f_1, V_{out} increases to 7.07 V.

(b) As the frequency increases from f_1 to f_r, V_{out} increases from 7.07 V to 10 V.

(c) As the frequency increases from f_r to f_2, V_{out} decreases from 10 V to 7.07 V.

(d) As the frequency increases above f_2, V_{out} decreases below 7.07 V.

FIGURE 14–20

Example of the frequeny response of a series resonant band-pass filter with the input voltage at a constant 10 V rms. The winding resistance of the coil is neglected.

Bandwidth of the Pass Band

The **bandwidth (BW)** of a band-pass filter is the range of frequencies for which the current (or output voltage) is equal to or greater than 70.7% of its value at the resonant frequency. Figure 14–21 shows the bandwidth on the response curve for the band-pass filter.

The frequencies at which the output of a filter is 70.7% of its maximum are the **cut-off frequencies**. Notice in Figure 14–21 that frequency f_1, which is below f_r, is the frequency at which I (or V_{out}) is 70.7% of the resonant value (I_{max}); f_1 is commonly called the *lower cutoff frequency*. At frequency f_2, above f_r, the current (or V_{out}) is again 70.7% of its maximum; f_2 is called the *upper cutoff frequency*. Other names for f_1 and f_2 are –3 dB frequencies, band frequencies, and half-power frequencies. (The term *decibel*, abbreviated dB, is defined on page 590).

The formula for calculating the bandwidth is as follows:

$$BW = f_2 - f_1$$ (14–7)

The unit of bandwidth is the hertz (Hz), the same as for frequency.

FIGURE 14–21

Generalized response curve of a series resonant band-pass filter.

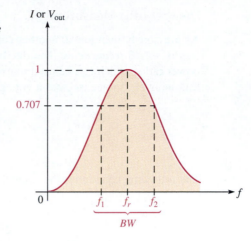

EXAMPLE 14–8

A certain series resonant band-pass filter has a maximum current of 100 mA at the resonant frequency. What is the value of the current at the cutoff frequencies?

Solution   Current at the cutoff frequencies is 70.7% of maximum:

$$I_{f1} = I_{f2} = 0.707I_{max} = 0.707(100 \text{ mA}) = 70.7 \text{ mA}$$

Related Exercise   Will a change in the cutoff frequencies affect the current at these frequencies if the maximum current remains 100 mA?

EXAMPLE 14–9

A resonant circuit has a lower cutoff frequency of 8 kHz and an upper cutoff frequency of 12 kHz. Determine the bandwidth.

Solution $$BW = f_2 - f_1 = 12 \text{ kHz} - 8 \text{ kHz} = 4 \text{ kHz}$$

Related Exercise   What is the bandwidth when $f_1 = 1$ MHz and $f_2 = 1.2$ MHz?

Half-Power Points of the Filter Response

As previously mentioned, the upper and lower cutoff frequencies are sometimes called the **half-power frequencies.** This term is derived from the fact that the true power from the source at these frequencies is one-half the power delivered at the resonant frequency. The following steps show that this relationship is true for a series resonant circuit.

At resonance,

$$P_{max} = I_{max}^2 R$$

The power at f_1 or f_2 is

$$P_{f1} = I_{f1}^2 R = (0.707I_{max})^2 R = (0.707)^2 I_{max}^2 R$$
$$= 0.5I_{max}^2 R = 0.5P_{max}$$

Decibel (dB) Measurement

As previously mentioned, another common term for the upper and lower cutoff frequencies is *−3 dB frequencies*. The **decibel** (dB) is a logarithmic measurement of voltage or power ratios which can be used to express the input-to-output relationship of a filter. The following equation expresses a voltage ratio in decibels:

$$dB = 20 \log\left(\frac{V_{out}}{V_{in}}\right) \qquad \qquad \textbf{(14–8)}$$

The following equation is the decibel formula for a power ratio:

$$dB = 10 \log\left(\frac{P_{out}}{P_{in}}\right) \qquad \qquad \textbf{(14–9)}$$

EXAMPLE 14–10 At a certain frequency, the output voltage of a filter is 5 V and the input is 10 V. Express the voltage ratio in decibels.

Solution
$$20 \log\left(\frac{V_{out}}{V_{in}}\right) = 20 \log\left(\frac{5 \text{ V}}{10 \text{ V}}\right) = 20 \log(0.5) = -6.02 \text{ dB}$$

The calculator sequence is

| 5 | ÷ | 1 | 0 | = | log | × | 2 | 0 | = |

Related Exercise Express 0.2 in decibels.

The −3 dB Frequencies The output of a filter is said to be down 3 dB at the cutoff frequencies. As you know, this frequency is the point at which the output voltage is 70.7% of the maximum voltage at resonance. We can show that the 70.7% point is the same as 3 dB below maximum (or −3 dB) as follows. The maximum voltage is the 0 dB reference:

$$20 \log\left(\frac{0.707 V_{max}}{V_{max}}\right) = 20 \log(0.707) = -3 \text{ dB}$$

Selectivity of a Band-Pass Filter

The response curve in Figure 14–21 is also called a *selectivity curve*. **Selectivity** defines how well a resonant circuit responds to a certain frequency and discriminates against all others. *The narrower the bandwidth, the greater the selectivity.*

We normally assume that a resonant circuit accepts frequencies within its bandwidth and completely eliminates frequencies outside the bandwidth. Such is not actually the case, however, because signals with frequencies outside the bandwidth are not completely eliminated. Their magnitudes, however, are greatly reduced. The further the frequencies are from the cutoff frequencies, the greater is the reduction, as illustrated in Figure 14–22(a). An ideal selectivity curve is shown in Figure 14–22(b).

As you can see in Figure 14–23, another factor that influences selectivity is the steepness of the slopes of the curve. The faster the curve drops off at the cutoff frequencies, the more selective the circuit is, because the frequencies outside the pass band are more quickly reduced (attenuated).

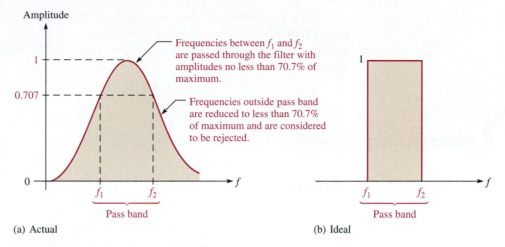

FIGURE 14–22
Generalized selectivity curve of a band-pass filter.

FIGURE 14–23
Comparative selectivity curves.

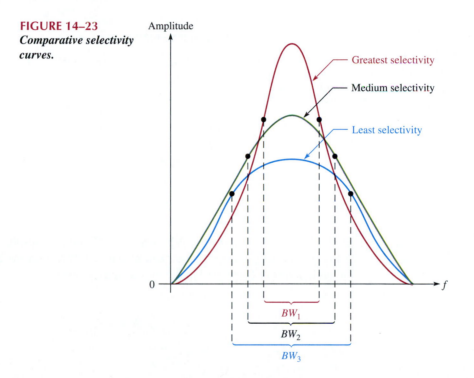

The Quality Factor (Q) of a Resonant Circuit and Its Effect on Bandwidth

The **quality factor (Q)** is the ratio of the reactive power in the inductor to the true power in the winding resistance of the coil or the resistance in series with the coil. It is a ratio of the power in L to the power in R. The quality factor is very important in resonant circuits. A formula for Q is developed as follows:

$$Q = \frac{\text{energy stored}}{\text{energy dissipated}} = \frac{\text{reactive power}}{\text{true power}} = \frac{I^2 X_L}{I^2 R}$$

In a series circuit, I is the same in L and R; thus, the I^2 terms cancel, leaving

$$Q = \frac{X_L}{R} \qquad\qquad \textbf{(14–10)}$$

When the resistance is just the winding resistance of the coil, the circuit Q and the coil Q are the same. Since Q varies with frequency because X_L varies, we are interested mainly in Q at resonance. Note that Q is a ratio of like units (ohms) and, therefore, has no unit itself.

EXAMPLE 14–11

Determine Q at resonance for the circuit in Figure 14–24.

FIGURE 14–24

Solution   Determine the inductive reactance:

$$X_L = 2\pi f_r\, L = 2\pi(1.007 \text{ kHz})(5 \text{ mH}) = 31.6\ \Omega$$

The quality factor is

$$Q = \frac{X_L}{R} = \frac{31.6\ \Omega}{10\ \Omega} = 3.16$$

Related Exercise   Calculate Q if R in Figure 14–24 is decreased to 2.7 Ω.

How Q Affects Bandwidth   A higher value of circuit Q results in a smaller bandwidth. A lower value of Q causes a larger bandwidth. A formula for the bandwidth of a resonant circuit in terms of Q is stated in the following equation:

$$BW = \frac{f_r}{Q} \tag{14–11}$$

EXAMPLE 14–12

What is the bandwidth of the filter in Figure 14–25?

FIGURE 14–25

Solution The total resistance is

$$R_T = R + R_W = 91\ \Omega + 9\ \Omega = 100\ \Omega$$

The bandwidth is found as follows:

$$f_r = \frac{1}{2\pi\sqrt{LC}} = \frac{1}{2\pi\sqrt{(10\ \text{mH})(0.005\ \mu\text{F})}} = 22.5\ \text{kHz}$$

$$X_L = 2\pi f_r L = 2\pi(22.5\ \text{kHz})(10\ \text{mH}) = 1.41\ \text{k}\Omega$$

$$Q = \frac{X_L}{R_T} = \frac{1.41\ \text{k}\Omega}{100\ \Omega} = 14.1$$

$$BW = \frac{f_r}{Q} = \frac{22.5\ \text{kHz}}{14.1} = 1.6\ \text{kHz}$$

Related Exercise Determine the bandwidth if L is changed to 50 mH in Figure 14–25.

The Band-Stop Filter

The basic series resonant **band-stop filter** is shown in Figure 14–26. Notice that the output voltage is taken across the LC portion of the circuit. This filter is still a series RLC circuit, just as the band-pass filter is. The difference is that in this case, the output voltage is taken across the combination of L and C rather than across R.

The band-stop filter rejects signals with frequencies between the upper and lower cutoff frequencies and passes those signals with frequencies below and above the cutoff values, as shown in the response curve of Figure 14–27. The range of frequencies between the lower and upper cutoff points is called the **stop band**. This type of filter is also referred to as a *band-elimination filter, band-reject filter,* or a *notch filter.*

FIGURE 14–26

A basic series resonant band-stop filter.

FIGURE 14–27

Generalized response curve for a band-stop filter.

At very low frequencies, the LC combination appears as a near open due to the high X_C, thus allowing most of the input voltage to pass through to the output. As the frequency increases, the impedance of the LC combination decreases until, at resonance, it is zero (ideally). Thus, the input signal is shorted to ground, and there is very little output voltage. As the frequency goes above its resonant value, the LC impedance increases, allowing an increasing amount of voltage to be dropped across it. The general frequency response of a series resonant band-stop filter is illustrated in Figure 14–28.

(a) As frequency increases to f_1, V_{out} decreases from 10 V to 7.07 V.

(b) As frequency increases from f_1 to f_r, V_{out} decreases from 7.07 V to 0 V.

(c) As frequency increases from f_r to f_2, V_{out} increases from 0 V to 7.07 V.

(d) As frequency increases above f_2, V_{out} increases toward 10 V.

FIGURE 14–28

Example of the frequency response of a series resonant band-stop filter with V_{in} at a constant 10 V rms. The winding resistance is neglected.

Characteristics of the Band-Stop Filter

All the characteristics that have been discussed in relation to the band-pass filter (current response, impedance characteristic, bandwidth, selectivity, and Q) apply equally to the band-stop filter, with the exception that the response curve of the output voltage is opposite. For the band-pass filter, V_{out} is maximum at resonance. For the band-stop filter, V_{out} is minimum at resonance.

EXAMPLE 14–13   Find the output voltage at f_r and the bandwidth in Figure 14–29.

FIGURE 14–29

Solution Since $X_L = X_C$ at resonance, then

$$V_{\text{out}} = \left(\frac{R_W}{R_1 + R_W} \right) V_{\text{in}} = \left(\frac{3 \ \Omega}{50 \ \Omega} \right) 100 \text{ mV} = 6.0 \text{ mV}$$

The bandwidth is found as follows:

$$f_r = \frac{1}{2\pi\sqrt{LC}} = \frac{1}{2\pi\sqrt{(100 \text{ mH})(0.01 \ \mu\text{F})}} = 5.03 \text{ kHz}$$

$$X_L = 2\pi f_r L = 2\pi(5.03 \text{ kHz})(100 \text{ mH}) = 3160 \ \Omega$$

$$Q = \frac{X_L}{R} = \frac{X_L}{R_1 + R_W} = \frac{3160 \ \Omega}{50 \ \Omega} = 63.2$$

$$BW = \frac{f_r}{Q} = \frac{5.03 \text{ kHz}}{63.2} = 79.6 \text{ Hz}$$

Related Exercise What happens to V_{out} if the frequency is increased above resonance? Below resonance?

SECTION 14–4 REVIEW

1. The output voltage of a certain band-pass filter is 15 V at the resonant frequency. What is its value at the cutoff frequencies?
2. For a certain band-pass filter, $f_r = 120$ kHz and $Q = 12$. What is the bandwidth of the filter?
3. In a band-stop filter, is the current minimum or maximum at resonance? Is the output voltage minimum or maximum at resonance?

14–5 ■ PARALLEL *RLC* CIRCUITS

In this section, you will learn how to determine the impedance and phase angle of a parallel RLC circuit. The current relationships are also covered.

After completing this section, you should be able

■ **Analyze parallel *RLC* circuits**
 □ Calculate the impedance
 □ Calculate the phase angle
 □ Determine all currents
 □ Convert from a series-parallel *RLC* circuit to an equivalent parallel form

Impedance and Phase Angle

The circuit in Figure 14–30 consists of the parallel combination of *R*, *L*, and *C*. To find the admittance, add the conductance (*G*) and the total susceptance (B_T) as phasor quanti-

FIGURE 14–30
Parallel RLC circuit.

ties. B_T is the difference of the capacitive susceptance and the inductive susceptance. Thus,

$$Y = \sqrt{G^2 + (B_C - B_L)^2} \qquad\qquad (14\text{--}12)$$

The total impedance is the reciprocal of the admittance:

$$Z_T = \frac{1}{Y}$$

The phase angle of the circuit is given by the following formula:

$$\theta = \tan^{-1}\left(\frac{B_T}{G}\right) \qquad\qquad (14\text{--}13)$$

When the frequency is above its resonant value ($X_C < X_L$), the impedance of the circuit in Figure 14–30 is predominantly capacitive because the capacitive current is greater, and the total current leads the source voltage. When the frequency is below its resonant value ($X_L < X_C$), the impedance of the circuit is predominantly inductive, and the total current lags the source voltage.

EXAMPLE 14–14   Determine the total impedance and the phase angle in Figure 14–31.

FIGURE 14–31

Solution   First determine the admittance as follows:

$$G = \frac{1}{R} = \frac{1}{10\ \Omega} = 100\ \text{mS}$$

$$B_C = \frac{1}{X_C} = \frac{1}{10\ \Omega} = 100\ \text{mS}$$

$$B_L = \frac{1}{X_L} = \frac{1}{5\ \Omega} = 200\ \text{mS}$$

$$Y = \sqrt{G^2 + (B_C - B_L)^2} = \sqrt{(100\ \text{mS})^2 + (100\ \text{mS} - 200\ \text{mS})^2}$$
$$= \sqrt{(100\ \text{mS})^2 + (-100\ \text{mS})^2} = 141.4\ \text{mS}$$

From Y, you can get Z_T:

$$Z_T = \frac{1}{Y} = \frac{1}{141.4\ \text{mS}} = 7.07\ \Omega$$

The phase angle is

$$\theta = \tan^{-1}\left(\frac{B_T}{G}\right) = \tan^{-1}\left(\frac{100\ \text{mS}}{100\ \text{mS}}\right) = 45°$$

Related Exercise   Is the circuit in Figure 14–31 predominately inductive or predominately capacitive?

Current Relationships

In a parallel *RLC* circuit, the current in the capacitive branch and the current in the inductive branch are always 180° out of phase with each other (neglecting any coil resistance). For this reason, I_C and I_L subtract from each other, and thus the total current into the parallel branches of L and C is always less than the largest individual branch current, as illustrated in Figure 14–32 and in the waveform diagram of Figure 14–33. Of course, the current in the resistive branch is always 90° out of phase with both reactive currents, as shown in the current phasor diagram of Figure 14–34. The total current can be expressed as

$$I_T = \sqrt{I_R^2 + (I_C - I_L)^2} \tag{14–14}$$

The phase angle can also be expressed in terms of the branch currents as

$$\theta = \tan^{-1}\left(\frac{I_{CL}}{I_R}\right) \tag{14–15}$$

where I_{CL} is $|I_C - I_L|$.

FIGURE 14–32
The total current into the parallel combination of C and L is the difference of the two branch currents ($I_{CL} = I_C - I_L$).

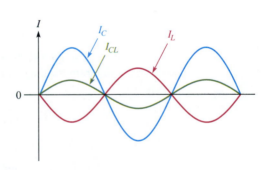

FIGURE 14–33
I_C and I_L effectively subtract.

FIGURE 14–34
Current phasor diagram for a parallel RLC circuit.

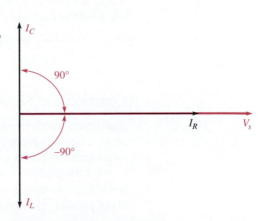

EXAMPLE 14–15   Find each branch current and the total current in Figure 14–35, and show their relationship.

FIGURE 14–35

Solution   Each branch current can be found using Ohm's law:

$$I_R = \frac{V_s}{R} = \frac{5 \text{ V}}{2.2 \text{ }\Omega} = 2.27 \text{ A}$$

$$I_C = \frac{V_s}{X_C} = \frac{5 \text{ V}}{5 \text{ }\Omega} = 1 \text{ A}$$

$$I_L = \frac{V_s}{X_L} = \frac{5 \text{ V}}{10 \text{ }\Omega} = 0.5 \text{ A}$$

The total current is the phasor sum of the branch currents:

$$I_T = \sqrt{I_R^2 + (I_C - I_L)^2} = \sqrt{(2.27 \text{ A})^2 + (1 \text{ A} - 0.5 \text{ A})^2}$$
$$= \sqrt{(2.27 \text{ A})^2 + (0.5 \text{ A})^2} = 2.32 \text{ A}$$

The phase angle is

$$\theta = \tan^{-1}\left(\frac{I_{CL}}{I_R}\right) = \tan^{-1}\left(\frac{0.5 \text{ A}}{2.27 \text{ A}}\right) = 12.4°$$

The total current is 2.32 A, leading V_s by 12.4°. Figure 14–36 is the current phasor diagram for the circuit.

FIGURE 14–36

Related Exercise   Will the total current increase or decrease if the frequency in Figure 14–35 is increased? Why?

Conversion of Series-Parallel to Parallel

The series-parallel configuration shown in Figure 14–37 is important because it represents a circuit having parallel *L* and *C* branches, with the winding resistance of the coil taken into account as a series resistance in the *L* branch.

It is helpful to view the series-parallel circuit in Figure 14–37 in an equivalent purely parallel form, as indicated in Figure 14–38. This equivalent form will simplify the analysis of parallel resonant characteristics as discussed in Section 14–6.

FIGURE 14–37
An important series-parallel RLC circuit
(Q = X$_L$/R$_W$).

FIGURE 14–38
Parallel equivalent form of the circuit in
Figure 14–37.

The equivalent inductance L_{eq} and the equivalent parallel resistance $R_{p(eq)}$ are given by the following formulas:

$$L_{eq} = L\left(\frac{Q^2 + 1}{Q^2}\right)$$

(14–16)

$$R_{p(eq)} = R_W(Q^2 + 1)$$

(14–17)

where Q is the quality factor of the coil, X_L/R_W. Derivations of these formulas are quite involved and thus are not given here. Notice in the equations that for a $Q \geq 10$, the value of L_{eq} is approximately the same as the original value of L. For example, if $L = 10$ mH, then

$$L_{eq} = 10 \text{ mH}\left(\frac{10^2 + 1}{10^2}\right) = 10 \text{ mH}(1.01) = 10.1 \text{ mH}$$

The equivalency of the two circuits means that at a given frequency, when the same value of voltage is applied to both circuits, the same total current flows in both circuits and the phase angles are the same. Basically, an equivalent circuit simply makes circuit analysis more convenient.

EXAMPLE 14–16   Convert the series-parallel circuit in Figure 14–39 to an equivalent parallel form at the given frequency.

FIGURE 14–39

Solution   Determine the inductive reactance:

$$X_L = 2\pi f L = 2\pi(15.9 \text{ kHz})(5 \text{ mH}) = 500 \text{ } \Omega$$

The Q of the coil is

$$Q = \frac{X_L}{R_W} = \frac{500 \ \Omega}{25 \ \Omega} = 20$$

Since $Q > 10$, then $L_{eq} \cong L = 5$ mH.

The equivalent parallel resistance is

$$R_{p(eq)} = R_W(Q^2 + 1) = (25 \ \Omega)(20^2 + 1) = 10.0 \ \text{k}\Omega$$

This equivalent resistance ($R_{p(eq)}$) appears in parallel with R_1 as shown in Figure 14–40(a). When combined, they give a total parallel resistance ($R_{p(tot)}$) of 3.38 kΩ, as indicated in Figure 14–40(b).

(a) Parallel equivalent of the circuit in Figure 14-39

(b) $R_{p(tot)} = R_1 \| R_{p(eq)} = 3.38$ kΩ

FIGURE 14–40

Related Exercise   Find the equivalent parallel circuit if $R_W = 10 \ \Omega$ in Figure 14–39.

SECTION 14–5 REVIEW

1. In a three-branch parallel circuit, $R = 150 \ \Omega$, $X_C = 100 \ \Omega$, and $X_L = 50 \ \Omega$ at a certain frequency. Determine the current in each branch when $V_s = 12$ V.
2. Is the circuit in Question 1 capacitive or inductive? Why?
3. Find the equivalent parallel inductance and resistance for a 20 mH coil with a winding resistance of 10 Ω at a frequency of 1 kHz.

14–6 ▪ PARALLEL RESONANCE

In this section, we will first look at the resonant condition in an ideal parallel LC circuit. Then, we will examine the more realistic case where the resistance of the coil is taken into account.

After completing this section, you should be able to

▪ **Analyze a circuit for parallel resonance**
   □ Describe parallel resonance in an ideal circuit
   □ Describe parallel resonance in an nonideal circuit
   □ Explain how impedance varies with frequency
   □ Determine current and phase angle at resonance
   □ Determine parallel resonant frequency
   □ Discuss the effects of loading a parallel resonant circuit

Condition for Ideal Parallel Resonance

Ideally, **parallel resonance** occurs when $X_L = X_C$. The frequency at which resonance occurs is called the *resonant frequency,* just as in the series case. When $X_L = X_C$, the two branch currents, I_C and I_L, are equal in magnitude, and, of course, they are always 180° out of phase with each other. Thus, the two currents cancel and the total current is zero, as presented in Figure 14–41. In this ideal case, the winding resistance of the coil is assumed to be zero.

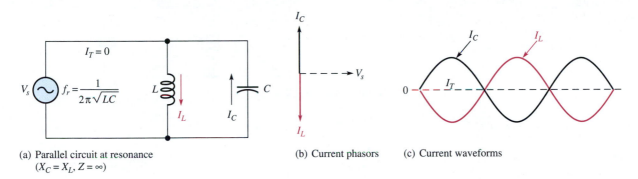

(a) Parallel circuit at resonance
    $(X_C = X_L, Z = \infty)$

(b) Current phasors

(c) Current waveforms

FIGURE 14–41
An ideal parallel LC circuit at resonance.

Since the total current is zero, the impedance of the parallel LC circuit is infinitely large (∞). These ideal resonant conditions are stated in the following equations:

$$X_L = X_C$$

$$Z_r = \infty$$

The Parallel Resonant Frequency

For an ideal parallel resonant circuit, the frequency at which resonance occurs is determined by the same formula as in series resonant circuits:

$$f_r = \frac{1}{2\pi\sqrt{LC}} \tag{14–18}$$

Currents in a Parallel Resonant Circuit

It is interesting to see how the currents in a parallel LC circuit vary as the frequency is increased from below the resonant value, through resonance, and then above the resonant value. Figure 14–42 illustrates the general response of an ideal circuit in terms of changes in the currents. Part (a) indicates the response as the source frequency is increased from zero toward f_r. At very low frequencies, X_C is very high and X_L is very low, so most of the current is through L. As the frequency increases, the current through L decreases and the current through C increases, causing the total current to decrease. At all times, I_L and I_C are 180° out of phase with each other, and thus the total current is the difference of the two branch currents. During this time, the impedance is increasing, as indicated by the decrease in total current.

When the frequency reaches its resonant value, f_r, as shown in Figure 14–42(b), X_C and X_L are equal, so I_C and I_L cancel because they are equal in magnitude but opposite in phase. At this point the total current is at its minimum value of zero. Since I_T is zero, Z is infinite. Thus, the ideal parallel LC circuit appears as an open at the resonant frequency.

FIGURE 14–42

An illustration of how the currents respond in a parallel LC circuit as the frequency is varied from below resonance to above resonance. The source voltage amplitude is constant.

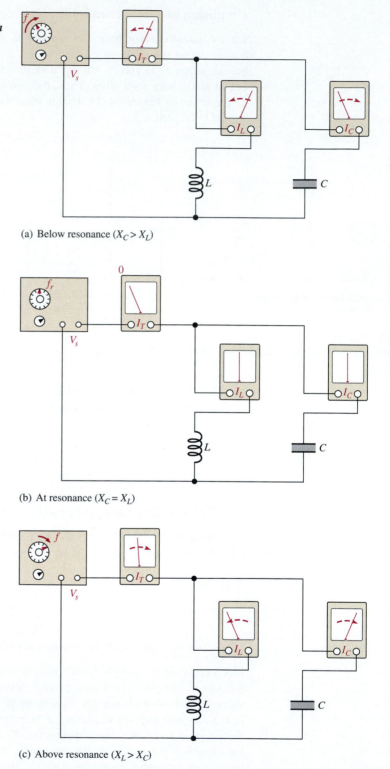

(a) Below resonance ($X_C > X_L$)

(b) At resonance ($X_C = X_L$)

(c) Above resonance ($X_L > X_C$)

    As the frequency is increased above resonance, as indicated in Figure 14–42(c), X_C continues to decrease and X_L continues to increase, causing the branch currents again to be unequal, with I_C being larger. As a result, total current increases and impedance decreases. As the frequency becomes very high, the impedance becomes very small due to the dominance of a very small X_C in parallel with a very large X_L.

In summary, the current dips to a minimum as the impedance peaks to a maximum at parallel resonance. The expression for the total current into the L and C branches is

$$I_T = |I_C - I_L| \qquad\qquad (14\text{–}19)$$

EXAMPLE 14–17   Find the resonant frequency and the branch currents in the ideal (winding resistance neglected) parallel LC circuit of Figure 14–43.

FIGURE 14–43

Solution   The resonant frequency is

$$f_r = \frac{1}{2\pi\sqrt{LC}} = \frac{1}{2\pi\sqrt{(100 \text{ mH})(0.047 \ \mu\text{F})}} = 2.32 \text{ kHz}$$

Determine the branch currents as follows:

$$X_L = 2\pi f_r L = 2\pi(2.32 \text{ kHz})(100 \text{ mH}) = 1.46 \text{ k}\Omega$$

$$X_C = X_L = 1.46 \text{ k}\Omega$$

$$I_L = \frac{V_s}{X_L} = \frac{10 \text{ V}}{1.46 \text{ k}\Omega} = 6.85 \text{ mA}$$

$$I_C = I_L = 6.85 \text{ mA}$$

The total current into the parallel LC circuit is

$$I_T = I_C - I_L = 0 \text{ A}$$

Related Exercise   Find f_r and all the currents at resonance in Figure 14–43 if C is changed to 0.022 μF.

The Parallel Resonant *LC* Circuit Is Often Called a Tank Circuit

The term **tank circuit** refers to the fact that the parallel resonant circuit stores energy in the magnetic field of the coil and in the electric field of the capacitor. The stored energy is transferred back and forth between the capacitor and the coil on alternate half-cycles as the current goes first one way and then the other when the indicator deenergizes and the capacitor charges, and vice versa. This concept is illustrated in Figure 14–44.

FIGURE 14–44

Energy storage in an ideal parallel resonant tank circuit.

(a) The coil deenergizes as the capacitor charges.

(b) The capacitor discharges as the coil energizes.

Parallel Resonant Conditions in a Nonideal Circuit

So far, the resonance of an ideal parallel *LC* circuit has been examined. Now we will consider resonance in a tank circuit with the resistance of the coil taken into account. Figure 14–45 shows a nonideal tank circuit and its parallel *RLC* equivalent.

FIGURE 14–45
A practical treatment of parallel resonant circuits includes the coil resistance.

(a) Nonideal tank circuit   (b) Parallel *RLC* equivalent

The *Q* of the circuit at resonance is simply the *Q* of the coil:

$$Q = \frac{X_L}{R_W}$$

The expressions for the equivalent parallel resistance and the equivalent inductance were given in Equations (14–17) and (14–16) as

$$R_{p(eq)} = R_W(Q^2 + 1)$$

$$L_{eq} = L\left(\frac{Q^2 + 1}{Q^2}\right)$$

Recall that for $Q \geq 10$, $L_{eq} \cong L$.

At parallel resonance,

$$X_{L(eq)} = X_C$$

In the parallel equivalent circuit, $R_{p(eq)}$ is in parallel with an ideal coil and a capacitor, so the *L* and *C* branches act as an ideal tank circuit which has an infinite impedance at resonance as shown in Figure 14–46. Therefore, the total impedance of the nonideal tank circuit at resonance can be expressed as simply the equivalent parallel resistance:

$$Z_r = R_W(Q^2 + 1) \tag{14–20}$$

FIGURE 14–46
At resonance, the parallel LC portion appears open and the source sees only $R_{p(eq)} = R_W(Q^2 = 1)$.

EXAMPLE 14–18 Determine the impedance of the circuit in Figure 14–47 at the resonant frequency ($f_r \cong$ 17,794 Hz).

FIGURE 14–47

Solution First calculate the quality factor Q as follows:

$$X_L = 2\pi f_r L = 2\pi(17{,}794 \text{ Hz})(8 \text{ mH}) = 894 \text{ }\Omega$$

$$Q = \frac{X_L}{R_W} = \frac{894 \text{ }\Omega}{50 \text{ }\Omega} = 17.9$$

The impedance is

$$Z_r = R_W(Q^2 + 1) = 50 \text{ }\Omega(17.9^2 + 1) = 16.1 \text{ k}\Omega$$

The calculator sequence is

[2] [×] [2nd] [π] [×] [1] [7] [7] [9] [4] [×] [8] [EXP] [+/−] [3] [=] [÷] [5] [0]
[=] [x²] [+] [1] [=] [×] [5] [0] [=]

Related Exercise Determine Z_r in Figure 14–47 if the winding resistance is 10 Ω.

Variation of the Impedance with Frequency

The impedance of a parallel resonant circuit is maximum at the resonant frequency and decreases at lower and higher frequencies, as indicated by the curve in Figure 14–48.

At very low frequencies, X_L is very small and X_C is very high, so the total impedance is essentially equal to that of the inductive branch. As the frequency goes up, the

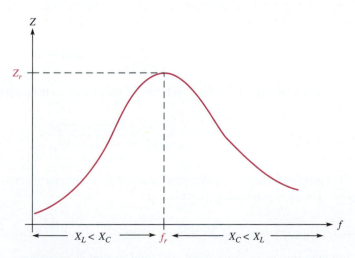

FIGURE 14–48
Generalized impedance curve for a parallel resonant circuit. The circuit is inductive below f_r, resistive at f_r, and capacitive above f_r.

impedance also increases, and the inductive reactance dominates (because it is less than X_C) until the resonant frequency is reached. At this point, of course, $X_L \cong X_C$ (for $Q > 10$) and the impedance is at its maximum. As the frequency goes above resonance, the capacitive reactance dominates (because it is less than X_L) and the impedance decreases.

Current and Phase Angle at Resonance

In the ideal tank circuit, the total current from the source at resonance is zero because the impedance is infinite. In the nonideal case, there is some total current at the resonant frequency, and it is determined by the impedance at resonance:

$$I_T = \frac{V_s}{Z_r} \qquad (14\text{–}21)$$

The phase angle of the parallel resonant circuit is 0° because the impedance is purely resistive at the resonant frequency.

Parallel Resonant Frequency in a Nonideal Circuit

As you know, when the coil resistance is considered, the resonant condition is

$$X_{L(eq)} = X_C$$

which can be expressed as

$$2\pi f_r L \left(\frac{Q^2 + 1}{Q^2} \right) = \frac{1}{2\pi f_r C}$$

Solving for f_r in terms of Q,

$$f_r = \frac{1}{2\pi\sqrt{LC}} \sqrt{\frac{Q^2}{Q^2 + 1}}$$

When $Q \geq 10$, the term with the Q factors is approximately 1:

$$\sqrt{\frac{Q^2}{Q^2 + 1}} = \sqrt{\frac{100}{101}} = 0.995 \cong 1$$

Therefore, the parallel resonant frequency is approximately the same as the series resonant frequency as long as Q is equal to or greater than 10:

$$f_r \cong \frac{1}{2\pi\sqrt{LC}} \qquad \text{for } Q \geq 10$$

A precise expression for f_r in terms of the circuit component values is

$$f_r = \frac{\sqrt{1 - (R_W^2 C/L)}}{2\pi\sqrt{LC}} \qquad (14\text{–}22)$$

This precise formula is seldom necessary and the simpler equation $f_r = 1/2\pi\sqrt{LC}$ is sufficient for most practical situations.

EXAMPLE 14–19   Find the frequency, impedance, and total current at resonance for the circuit in Figure 14–49 using Equation (14–22).

FIGURE 14–49

Solution   The precise resonant frequency is

$$f_r = \frac{\sqrt{1 - (R_W^2 C/L)}}{2\pi\sqrt{LC}} = \frac{\sqrt{1 - [(100\ \Omega)^2(0.05\ \mu F)/0.1\ H]}}{2\pi\sqrt{(0.05\ \mu F)(0.1\ H)}} = 2.25\ \text{kHz}$$

Calculate impedance as follows:

$$X_L = 2\pi f_r L = 2\pi(2.25\ \text{kHz})(0.1\ H) = 1.4\ \text{k}\Omega$$

$$Q = \frac{X_L}{R_W} = \frac{1.4\ \text{k}\Omega}{100\ \Omega} = 14$$

$$Z_r = R_W(Q^2 + 1) = 100\ \Omega(14^2 + 1) = 19.7\ \text{k}\Omega$$

The total current is

$$I_T = \frac{V_s}{Z_r} = \frac{10\ \text{V}}{19.7\ \text{k}\Omega} = 508\ \mu A$$

The calculator sequence for f_r is

1	−	(	1	0	0	x^2	×	•	0	5	EXP	+/−	6	÷	•	1
=	)	=	2nd	√x	÷	2	÷	2nd	π	÷	(	(	•	0	5
EXP	+/−	6	×	•	1	)	2nd	√x	=

Related Exercise   Repeat this example using the formula $f_r = 1/2\pi\sqrt{LC}$ and compare the results.

An External Parallel Load Resistance Affects a Tank Circuit

There are many practical situations in which an external load resistance appears in parallel with a tank circuit as shown in Figure 14–50(a). Obviously, the external resistor (R_L) will dissipate more of the energy delivered by the source and thus will lower the overall Q (Q_O) of the circuit. The external resistor effectively appears in parallel with the equiva-

(a)   (b)

FIGURE 14–50
Tank circuit with a parallel load resistor and its equivalent circuit.

lent parallel resistance of the coil, $R_{p(eq)}$, and both are combined to determine a total parallel resistance, $R_{p(tot)}$, as indicated in Figure 14–50(b):

$$R_{p(tot)} = R_L \parallel R_{p(eq)}$$

The overall parallel *RLC* circuit Q is expressed differently from the Q of a series circuit:

$$Q_O = \frac{R_{p(tot)}}{X_{L(eq)}} \qquad \textbf{(14–23)}$$

As you can see, the effect of loading the tank circuit is to reduce its overall Q (which is equal to the coil Q when unloaded).

SECTION 14–6 REVIEW

1. Is the impedance minimum or maximum at parallel resonance?
2. Is the current minimum or maximum at parallel resonance?
3. At ideal parallel resonance, $X_L = 1.5$ kΩ. What is X_C?
4. A tank circuit has the following values: $R_W = 4$ Ω, $L = 50$ mH, and $C = 10$ pF. Calculate f_r and Z_r.
5. If $Q = 25$, $L = 50$ mH, and $C = 1000$ pF, what is f_r?
6. In Question 5, if $Q = 2.5$, what is f_r?
7. In a certain tank circuit, the coil resistance is 20 Ω. What is the total impedance at resonance if $Q = 20$?

14–7 ■ PARALLEL RESONANT FILTERS

Parallel resonant circuits are commonly applied to band-pass and band-stop filters. In this section, we examine these applications.

After completing this section, you should be able to

■ **Analyze the operation of parallel resonant filters**
  □ Show how a band-pass filter is implemented
  □ Define *bandwidth*
  □ Explain how loading affects selectivity
  □ Show how a band-stop filter is implemented
  □ Determine the resonant frequency, bandwidth, and output voltage of band-pass and band-stop parallel resonant filters

---

The Band-Pass Filter

A basic parallel resonant band-pass filter is shown in Figure 14–51. Notice that the output is taken across the tank circuit in this application.

FIGURE 14–51
A basic parallel resonant band-pass filter.

The bandwidth and cutoff frequencies for a parallel resonant band-pass filter are defined in the same way as for the series resonant circuit, and the formulas given in Section 14–4 still apply. General band-pass frequency response curves showing both V_{out} and I_T versus frequency are given in Figures 14–52(a) and 14–52(b), respectively.

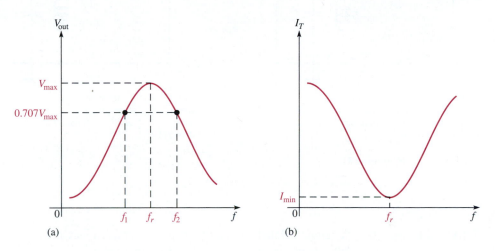

FIGURE 14–52
Generalized frequency response curves for a parallel resonant band-pass filter.

The filtering action is as follows: At very low frequencies, the impedance of the tank circuit is very low, and therefore only a small amount of voltage is dropped across it. As the frequency increases, the impedance of the tank circuit increases, and, as a result, the output voltage increases. When the frequency reaches its resonant value, the impedance is at its maximum and so is the output voltage. As the frequency goes above resonance, the impedance begins to decrease, causing the output voltage to decrease. The general response of a parallel resonant band-pass filter is illustrated in Figure 14–53. This illustration pictorially shows how the current and the output voltage change with frequency. (Parts (c) and (d) of Figure 14–53 are on p. 610.)

(a) As the frequency increases to f_1, V_{out} increases to 7.07 V, and I_T decreases.

(b) As the frequency increases from f_1 to f_r, V_{out} increases from 7.07 V to 10 V, and I_T decreases to its minimum value.

FIGURE 14–53
Example of the response of a parallel resonant band-pass filter with the input voltage at a constant 10 V rms.

(c) As the frequency increases from f_r to f_2, V_{out} decreases from 10 V to 7.07 V, and I_T increases from its minimum.

(d) As the frequency increases above f_2, V_{out} decreases below 7.07 V, and I_T continues to increase.

FIGURE 14–53 (*Continued*)

EXAMPLE 14–20

A certain parallel resonant band-pass filter has a maximum output voltage of 4 V at f_r. What is the value of V_{out} at the cutoff frequencies?

Solution V_{out} at the cutoff frequencies is 70.7% of maximum:

$$V_{out(1)} = V_{out(2)} = 0.707V_{out(max)} = 0.707(4 \text{ V}) = 2.828 \text{ V}$$

Related Exercise   What is V_{out} at the cutoff frequencies if V_{out} at the resonant frequency is 10 V?

EXAMPLE 14–21

A parallel resonant circuit has a lower cutoff frequency of 3.5 kHz and an upper cutoff frequency of 6 kHz. What is the bandwidth?

Solution $$BW = f_2 - f_1 = 6 \text{ kHz} - 3.5 \text{ kHz} = 2.5 \text{ kHz}$$

Related Exercise   A filter has a lower cutoff frequency of 520 kHz and a bandwidth of 10 kHz. What is the upper cutoff frequency?

EXAMPLE 14–22

A certain parallel resonant band-pass filter has a resonant frequency of 12 kHz and a Q of 10. What is its bandwidth?

Solution $$BW = \frac{f_r}{Q} = \frac{12 \text{ kHz}}{10} = 1.2 \text{ kHz}$$

Related Exercise   A certain parallel-resonant band-pass filter has a bandwidth of 4 MHz and a resonant frequency of 100 MHz. What is its Q?

Loading Affects the Selectivity of a Parallel Resonant Band-Pass Filter

When a resistive load is connected across the output of a filter as shown in Figure 14–54(a), the Q of the filter is reduced. Since $BW = f_r/Q$, the bandwidth is increased, thus reducing the selectivity. Also, the impedance of the filter at resonance is decreased because R_L effectively appears in parallel with $R_{p(eq)}$. Thus, the maximum output voltage is reduced by the voltage-divider effect of $R_{p(tot)}$ and the internal source resistance R_s, as illustrated in Figure 14–54(b). Figure 14–54(c) shows the general effect on the filter response curve.

(a) Tank circuit with load

(b) Equivalent circuit

R_s and $R_{p(tot)}$ act as a voltage divider.

$Z = \infty$ (open)

$V_{out} = \left(\dfrac{R_{p(tot)}}{R_{p(tot)} + R_s} \right) V_s$

(c) Loading widens the bandwidth and reduces the output.

FIGURE 14–54

Effects of loading on a parallel resonant band-pass filter.

EXAMPLE 14–23

(a) Determine f_r, BW, and V_{out} at resonance for the unloaded filter in Figure 14–55. The source resistance is 600 Ω.

(b) Repeat part (a) when the filter is loaded with a 50 kΩ resistance, and compare the results.

FIGURE 14–55

$R_W = 50\ \Omega$

V_s
5 V

$R_s = 600\ \Omega$

L
100 mH

C
0.1 μF

Solution

(a) $f_r \cong \dfrac{1}{2\pi\sqrt{LC}} = \dfrac{1}{2\pi\sqrt{(100 \text{ mH})(0.1 \text{ }\mu\text{F})}} = 1.59 \text{ kHz}$

At resonance, the bandwidth is calculated as follows:

$$X_L = 2\pi f_r L = 2\pi(1.59 \text{ kHz})(100 \text{ mH}) = 999 \text{ }\Omega$$

$$Q = \frac{X_L}{R_W} = \frac{999 \text{ }\Omega}{50 \text{ }\Omega} = 20$$

$$BW = \frac{f_r}{Q} = \frac{1.59 \text{ kHz}}{20} = 79.5 \text{ Hz}$$

The output voltage is calculated as follows:

$$R_{p(\text{eq})} = R_W(Q^2 + 1) = 50 \text{ }\Omega(20^2 + 1) = 20.1 \text{ k}\Omega$$

$$V_{\text{out}} = \left(\frac{R_{p(\text{eq})}}{R_{p(\text{eq})} + R_s}\right)V_s = \left(\frac{20.1 \text{ k}\Omega}{20.7 \text{ k}\Omega}\right)5 \text{ V} = 4.86 \text{ V}$$

(b) When a 50 kΩ load resistance is connected, the following values are obtained:

$$R_{p(\text{tot})} = R_{p(\text{eq})} \parallel R_L = 20.1 \text{ k}\Omega \parallel 50 \text{ k}\Omega = 14.3 \text{ k}\Omega$$

$$Q_O = \frac{R_{p(\text{tot})}}{X_L} = \frac{14.3 \text{ k}\Omega}{999 \text{ }\Omega} = 14.3$$

$$BW = \frac{f_r}{Q_O} = \frac{1.59 \text{ kHz}}{14.3} = 111 \text{ Hz}$$

$$V_{\text{out}} = \left(\frac{R_{p(\text{tot})}}{R_{p(\text{tot})} + R_s}\right)V_s = \left(\frac{14.3 \text{ k}\Omega}{14.9 \text{ k}\Omega}\right)5 \text{ V} = 4.80 \text{ V}$$

The result of adding the load resistance is an increased bandwidth and a decreased output voltage at resonance.

Related Exercise   How does load resistance affect the Q?

The Band-Stop Filter

A basic parallel resonant band-stop filter is shown in Figure 14–56. The output is taken across a load resistor in series with the tank circuit.

FIGURE 14–56
A basic parallel resonant band-stop filter.

The variation of the tank circuit impedance with frequency produces the familiar current response that has been previously discussed; that is, the current is minimum at resonance and increases on both sides of resonance. Since the output voltage is across the series load resistor, it follows the current, thus creating the band-stop response characteristic as indicated in Figure 14–57.

FIGURE 14–57
Band-stop filter response.

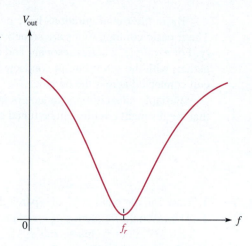

Actually, the band-stop filter in Figure 14–56 can be viewed as a voltage divider created by Z_r of the tank and the load resistance. Thus, the output voltage at f_r is

$$V_{out} = \left(\frac{R_L}{R_L + Z_r} \right) V_{in}$$

EXAMPLE 14–24   Find f_r and the output voltage at resonance for the band-stop filter in Figure 14–58.

FIGURE 14–58

Solution   The resonant frequency and impedance are calculated as follows:

$$f_r = \frac{\sqrt{1 - (R_W^2 C/L)}}{2\pi\sqrt{LC}} = 249 \text{ Hz}$$

$$X_L = 2\pi f_r L = 516 \ \Omega$$

$$Q = \frac{X_L}{R_W} = \frac{516 \ \Omega}{250 \ \Omega} = 2.06$$

$$Z_r = R_W(Q^2 + 1) = 1.31 \text{ k}\Omega$$

At resonance,

$$V_{out} = \left(\frac{R_L}{R_L + Z_r} \right) V_s = \left(\frac{1 \text{ k}\Omega}{2.31 \text{ k}\Omega} \right) 12 \text{ V} = 5.19 \text{ V}$$

Related Exercise   Why wouldn't the formula for ideal resonant frequency ($f_r = 1/2\pi \sqrt{LC}$) work in this particular case?

Basic filter configurations have been introduced in this section and in Section 14–4. These basic configurations are sometimes used in combination to increase filter selectivity. For example, a series resonant and a parallel resonant band-pass filter can be used together, with the series circuit connected between the input and output and the parallel circuit connected across the output.

Variable capacitors or inductors are sometimes used in certain filter applications so that the resonant circuits can be tuned over a range of resonant frequencies.

SECTION 14–7 REVIEW

1. How can the bandwidth of a parallel resonant filter be increased?
2. The resonant frequency of a certain high-Q ($Q > 10$) filter is 5 kHz. If the Q is lowered to 2, does f_r change and, if so, to what value?
3. What is the impedance of a tank circuit at its resonant frequency if $R_W = 75 \, \Omega$ and $Q = 25$?

14–8 ■ SYSTEM APPLICATIONS

Resonant circuits are used in a wide variety of applications, particularly in communication systems. In this section, we will look briefly at a few common communication systems applications. The purpose in this section is not to explain how the systems work, but to illustrate the importance of resonant circuits in electronic communication.

After completing this section, you should be able to

■ **Discuss some system applications of resonant circuits**
  □ Describe a tuned amplifier application
  □ Describe antenna coupling
  □ Describe tuned amplifiers
  □ Describe signal separation in a receiver
  □ Describe a radio receiver

Tuned Amplifiers

A tuned amplifier is a circuit that amplifies signals within a specified band. Typically, a parallel resonant circuit is used in conjunction with an amplifier to achieve the selectivity. In terms of the general operation, input signals with frequencies that range over a wide band are accepted on the amplifier's input and are amplified. The resonant circuit allows only a relatively narrow band of those frequencies to be passed on. The variable capacitor allows tuning over the range of input frequencies so that a desired frequency can be selected, as indicated in Figure 14–59.

Antenna Input to a Receiver

Radio signals are sent out from a transmitter via electromagnetic waves which propagate through the atmosphere. When the electromagnetic waves cut across the receiving antenna, small voltages are induced. Out of all the wide range of electromagnetic frequencies, only one frequency or a limited band of frequencies must be extracted. Figure 14–60 shows a typical arrangement of an antenna coupled to the receiver input by a transformer. A variable capacitor is connected across the transformer secondary to form a parallel resonant circuit.

FIGURE 14–59
A basic tuned band-pass amplifier.

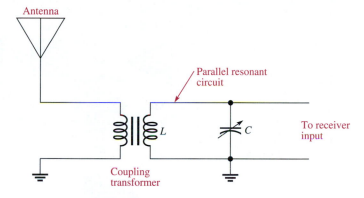

FIGURE 14–60
Resonant coupling from an antenna.

Double-Tuned Transformer Coupling in a Receiver

In some types of communication receivers, tuned amplifiers are transformer-coupled together to increase the amplification. Capacitors can be placed in parallel with the primary and secondary windings of the transformer, effectively creating two parallel resonant band-pass filters that are coupled together. This technique, illustrated in Figure 14–61, can result in a wider bandwidth and steeper slopes on the response curve, thus increasing the selectivity for a desired band of frequencies.

FIGURE 14–61
Double-tuned amplifiers.

Signal Reception and Separation in a TV Receiver

A television receiver must handle both video (picture) signals and audio (sound) signals. Each TV transmitting station is allotted a 6 MHz bandwidth. Channel 2 is allotted a band from 54 MHz through 59 MHz, channel 3 is allotted a band from 60 MHz through 65 MHz, on up to channel 13, which has a band from 210 MHz through 215 MHz. You can tune the front end of the TV receiver to select any one of these channels by using tuned amplifiers. The signal output of the front end of the receiver has a bandwidth from 41 MHz through 46 MHz, regardless of the channel that is tuned in. This band, called the *intermediate frequency* (IF) band, contains both video and audio. Amplifiers tuned to the IF band boost the signal and feed it to the video amplifier.

Before the output of the video amplifier is applied to the cathode-ray tube, the audio signal is removed by a 4.5 MHz band-stop filter (called a *wave trap*), as shown in Figure 14–62. This trap keeps the sound signal from interfering with the picture. The video amplifier output is also applied to band-pass circuits that are tuned to the sound carrier frequency of 4.5 MHz. The sound signal is then processed and applied to the speaker as indicated in Figure 14–62.

FIGURE 14–62
A simplified portion of a TV receiver showing filter usage.

Superheterodyne Receiver

Another good example of filter applications is in the common AM (amplitude modulation) receiver. The AM broadcast band ranges from 535 kHz to 1605 kHz. Each AM station is assigned a certain narrow bandwidth within that range. A simplified block diagram of a superheterodyne AM receiver is shown in Figure 14–63.

There are basically three parallel resonant band-pass filters in the front end of the receiver. Each of these filters is gang-tuned by capacitors; that is, the capacitors are mechanically or electronically linked together so that they change together as a station is selected. The front end is tuned to receive a desired station, for example, one that transmits at 600 kHz. The input filter from the antenna and the RF (radio frequency) amplifier filter select only a frequency of 600 kHz out of all the frequencies crossing the antenna. The actual audio (sound) signal is carried by the 600 kHz carrier frequency by modulating the amplitude of the carrier so that it follows the audio signal as indicated. The variation in the amplitude of the carrier corresponding to the audio signal is called the *envelope*. The 600 kHz is then applied to a circuit called the *mixer*. The local oscillator (LO) is tuned to a frequency that is 455 kHz above the selected frequency (1055 kHz, in this case). By a process called *heterodyning* or *beating,* the AM signal and the local oscillator signal

FIGURE 14–63

A simplified diagram of a superheterodyne AM radio broadcast receiver showing the application of tuned resonant circuits.

are mixed together, and the 600 kHz AM signal is converted to a 455 kHz AM signal (1055 kHz − 600 kHz = 455 kHz). The 455 kHz is the intermediate frequency (IF) for standard AM receivers. No matter which station within the broadcast band is selected, its frequency is always converted to the 455 kHz IF. The amplitude-modulated IF is applied to an audio detector which removes the IF, leaving only the envelope or audio signal. The audio signal is then amplified and applied to the speaker.

SECTION 14–8 REVIEW	**1.** Generally, why is a tuned filter necessary when a signal is coupled from an antenna to the input of a receiver? **2.** What is a wave trap? **3.** What is meant by *ganged tuning*?

14–9 ■ APPLICATION ASSIGNMENT

In the application assignment from the last chapter, you identified two types of filters in sealed housings. In this application assignment, you have another filter to evaluate which is some type of resonant filter with no specifications available. Your supervisor has requested that you come up with a frequency response curve and determine the resonant frequency and the bandwidth.

In this assignment you will use oscilloscope measurements to determine the frequency response characteristics of the filter. The internal circuitry and component values are of no concern in this case.

Step 1: Measure the Frequency Response

Based on the series of five oscilloscope measurements in Figure 14–64, create a frequency response curve (plot of output voltage versus frequency) for the filter.

FIGURE 14–64
Frequency response measurements.

Step 2: Analyze the Response Curve

Specify the type of filter and determine the resonant frequency and the bandwidth.

| **SECTION 14–9 REVIEW** | **1.** What are the half-power frequencies for the filter in this assignment? |
| | **2.** From the measurements in Figure 14–64, can you determine the circuit arrangement or component values? |

▪ SUMMARY

- X_L and X_C have opposing effects in an *RLC* circuit.
- In a series *RLC* circuit, the larger reactance determines the net reactance of the circuit.
- In a parallel *RLC* circuit, the smaller reactance determines the net reactance of the circuit.

For series resonance:
- The reactances are equal.
- The impedance is minimum and equal to the resistance.
- The current is maximum.
- The phase angle is zero.
- The voltages across *L* and *C* are equal in magnitude and, as always, 180° out of phase with each other and thus they cancel.

For parallel resonance:
- The reactances are approximately equal for $Q \geq 10$.
- The impedance is maximum.
- The current is minimum and, ideally, equal to zero.
- The phase angle is zero.
- The currents in the *L* and *C* branches are equal in magnitude and, as always, 180° out of phase with each other and thus they cancel.

▪ GLOSSARY

Band-pass filter A filter that passes frequencies within a certain range and blocks all others.

Band-stop filter A filter that blocks frequencies within a certain range.

Bandwidth (BW) The characteristic of certain electronic circuits that specifies the usable range of frequencies for which signals pass from input to output without significant reduction in amplitude.

Cutoff frequencies The frequencies at which the output of a filter is 70.7% of its maximum.

Decibel (dB) The unit of the logarithmic expression of a ratio, such as power or voltage ratios.

Half-power frequency The frequency at which the output of a filter is 70.7% of maximum.

Parallel resonance In a parallel *RLC* circuit, the condition where the impedance is maximum and the reactances are equal.

Pass band The range of frequencies passed by a filter.

Quality factor (Q) The ratio of reactive power to true power in a coil or a resonant circuit.

Resonant frequency The frequency at which a resonant condition occurs in a series or parallel *RLC* circuit.

Selectivity A measure of how well a filter passes certain frequencies and rejects others. The narrower the bandwidth, the greater the selectivity.

Series resonance In a series *RLC* circuit, the condition where the impedance is minimum and the reactances are equal.

Stop band The range of frequencies between the upper and lower cutoff points.

Tank circuit A parallel resonant circuit.

▪ FORMULAS

(14–1)	$X_T = \lvert X_L - X_C \rvert$	Total series reactance (absolute value)
(14–2)	$Z_T = \sqrt{R^2 + (X_L - X_C)^2}$	Total series *RLC* impedance
(14–3)	$\theta = \tan^{-1}\left(\dfrac{X_T}{R}\right)$	Series *RLC* phase angle

(14–4)	$X_L = X_C$	Condition for series resonance
(14–5)	$Z_r = R$	Series resonant impedance
(14–6)	$f_r = \dfrac{1}{2\pi\sqrt{LC}}$	Series resonant frequency
(14–7)	$BW = f_2 - f_1$	Bandwidth
(14–8)	$\text{dB} = 20 \log\left(\dfrac{V_{out}}{V_{in}}\right)$	Decibel formula for voltage ratio
(14–9)	$\text{dB} = 10 \log\left(\dfrac{P_{out}}{P_{in}}\right)$	Decibel formula for power ratio
(14–10)	$Q = \dfrac{X_L}{R}$	Series resonant Q
(14–11)	$BW = \dfrac{f_r}{Q}$	Bandwidth
(14–12)	$Y = \sqrt{G^2 + (B_C - B_L)^2}$	Parallel *RLC* admittance
(14–13)	$\theta = \tan^{-1}\left(\dfrac{B_T}{G}\right)$	Parallel *RLC* phase angle
(14–14)	$I_T = \sqrt{I_R^2 + (I_C - I_L)^2}$	Total parallel *RLC* current
(14–15)	$\theta = \tan^{-1}\left(\dfrac{I_{CL}}{I_R}\right)$	Parallel *RLC* phase angle
(14–16)	$L_{eq} = L\left(\dfrac{Q^2 + 1}{Q^2}\right)$	Equivalent parallel inductance
(14–17)	$R_{p(eq)} = R_W(Q^2 + 1)$	Equivalent parallel resistance
(14–18)	$f_r = \dfrac{1}{2\pi\sqrt{LC}}$	Ideal parallel resonant frequency
(14–19)	$I_T = |I_C - I_L|$	Total parallel *LC* current (absolute value)
(14–20)	$Z_r = R_W(Q^2 + 1)$	Impedance at parallel resonance
(14–21)	$I_T = \dfrac{V_s}{Z_r}$	Total current at parallel resonance
(14–22)	$f_r = \dfrac{\sqrt{1 - (R_W^2 C/L)}}{2\pi\sqrt{LC}}$	Parallel resonant frequency (exact)
(14–23)	$Q_o = \dfrac{R_{p(tot)}}{X_{L(eq)}}$	Parallel *RLC* Q

▪ **SELF-TEST**

1. The total reactance of a series *RLC* circuit at resonance is
   (a) zero   (b) equal to the resistance   (c) infinity   (d) capacitive

2. The phase angle of a series *RLC* circuit at resonance is
   (a) −90°   (b) +90°   (c) 0°   (d) is dependent on the reactance

3. The impedance at the resonant frequency of a series *RLC* circuit with $L = 15$ mH, $C = 0.015$ μF, and $R_W = 80$ Ω is
   (a) 15 kΩ   (b) 80 Ω   (c) 30 Ω   (d) 0 Ω

4. In a series *RLC* circuit that is operating below the resonant frequency, the current
   (a) is in phase with the applied voltage   (b) lags the applied voltage
   (c) leads the applied voltage

5. If the value of C in a series *RLC* circuit is increased, the resonant frequency
   (a) is not affected   (b) increases   (c) remains the same   (d) decreases

6. In a certain series resonant circuit, $V_C = 150$ V, $V_L = 150$ V, and $V_R = 50$ V. The value of the source voltage is
   (a) 150 V   (b) 300 V   (c) 50 V   (d) 350 V

7. A certain series resonant band-pass filter has a bandwidth of 1 kHz. If the existing coil is replaced with one having a lower value of Q, the bandwidth will
(a) increase   (b) decrease   (c) remain the same   (d) be more selective

8. At frequencies below resonance in a parallel RLC circuit, the current
(a) leads the source voltage   (b) lags the source voltage
(c) is in phase with the source voltage

9. The total current into the L and C branches of a parallel circuit at resonance is ideally
(a) maximum   (b) low   (c) high   (d) zero

10. To tune a parallel resonant circuit to a lower frequency, the capacitance should be
(a) increased   (b) decreased   (c) left alone   (d) replaced with inductance

11. The resonant frequency of a parallel circuit is the same as a series circuit when
(a) the Q is very low   (b) the Q is very high   (c) there is no resistance

12. If the resistance in parallel with a parallel resonant filter is reduced, the bandwidth
(a) disappears   (b) decreases   (c) becomes sharper   (d) increases

▪ BASIC PROBLEMS

SECTION 14–1   Impedance and Phase Angle of Series *RLC* Circuits

1. A certain series RLC circuit has the following values: $R = 10\ \Omega$, $C = 0.05\ \mu F$, and $L = 5\ mH$. Determine the impedance and phase angle. What is the total reactance? $f = 5\ kHz$.

2. Find the impedance in Figure 14–65.

3. If the frequency of the source voltage in Figure 14–65 is doubled from the value that produces the indicated reactances, how does the impedance change?

SECTION 14–2   Analysis of Series *RLC* Circuits

4. For the circuit in Figure 14–65, find I_T, V_R, V_L, and V_C.

5. Sketch the voltage phasor diagram for the circuit in Figure 14–65.

6. Analyze the circuit in Figure 14–66 for the following ($f = 25\ kHz$):
(a) I_T (b) P_{true} (c) P_r (d) P_a

FIGURE 14–65

FIGURE 14–66

SECTION 14–3   Series Resonance

7. Find X_L, X_C, Z, and I at the resonant frequency in Figure 14–67.

8. A certain series resonant circuit has a maximum current of 50 mA and a V_L of 100 V. The applied voltage is 10 V. What is Z? What are X_L and X_C?

9. For the RLC circuit in Figure 14–68, determine the resonant frequency and the cutoff frequencies.

10. What is the value of the current at the half-power points in Figure 14–68?

FIGURE 14–67

FIGURE 14–68

SECTION 14–4 Series Resonant Filters

11. Determine the resonant frequency for each filter in Figure 14–69. Are these filters band-pass or band-stop types?

12. Assuming that the coils in Figure 14–69 have a winding resistance of 10 Ω, find the bandwidth for each filter.

FIGURE 14–69

(a)   (b)

13. Determine the resonant frequency and bandwidth for each filter in Figure 14–70.

(a)   (b)

FIGURE 14–70

SECTION 14–5 Parallel *RLC* Circuits

14. Find the total impedance of the circuit in Figure 14–71.

15. Is the circuit in Figure 14–71 capacitive or inductive? Explain.

16. For the circuit in Figure 14–71, find all the currents and voltages.

FIGURE 14–71

17. Find the total impedance for the circuit in Figure 14–72.

FIGURE 14–72

SECTION 14–6   Parallel Resonance

18. What is the impedance of an ideal parallel resonant circuit (no resistance in either branch)?

19. Find Z at resonance and f_r for the tank circuit in Figure 14–73.

20. How much current is drawn from the source in Figure 14–73 at resonance? What are the inductive current and the capacitive current at the resonant frequency?

FIGURE 14–73

SECTION 14–7   Parallel Resonant Filters

21. At resonance, $X_L = 2$ kΩ and $R_W = 25$ Ω in a parallel resonant band-pass filter. The resonant frequency is 5 kHz. Determine the bandwidth.

22. If the lower cutoff frequency is 2400 Hz and the upper cutoff frequency is 2800 Hz, what is the bandwidth?

23. In a certain resonant circuit, the power at resonance is 2.75 W. What is the power at the lower and upper cutoff frequencies?

24. What values of L and C should be used in a tank circuit to obtain a resonant frequency of 8 kHz? The bandwidth must be 800 Hz. The winding resistance of the coil is 10 Ω.

25. A parallel resonant circuit has a Q of 50 and a BW of 400 Hz. If Q is doubled, what is the bandwidth for the same f_r?

■ ADVANCED PROBLEMS

26. For each following case, express the voltage ratio in decibels:
   (a) $V_{in} = 1$ V, $V_{out} = 1$ V   **(b)** $V_{in} = 5$ V, $V_{out} = 3$ V
   (c) $V_{in} = 10$ V, $V_{out} = 7.07$ V   **(d)** $V_{in} = 25$ V, $V_{out} = 5$ V

27. Find the current through each component in Figure 14–74. Find the voltage across each component.

FIGURE 14–74

28. Determine whether there is a value of C that will make $V_{ab} = 0$ V in Figure 14–75. If not, explain.

29. If the value of C is 0.2 μF, how much current flows through each branch in Figure 14–75? What is the total current?

30. Determine the resonant frequencies in Figure 14–76 and find V_{out} at each resonant frequency.

FIGURE 14–75   **FIGURE 14–76**

31. Design a band-pass filter using a parallel resonant circuit to meet the following specifications: $BW = 500$ Hz, $Q = 40$, $I_{C(max)} = 20$ mA, $V_{C(max)} = 2.5$ V.

32. Design a circuit in which the following series resonant frequencies are switch-selectable: 500 kHz, 1000 kHz, 1500 kHz, 2000 kHz.

33. Design a parallel-resonant network using a single coil and switch-selectable capacitors to produce the following resonant frequencies: 8 MHz, 9 MHz, 10 MHz, and 11 MHz. Assume a 10 μH coil with a winding resistance of 5 Ω.

▪ **ANSWERS TO SECTION REVIEWS**

Section 14–1

1. The circuit is inductive if $X_L > X_C$ and capacitive if $X_C > X_L$.

2. $X_T = X_C - X_L = 70 \, \Omega$; The circuit is capacitive.

3. $Z = \sqrt{R^2 + X_T^2} = 83.2 \, \Omega$; $\theta = \tan^{-1}(X_T/R) = 57.3°$; Current is leading V_s.

Section 14–2

1. $V_s = \sqrt{V_R^2 + (V_C - V_L)^2} = 38.4$ V

2. Current leads V_s because the circuit is capacitive.

3. $X_T = X_C - X_L = 6 \, \Omega$

Section 14–3

1. Series resonance occurs when $X_L = X_C$.

2. The current is maximum because the impedance is minimum.

3. $f_r = 1/2\pi\sqrt{LC} = 159$ kHz

4. It is capacitive because $X_C > X_L$.

Section 14–4

1. $V_{out} = 0.707(15 \text{ V}) = 10.6$ V

2. $BW = f_r/Q = 10$ kHz

3. Current is maximum; output voltage is minimum.

Section 14–5

1. $I_R = V_s/R = 80$ mA; $I_C = V_s/X_C = 120$ mA; $I_L = V_s/X_L = 240$ mA

2. The circuit is inductive ($X_L < X_C$).

3. $L_{eq} = L[(Q^2 + 1)/Q^2] = 20.1$ mH; $R_{eq} = R_W(Q^2 + 1) = 1598 \, \Omega$

Section 14–6

1. The impedance is maximum.
2. The current is minimum.
3. At ideal parallel resonance, $X_C = X_L = 1.5$ kΩ.
4. $f_r = \sqrt{1 - R_W^2\, C/L}/2\pi\sqrt{LC} = 225$ kHz; $Z_r = R_W(Q^2 + 1) = 1249$ MΩ
5. $f_r = 1/2\pi\sqrt{LC} = 22.5$ kHz
6. $f_r = \sqrt{Q^2/(Q^2 + 1)}/2\pi\sqrt{LC} = 20.9$ kHz
7. $Z_r = R_W(Q^2 + 1) = 8.02$ kΩ

Section 14–7

1. The bandwidth can be increased by reducing the parallel resistance.
2. f_r changes to 4.47 kHz.
3. $Z_r = R_W(Q^2 + 1) = 47.0$ kΩ

Section 14–8

1. A tuned filter is used to select a narrow band of frequencies.
2. A wave trap is a band-stop filter.
3. Several variable capacitors (or inductors) whose values can be varied simultaneously with a common control is an example of ganged tuning.

Section 14–9

1. $f_{c(1)} = 8$ kHz; $f_{c(2)} = 11.6$ kHz
2. The circuit or component values cannot be determined from the data given.

■ **ANSWERS TO RELATED EXERCISES FOR EXAMPLES**

14–1	12.3 Ω; 63.0°
14–2	4.64 kΩ
14–3	Current increases, then decreases.
14–4	More capacitive
14–5	22.5 kHz
14–6	2.27 A
14–7	Z increases; Z increases
14–8	No
14–9	200 kHz
14–10	−14 dB
14–11	11.7
14–12	319 Hz
14–13	V_{out} increases; V_{out} increases
14–14	Inductive
14–15	Increase
14–16	$R_{p(\text{eq})} = 25$ kΩ; $L_{\text{eq}} = 5$ mH; $C = 0.02$ μF
14–17	$f_r = 3.39$ kHz; $I_L = 6.85$ mA; $I_C = 6.85$ mA; $I_T = 0$ A
14–18	79.9 kΩ
14–19	The differences are negligible.
14–20	7.07 V
14–21	530 kHz
14–22	25
14–23	Q decreases with less load resistance.
14–24	The Q is less than 10.

15

PULSE RESPONSE OF REACTIVE CIRCUITS

In Chapters 12 and 13, the frequency (sinusoidal) response of *RC* and *RL* circuits was covered. In this chapter, the response of *RC* and *RL* filters to pulse inputs is examined. With pulse inputs, the time response of the circuits is of primary importance. In the areas of pulse and digital circuits, the technician is often concerned with how a circuit responds over an interval of time to rapid changes in voltage or current. The relationship of the circuit time constant to the input pulse characteristics, such as pulse width and period, determines the wave shapes of the circuit voltages.

Before starting this chapter, you should review the material in Sections 9–5 and 10–5. Understanding exponential changes in voltages and currents in capacitors and inductors is crucial to the study of pulse response. Exponential formulas that were given in Chapter 9 are used throughout this chapter.

Integrator and *differentiator,* terms used throughout this coverage, refer to mathematical functions that are approximated by these circuits under certain conditions. Integration is an averaging process, and differentiation is a process for establishing a rate of change of a function. We will not deal with the purely mathematical aspects at this level of coverage.

■ CHAPTER OBJECTIVES

☐ Explain the operation of an *RC* integrator

☐ Analyze an *RC* integrator with a single input pulse

☐ Analyze an *RC* integrator with repetitive input pulses

☐ Analyze an *RC* differentiator with a single input pulse

☐ Analyze an *RC* differentiator with repetitive input pulses

☐ Analyze the operation of an *RL* integrator

☐ Analyze the operation of an *RL* differentiator

☐ Discuss some basic applications of integrators and differentiators

☐ Troubleshoot integrators and differentiators

■ APPLICATION ASSIGNMENT

For the application assignment in this chapter, you will breadboard a time-delay circuit and determine the component values required to meet certain specifications. You will also determine how to test the circuit for proper operation using a pulse generator and an oscilloscope. After studying this chapter, you should be able to complete the application assignment in Section 15–10.

15–1 ▪ THE *RC* INTEGRATOR

A series RC circuit in which the output voltage is taken across the capacitor is known as an integrator in terms of pulse response. Recall that in terms of frequency response, it is a low-pass filter. The term, integrator, is derived from a mathematical function which this type of circuit approximates under certain conditions.

After completing this section, you should be able to

▪ **Explain the operation of an *RC* integrator**
  ☐ Describe how the capacitor charges and discharges
  ☐ Explain how a capacitor reacts to an instantaneous change in voltage or current
  ☐ Describe the basic output voltage waveform

The Capacitor Charges and Discharges with a Pulse Input

When a pulse generator is connected to the input of an *RC* **integrator,** as shown in Figure 15–1, the capacitor will charge and discharge in response to the pulses. When the input goes from its low level to its high level, the capacitor charges toward the high level of the pulse through the resistor. This changing action is analogous to connecting a battery through a switch to the *RC* network, as illustrated in Figure 15–2(a). When the pulse goes from its high level back to its low level, the capacitor discharges back through the source. The resistance of the source is assumed to be negligible compared to *R*. This discharging action is analogous to replacing the source with a closed switch, as illustrated in Figure 15–2(b).

The capacitor will charge and discharge following an exponential curve. Its rate of charging and discharging, of course, depends on the *RC* **time constant** ($\tau = RC$).

FIGURE 15–1
An RC integrating circuit.

(a) When the input pulse goes high, the source effectively acts as a battery in series with a switch closure, thereby charging the capacitor.

(b) When the input pulse goes back low, the source effectively acts as a closed switch, providing a discharge path for the capacitor.

FIGURE 15–2
The equivalent action when a pulse source charges and discharges the capacitor.

For an ideal pulse, each edge is considered to occur instantaneously. Two basic rules of capacitor behavior help in understanding the **pulse response** of *RC* circuits:

1. The capacitor appears as a short to an instantaneous change in current and as an open to dc.

2. The voltage across the capacitor cannot change instantaneously—it can change only exponentially.

The Capacitor Voltage

In an *RC* integrator, the output is the capacitor voltage. The capacitor charges during the time that the input pulse is high. If the pulse is at its high level long enough, the capacitor will fully charge to the voltage amplitude of the pulse, as illustrated in Figure 15–3. The capacitor discharges during the time that the pulse is low. If the low time between pulses is long enough, the capacitor will fully discharge to zero, as shown in the figure. Then when the next pulse occurs, it will charge again.

FIGURE 15–3

Illustration of a capacitor fully charging and discharging in response to a pulse input.

15–2 ■ SINGLE-PULSE RESPONSE OF *RC* INTEGRATORS

From the previous section, you have a general idea of how an RC integrator responds to a pulse input. In this section, the response to a single pulse is examined in detail.

After completing this section, you should be able to

■ **Analyze an *RC* integrator with a single input pulse**
  □ Discuss the importance of the circuit time constant
  □ Define *transient time*
  □ Determine the response when the pulse width is equal to or greater than five time constants
  □ Determine the response when the pulse width is less than five time constants

Two conditions of pulse response must be considered:

1. When the input pulse width (t_W) is equal to or greater than five time constants ($t_W \geq 5\tau$)

2. When the input pulse width is less than five time constants ($t_W < 5\tau$)

Recall that five time constants is accepted as the time a capacitor needs to fully charge or fully discharge; this time is often called the **transient time.**

When the Pulse Width Is Equal to or Greater Than Five Time Constants

The capacitor will fully charge if the pulse width is equal to or greater than five time constants (5τ). This condition is expressed as $t_W \geq 5\tau$. At the end of the pulse, the capacitor fully discharges back through the source.

Figure 15–4 illustrates the output waveforms for various *RC* time constants and a fixed input pulse width. Notice that the shape of the output pulse approaches that of the input as the transient time is made small compared to the pulse width. In each case the output reaches the full amplitude of the input.

FIGURE 15–4

Variation of an integrator's output pulse shape with time constant. The shaded areas indicate when the capacitor is charging and discharging.

Figure 15–5 shows how a fixed time constant and a variable input pulse width affect the integrator output. Notice that as the pulse width is increased, the shape of the output pulse approaches that of the input. Again, this means that the transient time is short compared to the pulse width.

(a) $t_W = 5\tau$

(b) $t_W = 10\tau$

(c) $t_W = 20\tau$

(d) $t_W = 40\tau$

FIGURE 15–5

Variation of an integrator's output pulse shape with input pulse width (the time constant is fixed). Black is input and color is output.

When the Pulse Width Is Less Than Five Time Constants

Now let's examine the case in which the width of the input pulse is less than five time constants of the integrator. This condition is expressed as $t_W < 5\tau$.

As before, the capacitor charges for the duration of the pulse. The pulse width is the time the capacitor has for charging. However, because the pulse width is less than the time the capacitor needs to fully charge (5τ), the output voltage will *not* reach the full input voltage before the end of the pulse. The capacitor only partially charges, as illustrated in Figure 15–6 for several values of *RC* time constants. Notice that for longer time constants, the output reaches a lower voltage because the capacitor cannot charge as much during the pulse width. Of course, in our examples with a single-pulse input, the capacitor fully discharges after the pulse ends.

FIGURE 15–6

Capacitor voltage for various time constants which are longer than the input pulse width. Black is input and color is output.

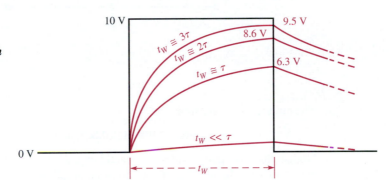

When the time constant is much greater than the input pulse width, the capacitor charges very little, and, as a result, the output voltage becomes almost negligible, as indicated in Figure 15–6.

Figure 15–7 illustrates the effect of reducing the input pulse width for a fixed time constant value. As the width is reduced, the output voltage becomes smaller because the capacitor has less time to charge. However, it takes the capacitor the same length of time (5τ) to discharge back to zero for each condition after the pulse is removed.

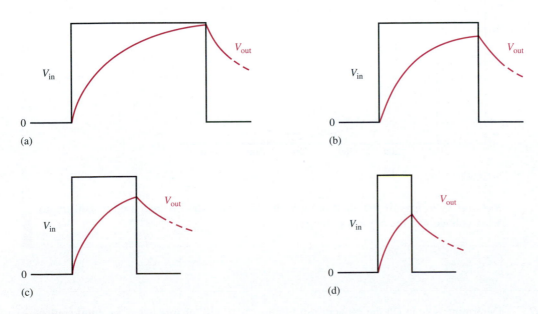

FIGURE 15–7

The capacitor charges less and less as the input pulse width is reduced. The time constant is fixed.

EXAMPLE 15–1

A single 10 V pulse with a width of 100 μs is applied to the integrator in Figure 15–8.
(a) To what voltage will the capacitor charge?
(b) How long will it take the capacitor to discharge?
(c) Sketch the output voltage.

FIGURE 15–8

Solution
(a) The circuit time constant is

$$\tau = RC = (100 \text{ k}\Omega)(0.001 \ \mu\text{F}) = 100 \ \mu\text{s}$$

Notice that the pulse width is exactly equal to one time constant. Thus, the capacitor will charge 63% of the full input amplitude in one time constant, so the output will reach a maximum voltage of 6.3 V.

(b) The capacitor discharges back through the source when the pulse ends. The total discharge time is

$$5\tau = 5(100 \ \mu\text{s}) = 500 \ \mu\text{s}$$

(c) The output charging and discharging curve is shown in Figure 15–9.

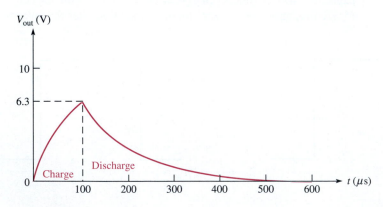

FIGURE 15–9

Related Exercise If the input pulse width in Figure 15–8 is increased to 200 μs, to what voltage will the capacitor charge?

EXAMPLE 15–2

Determine how much the capacitor in Figure 15–10 will charge when the single pulse is applied to the input.

FIGURE 15–10

Solution Calculate the time constant:

$$\tau = RC = (2.2 \text{ k}\Omega)(1 \text{ }\mu\text{F}) = 2.2 \text{ ms}$$

Because the pulse width is 5 ms, the capacitor charges for approximately 2.27 time constants (5 ms/2.2 ms = 2.27). Use the exponential formula, Equation (9–18), to find the voltage to which the capacitor will charge. With $V_F = 25$ V and $t = 5$ ms, the calculation is done as follows:

$$v = V_F(1 - e^{-t/RC})$$
$$= (25 \text{ V})(1 - e^{-5/2.2}) = (25 \text{ V})(1 - e^{-2.27})$$
$$= (25 \text{ V})(1 - 0.103) = (25 \text{ V})(0.897) = 22.4 \text{ V}$$

These calculations show that the capacitor charges to 22.4 V during the 5 ms duration of the input pulse. It will discharge back to zero when the pulse goes back to zero.

The calculator sequence for v is

Related Exercise Determine how much *C* will charge if the pulse width is increased to 10 ms.

SECTION 15–2 REVIEW

1. When an input pulse is applied to an *RC* integrator, what condition must exist in order for the output voltage to reach the full amplitude of the input?
2. For the circuit in Figure 15–11, which has a single input pulse, find the maximum output voltage and determine how long the capacitor will discharge.

FIGURE 15–11

3. For Figure 15–11, sketch the approximate shape of the output voltage with respect to the input pulse.

4. If the integrator time constant equals the input pulse width, will the capacitor fully charge?

5. Describe the condition in an integrator under which the output voltage has the approximate shape of a rectangular input pulse.

15–3 ■ REPETITIVE-PULSE RESPONSE OF *RC* INTEGRATORS

In the last section, you learned how an RC integrator responds to a single-pulse input. These basic ideas are extended in this section to include the integrator response to repetitive pulses. In electronic systems, you will encounter repetitive-pulse waveforms much more often than single pulses. However, an understanding of the integrator's response to single pulses is necessary in order to understand how these circuits respond to repeated pulses.

After completing this section, you should be able to

■ **Analyze an *RC* integrator with repetitive input pulses**
  □ Determine the response when the capacitor does not fully charge or discharge
  □ Define *steady state*
  □ Describe the effect of a change in time constant on circuit response

If a **periodic** pulse waveform is applied to an *RC* integrator, as shown in Figure 15–12, *the output waveshape depends on the circuit time constant and the frequency and duty cycle of the input pulses.* The capacitor, of course, charges and discharges in response to a pulse input. The amount of charge and discharge of the capacitor depends both on the circuit time constant and on the input frequency and duty cycle, as mentioned.

If the pulse width and the time between pulses are each equal to or greater than five time constants, the capacitor will fully charge and fully discharge during each period of the input waveform. This case is shown in Figure 15–12.

FIGURE 15–12
RC integrator with a repetitive-pulse waveform input.

When the Capacitor Does Not Fully Charge and Discharge

When the pulse width and the time between pulses are shorter than 5 time constants, as illustrated in Figure 15–13 for the case of a square wave, the capacitor will *not* completely charge or discharge. We will now examine the effects of this situation on the output voltage of the *RC* integrator.

For convenience, let's take an example of an *RC* integrator with a charging and discharging time constant equal to the pulse width of a 10 V square wave input, as in Figure 15–14. This choice will simplify the analysis and will demonstrate the basic action of the

FIGURE 15–13

Waveform that does not allow full charge or discharge of the integrator capacitor.

FIGURE 15–14

Integrator with a square wave input having a period equal to two time constants (T = 2τ).

integrator under these conditions. At this point, we really do not care what the exact time constant value is because we know that an *RC* circuit charges 63% during one time constant interval.

We will assume that the capacitor in Figure 15–14 begins initially uncharged and examine the output voltage on a pulse-by-pulse basis. The results of this analysis are shown in Figure 15–15.

First Pulse During the first pulse, the capacitor charges. The output voltage reaches 6.3 V (63% of 10 V), as shown in Figure 15–15.

Between First and Second Pulses The capacitor discharges, and the voltage decreases to 37% of the voltage at the beginning of this interval: 0.37(6.3 V) = 2.33 V.

Second Pulse The capacitor voltage begins at 2.33 V and increases 63% of the way to 10 V. The total charging range is 10 V − 2.33 V = 7.67 V. The capacitor voltage will increase an additional 63% of 7.67 V, which is 4.83 V. Thus, at the end of the second pulse, the output voltage is 2.33 V + 4.83 V = 7.16 V, as indicated in Figure 15–15. Notice that the average is building up.

Between Second and Third Pulses The capacitor discharges during this time, and therefore the voltage decreases to 37% of the initial voltage by the end of the second pulse: 0.37(7.16 V) = 2.65 V.

Third Pulse At the start of the third pulse, the capacitor voltage is 2.65 V. The capacitor charges 63% of the way from 2.65 V to 10 V: 0.63(10 V − 2.65 V) = 4.63 V. Therefore, the voltage at the end of the third pulse is 2.65 V + 4.63 V = 7.28 V.

Between Third and Fourth Pulses The voltage during this interval decreases due to capacitor discharge. It will decrease to 37% of its value by the end of the third pulse. The final voltage in this interval is 0.37(7.28 V) = 2.69 V.

Fourth Pulse At the start of the fourth pulse, the capacitor voltage is 2.69 V. The voltage increases by 0.63(10 V − 2.69 V) = 4.605 V. Therefore, at the end of the fourth pulse, the capacitor voltage is 2.69 V + 4.605 V = 7.295 V. Notice that the values are leveling off as the pulses continue.

Between Fourth and Fifth Pulses   Between these pulses, the capacitor voltage drops to $0.37(7.295 \text{ V}) = 2.7 \text{ V}$.

Fifth Pulse   During the fifth pulse, the capacitor charges $0.63(10 \text{ V} - 2.7 \text{ V}) = 4.6 \text{ V}$. Since it started at 2.7 V, the voltage at the end of the pulse is $2.7 \text{ V} + 4.6 \text{ V} = 7.3 \text{ V}$.

FIGURE 15–15
Input and output for the initially uncharged integrator in Figure 15–14.

Steady-State Response

In the preceding discussion, the output voltage gradually built up and then began leveling off. It takes approximately 5τ for the output voltage to build up to a constant average value, regardless of the number of pulses that may occur during that time interval. This interval is the transient time of the circuit. Once the output voltage reaches the average value of the input voltage, a **steady-state** condition is reached, which continues as long as the periodic input continues. This condition is illustrated in Figure 15–16 based on the values obtained in the preceding discussion.

The transient time for our example circuit is the time from the beginning of the first pulse to the end of the third pulse. The reason for this interval is that the capacitor voltage at the end of the third pulse is 7.28 V, which is about 99% of the final voltage.

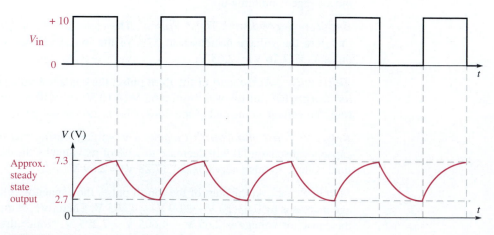

FIGURE 15–16
Output reaches steady state after 5τ.

The Effect of an Increase in Time Constant

What happens to the output voltage if the *RC* time constant of the integrator is increased with a variable resistor, as indicated in Figure 15–17? As the time constant is increased, the capacitor charges less during a pulse and discharges less between pulses. The result is a smaller fluctuation in the output voltage for increasing values of time constant, as shown in Figure 15–18.

FIGURE 15–17
Integrator with a variable time constant.

FIGURE 15–18
Effect of longer time constants on the output of an integrator ($\tau_3 > \tau_2 > \tau_1$).

As the time constant becomes extremely long compared to the pulse width, the output voltage approaches a constant dc voltage, as shown in Figure 15–18(c). This value is the average value of the input. For a square wave, it is one-half the amplitude.

EXAMPLE 15–3
Determine the output voltage waveform for the first two pulses applied to the integrator circuit in Figure 15–19. Assume that the capacitor is initially uncharged.

FIGURE 15–19

V$_{in}$ 5 V, 0, 10 µs, 15 µs, 10 µs

R 4.7 kΩ, *C* 0.01 µF, *V*$_{out}$

Solution   First calculate the circuit time constant:

$$\tau = RC = (4.7 \text{ k}\Omega)(0.01 \text{ µF}) = 47 \text{ µs}$$

Obviously, the time constant is much longer than the input pulse width or the interval between pulses (notice that the input is not a square wave). Thus, in this case, the exponential formulas must be applied, and the analysis is relatively difficult. Follow the solution carefully.

1. *Calculation for first pulse:* Use Equation (9–18) because C is charging. Note that V_F is 5 V, and t equals the pulse width of 10 μs. Therefore,

$$v_C = V_F(1 - e^{-t/RC}) = (5\text{ V})(1 - e^{-10\mu s/47\mu s})$$
$$= (5\text{ V})(1 - 0.808) = 960\text{ mV}$$

This result is plotted in Figure 15–20(a).

2. *Calculation for interval between first and second pulse:* Use Equation (9–19) because C is discharging. Note that V_i is 960 mV because C begins to discharge from this value at the end of the first pulse. The discharge time is 15 μs. Therefore,

$$v_C = V_i e^{-t/RC} = (960\text{ mV})^{-15\mu s/47\mu s}$$
$$= (960\text{ mV})(0.727) = 698\text{ mV}$$

This result is shown in Figure 15–20(b).

3. *Calculation for second pulse:* At the beginning of the second pulse, the output voltage is 698 mV. During the second pulse, the capacitor will again charge. In this case it does not begin at zero volts. It already has 698 mV from the previous charge and discharge. To handle this situation, you must use the general formula of Equation (9–16):

$$v = V_F + (V_i - V_F)e^{-t/\tau}$$

Using this equation, calculate the voltage across the capacitor at the end of the second pulse as follows:

$$v_C = V_F + (V_i - V_F)e^{-t/RC}$$
$$= 5\text{ V} + (698\text{ mV} - 5\text{ V})e^{-10\mu s/47\mu s}$$
$$= 5\text{ V} + (-4.30\text{ V})(0.808) = 5\text{ V} - 3.47\text{ V} = 1.53\text{ V}$$

This result is shown in Figure 15–20(c).

FIGURE 15–20

(a)

(b)

(c)

Notice that the output waveform builds up on successive input pulses. After approximately 5τ, it will reach its steady state and will fluctuate between a constant maximum and a constant minimum, with an average equal to the average value of the input. You can demonstrate this pattern by carrying the analysis in this example further.

Related Exercise Determine V_{out} at the beginning of the third pulse.

SECTION 15–3 REVIEW

1. What conditions allow an *RC* integrator capacitor to fully charge and discharge when a periodic pulse waveform is applied to the input?
2. What will the output waveform look like if the circuit time constant is extremely small compared to the pulse width of a square wave input?
3. When 5τ is greater than the pulse width of an input square wave, the time required for the output voltage to build up to a constant average value is called _____ .
4. Define *steady-state response*.
5. What does the average value of the output voltage of an integrator equal during steady state?

15–4 ▪ SINGLE-PULSE RESPONSE OF *RC* DIFFERENTIATORS

A series RC circuit in which the output voltage is taken across the resistor is known as a differentiator in terms of pulse response. Recall that in terms of frequency response, a series RC circuit is a high-pass filter. The term, differentiator, is derived from a mathematical function which this type of circuit approximates under certain conditions.

After completing this section, you should be able to

▪ **Analyze an *RC* differentiator with a single input pulse**
  ☐ Describe the response at the rising edge of the input pulse
  ☐ Determine the response during and at the end of a pulse for various pulse width–time constant relationships

---

Figure 15–21 shows an *RC* **differentiator** with a pulse input. The same action occurs in the differentiator as in the integrator, except the output voltage is taken across the resistor rather than the capacitor. The capacitor charges exponentially at a rate depending on the *RC* time constant. The shape of the differentiator's resistor voltage is determined by the charging and discharging action of the capacitor.

FIGURE 15–21
An RC differentiating circuit.

Pulse Response

To understand how the output voltage is shaped by a differentiator, you must consider the following:

1. The response to the rising pulse edge
2. The response between the rising and falling edges
3. The response to the falling pulse edge

Response to the Rising Edge of the Input Pulse Assume that the capacitor is initially uncharged prior to the rising pulse edge. Prior to the pulse, the input is zero volts. Thus, there are zero volts across the capacitor and also zero volts across the resistor, as indicated in Figure 15–22(a).

Now asssume that a 10 V pulse is applied to the input. When the rising edge occurs, point A goes to +10 V. Recall that the voltage across a capacitor cannot change instantaneously, and thus the capacitor appears instantaneously as a short. Therefore, if point A

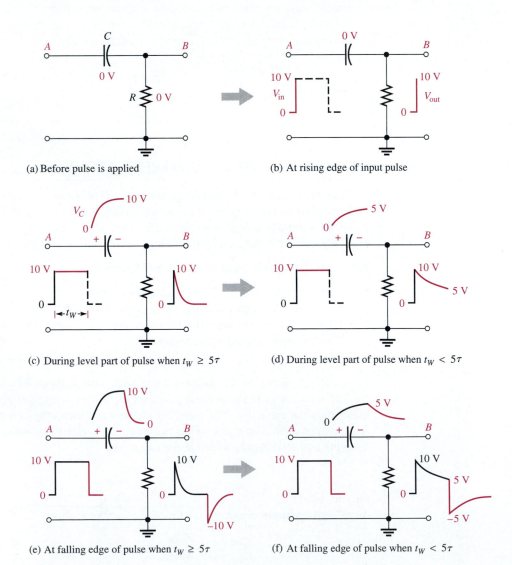

(a) Before pulse is applied

(b) At rising edge of input pulse

(c) During level part of pulse when $t_W \geq 5\tau$

(d) During level part of pulse when $t_W < 5\tau$

(e) At falling edge of pulse when $t_W \geq 5\tau$

(f) At falling edge of pulse when $t_W < 5\tau$

FIGURE 15–22

Examples of the response of a differentiator to a single input pulse under two conditions:
$t_W \geq 5\tau$ and $t_W < 5\tau$.

goes instantly to +10 V, then point *B must* also go instantly to +10 V, keeping the capacitor voltage zero for the instant of the rising edge. The capacitor voltage is the voltage from point *A* to point *B*.

The voltage at point *B* with respect to ground is the voltage across the resistor (and the output voltage). Thus, the output voltage suddenly goes to +10 V in response to the rising pulse edge, as indicated in Figure 15–22(b).

Response During Pulse When $t_W \geq 5\tau$ While the pulse is at its high level between the rising edge and the falling edge, the capacitor is charging. When the pulse width is equal to or greater than five time constants ($t_W \geq 5\tau$), the capacitor has time to fully charge.

As the voltage across the capacitor builds up exponentially, the voltage across the resistor decreases exponentially until it reaches zero volts at the time the capacitor reaches full charge (+10 V in this case). This decrease in the resistor voltage occurs because the sum of the capacitor voltage and the resistor voltage at any instant must be equal to the applied voltage, in compliance with Kirchhoff's voltage law ($v_C + v_R = v_{in}$). This part of the response is illustrated in Figure 15–22(c).

Response During Pulse When $t_W < 5\tau$ When the pulse width is less than five time constants ($t_W < 5\tau$), the capacitor does not have time to fully charge. Its partial charge depends on the relation of the time constant and the pulse width.

Because the capacitor does not reach the full +10 V, the resistor voltage will not reach zero volts by the end of the pulse. For example, if the capacitor charges to +5 V during the pulse interval, the resistor voltage will decrease to +5 V, as illustrated in Figure 15–22(d).

Response to Falling Edge When $t_W \geq 5\tau$ Let's first examine the case in which the capacitor is fully charged at the end of the pulse ($t_W \geq 5\tau$). Refer to Figure 15–22(e). On the falling edge, the input pulse suddenly goes from +10 V back to zero. An instant before the falling edge, the capacitor is charged to 10 V, so point *A* is +10 V and point *B* is 0 V. Since the voltage across a capacitor cannot change instantaneously, when point *A* makes a transition from +10 V to zero on the falling edge, point *B* *must* also make a 10 V transition from zero to −10 V. This keeps the voltage across the capacitor at 10 V for the instant of the falling edge.

The capacitor now begins to discharge exponentially. As a result, the resistor voltage goes from −10 V to zero in an exponential curve, as indicated in Figure 15–22(e).

Response to Falling Edge When $t_W < 5\tau$ Next, let's examine the case in which the capacitor is only partially charged at the end of the pulse ($t_W < 5\tau$). For example, if the capacitor charges to +5 V, the resistor voltage at the instant before the falling edge is also +5 V, because the capacitor voltage plus the resistor voltage must add up to +10 V, as illustrated in Figure 15–22(d).

When the falling edge occurs, point *A* goes from +10 V to zero. As a result, point *B* goes from +5 V to −5 V, as illustrated in Figure 15–22(f). This decrease occurs, of course, because the capacitor voltage cannot change at the instant of the falling edge. Immediately after the falling edge, the capacitor begins to discharge to zero. As a result, the resistor voltage goes from −5 V to zero, as shown.

Summary of Differentiator Response to a Single Pulse

Perhaps a good way to summarize this section is to look at the general output waveforms of a differentiator as the time constant is varied from one extreme, when 5τ is much less than the pulse width, to the other extreme, when 5τ is much greater than the pulse width. These situations are illustrated in Figure 15–23. In part (a), the output consists of narrow positive and negative "spikes." In part (f), the output approaches the shape of the input. Various conditions between these extremes are illustrated in parts (b) through (e).

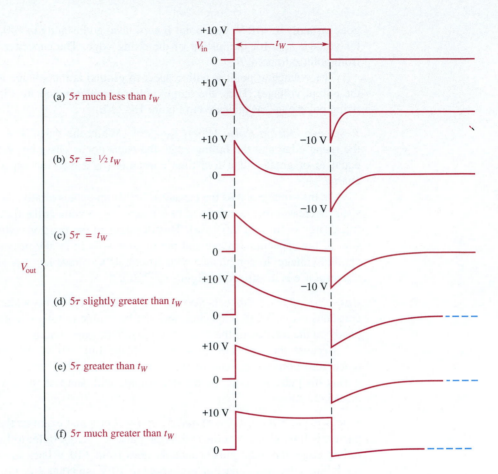

(a) 5τ much less than t_W

(b) $5\tau = \frac{1}{2} t_W$

(c) $5\tau = t_W$

V_{out}

(d) 5τ slightly greater than t_W

(e) 5τ greater than t_W

(f) 5τ much greater than t_W

FIGURE 15–23

Effects of a change in time constant on the shape of the output voltage of a differentiator.

EXAMPLE 15–4   Sketch the output voltage for the circuit in Figure 15–24.

FIGURE 15–24

Solution   First calculate the time constant:

$$\tau = RC = (180 \text{ k}\Omega)(10 \text{ pF}) = 1.8 \ \mu s$$

In this case, $t_W > 5\tau$, so the capacitor reaches full charge in 9 μs (before the end of the pulse).

On the rising edge, the resistor voltage jumps to +5 V and then decreases exponentially to zero before the end of the pulse. On the falling edge, the resistor voltage jumps to −5 V and then goes back to zero exponentially. The resistor voltage is, of course, the output, and its shape is shown in Figure 15–25.

FIGURE 15–25

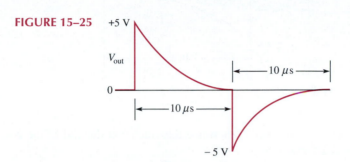

Related Exercise Sketch the output voltage if $R = 18$ kΩ and $C = 47$ pF in Figure 15–24.

EXAMPLE 15–5 Determine the output voltage waveform for the differentiator in Figure 15–26.

FIGURE 15–26

Solution First calculate the time constant:

$$\tau = (2.2 \text{ k}\Omega)(1 \text{ } \mu\text{F}) = 2.2 \text{ ms}$$

On the rising edge, the resistor voltage immediately jumps to +25 V. Because the pulse width is 5 ms, the capacitor charges for approximately 2.27 time constants during the pulse, and therefore does not reach full charge. Thus, you must use the formula for a decreasing exponential, Equation (9–19), in order to calculate to what voltage the output decreases by the end of the pulse:

$$v_{out} = V_i e^{-t/RC} = 25e^{-5ms/2.2ms} = 25(0.103) = 2.58 \text{ V}$$

where $V_i = 25$ V and $t = 5$ ms. This calculation gives the resistor voltage at the end of the 5 ms pulse width interval.

On the falling edge, the resistor voltage immediately jumps from + 2.58 V down to −22.4 V (a 25 V transition). The resulting waveform of the output voltage is shown in Figure 15–27.

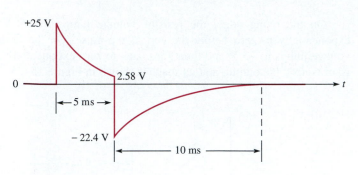

FIGURE 15–27

Related Exercise Determine the voltage at the end of the pulse in Figure 15–26 if R is 1.5 kΩ.

SECTION 15–4 REVIEW

1. Sketch the output of a differentiator for a 10 V input pulse when $5\tau = 0.5t_W$.
2. Under what condition does the output pulse shape most closely resemble the input pulse for a differentiator?
3. What does the differentiator output look like when 5τ is much less than the pulse width of the input?
4. If the resistor voltage in a differentiating circuit is down to +5 V at the end of a 15 V input pulse, to what negative value will the resistor voltage go in response to the falling edge of the input?

15–5 ▪ REPETITIVE-PULSE RESPONSE OF *RC* DIFFERENTIATORS

The RC differentiator response to a single pulse, covered in the last section, is extended in this section to repetitive pulses.

After completing this section, you should be able to

▪ **Analyze an *RC* differentiator with repetitive input pulses**
 ☐ Determine the response when the pulse width is less than five time constants

If a periodic pulse waveform is applied to an *RC* differentiating circuit, two conditions again are possible: $t_W \geq 5\tau$ or $t_W < 5\tau$. Figure 15–28 shows the output when $t_W = 5\tau$. As the time constant is reduced, both the positive and the negative portions of the output become narrower. Notice that the average value of the output is zero.

FIGURE 15–28
Example of differentiator response when $t_W = 5\tau$.

Figure 15–29 shows the steady-state output when $t_W < 5\tau$. As the time constant is increased, the positively and negatively sloping portions become flatter. For a very long time constant, the output approaches the shape of the input, but with an average value of zero. The average value of a waveform is its **dc component.** Because a capacitor blocks dc, the dc component of the input is prevented from passing through to the output.

FIGURE 15–29
Example of differentiator response when $t_W < 5\tau$.

Like the integrator, the differentiator output takes time (5τ) to reach steady state. To illustrate the response, let's take an example in which the time constant equals the input pulse width.

Analysis of a Repetitive Waveform

At this point, we do not care what the circuit time constant is because we know that the resistor voltage will decrease to 37% of its maximum value during one pulse (1τ). Assume that the capacitor in Figure 15–30 begins initially uncharged and examine the output voltage on a pulse-by-pulse basis. The results of the analysis to follow are shown in Figure 15–31.

FIGURE 15–30
RC differentiator with $\tau = t_W$.

First Pulse   On the rising edge, the output instantaneously jumps to +10 V. Then the capacitor partially charges to 63% of 10 V, which is 6.3 V. Thus, the output voltage must decrease to 3.7 V, as shown in Figure 15–31.

On the falling edge, the output instantaneously makes a negative-going 10 V transition to −6.3 V (−10 V + 3.7 V = −6.3 V).

Between First and Second Pulses   The capacitor discharges to 37% of 6.3 V, which is 2.33 V. Thus, the resistor voltage, which starts at −6.3 V, must increase to −2.33 V. Why? Because at the instant prior to the next pulse, the input voltage is zero. Therefore, the sum of v_C and v_R must be zero (+2.33 V − 2.33 V = 0). Remember that $v_C + v_R = v_{in}$ at all times.

Second Pulse   On the rising edge, the output makes an instantaneous, positive-going, 10 V transition from −2.33 V to 7.67 V. Then by the end of the pulse the capacitor charges 0.63 × (10 V − 2.33 V) = 4.83 V. Thus, the capacitor voltage increases from 2.33 V to 2.33 V + 4.83 V = 7.16 V. The output voltage drops to 0.37 × (7.67 V) = 2.84 V.

On the falling edge, the output instantaneously makes a negative-going transition from 2.84 V to −7.16 V, as shown in Figure 15–31.

Between Second and Third Pulses   The capacitor discharges to 37% of 7.16 V, which is 2.65 V. Thus, the output voltage starts at −7.16 V and increases to −2.65 V, because the capacitor voltage and the resistor voltage must add up to zero at the instant prior to the third pulse (the input is zero).

Third Pulse   On the rising edge, the output makes an instantaneous 10 V transition from −2.65 V to +7.35 V. Then the capacitor charges 0.63 × (10 V − 2.65 V) = 4.63 V to 2.65 V + 4.63 V = +7.28 V. As a result, the output voltage drops to 0.37 × 7.35 V = 2.72 V.   On the falling edge, the output instantly goes from +2.72 V down to −7.28 V.

After the third pulse, five time constants have elapsed, and the output voltage is close to its steady state. Thus, it will continue to vary from a positive maximum of about +7.3 V to a negative maximum of about −7.3 V, with an average value of zero.

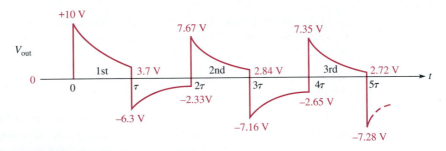

FIGURE 15–31
Differentiator output waveform during transient time for the circuit in Figure 15–30.

SECTION 15–5 REVIEW

1. What conditions allow an *RC* differentiator to fully charge and discharge when a periodic pulse waveform is applied to the input?
2. What will the output waveform look like if the circuit time constant is extremely small compared to the pulse width of a square wave input?
3. What does the average value of the differentiator output voltage equal during steady state?

15–6 ■ PULSE RESPONSE OF *RL* INTEGRATORS

A series RL circuit in which the output voltage is taken across the resistor is known as an integrator in terms of pulse response.

After completing this section, you should be able to

■ **Analyze the operation of an *RL* integrator**
   □ Determine the response to a single input pulse

Figure 15–32 shows an *RL* integrator. The output waveform is taken across the resistor and, under equivalent conditions, is the same shape as that for the *RC* integrator. Recall that in the *RC* case, the output was across the capacitor.

FIGURE 15–32
An RL integrating circuit.

As you know, both edges of an ideal pulse are considered to occur instantaneously. Two basic rules for inductor behavior will aid in analyzing *RL* circuit responses to pulse inputs:

1. The inductor appears as an open to an instantaneous change in current and as a short (ideally) to dc.

2. The current in an inductor cannot change instantaneously—it can change only exponentially.

Response of the Integrator to a Single Pulse

When a pulse generator is connected to the input of the integrator and the voltage pulse goes from its low level to its high level, the inductor prevents a sudden change in current. As a result, the inductor acts as an open, and all the input voltage is across it at the instant of the rising pulse edge. This situation is indicated in Figure 15–33(a).

(a) At rising edge of pulse (*i* = 0)

(b) During flat portion of pulse

(c) At falling edge of pulse and after

FIGURE 15–33
Illustration of the pulse response of an RL integrator (t$_W$ > 5τ).

After the rising edge, the current builds up, and the output voltage follows the current as it increases exponentially, as shown in Figure 15–33(b). The current can reach a maximum of V_p/R if the transient time is shorter than the pulse width ($V_p = 10$ V in this example).

When the pulse goes from its high level to its low level, an induced voltage with reversed polarity is created across the coil in an effort to keep the current equal to V_p/R. The output voltage begins to decrease exponentially, as shown in Figure 15–33(c).

The exact shape of the output depends on the L/R time constant as summarized in Figure 15–34 for various relationships between the time constant and the pulse width. You should note that the response of this RL circuit in terms of the shape of the output is identical to that of the RC integrator. The relationship of the L/R time constant to the input pulse width has the same effect as the RC time constant that we discussed earlier in this chapter. For example, when $t_W < 5\tau$, the output voltage will not reach its maximum possible value.

FIGURE 15–34
Illustration of the variation in integrator output pulse shape with time constant.

EXAMPLE 15–6   Determine the maximum output voltage for the integrator in Figure 15–35 when a single pulse is applied as shown.

FIGURE 15–35

Solution   Calculate the time constant:

$$\tau = \frac{L}{R} = \frac{100 \text{ mH}}{47 \text{ }\Omega} = 2.13 \text{ ms}$$

Because the pulse width is 5 ms, the inductor charges for approximately 2.35τ (5 ms/2.13 ms = 2.35). Use the exponential formula to calculate the voltage:

$$v_{\text{out}} = V_F(1 - e^{-t/\tau}) = 25(1 - e^{-5/2.13})$$
$$= 25(1 - e^{-2.35}) = 25(1 - 0.095) = 25(0.905) = 22.6 \text{ V}$$

Related Exercise What value must *R* be for the output voltage to reach 25 V by the end of the input pulse in Figure 15–35?

EXAMPLE 15–7 A pulse is applied to the *RL* integrator circuit in Figure 15–36. Determine the complete waveshapes and the values for *I*, V_R, and V_L.

FIGURE 15–36

Solution The circuit time constant is

$$\tau = \frac{L}{R} = \frac{5 \text{ mH}}{1.5 \text{ k}\Omega} = 3.33 \text{ }\mu\text{s}$$

Since $5\tau = 16.7$ μs is less than t_W, the current will reach its maximum value and remain there until the end of the pulse.
   At the rising edge of the pulse,

$$i = 0 \text{ A}$$
$$v_R = 0 \text{ V}$$
$$v_L = 10 \text{ V}$$

Since the inductor is initially an open, all of the input voltage appears across *L*.
   During the pulse,

$$i \text{ increases exponentially to } \frac{V_p}{R} = \frac{10 \text{ V}}{1.5 \text{ k}\Omega} = 6.67 \text{ mA in 16.7 } \mu\text{s}$$

v_R increases exponentially to 10 V in 16.7 μs

v_L decreases exponentially to zero in 16.7 μs

At the falling edge of the pulse,

$$i = 6.67 \text{ mA}$$
$$v_R = 10 \text{ V}$$
$$v_L = -10 \text{ V}$$

After the pulse,

$$i \text{ decreases exponentially to zero in } 16.7 \ \mu s$$
$$v_R \text{ decreases exponentially to zero in } 16.7 \ \mu s$$
$$v_L \text{ decreases exponentially to zero in } 16.7 \ \mu s$$

The waveforms are shown in Figure 15–37.

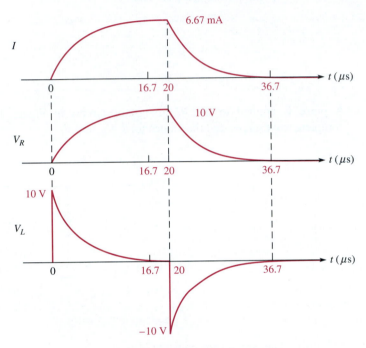

FIGURE 15–37

Related Exercise   What will be the maximum output voltage if the amplitude of the input pulse is increased to 20 V in Figure 15–36?

EXAMPLE 15–8   A 10 V pulse with a width of 1 ms is applied to the integrator in Figure 15–38. Determine the voltage level that the output will reach during the pulse. If the source has an internal resistance of 30 Ω, how long will it take the output to decay to zero? Sketch the output voltage waveform.

FIGURE 15–38

Solution The coil charges through the 30 Ω source resistance plus the 470 Ω external resistor. The time constant is

$$\tau = \frac{L}{R_T} = \frac{500 \text{ mH}}{470 \text{ } \Omega + 30 \text{ } \Omega} = \frac{500 \text{ mH}}{500 \text{ } \Omega} = 1 \text{ ms}$$

Notice that in this case the pulse width is exactly equal to τ. Thus, the output V_R will reach 63% of the full input amplitude in 1τ. Therefore, the output voltage gets to 6.3 V at the end of the pulse.

After the pulse is gone, the inductor discharges back through the 30 Ω source and the 470 Ω resistor. The source takes 5τ to completely discharge.

$$5\tau = 5(1 \text{ ms}) = 5 \text{ ms}$$

The output voltage is shown in Figure 15–39.

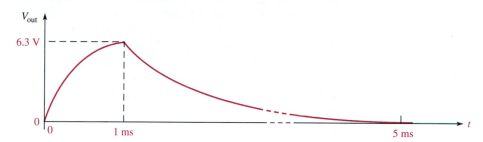

FIGURE 15–39

Related Exercise To what value must R be changed in Figure 15–38 to allow the output voltage to reach the input level during the pulse?

SECTION 15–6 REVIEW

1. In an *RL* integrator, across which component is the output voltage taken?
2. When a pulse is applied to an *RL* integrator, what condition must exist in order for the output voltage to reach the amplitude of the input?
3. Under what condition will the output voltage have the approximate shape of the input pulse?

15–7 ■ PULSE RESPONSE OF *RL* DIFFERENTIATORS

A series RL circuit in which the output voltage is taken across the inductor is known as a differentiator.

After completing this section, you should be able to

■ **Analyze the operation of an *RL* differentiator**
  □ Determine the response to a single input pulse

Figure 15–40 shows an *RL* differentiator. Compare this circuit to the *RC* differentiator in which the output voltage is across the resistor.

FIGURE 15–40
An RL differentiating circuit.

Response of the Differentiator to a Single Pulse

A pulse generator is connected to the input of the differentiator. Initially, before the pulse, there is no current in the circuit. When the input pulse goes from its low level to its high level, the inductor prevents a sudden change in current. It does this, as you know, with an induced voltage equal and opposite to the input. As a result, L looks like an open, and all of the input voltage appears across it at the instant of the rising edge, as shown in Figure 15–41(a) with a 10 V pulse.

During the pulse, the current exponentially builds up. As a result, the inductor voltage decreases, as shown in Figure 15–41(b). The rate of decrease, as you know, depends on the L/R time constant. When the falling edge of the input appears, the inductor reacts to keep the current as is, by creating an induced voltage in a direction as indicated in Figure 15–41(c). This reaction is seen as a sudden negative-going transition of the inductor voltage, as indicated in parts (c) and (d).

(a) At rising edge of pulse

(b) During flat portion of pulse

(c) At falling edge when $t_W < 5\tau$

(d) At falling edge when $t_W \geq 5\tau$

FIGURE 15–41
Illustration of the response of an RL differentiator for both time constant conditions.

Two conditions are possible, as indicated in Figure 15–41(c) and (d). In part (c), 5τ is greater than the input pulse width, and the output voltage does not have time to decay to zero. In part (d), 5τ is less than the pulse width, and so the output decays to zero before the end of the pulse. In this case a full, negative, 10 V transition occurs at the trailing edge.

Keep in mind that as far as the input and output waveforms are concerned, the *RL* integrator and differentiator perform the same as their *RC* counterparts.

A summary of the *RL* differentiator response for relationships of various time constants and pulse widths is shown in Figure 15–42.

FIGURE 15–42

Illustration of the variation in output pulse shape with the time constant.

EXAMPLE 15–9   Sketch the output voltage for the circuit in Figure 15–43.

FIGURE 15–43

Solution   First calculate the time constant:

$$\tau = \frac{L}{R} = \frac{200 \ \mu\text{H}}{100 \ \Omega} = 2 \ \mu\text{s}$$

In this case, $t_W = 5\tau$, so the output will decay to zero at the end of the pulse.

On the rising edge, the inductor voltage jumps to +5 V and then decays exponentially to zero. It reaches approximately zero at the instant of the falling edge. On the falling edge of the input, the inductor voltage jumps to −5 V and then goes back to zero. The output waveform is shown in Figure 15–44.

FIGURE 15–44

Related Exercise   Sketch the output voltage if the input pulse width is reduced to 5 μs in Figure 15–43.

EXAMPLE 15–10   Determine the output voltage waveform for the differentiator in Figure 15–45.

FIGURE 15–45

Solution   First calculate the time constant:

$$\tau = \frac{L}{R} = \frac{20 \text{ mH}}{10 \ \Omega} = 2 \text{ ms}$$

On the rising edge, the inductor voltage immediately jumps to +25 V. Because the pulse width is 5 ms, the inductor charges for only two and a half time constants (2.5τ), so you must use the formula for a decreasing exponential:

$$v_L = V_i e^{-t/\tau} = 25e^{-5\text{ms}/2\text{ms}} = 25e^{-2.5} = 25(0.082) = 2.05 \text{ V}$$

This result is the inductor voltage at the end of the 5 ms input pulse.

On the falling edge, the output immediately drops from +2.05 V to −22.95 V (a 25 V negative-going transition). The complete output waveform is sketched in Figure 15–46.

FIGURE 15–46

Related Exercise   What must be the value of R for V_{out} to reach zero by the end of the input pulse in Figure 15–45?

15–8 ▪ APPLICATIONS

Although integrators and differentiators are found in many different kinds of applications, only three basic ones are discussed in this section to give you a general idea of how these circuits can be used.

After completing this section, you should be able to

▪ **Discuss some basic applications of integrators and differentiators**
- ☐ Describe how *RC* integrators are used in timing circuits
- ☐ Describe how an integrator can be used for converting pulse waveforms to dc
- ☐ Describe how a differentiator can be used as a trigger pulse generator

Timing Circuits

RC integrators can be used in timing circuits to set a specified time interval for various purposes. By changing the time constant, the interval can be adjusted. For example, the *RC* integrator circuit in Figure 15–47(a) is used to provide a delay between the time a

(a) Basic delay circuit

(b) Timing diagram

FIGURE 15–47
A basic time delay application of an RC integrator.

power-on switch is closed and the time a certain event is activated by a threshold circuit. The threshold circuit is designed to respond to the input voltage only when the input reaches a certain specified level. Don't worry how the threshold circuit works at this point; you will learn about this type of circuit in a later course.

When the switch is thrown, the capacitor begins to charge at a rate set by the RC time constant. After a certain time, the capacitor voltage reaches the threshold value and triggers (turns on) the circuit to activate a device such as a motor, relay, lamp, etc., depending on the particular application. This action is illustrated by the waveforms in Figure 15–47(b).

EXAMPLE 15–11   What is the time delay in the circuit of Figure 15–47(a) with $V_{in} = 9$ V, $R = 10$ MΩ, and $C = 0.47$ μF? Assume the threshold voltage is 5 V.

Solution   Use the exponential formula, Equation (9–18), and solve for time as follows:

$$v = V_F(1 - e^{-t/RC})$$
$$v = V_F - V_F e^{-t/RC}$$
$$V_F - v = V_F e^{-t/RC}$$
$$e^{-t/RC} = \frac{V_F - v}{V_F}$$

Taking the natural log (ln) of both sides of the equation yields

$$-\frac{t}{RC} = \ln\left(\frac{V_F - v}{V_F}\right)$$
$$t = -RC \ln\left(\frac{V_F - v}{V_F}\right)$$

V_F is the final voltage to which C will charge and equals V_{in}. Substituting values and solving for t:

$$t = -(10 \text{ MΩ})(0.47 \text{ μF})\ln\left(\frac{9 \text{ V} - 5 \text{ V}}{9 \text{ V}}\right) = -(10 \text{ MΩ})(0.47 \text{ μF})\ln\left(\frac{4 \text{ V}}{9 \text{ V}}\right) = 3.8 \text{ s}$$

Related Exercise   Find the time delay for the circuit in Figure 15–47 for $V_{in} = 24$ V, $R = 10$ kΩ, and $C = 1500$ pF.

A Pulse Waveform-to-DC Converter

An integrating circuit can be used to convert a pulse waveform to a constant dc value equal to the average of the waveform. This is accomplished by using a time constant that is extremely large compared to the period of the pulse waveform, as illustrated in Figure 15–48(a). Actually, there will be a slight ripple on the output as the capacitor charges and discharges a little. This ripple is less the larger the time constant and can approach a constant level, as shown in Figure 15–48(b). In terms of sinusoidal response, this circuit is a low-pass filter.

A Positive and Negative Trigger Pulse Generator

A differentiator circuit can be used to produce very short duration pulses (spikes) with both positive and negative polarities, as illustrated in Figure 15–49(a). When the time

(a)

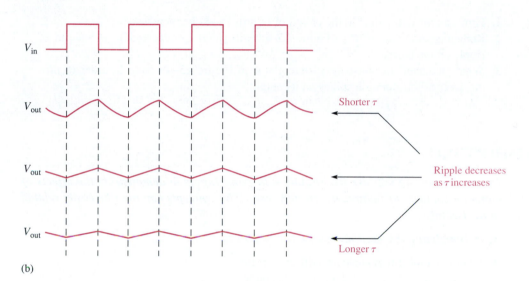

(b)

FIGURE 15–48

A long time constant integrator acts as a pulse-to-dc converter.

(a)

(b)

FIGURE 15–49

A very short time constant differentiator generates positive and negative spikes.

constant is small compared to the pulse width of the input, the differentiator produces a positive spike on each positive-going edge of the input and a negative spike on each negative-going edge.

To separate the positive and negative spikes, two clipping circuits are used, as shown in Figure 15–49(b). Each clipping circuit consists of a diode, which is a semiconductor device. The negative clipper removes the negative spike and passes the positive spike. The positive clipper does just the opposite.

<table>
<tr><td>SECTION 15–8
REVIEW</td><td>1. How can the delay time in the circuit of Figure 15–47 be increased?
2. Reducing the value of R in Figure 15–48 will (increase, decrease) the amount of ripple on the output.
3. What characteristic must the differentiator in Figure 15–49 have in order to produce very short duration spikes on its output?</td></tr>
</table>

15–9 ▪ TROUBLESHOOTING

In this section, we will use RC circuits with pulse inputs to demonstrate the effects of common component failures in selected cases. The concepts can then be easily related to RL circuits.

After completing this section, you should be able to

▪ **Troubleshoot integrators and differentiators**
   ☐ Recognize the effect of an open capacitor
   ☐ Recognize the effect of a leaky capacitor
   ☐ Recognize the effect of a shorted capacitor
   ☐ Recognize the effect of an open resistor

Open Capacitor

If the capacitor in an integrator opens, the output has the same waveshape as the input, as shown in Figure 15–50(a). If the capacitor in a differentiator opens, the output is zero because it is held at ground through the resistor, as illustrated in part (b).

FIGURE 15–50
Examples of the effect of an open capacitor.

(a) Integrator

(b) Differentiator

Leaky Capacitor

If the capacitor in an integrator becomes leaky, the time constant will be effectively reduced by the leakage resistance. When Thevenized, looking from C, R_{leak} appears in parallel with R through the zero resistance of the input source. The waveshape of the output voltage (across C) is altered from normal by a shorter charging time. Also, the amplitude of the output is reduced because R and R_{leak} effectively act as a voltage divider. These effects are illustrated in Figure 15–51(a).

If the capacitor in a differentiator becomes leaky, the time constant is reduced, just as in the integrator (they are both simply series RC circuits). When the capacitor reaches full charge, the output voltage (across R) is set by the effective voltage-divider action of R and R_{leak}, as shown in Figure 15–51(b).

(a)

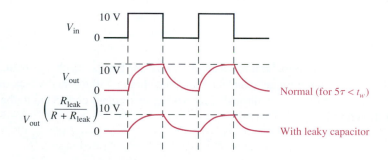

(b)

FIGURE 15–51
Examples of the effect of a leaky capacitor.

Shorted Capacitor

If the capacitor in an integrator shorts, the output is at ground as shown in Figure 15–52(a). If the capacitor in a differentiator shorts, the output is the same as the input, as shown in part (b).

(a)

(b)

FIGURE 15–52
Examples of the effect of a shorted capacitor.

Open Resistor

If the resistor in an integrator opens, the capacitor has no discharge path, and, ideally, it will hold its charge. In an actual situation, the charge will gradually leak off or the capacitor will discharge slowly through a measuring instrument connected to the output. This is illustrated in Figure 15–53(a).

FIGURE 15–53

Examples of the effect of an open resistor.

(a)

(b)

If the resistor in a differentiator opens, the output looks like the input except for the dc level because the capacitor now must charge and discharge through the extremely high resistance of the oscilloscope, as shown in Figure 15–53(b).

SECTION 15–9 REVIEW

1. An *RC* integrator has a zero output with a square wave input. What are the possible causes of this problem?
2. If the integrator capacitor is open, what is the output for a square wave input?
3. If the capacitor in a differentiator is shorted, what is the output for a square wave input?

15–10 ▪ APPLICATION ASSIGNMENT

Your supervisor has given you the assignment of breadboarding and testing a time-delay circuit that will provide five switch-selectable delay times. An RC integrator is selected for this application. The input is a 5 V pulse of long duration which begins the time delay. The exponential output voltage goes to a threshold trigger circuit that is used to turn the power on to a portion of a system after any one of the five selectable time delays.

A schematic of the selectable time-delay integrator is shown in Figure 15–54. The *RC* integrator is driven by a positive pulse input and the output is the exponential voltage across the selected capacitor. The output voltage triggers a threshold circuit at the 3.5 V level, which then turns on the power to a portion of a system. The basic concept of operation is shown in Figure 15–55. In this application, the delay time of the integrator is spec-

FIGURE 15–54
Integrator delay circuit.

FIGURE 15–55
Illustration of the time-delay operation.

TABLE 15–1

SWITCH POSITION	DELAY TIME
A	10 ms
B	25 ms
C	40 ms
D	65 ms
E	85 ms

ified as the time from the rising edge of the input pulse to the point where the output voltage reaches 3.5 V. The specified time delays are listed in Table 15–1.

Step 1: Capacitor Values

Determine a value for each of the five capacitors in the time-delay circuit that will provide a delay time within 10% of the specified delay time. Select from the following list of standard values: 0.1 μF, 0.12 μF, 0.15 μF, 0.18 μF, 0.22 μF, 0.27 μF, 0.33 μF, 0.39 μF,

FIGURE 15–56
Time-delay circuit breadboard and test instruments.

0.47 μF, 0.56 μF, 0.68 μF, 0.82 μF, 1.0 μF, 1.2 μF, 1.5 μF, 1.8 μF, 2.2 μF, 2.7 μF, 3.3 μF, 3.9 μF, 4.7 μF, 5.6 μF, 6.8 μF, and 8.2 μF.

Step 2: Circuit Connections

Refer to Figure 15–56. The components for the time-delay circuit are assembled, but not connected, on the breadboard. Using the circled numbers, develop a point-to-point wiring list to show how the circuit and measurement instruments should be connected.

Step 3: Test Procedure and Instrument Settings

Develop a procedure for fully testing the time-delay circuit. Specify the amplitude, frequency, and duty cycle settings for the function generator in order to test each delay time. Specify the oscilloscope sec/div settings for measuring each of the five specified delay times.

Step 4: Measurement

Explain how you will verify that each switch setting produces the proper output delay time.

SECTION 15–10 REVIEW	**1.** Explain why the input pulse to the time-delay circuit must be of sufficiently long duration. **2.** Indicate how you would modify the time-delay circuit so that each selected time delay had some adjustment range.

■ SUMMARY

- In an *RC* integrating circuit, the output voltage is taken across the capacitor.
- In an *RC* differentiating circuit, the output voltage is taken across the resistor.
- In an *RL* integrating circuit, the output voltage is taken across the resistor.
- In an *RL* differentiating circuit, the output voltage is taken across the inductor.
- In an integrator, when the pulse width (t_W) of the input is much less than the transient time, the output voltage approaches a constant level equal to the average value of the input.
- In an integrator, when the pulse width of the input is much greater than the transient time, the output voltage approaches the shape of the input.
- In a differentiator, when the pulse width of the input is much less than the transient time, the output voltage approaches the shape of the input but with an average value of zero.
- In a differentiator, when the pulse width of the input is much greater than the transient time, the output voltage consists of narrow, positive-going and negative-going spikes occurring on the leading and trailing edges of the input pulses.

■ GLOSSARY

DC component   The average value of a pulse waveform.

Differentiator   A circuit producing an output that approaches the mathematical derivative of the input.

Integrator   A circuit producing an output that approaches the mathematical derivative of the input.

Periodic   Characterized by a repetition at fixed intervals.

Pulse response   The reaction of a circuit to a given input.

Steady state   The equilibrium condition of a circuit that occurs after an initial transient time.

Time constant   A fixed time interval, set by *R, C,* and *L* values, that determines the time response of a circuit.

Transient time   An interval equal to approximately five time constants.

■ **SELF-TEST**

1. The output of an *RC* integrator is taken across the
   (a) resistor **(b)** capacitor **(c)** source **(d)** coil

2. When a 10 V input pulse with a width equal to one time constant is applied to an *RC* integrator, the capacitor charges to
   (a) 10 V **(b)** 5 V **(c)** 6.3 V **(d)** 3.7 V

3. When a 10 V pulse with a width equal to one time constant is applied to an *RC* differentiator, the capacitor charges to
   (a) 6.3 V **(b)** 10 V **(c)** 0 V **(d)** 3.7 V

4. In an *RC* integrator, the output pulse closely resembles the input pulse when
   (a) τ is much larger than the pulse width **(b)** τ is equal to the pulse width
   (c) τ is less than the pulse width **(d)** τ is much less than the pulse width

5. In an *RC* differentiator, the output pulse closely resembles the input pulse when
   (a) τ is much larger than the pulse width **(b)** τ is equal to the pulse width
   (c) τ is less than the pulse width **(d)** τ is much less than the pulse width

6. The positive and negative portions of a differentiator's output voltage are equal when
   (a) $5\tau < t_W$ **(b)** $5\tau > t_W$ **(c)** $5\tau = t_W$ **(d)** $5\tau > 0$

7. The output of an *RL* integrator is taken across the
   (a) resistor **(b)** coil **(c)** source **(d)** capacitor

8. The maximum current in an *RL* differentiator is
   (a) $I = \dfrac{V_p}{X_L}$ **(b)** $I = \dfrac{V_p}{Z}$ **(c)** $I = \dfrac{V_p}{R}$

9. The current in an *RL* differentiator reaches its maximum possible value when
   (a) $5\tau = t_W$ **(b)** $5\tau < t_W$ **(c)** $5\tau > t_W$ **(d)** $\tau = 0.5 t_W$

10. If you have an *RC* and an *RL* differentiator with equal time constants sitting side-by-side and you apply the same input pulse to both,
    (a) the *RC* has the widest output pulse
    (b) the *RL* has the most narrow spikes on the output
    (c) the output of one is an increasing exponential and the output of the other is a decreasing exponential
    (d) you can't tell the difference by observing the output waveforms

---

■ **BASIC PROBLEMS**

SECTION 15–1   The *RC* Integrator

1. An integrating circuit has $R = 2.2$ kΩ in series with $C = 0.05$ μF. What is the time constant?

2. Determine how long it takes the capacitor in an integrating circuit to reach full charge for each of the following series *RC* combinations:
   (a) $R = 50$ Ω, $C = 50$ μF **(b)** $R = 3300$ Ω, $C = 0.015$ μF
   (c) $R = 22$ kΩ, $C = 100$ pF **(d)** $R = 5$ MΩ, $C = 10$ pF

SECTION 15–2   Single-Pulse Response of *RC* Integrators

3. A 20 V pulse is applied to an *RC* integrator. The pulse width equals one time constant. To what voltage does the capacitor charge during the pulse? Assume that it is initially uncharged.

4. Repeat Problem 3 for the following values of t_W:
   (a) 2τ **(b)** 3τ **(c)** 4τ **(d)** 5τ

5. Sketch the approximate shape of an integrator output voltage where 5τ is much less than the pulse width of a 10 V square wave input. Repeat for the case in which 5τ is much larger than the pulse width.

6. Determine the output voltage for an integrator with a single input pulse, as shown in Figure 15–57. For repetitive pulses, how long will it take this circuit to reach steady state?

FIGURE 15–57

SECTION 15–3   Repetitive-Pulse Response of *RC* Integrators

7. Sketch the integrator output in Figure 15–58 showing maximum voltages.

8. A 1 V, 10 kHz pulse waveform with a duty cycle of 25% is applied to an integrator with $\tau = 25$ μs. Graph the output voltage for three initial pulses. *C* is initially uncharged.

FIGURE 15–58

9. What is the steady-state output voltage of the integrator with a square wave input shown in Figure 15–59?

FIGURE 15–59

SECTION 15–4   Single-Pulse Response of *RC* Differentiators

10. Repeat Problem 5 for an *RC* differentiator.

11. Redraw the circuit in Figure 15–57 to make it a differentiator, and repeat Problem 6.

SECTION 15–5   Repetitive-Pulse Response of *RC* Differentiators

12. Sketch the differentiator output in Figure 15–60, showing maximum voltages.

13. What is the steady-state output voltage of the differentiator with the square wave input shown in Figure 15–61?

FIGURE 15–60   **FIGURE 15–61**

SECTION 15–6   Pulse Response of *RL* Integrators

14. Determine the output voltage for the circuit in Figure 15–62. A single input pulse is applied as shown.

15. Sketch the integrator output in Figure 15–63, showing maximum voltages.

FIGURE 15–62   **FIGURE 15–63**

SECTION 15–7   Pulse Response of *RL* Differentiators

16. (a) What is τ in Figure 15–64?
    (b) Sketch the output voltage.

17. Draw the output waveform if a periodic pulse waveform with $t_W = 25$ μs and $T = 60$ μs is applied to the circuit in Figure 15–64.

FIGURE 15–64

SECTION 15–8   Applications

18. What is the instantaneous voltage at point B in Figure 15–47 440 μs after the switch is closed for $R = 22$ kΩ and $C = 0.001$ μF? $V_{in} = 10$ V.

19. Ideally, what is the output of an *RC* integrator when the input is a square wave with an amplitude of 12 V? Assume that the time constant is much greater than the period of the input signal?

SECTION 15–9   Troubleshooting

20. Determine the most likely fault(s), in the circuit of Figure 15–65(a) for each set of waveforms in parts (b) through (d). V_{in} is a square wave with a period of 8 ms.

(a)

(b)   (c)   (d)

FIGURE 15–65

21. Determine the most likely fault(s), if any, in the circuit of Figure 15–66(a) for each set of waveforms in parts (b) through (d). V_{in} is a square wave with a period of 8 ms.

(a)

(b)   (c)   (d)

FIGURE 15–66

▪ **ADVANCED PROBLEMS**

22. (a) What is τ in Figure 15–67?
   (b) Sketch the output voltage.

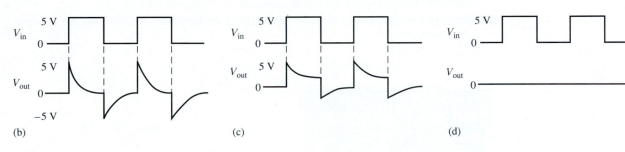

FIGURE 15–67

23. (a) What is τ in Figure 15–68?
   (b) Sketch the output voltage.

FIGURE 15–68

24. Determine the time constant in Figure 15–69. Is this circuit an integrator or a differentiator?

FIGURE 15–69

25. In a delay circuit like that of Figure 15–47, what time constant will produce a 1 s delay if the threshold of the circuit is 2.5 V and the amplitude of the input is 5 V?

26. Draw the schematic for the circuit in Figure 15–70 and determine if the oscilloscope presentation is correct.

(a) Oscilloscope display

(b) Circuit board with instrument leads connected

FIGURE 15–70

■ **ANSWERS TO SECTION REVIEWS**

Section 15–1

1. An integrator is a series *RC* circuit with a pulse input in which the output is across the capacitor.

2. A voltage applied to the input causes the capacitor to charge. Zero volts across the input causes the capacitor to discharge.

Section 15–2

1. The output of an integrator reaches full amplitude when $5\tau \leq t_W$.
2. $V_{out} = (0.63)1$ V $= 0.63$ V; $t_{disch} = 5\tau = 51.5$ ms.
3. See Figure 15–71.

FIGURE 15–71

0.63 V

0

0   10.3 ms   61.8 ms

4. No, C will not fully charge.
5. The output is approximately the shape of the input when $5\tau << t_W$ (5τ much less than t_W).

Section 15–3

1. An integrator capacitor will fully charge and discharge when $5\tau \leq t_W$ and $5\tau \leq$ time between pulses.
2. The output looks like the input.
3. transient time
4. Steady state is the response after the transient time has passed.
5. The average value of the output voltage equals the average value of the input voltage.

Section 15–4

1. See Figure 15–72.

FIGURE 15–72

+ 10 V

− 10 V

2. The output approximates the input when $5\tau >> t_W$.
3. The output consists of a positive and a negative spike.
4. $V_R = +5$ V $- 15$ V $= -10$ V

Section 15–5

1. C fully charges and discharges when $5\tau \leq t_W$ and $5\tau \leq$ time between pulses.
2. The output consists of positive and negative spikes.
3. $V_{out} = 0$ V

Section 15–6

1. The output is across the resistor.
2. The output reaches the input amplitude when $5\tau \leq t_W$.
3. The output approximates the input when $5\tau << t_W$.

Section 15–7

1. The output is across the inductor.
2. The output approximates the input when $5\tau >> t_W$.
3. $V_{out} = 2$ V $- 10$ V $= -8$ V

Section 15–8

1. Increase the time constant to increase delay time.
2. Reducing R will increase ripple.
3. A very short time constant produces very short duration spikes.

Section 15–9

1. Zero output of *RC* integrator indicates shorted capacitor or open resistor.

2. An open integrator capacitor results in an output exactly like the input.

3. A shorted differentiator capacitor results in an output exactly like the input.

Section 15–10

1. The input pulse must be sufficiently long to allow the output to reach 3.5 V for the longest time-constant setting.

2. Add a series potentiometer to adjust the total resistance.

▪ **ANSWERS TO RELATED EXERCISES FOR EXAMPLES**

15–1   8.65 V

15–2   24.7 V

15–3   1.11 V

15–4   See Figure 15–73.

FIGURE 15–73

15–5   892 mV

15–6   100 Ω

15–7   20 V

15–8   2.5 kΩ

15–9   See Figure 15–74.

FIGURE 15–74

15–10 20 Ω

15–11 3.5 μs

A TABLE OF STANDARD RESISTOR VALUES

Resistance Tolerance (± %)

0.1% 0.25% 0.5%	1%	2% 5%	10%	0.1% 0.25% 0.5%	1%	2% 5%	10%	0.1% 0.25% 0.5%	1%	2% 5%	10%	0.1% 0.25% 0.5%	1%	2% 5%	10%	0.1% 0.25% 0.5%	1%	2% 5%	10%	0.1% 0.25% 0.5%	1%	2% 5%	10%
10.0	10.0	10	10	14.7	14.7	—	—	21.5	21.5	—	—	31.6	31.6	—	—	46.4	46.4	—	—	68.1	68.1	68	68
10.1	—	—	—	14.9	—	—	—	21.8	—	—	—	32.0	—	—	—	47.0	—	47	47	69.0	—	—	—
10.2	10.2	—	—	15.0	15.0	15	15	22.1	22.1	22	22	32.4	32.4	—	—	47.5	47.5	—	—	69.8	69.8	—	—
10.4	—	—	—	15.2	—	—	—	22.3	—	—	—	32.8	—	—	—	48.1	—	—	—	70.6	—	—	—
10.5	10.5	—	—	15.4	15.4	—	—	22.6	22.6	—	—	33.2	33.2	33	33	48.7	48.7	—	—	71.5	71.5	—	—
10.6	—	—	—	15.6	—	—	—	22.9	—	—	—	33.6	—	—	—	49.3	—	—	—	72.3	—	—	—
10.7	10.7	—	—	15.8	15.8	—	—	23.2	23.2	—	—	34.0	34.0	—	—	49.9	49.9	—	—	73.2	73.2	—	—
10.9	—	—	—	16.0	—	16	—	23.4	—	—	—	34.4	—	—	—	50.5	—	—	—	74.1	—	—	—
11.0	11.0	11	—	16.2	16.2	—	—	23.7	23.7	—	—	34.8	34.8	—	—	51.1	51.1	51	—	75.0	75.0	75	—
11.1	—	—	—	16.4	—	—	—	24.0	—	24	—	35.2	—	—	—	51.7	—	—	—	75.9	—	—	—
11.3	11.3	—	—	16.5	16.5	—	—	24.3	24.3	—	—	35.7	35.7	—	—	52.3	52.3	—	—	76.8	76.8	—	—
11.4	—	—	—	16.7	—	—	—	24.6	—	—	—	36.1	—	36	—	53.0	—	—	—	77.7	—	—	—
11.5	11.5	—	—	16.9	16.9	—	—	24.9	24.9	—	—	36.5	36.5	—	—	53.6	53.6	—	—	78.7	78.7	—	—
11.7	—	—	—	17.2	—	—	—	25.2	—	—	—	37.0	—	—	—	54.2	—	—	—	79.6	—	—	—
11.8	11.8	—	—	17.4	17.4	—	—	25.5	25.5	—	—	37.4	37.4	—	—	54.9	54.9	—	—	80.6	80.6	—	—
12.0	—	12	12	17.6	—	—	—	25.8	—	—	—	37.9	—	—	—	56.2	—	—	—	81.6	—	—	—
12.1	12.1	—	—	17.8	17.8	—	—	26.1	26.1	—	—	38.3	38.3	—	—	56.6	56.6	56	56	82.5	82.5	82	82
12.3	—	—	—	18.0	—	18	18	26.4	—	—	—	38.8	—	—	—	56.9	—	—	—	83.5	—	—	—
12.4	12.4	—	—	18.2	18.2	—	—	26.7	26.7	—	—	39.2	39.2	39	39	57.6	57.6	—	—	84.5	84.5	—	—
12.6	—	—	—	18.4	—	—	—	27.1	—	27	27	39.7	—	—	—	58.3	—	—	—	85.6	—	—	—
12.7	12.7	—	—	18.7	18.7	—	—	27.4	27.4	—	—	40.2	40.2	—	—	59.0	59.0	—	—	86.6	86.6	—	—
12.9	—	—	—	18.9	—	—	—	27.7	—	—	—	40.7	—	—	—	59.7	—	—	—	87.6	—	—	—
13.0	13.0	13	—	19.1	19.1	—	—	28.0	28.0	—	—	41.2	41.2	—	—	60.4	60.4	—	—	88.7	88.7	—	—
13.2	—	—	—	19.3	—	—	—	28.4	—	—	—	41.7	—	—	—	61.2	—	—	—	89.8	—	—	—
13.3	13.3	—	—	19.6	19.6	—	—	28.7	28.7	—	—	42.2	42.2	—	—	61.9	61.9	62	—	90.9	90.9	91	—
13.5	—	—	—	19.8	—	—	—	29.1	—	—	—	42.7	—	—	—	62.6	—	—	—	92.0	—	—	—
13.7	13.7	—	—	20.0	20.0	20	—	29.4	29.4	—	—	43.2	43.2	43	—	63.4	63.4	—	—	93.1	93.1	—	—
13.8	—	—	—	20.3	—	—	—	29.8	—	—	—	43.7	—	—	—	64.2	—	—	—	94.2	—	—	—
14.0	14.0	—	—	20.5	20.5	—	—	30.1	30.1	30	—	44.2	44.2	—	—	64.9	64.9	—	—	95.3	95.3	—	—
14.2	—	—	—	20.8	—	—	—	30.5	—	—	—	44.8	—	—	—	65.7	—	—	—	96.5	—	—	—
14.3	14.3	—	—	21.0	21.0	—	—	30.9	30.9	—	—	45.3	45.3	—	—	66.5	66.5	—	—	97.6	97.6	—	—
14.5	—	—	—	21.3	—	—	—	31.2	—	—	—	45.9	—	—	—	67.3	—	—	—	98.8	—	—	—

NOTE: These values are generally available in multiples of 0.1, 1, 10, 100, 1 k, and 1 M.

B BATTERIES

Batteries are an important source of dc voltage. They are available in two basic categories: the wet cell and the dry cell. A battery generally is made up of several individual cells.

A cell consists basically of two electrodes immersed in an electrolyte. A voltage is developed between the electrodes as a result of the chemical action between the electrodes and the electrolyte. The electrodes typically are two dissimilar metals, and the electrolyte is a chemical solution or substance.

Wet Cells and Dry Cells

In a wet cell, the electrolyte is in liquid form, as you might guess. The automotive battery is a common type of wet-cell battery. In a dry cell, the electrolyte is typically in a paste-like form. There are many more varieties of dry-cell batteries than wet-cell batteries.

Primary and Secondary Cells

Primary cells cannot be recharged and thus are used once and then discarded. The 1.5 V carbon-zinc or alkaline flashlight battery is an example of a primary cell.

Secondary cells are rechargable and can be reused many times. Examples are nickel-cadmium (Ni-Cad) cells and lead-acid automotive cells.

Charging Rate

The charging rate for secondary cells specifies the amount of reverse current recommended over a given period of time. For example, the charging rate might be specified as 100 mA for 10 h. The same battery could be charged at a faster equivalent rate of, say, 200 mA for 5 h or at a slower equivalent rate of 50 mA for 20 h.

Charge/Discharge Cycle

When a battery charges and is then discharged to below its useful output voltage, it has gone through one charge/discharge cycle. The total number of charge/discharge cycles that a battery can go through before it becomes no longer usable is called the *cycle life*. The cycle life of a typical rechargable battery can be several hundred cycles.

Capacity

The capacity of any type of battery specifies how much current the battery can deliver to a load over a certain period of time before its output voltage decreases to an unusable level. Capacity is expressed in ampere-hours (Ah). For example, if a certain battery has a capacity of 50 Ah, it can deliver 50 A for 1 h or 100 A for $\frac{1}{2}$ h, 25 A for 2 h, and so on.

TYPES OF BATTERIES

Lead-Acid Cell

The positive electrode of a lead-acid cell is lead peroxide (PbO_2), and the negative electrode is spongy lead (Pb). The electrolyte is sulfuric acid (H_2SO_4) in water. Thus, the lead-acid cell is classified as a wet cell.

Two positive hydrogen ions (2H$^+$) and one negative sulfate ion (SO$_4^{--}$) are formed when the sulfuric acid ionizes in the water. Lead ions (Pb^{++}) from both electrodes displace the hydrogen ions in the electrolyte solution. When the lead ion from the spongy lead electrode enters the solution, it combines with a sulfate ion (SO$_4^{--}$) to form lead sulfate (PbSO$_4$), and it leaves two excess electrons on the electrode.

When a lead ion from the lead peroxide electrode enters the solution, it also leaves two excess electrons on the electrode and forms lead sulfate in the solution. However, because this electrode is lead peroxide, two free oxygen atoms are created when a lead atom leaves and enters the solution as a lead ion. These two oxygen atoms take four electrons from the lead peroxide electrode and become oxygen ions (O^{--}). This process creates a deficiency of two electrons on this electrode (there were initially two excess electrons).

The two oxygen ions (2O^{--}) combine in the solution with four hydrogen ions (4H$^+$) to produce two molecules of water (2H$_2$O). This process dilutes the electrolyte over a period of time. Also, there is a buildup of lead sulfate on the electrodes. These two factors result in a reduction in the voltage produced by the cell and necessitate periodic recharging.

As you have seen, for each departing lead ion, there is an excess of two electrons on the spongy lead electrode, and there is a deficiency of two electrons on the lead peroxide electrode. Therefore, the lead peroxide electrode is positive, and the spongy lead electrode is negative. This chemical reaction is pictured in Figure B–1.

FIGURE B–1
Chemical reaction in a discharging lead-acid cell.

As mentioned, the dilution of the electrolyte by the formation of water and lead sulfate requires that the lead-acid cell be recharged to reverse the chemical process. The cell is recharged by connection of an external voltage source to the electrodes, as shown in Figure B–2. The formula for the chemical reaction in a lead-acid cell is as follows:

$$Pb + PbO_2 + 2H_2SO_4 \longrightarrow 2PbSO_4 + 2H_2O$$

Carbon-Zinc Cell

A typical carbon-zinc dry cell is illustrated in Figure B–3. The zinc container or can is dissolved by the electrolyte. As a result of this reaction, an excess of electrons accumulates on the container, making it the negative electrode. The hydrogen ions in the electrolyte take electrons from the carbon rod, making it the positive electrode. Hydrogen gas is formed near the carbon electrode, but his gas is eliminated by reaction with manganese dioxide (called a *depolarizing agent*). This depolarization prevents bursting of the container due to gas formation. Because the chemical reaction is not reversible, the carbon-zinc cell is a primary cell.

Several common types of dry cells are summarized in Table B–1.

FIGURE B–2
Recharging a lead-acid cell.

FIGURE B–3
Cutaway view of a typical carbon-zinc cell.

TABLE B–1
Types of battery cells

TYPE	+ ELECTRODE	–ELECTRODE	ELECTROLYTE	VOLTS	COMMENTS
Carbon-zinc	Carbon	Zinc	Ammonium and zinc chloride	1.5	Dry, primary
Lead-acid	Lead peroxide	Spongy lead	Sulfuric acid	2.0	Wet, secondary
Manganese-alkaline	Manganese dioxide	Zinc	Potassium hydroxide	1.5	Dry, primary or secondary
Mercury	Zinc	Mercuric oxide	Potassium hydroxide	1.3	Dry, primary
Nickel-cadmium	Nickel	Cadmium hydroxide	Potassium hydroxide	1.25	Dry, secondary
Nickel-iron (Edison cell)	Nickel oxide	Iron	Potassium hydroxide	1.36	Wet, secondary

C

CIRCUIT BREADBOARDING AND MEASUREMENTS

Breadboarding a Circuit

The term *breadboarding* refers to the process of installing components on a circuit board and interconnecting them to form a specified circuit. One common type of circuit board used for constructing circuits in the laboratory is shown in Figure C–1. It consists of rows of small sockets into which component leads and jumper wires are inserted.

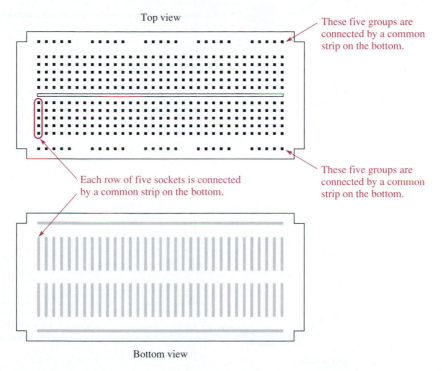

Top view

These five groups are connected by a common strip on the bottom.

These five groups are connected by a common strip on the bottom.

Each row of five sockets is connected by a common strip on the bottom.

Bottom view

FIGURE C–1
A typical circuit board used for breadboarding.

In this particular configuration, all five sockets in each row are connected together and are effectively one electrical point as shown in the bottom view. All the sockets arranged on the outer edges of the board are typically connected together as shown. These two strips of sockets are normally used to distribute the supply voltages and ground to the circuit.

As examples of breadboarding a circuit, Figure C–2 shows how both simple series and parallel resistive circuits might be connected on the circuit board.

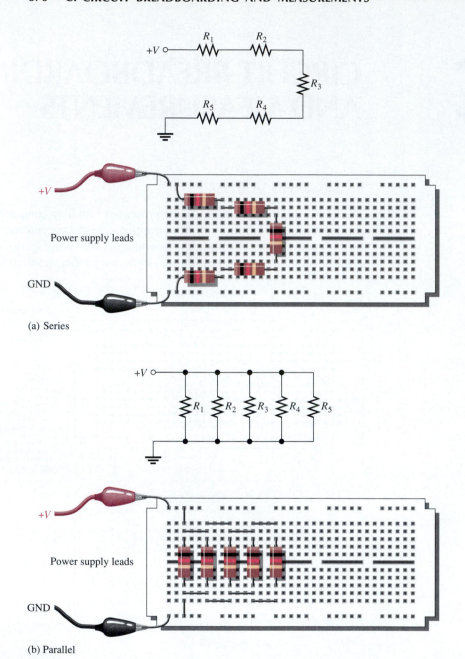

FIGURE C–2
Examples of breadboarding a circuit.

FIGURE C–3
Example of measuring voltage in a circuit.

Voltage Measurement

To measure voltage in a circuit using a multimeter, the following general steps should be taken:

1. Insert the red meter lead into the connector labeled for *voltage* (V) on the meter and insert the black meter lead into the connector labeled *common* (COM).

2. Set the function switch to the *volts* position (select either dc or ac).

3. Set the range switch (the range switch and function switch are often the same) to the lowest setting that is higher than the voltage you expect to measure.

4. Select the two points in the circuit between which you want to measure the voltage.

5. Place the red meter lead on the point closest to the positive side of the circuit and the black lead on the point closest to the negative or ground side of the circuit as shown in Figure C–3.

6. Read the voltage.

FIGURE C–4
Example of measuring current in a circuit.

Current Measurement

To measure current in a circuit using a multimeter, the following general steps are taken:

1. Insert the red meter lead into the connector labeled for *current* (A or mA) on the meter and insert the black meter lead into the connector labeled *common* (COM).

2. Set the function switch to the *amps* position (select either dc or ac).

3. Set the range switch to the lowest setting that is higher than the current you expect to measure.

4. Select the point in the circuit at which you want to measure the current.

5. Turn off the power and break the circuit at the selected point by simply lifting a component lead or one end of a jumper wire from the socket. This permits the ammeter to be placed in series with the current path.

6. Insert the ammeter in the break by connecting the red lead to the point closest to the positive side of the circuit and the black lead to the point closest to the negative or ground side of the circuit as shown in Figure C–4.

7. Turn the power back on and read the current.

Read resistance value

Select ohms function

Select multiplier

Disconnect power

Remove resistor from circuit by lifting one lead

+V

Power supply leads

GND

All 1 kΩ resistors

FIGURE C–5
Example of measuring resistance in a circuit.

Resistance Measurement

To measure resistance in a circuit using a multimeter, the following general steps are taken:

1. Insert one meter lead into the connector labeled for *ohms* (Ω) on the meter and the other lead into the connector labeled *common* (COM). Zero the ohmmeter if necessary.

2. Set the function switch to the *ohms* (Ω) position.

3. Select an appropriate multiplier setting (×1, ×10, etc.).

4. Disconnect power from the circuit.

5. Remove the resistor to be measured or lift one lead so that it is disconnected from the rest of the circuit. If you wish to measure total resistance, none of the components has to be removed.

6. Connect the ohmmeter leads across the selected resistor as shown in Figure C–5. Polarity does not matter.

7. Read the resistance and reconnect the resistor and the power.

D CAPACITOR COLOR CODING

Some capacitors have color-coded designations. The color code used for capacitors is basically the same as that used for resistors. Some variations occur in tolerance designation. The basic color codes are shown in Table D–1, and some typical color-coded capacitors are illustrated in Figure D–1.

TABLE D–1

Typical composite color codes for capacitors (picofarads)

COLOR	DIGIT	MULTIPLIER	TOLERANCE
Black	0	1	20%
Brown	1	10	1%
Red	2	100	2%
Orange	3	1000	3%
Yellow	4	10000	
Green	5	100000	5% (EIA)
Blue	6	1000000	
Violet	7		
Gray	8		
White	9		
Gold		0.1	5% (JAN)
Silver		0.01	10%
No color			20%

NOTE: EIA stands for Electronic Industries Association, and JAN stands for Joint Army-Navy, a military standard.

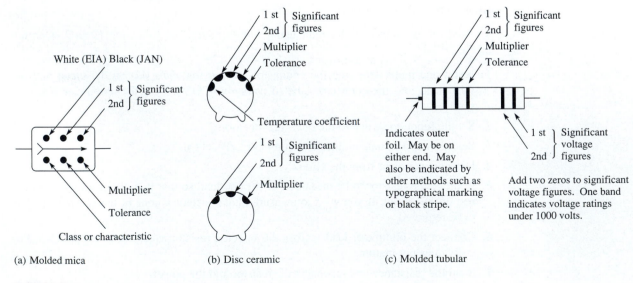

(a) Molded mica

(b) Disc ceramic

(c) Molded tubular

FIGURE D–1

Typical color-coded capacitors.

NORTON'S THEOREM, MILLMAN'S THEOREM, AND Δ-TO-Y/Y-TO-Δ NETWORK CONVERSIONS

E

NORTON'S THEOREM

Like Thevenin's theorem, Norton's theorem provides a method of reducing a more complex circuit to a simpler form. The basic difference is that Norton's theorem gives an equivalent current source in parallel with an equivalent resistance. The form of Norton's equivalent circuit is shown in Figure E–1. Regardless of how complex the original circuit is, it can always be reduced to this equivalent form. The equivalent current source is designated I_N, and the equivalent resistance, R_N.

FIGURE E–1

The form of Norton's equivalent circuit.

To apply Norton's theorem, you must know how to find the two quantities I_N and R_N. Once you know them for a given circuit, simply connect them in parallel to get the complete Norton circuit.

Norton's Equivalent Current (I_N)

As stated, I_N is one part of the complete Norton equivalent circuit; R_N is the other part. I_N is defined to be the short circuit current between two points in a circuit. Any component connected between these two points effectively "sees" a current source of value I_N in parallel with R_N.

To illustrate, suppose that a resistive circuit of some kind has a resistor connected between two points in the circuit, as shown in Figure E–2(a). We wish to find the Norton circuit that is equivalent to the one shown as "seen" by R_L. To find I_N, calculate the current between points A and B with these two points shorted, as shown in Figure E–2(b). Example E–1 demonstrates how to find I_N.

(a) Original circuit   (b) Short the terminals to get I_N.

FIGURE E–2

Determining the Norton equivalent current, I_N.

681

EXAMPLE E–1   Determine I_N for the circuit within the shaded area in Figure E–3(a).

FIGURE E–3

Solution   Short terminals A and B as shown in Figure E–3(b). I_N is the current through the short and is calculated as follows: First, the total resistance seen by the voltage source is

$$R_T = R_1 + \frac{R_2 R_3}{R_2 + R_3} = 47\ \Omega + \frac{(47\ \Omega)(100\ \Omega)}{147\ \Omega} = 79\ \Omega$$

The total current from the source is

$$I_T = \frac{V_S}{R_T} = \frac{83.3\ \text{V}}{79\ \Omega} = 1.05\ \text{A}$$

Now apply the current-divider formula to find I_N (the current through the short):

$$I_N = \left(\frac{R_2}{R_2 + R_3} \right) I_T = \left(\frac{47\ \Omega}{147\ \Omega} \right) 1.05\ \text{A} = 336\ \text{mA}$$

This is the value for the equivalent Norton current source.

Norton's Equivalent Resistance (R_N)

We define R_N in the same way as R_{TH}: it is the total resistance appearing between two terminals in a given circuit with all sources replaced by their internal resistances. Example E–2 demonstrates how to find R_N.

EXAMPLE E–2   Find R_N for the circuit within the shaded area of Figure E–3 (see Example E–1).

Solution   First reduce V_S to zero by shorting it, as shown in Figure E–4.
Looking in at terminals A and B, you can see that the parallel combination of R_1 and R_2 is in series with R_3. Thus,

$$R_N = R_3 + \frac{R_1}{2} = 100\ \Omega + \frac{47\ \Omega}{2} = 124\ \Omega$$

FIGURE E–4

The two examples have shown how to find the two equivalent components of a Norton equivalent circuit, I_N and R_N. Keep in mind that these values can be found for any linear circuit. Once these are known, they must be connected in parallel to form the Norton equivalent circuit, as illustrated in Example E–3.

EXAMPLE E–3

Draw the complete Norton circuit for the original circuit in Figure E–3 (Example E–1).

Solution In Examples E–1 and E–2 you found that $I_N = 336$ mA and $R_N = 124\ \Omega$. The Norton equivalent circuit is shown in Figure E–5.

FIGURE E–5

Summary of Norton's Theorem

Any load resistor connected between the terminals of a Norton equivalent circuit will have the same current through it and the same voltage across it as if it were connected to the terminals of the original circuit. A summary of steps for theoretically applying Norton's theorem is as follows:

1. Short the two terminals between which you want to find the Norton equivalent circuit.
2. Determine the current (I_N) through the shorted terminals.
3. Determine the resistance (R_N) between the two terminals (opened) with all voltage sources shorted and all current sources opened ($R_N = R_{TH}$).
4. Connect I_N and R_N in parallel to produce the complete Norton equivalent for the original circuit.

Norton's equivalent circuit can also be derived from Thevenin's equivalent circuit by use of the source conversion method.

MILLMAN'S THEOREM

Millman's theorem permits any number of parallel voltage sources to be reduced to a single equivalent voltage source. It simplifies finding the voltage across or current through a load. Millman's theorem gives the same results as Thevenin's theorem for the special case of parallel voltage sources. A conversion by Millman's theorem is illustrated in Figure E–6.

FIGURE E–6
Reduction of parallel voltage sources to a single equivalent voltage source.

Millman's Equivalent Voltage (V_{EQ}) and Equivalent Resistance (R_{EQ})

Millman's theorem gives us a formula for calculating the equivalent voltage, V_{EQ}. To find V_{EQ}, convert each of the parallel voltage sources into the current sources, as shown in Figure E–7.

In Figure E–7(b), the total current from the parallel current sources is

$$I_T = I_1 + I_2 + I_3 + \cdots + I_n$$

The total conductance between terminals A and B is

$$G_T = G_1 + G_2 + G_3 + \cdots + G_n$$

where $G_T = 1/R_T$, $G_1 = 1/R_1$, and so on. Remember, the current sources are effectively open. Therefore, by Millman's theorem, the equivalent resistance, R_{EQ}, is the total resistance R_T.

$$R_{EQ} = \frac{1}{G_T} = \frac{1}{(1/R_1) + (1/R_2) + (1/R_3) + \cdots + (1/R_n)} \tag{E–1}$$

FIGURE E–7
Parallel voltage sources converted to current sources.

By Millman's theorem, the equivalent voltage is $I_T R_{EQ}$, where I_T is expressed as follows:

$$I_T = I_1 + I_2 + I_3 + \cdots + I_n = \frac{V_1}{R_1} + \frac{V_2}{R_2} + \frac{V_3}{R_3} + \cdots + \frac{V_n}{R_n}$$

The following is the formula for the equivalent voltage:

$$V_{EQ} = \frac{(V_1/R_1) + (V_2/R_2) + (V_3/R_3) + \cdots + (V_n/R_n)}{(1/R_1) + (1/R_2) + (1/R_3) + \cdots + (1/R_n)} \tag{E–2}$$

Equations (E–1) and (E–2) are the two Millman formulas. The equivalent voltage source has a polarity such that the total current through a load will flow in the same direction as in the original circuit.

EXAMPLE E–4

Use Millman's theorem to find the voltage across R_L and the current through R_L in Figure E–8.

FIGURE E–8

Solution   Apply Millman's theorem as follows:

$$R_{EQ} = \frac{1}{(1/R_1) + (1/R_2) + (1/R_3)}$$

$$= \frac{1}{(1/22\ \Omega) + (1/22\ \Omega) + (1/10\ \Omega)} = \frac{1}{0.19} = 5.24\ \Omega$$

$$V_{EQ} = \frac{(V_1/R_2) + (V_2/R_2) + (V_3/R_3)}{(1/R_1) + (1/R_2) + (1/R_3)}$$

$$= \frac{(10\ V/22\ \Omega) + (5\ V/22\ \Omega) + (15\ V/10\ \Omega)}{(1/22\ \Omega) + (1/22\ \Omega) + (1/10\ \Omega)} = \frac{2.18\ A}{0.19\ S} = 11.5\ V$$

The single equivalent voltage source is shown in Figure E–9.

FIGURE E–9

Now calculate I_L and V_L for the load resistor:

$$I_L = \frac{V_{EQ}}{R_{EQ} + R_L} = \frac{11.5\ V}{52.2\ \Omega} = 220\ mA$$

$$V_L = I_L R_L = (220\ mA)(47\ \Omega) = 10.3\ V$$

DELTA-WYE (Δ-TO-Y) AND WYE-DELTA (Y-TO-Δ) NETWORK CONVERSIONS

A resistive delta (Δ) network has the form shown in Figure E–10(a). A wye (Y) network is shown in Figure E–10(b). Notice that letter subscripts are used to designate resistors in the delta network and that numerical subscripts are used to designate resistors in the wye.

FIGURE E–10
Delta and wye networks.

(a) Delta   (b) Wye

Conversion between these two forms of circuits is sometimes helpful in areas such as bridge analysis and three-phase power systems. In this section, the conversion formulas and rules for remembering them are given.

Δ-to-Y Conversion

It is convenient to think of the wye positioned within the delta, as shown in Figure E–11. To convert from delta to wye, we need R_1, R_2, and R_3 in terms of R_A, R_B, and R_C. The conversion rule is as follows: *Each resistor in the wye is equal to the product of the resistors in two adjacent delta branches, divided by the sum of all three delta resistors.*

FIGURE E–11
"Y within Δ" aid for conversion formulas.

In Figure E–11, R_A and R_C are "adjacent" to R_1:

$$R_1 = \frac{R_A R_C}{R_A + R_B + R_C} \qquad \text{(E–3)}$$

Also, R_B and R_C are "adjacent" to R_2:

$$R_2 = \frac{R_B R_C}{R_A + R_B + R_C} \qquad \text{(E–4)}$$

and R_A and R_B are "adjacent" to R_3:

$$R_3 = \frac{R_A R_B}{R_A + R_B + R_C} \qquad \text{(E–5)}$$

Y-to-Δ Conversion

To convert from wye to delta, we need R_A, R_B, and R_C in terms of R_1, R_2 and R_3. The conversion rule is as follows: *Each resistor in the delta is equal to the sum of all possible products of wye resistors taken two at a time, divided by the opposite wye resistor.*

In Figure E–11, R_2 is "opposite" to R_A:

$$R_A = \frac{R_1R_2 + R_1R_3 + R_2R_3}{R_2} \qquad \text{(E–6)}$$

Also, R_1 is "opposite" to R_B:

$$R_B = \frac{R_1R_2 + R_1R_3 + R_2R_3}{R_1} \qquad \text{(E–7)}$$

and R_3 is "opposite" to R_C:

$$R_C = \frac{R_1R_2 + R_1R_3 + R_2R_3}{R_3} \qquad \text{(E–8)}$$

The following two examples illustrate conversion between these two forms of circuits.

EXAMPLE E–5

Convert the delta network in Figure E–12 to a wye network.

FIGURE E–12

Solution   Use Equations (E–3), (E–4), and (E–5):

$$R_1 = \frac{R_A R_C}{R_A + R_B + R_C} = \frac{(220\ \Omega)(100\ \Omega)}{220\ \Omega + 560\ \Omega + 100\ \Omega} = 25\ \Omega$$

$$R_2 = \frac{R_B R_C}{R_A + R_B + R_C} = \frac{(560\ \Omega)(100\ \Omega)}{880\ \Omega} = 63.6\ \Omega$$

$$R_3 = \frac{R_A R_B}{R_A + R_B + R_C} = \frac{(220\ \Omega)(560\ \Omega)}{880\ \Omega} = 140\ \Omega$$

The resulting wye network is shown in Figure E–13.

FIGURE E–13

EXAMPLE E–6   Convert the wye network in Figure E–14 to a delta network.

FIGURE E–14

$$
\begin{array}{c}
R_1 \quad R_2 \\
1\ \text{k}\Omega \quad 2.2\ \text{k}\Omega \\
R_3 \quad 5.6\ \text{k}\Omega
\end{array}
$$

Solution   Use Equations (E–6), (E–7), and (E–8):

$$R_A = \frac{R_1 R_2 + R_1 R_3 + R_2 R_3}{R_2}$$

$$= \frac{(1\ \text{k}\Omega)(2.2\ \text{k}\Omega) + (1\ \text{k}\Omega)(5.6\ \text{k}\Omega) + (2.2\ \text{k}\Omega)(5.6\ \text{k}\Omega)}{2.2\ \text{k}\Omega} = 9.15\ \text{k}\Omega$$

$$R_B = \frac{R_1 R_2 + R_1 R_3 + R_2 R_3}{R_1}$$

$$= \frac{(1\ \text{k}\Omega)(2.2\ \text{k}\Omega) + (1\ \text{k}\Omega)(5.6\ \text{k}\Omega) + (2.2\ \text{k}\Omega)(5.6\ \text{k}\Omega)}{1\ \text{k}\Omega} = 20.1\ \text{k}\Omega$$

$$R_C = \frac{R_1 R_2 + R_1 R_3 + R_2 R_3}{R_3}$$

$$= \frac{(1\ \text{k}\Omega)(2.2\ \text{k}\Omega) + (1\ \text{k}\Omega)(5.6\ \text{k}\Omega) + (2.2\ \text{k}\Omega)(5.6\ \text{k}\Omega)}{5.6\ \text{k}\Omega} = 3.59\ \text{k}\Omega$$

The resulting delta network is shown in Figure E–15.

FIGURE E–15

$$
\begin{array}{c}
R_C \\
3.59\ \text{k}\Omega \\
R_A \qquad R_B \\
9.15\ \text{k}\Omega \qquad 20.1\ \text{k}\Omega
\end{array}
$$

Application of Δ-to-Y Conversion to the Simplification of a Bridge Circuits

In Chapter 6, you saw how Thevenin's theorem can be used to simplify a bridge circuit. Now you will see how Δ-to-Y conversion can be used for conversion of a bridge circuit to a series parallel form for easier analysis.

Figure E–16 illustrates how the delta (Δ) formed by R_A, R_B, and R_C can be converted to a wye (Y), thus creating an equivalent series-parallel circuit. Equations (E–3), (E–4), and (E–5) are used in this conversion.

In a bridge circuit, the load is connected across points C and D. In Figure E–16(a), R_C represents the load resistor. When voltage is applied across points A and B, the voltage

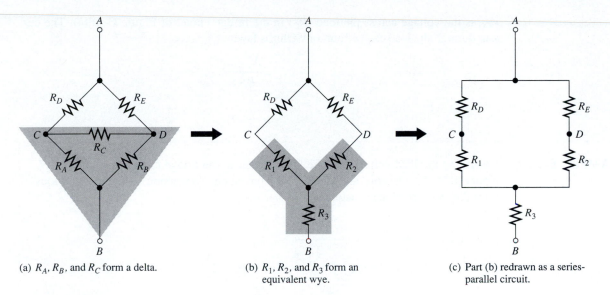

FIGURE E–16
Conversion of a bridge circuit to a series-parallel configuration.

(a) R_A, R_B, and R_C form a delta.

(b) R_1, R_2, and R_3 form an equivalent wye.

(c) Part (b) redrawn as a series-parallel circuit.

from C to D (V_{CD}) can be determined using the equivalent series-parallel circuit in Figure E–16(c) as follows. The total resistance from point A to point B is

$$R_T = \frac{(R_1 + R_D)(R_2 + R_E)}{(R_1 + R_D) + (R_2 + R_E)} + R_3$$

Then

$$I_T = \frac{V_{AB}}{R_T}$$

The resistance of the parallel portion of the circuit in Figure E–16(c) is

$$R_{T(p)} = \frac{(R_1 + R_D)(R_2 + R_E)}{(R_1 + R_D) + (R_2 + R_E)}$$

The current through the left branch is

$$I_{AC} = \left(\frac{R_{T(p)}}{R_1 + R_D}\right)I_T$$

The current through the right branch is

$$I_{AD} = \left(\frac{R_{T(p)}}{R_2 + R_E}\right)I_T$$

The voltage at point C with respect to point A is

$$V_{CA} = V_A - I_{AC}R_D$$

The voltage at point D with respect to point A is

$$V_{DA} = V_A - I_{AD}R_E$$

The voltage from point C to point D is

$$V_{CD} = V_{CA} - V_{DA} = (V_A - I_{AC}R_D) - (V_A - I_{AD}R_E)$$
$$= I_{AD}R_E - I_{AC}R_D$$

V_{CD} is the voltage across the load (R_C) in the bridge circuit of Figure E–16(a). The current through the load can be found by Ohm's law:

$$I_C = \frac{V_{CD}}{R_C}$$

EXAMPLE E–7

Determine the load voltage and the load current in the bridge circuit in Figure E–17. Notice that the resistors are labeled for convenient conversion using the previous formulas. R_C is the load resistor.

FIGURE E–17

Solution   First convert the delta formed by R_A, R_B, and R_C to a wye:

$$R_1 = \frac{R_A R_C}{R_A + R_B + R_C} = \frac{(2.2\ \text{k}\Omega)(18\ \text{k}\Omega)}{2.2\ \text{k}\Omega + 2.7\ \text{k}\Omega + 18\ \text{k}\Omega} = 1.73\ \text{k}\Omega$$

$$R_2 = \frac{R_B R_C}{R_A + R_B + R_C} = \frac{(2.7\ \text{k}\Omega)(18\ \text{k}\Omega)}{22.9\ \text{k}\Omega} = 2.12\ \text{k}\Omega$$

$$R_3 = \frac{R_A R_B}{R_A + R_B + R_C} = \frac{(2.2\ \text{k}\Omega)(2.7\ \text{k}\Omega)}{22.9\ \text{k}\Omega} = 259\ \Omega$$

The resulting equivalent series-parallel circuit is shown in Figure E–18.

FIGURE E–18

Now determine R_T and the branch currents in Figure E–18.

$$R_T = \frac{(R_1 + R_D)(R_2 + R_E)}{(R_1 + R_D) + (R_2 + R_E)} + R_3 = \frac{(6.43 \text{ k}\Omega)(6.02 \text{ k}\Omega)}{6.43 \text{ k}\Omega + 6.02 \text{ k}\Omega} + 259 \ \Omega = 3.37 \text{ k}\Omega$$

$$I_T = \frac{V_{AB}}{R_T} = \frac{120 \text{ V}}{3.37 \text{ k}\Omega} = 35.6 \text{ mA}$$

The total resistance of the parallel part of the circuit, $R_{T(p)}$, is 3.11 kΩ.

$$I_{AC} = \left(\frac{R_{T(p)}}{R_1 + R_D} \right) I_T = \left(\frac{3.11 \text{ k}\Omega}{1.73 \text{ k}\Omega + 4.7 \text{ k}\Omega} \right) 35.6 \text{ mA} = 17.2 \text{ mA}$$

$$I_{AD} = \left(\frac{R_{T(p)}}{R_2 + R_E} \right) I_T = \left(\frac{3.11 \text{ k}\Omega}{2.12 \text{ k}\Omega + 3.9 \text{ k}\Omega} \right) 35.6 \text{ mA} = 18.4 \text{ mA}$$

The voltage across the load is

$$
\begin{aligned}
V_{CD} &= I_{AD}R_E - I_{AC}R_D \\
&= (18.4 \text{ mA})(3.9 \text{ k}\Omega) - (17.2 \text{ mA})(4.7 \text{ k}\Omega) \\
&= 71.8 \text{ V} - 80.8 \text{ V} = -9 \text{ V}
\end{aligned}
$$

The load current is

$$I_C = \frac{V_{CD}}{R_C} = \frac{-9 \text{ V}}{18 \text{ k}\Omega} = -500 \ \mu\text{A}$$

ANSWERS TO SELF-TESTS

Chapter 1
1. (c) **2.** (c) **3.** (c) **4.** (d) **5.** (b) **6.** (b) **7.** (a) **8.** (d)
9. (d) **10.** (d) **11.** (c) **12.** (c)

Chapter 2
1. (b) **2.** (a) **3.** (c) **4.** (c) **5.** (b) **6.** (b) **7.** (c) **8.** (d)
9. (b) **10.** (b) **11.** (e) **12.** (b) **13.** (b) **14.** (c)

Chapter 3
1. (b) **2.** (c) **3.** (b) **4.** (d) **5.** (a) **6.** (d) **7.** (b) **8.** (c)
9. (d) **10.** (a) **11.** (a) **12.** (d) **13.** (c) **14.** (a) **15.** (c)

Chapter 4
1. (a) **2.** (d) **3.** (b) **4.** (d) **5.** (d) **6.** (a) **7.** (b) **8.** (c)
9. (b) **10.** (c) **11.** (a) **12.** (d) **13.** (d) **14.** (d)

Chapter 5
1. (b) **2.** (c) **3.** (b) **4.** (d) **5.** (a) **6.** (c) **7.** (c) **8.** (c)
9. (d) **10.** (b) **11.** (a) **12.** (d) **13.** (c) **14.** (b)

Chapter 6
1. (e) **2.** (c) **3.** (c) **4.** (c) **5.** (b) **6.** (a) **7.** (b) **8.** (b)
9. (d) **10.** (b) **11.** (b) **12.** (c) **13.** (d) **14.** (a) **15.** (d)

Chapter 7
1. (b) **2.** (c) **3.** (a) **4.** (d) **5.** (b) **6.** (c) **7.** (a) **8.** (d)
9. (c) **10.** (c) **11.** (b) **12.** (c)

Chapter 8
1. (b) **2.** (b) **3.** (c) **4.** (b) **5.** (d) **6.** (a) **7.** (a) **8.** (a)
9. (c) **10.** (b) **11.** (a) **12.** (d) **13.** (c) **14.** (b) **15.** (d)

Chapter 9
1. (g) **2.** (b) **3.** (c) **4.** (d) **5.** (a) **6.** (d) **7.** (a) **8.** (f)
9. (c) **10.** (b) **11.** (d) **12.** (a) **13.** (b) **14.** (a) **15.** (b) **16.** (d)

Chapter 10
1. (c) **2.** (d) **3.** (c) **4.** (b) **5.** (d) **6.** (a) **7.** (d) **8.** (c)
9. (b) **10.** (a) **11.** (d) **12.** (b)

Chapter 11
1. (b) **2.** (c) **3.** (d) **4.** (a) **5.** (b) **6.** (c) **7.** (d) **8.** (b)
9. (a) **10.** (c) **11.** (d) **12.** (c) **13.** (a) **14.** (c)

Chapter 12
1. (c) **2.** (b) **3.** (b) **4.** (a) **5.** (d) **6.** (b) **7.** (a) **8.** (d)
9. (c) **10.** (b) **11.** (d) **12.** (d) **13.** (c) **14.** (b) **15.** (d)

Chapter 13

1. (f)　　**2.** (b)　　**3.** (a)　　**4.** (d)　　**5.** (a)　　**6.** (d)　　**7.** (b)　　**8.** (b)
9. (a)　　**10.** (d)　　**11.** (c)　　**12.** (d)　　**13.** (b)　　**14.** (c)　　**15.** (a)

Chapter 14

1. (a)　　**2.** (c)　　**3.** (b)　　**4.** (c)　　**5.** (d)　　**6.** (c)　　**7.** (a)　　**8.** (b)
9. (d)　　**10.** (a)　　**11.** (b)　　**12.** (c)

Chapter 15

1. (b)　　**2.** (c)　　**3.** (a)　　**4.** (d)　　**5.** (a)　　**6.** (a)　　**7.** (a)　　**8.** (c)
9. (b)　　**10.** (d)

ANSWERS TO ODD-NUMBERED PROBLEMS

Chapter 1

1. (a) 3×10^3 (b) 75×10^3 (c) 2×10^6

3. (a) $8.4 \times 10^3 = 0.84 \times 10^4 = 0.084 \times 10^5$ (b) $99 \times 10^3 = 9.9 \times 10^4 = 0.99 \times 10^5$
(c) $200 \times 10^3 = 20 \times 10^4 = 2 \times 10^5$

5. (a) 0.0000025   (b) 5000   (c) 0.39

7. (a) 126×10^6 (b) 5.00085×10^3 (c) 6060×10^{-9}

9. (a) 20×10^8 (b) 36×10^{14} (c) 15.4×10^{-15}

11. (a) 2370×10^{-6} (b) 18.56×10^{-6}
(c) $0.00574389 \times 10^{-6}$ (d) $100{,}000{,}000{,}000 \times 10^{-6}$

13. (a) 12.25×10^{14} (b) 5×10^3 (c) 10.575×10^{-6} (d) 2×10^{10}

15. (a) $3 \, \mu F$ (b) $3.3 \, M\Omega$ (c) $350 \, nA$

17. (a) $5000 \, \mu A$ (b) $3.2 \, mW$ (c) $5 \, MV$ (d) $10{,}000 \, kV$

Chapter 2

1. $80 \times 10^{12} \, C$

3. (a) 10 V   (b) 2.5 V   (c) 4 V

5. 20 V

7. (a) 75 A   (b) 20 A   (c) 2.5 A

9. 2 s

11. (1) $6800 \, \Omega \pm 10\%$ (2) $33 \, \Omega \pm 10\%$ (3) $47{,}000 \, \Omega \pm 20\%$

13. (a) Red, violet, brown   (b) 2—330 Ω, 4—2.2 kΩ, 12—56 kΩ, 6—100 kΩ, 1—39 kΩ

15. Current flows through lamp 2.

17. Ammeter in series with resistors, with its negative terminal to the negative terminal of source and its positive terminal to one side of R_1. Voltmeter placed across (in parallel with) the source (negative to negative, positive to positive).

19. Position 1: V1 = 0, V2 = V_S
Position 2: V1 = V_S, V2 = 0

21. (a) 0.25 V   (b) 250 V

23. (a) 200 Ω (b) 150 MΩ (c) 4500 Ω

25. 33.3 V

27. AWG #27

29. Circuit (b)

31. One ammeter in series with battery. One ammeter in series with each resistor (seven total).

33. See Figure P–1.

FIGURE P–1

Chapter 3

1. (a) 3 A   (b) 0.2 A   (c) 1.5 A

3. 15 mA

5. (a) 3.33 mA   (b) 550 μA   (c) 588 μA   (d) 500 mA   (e) 6.60 mA

7. (a) 2.50 mA   (b) 2.27 mA   (c) 8.33 mA

9. (a) 10 mV   (b) 1.65 V   (c) 15 kV   (d) 3.52 V
(e) 0.25 V   (f) 750 kV   (g) 8.5 kV   (h) 3.75 mV

11. (a) 81 V   (b) 500 V   (c) 125 V

13. (a) 2 kΩ (b) 3.5 kΩ (c) 2 kΩ (d) 100 kΩ (e) 1 MΩ

15. (a) 4 Ω (b) 3 kΩ (c) 200 kΩ

17. 417 mW

19. (a) 1 MW   (b) 3 MW   (c) 150 MW   (d) 8.7 MW

21. (a) 2,000,000 μW   (b) 500 μW   (c) 250 μW   (d) 6.67 μW

23. 16.5 mW

25. 1.18 kW

27. 6 W

29. 25 Ω

31. 0.00186 kWh

33. 1 W

35. (a) positive at top   (b) positive at bottom   (c) positive at right

37. 36 Ah

39. 13.5 mA

41. 4.25 W

43. 150 Ω

45. $V = 0$ V, $I = 0$ A; $V = 10$ V, $I = 100$ mA; $V = 20$ V, $I = 200$ mA; $V = 30$ V, $I = 300$ mA;
$V = 40$ V, $I = 400$ mA; $V = 50$ V, $I = 500$ mA; $V = 60$ V, $I = 600$ mA; $V = 70$ V, $I = 700$ mA;
$V = 80$ V, $I = 800$ mA; $V = 90$ V, $I = 900$ mA; $V = 100$ V, $I = 1$ A
The graph is a straight line. There is a linear relationship between V and I.

47. $R_1 = 0.5$ Ω; $R_2 = 1$ Ω; $R_3 = 2$ Ω

49. 10 V; 30 V

51. $I_{MAX} = 3.83$ mA; $I_{MIN} = 3.46$ mA

53. 216 kWh

55. 12 W

57. 2 A

Chapter 4

1. See Figure P–2.

FIGURE P–2

3. 0.1 A

5. 138 Ω

7. (a) 7.9 kΩ (b) 33 Ω (c) 13.24 MΩ
The series circuit is disconnected from the source, and the ohmmeter is connected across the circuit terminals.

9. 1126 Ω

11. (a) 170 kΩ (b) 50 Ω (c) 12.4 kΩ (d) 1.97 kΩ

13. (a) 625 μA   (b) 4.26 μA. The ammeter is connected in series.

15. 355 mA

17. 14 V

19. 26 V

21. (a) $V_2 = 6.8$ V   (b) $V_R = 8$ V, $V_{2R} = 16$ V, $V_{3R} = 24$ V, $V_{4R} = 32$ V
The voltmeter is connected across (in parallel with) each resistor for which the voltage is unknown.

23. (a) 3.84 V   (b) 6.77 V

25. $V_{5.6k\Omega} = 10$ V; $V_{1k\Omega} = 1.79$ V; $V_{560\Omega} = 1$ V; $V_{10k\Omega} = 17.9$ V

27. 55 mW

29. Measure V_A and V_B with respect to ground; then $V_{R2} = V_A - V_B$.

31. (a) R_4 open   (b) R_4 and R_5 shorted

33. 780 Ω

35. $V_{AF} = 100$ V; $V_{BF} = 77.2$ V; $V_{CF} = 66.9$ V; $V_{DF} = 18.2$ V; $V_{EF} = 5.81$ V; $V_F = 0$ V

37. 500 Ω

39. (a) 19.1 mA   (b) 45.8 V   (c) $R(\frac{1}{8}$ W$) = 343$ Ω, $R(\frac{1}{4}$ W$) = 686$ Ω, $R(\frac{1}{2}$ W$) = 1372$ Ω

41. See Figure P–3.

FIGURE P–3

43. $R_1 + R_7 + R_8 + R_{10} = 4.23$ kΩ; $R_2 + R_4 + R_6 + R_{11} = 23.6$ kΩ;
$R_3 + R_5 + R_9 + R_{12} = 19.9$ kΩ

45. A: 5.45 mA; B: 6.06 mA; C: 7.95 mA; D: 12 mA

47. A: $V_1 = 6.03$ V, $V_2 = 3.35$ V, $V_3 = 2.75$ V, $V_4 = 1.88$ V, $V_5 = 4.0$ V;
B: $V_1 = 6.71$ V, $V_2 = 3.73$ V, $V_3 = 3.06$ V, $V_5 = 4.5$ V;
C: $V_1 = 8.1$ V, $V_2 = 4.5$ V, $V_5 = 5.4$ V;
D: $V_1 = 10.8$ V, $V_5 = 7.2$ V

49. Yes, R_3 and R_5 are shorted.

51. (a) R_{11} has burned open due to excessive power.   (b) Replace R_{11} (10 kΩ).   (c) 338 V

Chapter 5

1. See Figure P–4.

FIGURE P–4

3. 12 V; 5 mA

5. 1350 mA

7. $I_2 = I_3 = 7.5$ mA. An ammeter in series with each resistor in each branch.

9. (a) 359 Ω (b) 25.6 Ω (c) 819 Ω (d) 996 Ω

11. 2 kΩ

13. (a) 909 mA   **(b)** 76 mA

15. Circuit (a)

17. $I_1 = 2.19$ A; $I_2 = 811$ mA

19. 200 mW

21. 682 mA; 4.09 A

23. The 1 kΩ resistor is open.

25. $R_2 = 25$ Ω; $R_3 = 100$ Ω; $R_4 = 12.5$ Ω

27. $I_R = 4.8$ A; $I_{2R} = 2.4$ A; $I_{3R} = 1.6$ A; $I_{4R} = 1.2$ A

29. (a) $R_1 = 100$ Ω, $R_2 = 200$ Ω, $I_2 = 50$ mA
     (b) $I_1 = 125$ mA, $I_2 = 74.9$ mA, $R_1 = 79.9$ Ω, $R_2 = 134$ mA, $V_s = 10$ V
     (c) $I_1 = 253$ mA, $I_2 = 147$ mA, $I_3 = 100$ mA, $R_1 = 395$ Ω

31. 53.7 Ω

33. 1.08 A

35. $R_1, R_2, R_5, R_9, R_{10}$, and R_{12}; $R_T = 2.57$ MΩ
     R_4, R_6, R_7, and R_8; $R_T = 2.77$ MΩ
     R_3 and R_{11}; $R_T = 2.13$ MΩ

37. $R_2 = 750$ Ω; $R_4 = 423$ Ω

39. The 4.7 kΩ resistor is open.

41. (a) One of the resistors has opened due to excessive power dissipation.   **(b)** 30 V
     (c) Replace the 1.8 kΩ resistor.

43. (a) 940 Ω   **(b)** 518 Ω   **(c)** 518 Ω   **(d)** 422 Ω

Chapter 6

1. R_2, R_3, and R_4 are in parallel, and this parallel combination is in series with both R_1 and R_5.

3. See Figure P–5.

(a)   (b)

FIGURE P–5

5. 2003 Ω

7. (a) 128 Ω   **(b)** 791 Ω

9. (a) $I_1 = I_4 = 11.7$ mA, $I_2 = I_3 = 5.85$ mA, $V_1 = 655$ mV, $V_2 = V_3 = 585$ mV, $V_4 = 257$ mV
     (b) $I_1 = 3.8$ mA, $I_2 = 618$ µA, $I_3 = 1.27$ mA, $I_4 = 1.91$ mA, $V_1 = 2.58$ V,
         $V_2 = V_3 = V_4 = 420$ mV

11. 7.5 V unloaded; 7.29 V loaded

13. 56 kΩ load

15. 1000 V

17. (a) 0.5 V range   **(b)** Approximately 1 mV

19. 360 Ω

21. 21 mA

23. $R_{TH} = 18$ kΩ; $V_{TH} = 2.7$ V

25. No, the meter should read 4.39 V. The 680 Ω is open.

27. The 7.62 V and 5.24 V readings are incorrect, indicating that the 3.3 kΩ resistor is open.

29. (a) $V_1 = -10$ V, all others 0 V   (b) $V_1 = -2.33$ V, $V_4 = -7.67$ V, $V_2 = -2.33$ V, $V_3 = 0$ V
   (c) $V_1 = -2.33$ V, $V_4 = -7.67$ V, $V_2 = V_3 = 0$ V   (d) $V_1 = -10$ V, all others 0 V

31. See Figure P–6.

FIGURE P–6

33. $R_T = 5.76$ kΩ; $V_A = 3.3$ V; $V_B = 1.7$ V; $V_C = 850$ mV

35. $V_1 = 1.61$ V; $V_2 = 6.7$ V; $V_3 = 1.77$ V; $V_4 = 3.4$ V; $V_5 = 387$ mV; $V_6 = 2.63$ V;
$V_7 = 387$ mV; $V_8 = 1.77$ V; $V_9 = 1.61$ V

37. 110 Ω

39. $R_1 = 180$ Ω; $R_2 = 60$ Ω. Output across R_2.

41. 842 μA

43. 11.7 V

45. See Figure P–7.

FIGURE P–7

47. Pos 1: $V_1 = 81.0$ V, $V_2 = 54.0$ V, $V_3 = 27.0$ V;
   Pos 2: $V_1 = 86.2$ V, $V_2 = 52.4$ V, $V_3 = 26.2$ V;
   Pos 3: $V_1 = 89.0$ V, $V_2 = 58.0$ V, $V_3 = 27.0$ V

49. One of the 12 kΩ resistors is open.

51. The 2.2 kΩ resistor is open.

53. $V_A = 0$ V; $V_B = 11.1$ V

Chapter 7

1. Decrease

3. 37.5 μWb

5. 597

7. 150 At

9. (a) Electromagnetic force   (b) Spring force

11. Electromagnetic force

13. Change the current.

15. Material A

17. 1 mA

19. To electrically connect the loop to the external circuit

21. The output voltage appears as the solid curve in Figure 7–38 with a frequency of 120 Hz and a peak of 10 V.

Chapter 8

1. (a) 1 Hz (b) 5 Hz (c) 20 Hz (d) 1 kHz (e) 2 kHz (f) 100 kHz

3. 2 μs

5. 250 Hz

7. 200 rps

9. (a) 7.07 mA (b) 4.5 mA (c) 14.14 mA

11. 15°, A leads.

13. See Figure P–8.

FIGURE P–8

15. (a) 22.5° (b) 60° (c) 90° (d) 108° (e) 216° (f) 324°

17. (a) 57.4 mA (b) 99.6 mA (c) −17.4 mA
   (d) −57.4 mA (e) −99.6 mA (f) 0 mA

19. 30°: 13.0 V; 45°: 14.5 V; 90°: 13.0 V; 180°: −7.5 V; 200°: −11.5 V; 300°: −7.5 V

21. (a) 7.07 mA (b) 6.37 mA (c) 10 mA (d) 20 mA (e) 10 mA

23. 7.38 V

25. 4.24 V

27. $V_p = 0.6$ V; $T = 500$ ms

29. $V_{rms} = 1.56$ V; $f = 1.47$ MHz

31. $t_r \cong 3.5$ ms; $t_f \cong 3.5$ ms; $t_W \cong 12.5$ ms; $V = 5$ V

33. (a) −0.375 V (b) 3.01 V

35. (a) 50 kHz (b) 10 Hz

37. 25 kHz

39. $I_{max} = 3.5$ A; $V_{avg} = 200$ V; See Figure P–9.

FIGURE P–9

41. (a) 2.5 (b) 3.54 V (c) 12.5 kHz

43. See Figure P–10.

45. $V_p = 1.89$ V; $f = 2$ Hz

FIGURE P–10

Chapter 9

1. **(a)** 5 μF   **(b)** 1 μC   **(c)** 10 V

3. **(a)** 0.001 μF   **(b)** 0.0035 μF   **(c)** 0.00025 μF

5. 2 μF

7. 88.5 pF

9. A 12.5 pF increase

11. Ceramic

13. **(a)** 0.02 μF   **(b)** 0.047 μF   **(c)** 0.001 μF   **(d)** 220 pF   **(e)** 100,000 pF

15. **(a)** Encapsulation   **(b)** Dielectric (ceramic disk)   **(c)** Plate (metal disk)   **(d)** Leads

17. **(a)** 0.67 μF   **(b)** 69 pF   **(c)** 2.7 μF

19. $V_1 = 2$ V; $V_2 = 10$ V; $V_3 = 5$ V; $V_4 = 1$ V

21. **(a)** 2.5 μF   **(b)** 717 pF   **(c)** 1.6 μF

23. **(a)** 100 μs   **(b)** 500 μs   **(c)** 23.5 μs   **(d)** 15 ms

25. **(a)** 9.48 V   **(b)** 13.0 V   **(c)** 14.3 V   **(d)** 14.7 V   **(e)** 14.9 V

27. **(a)** 2.72 V   **(b)** 5.90 V   **(c)** 11.7 V

29. **(a)** 339 kΩ **(b)** 13.6 kΩ **(c)** 677 Ω **(d)** 33.9 Ω

31. **(a)** 31.8 Ω **(b)** 111 kΩ **(c)** 49.7 Ω

33. 200 Ω

35. $P_{\text{true}} = 0$ W; $P_r = 3.39$ mVAR

37. 0 Ω

39. Shorted capacitor

41. 3.18 ms

43. 2.94 μs

45. **(a)** Charges to 3.32 V in 10 ms, then discharges to 0 V in 215 ms.
    (b) Charges to 3.32 V in 10 ms, then discharges to 2.85 V in 5 ms, then charges toward 20 V.

47. 0.0052 μF

49. $V_1 = 7.25$ V; $V_2 = 2.76$ V; $V_3 = 0.79$ V; $V_4 = 1.98$ V; $V_5 = 1.19$ V; $V_6 = 0.79$ V

Chapter 10

1. **(a)** 1000 mH   **(b)** 0.25 mH   **(c)** 0.01 mH   **(d)** 0.5 mH

3. 3536 turns

5. 50 mJ

7. 155 μH

9. 7.14 μH

11. **(a)** 4.33 H   **(b)** 50 mH   **(c)** 0.57 μH

13. **(a)** 1 μs   **(b)** 2.13 μs   **(c)** 2 μs

15. **(a)** 5.52 V   **(b)** 2.03 V   **(c)** 0.75 V   **(d)** 0.27 V   **(e)** 0.1 V

17. **(a)** 136 kΩ **(b)** 1.57 kΩ **(c)** 179 mΩ

19. $I_T = 10.1$ mA; $I_{L2} = 6.7$ mA; $I_{L3} = 3.35$ mA

21. 101 VAR

23. **(a)** Infinite resistance   **(b)** Zero resistance   **(c)** Lower R_W

25. 2.86 mA

27. 26.1 mA

Chapter 11

1. $1.5\ \mu\text{H}$

3. 3

5. (a) Positive at top   (b) Positive at bottom   (c) Positive at bottom

7. 500 turns

9. (a) Same polarity, 100 V rms   (b) Opposite polarity, 100 V rms

11. 240 V

13. 33.3 mA

15. $27.3\ \Omega$

17. 5.75 mA

19. 0.5

21. $5\ \text{k}\Omega$

23. No

25. 94.5 W

27. 0.98

29. 25 kVA

31. Secondary 1: 2; Secondary 2: 0.5; Secondary 3: 0.25

33. (a) 48 V   (b) 25 V

35. Excessive primary current is drawn, potentially burning out the source and/or the transformer, unless the primary is protected by a fuse.

37. (a) $V_{L1} = 35$ V, $I_{L1} = 2.92$ A, $V_{L2} = 15$ V, $I_{L2} = 1.5$ A   (b) $28.9\ \Omega$

39. (a) 20 V   (b) 10 V

41. 0.0145 (69.2:1)

Chapter 12

1. 8 kHz; 8 kHz

3. (a) $288\ \Omega$ (b) $1209\ \Omega$

5. (a) $798\ \text{k}\Omega$ (b) $169\ \text{k}\Omega$ (c) $97.3\ \text{k}\Omega$ (d) $64.4\ \text{k}\Omega$

7. (a) 34.7 mA   (b) 4.14 mA

9. $I_T = 12$ mA; $V_{C(0.1)} = 1.27$ V; $V_{C(0.2)} = 0.637$ V; $V_R = 0.6$ V; $\theta = 72.5°$ (V_s lagging I_T)

11. $807\ \Omega$; $-36.1°$

13. $353\ \Omega$; $62°$

15. $261\ \Omega$; $-79.8°$

17. $I_{C1} = 118$ mA; $I_{C2} = 50.3$ mA; $I_{R1} = 36.4$ mA; $I_{R2} = 44.4$ mA; $I_T = 187$ mA; $\theta = 64.4°$ (V_s lagging I_T)

19. (a) $3.9\ \text{k}\Omega$ (b) $20\ \mu\text{A}$ (c) $15.7\ \mu\text{A}$ (d) $25.6\ \mu\text{A}$ (e) $38.1°$ (V_s lagging I_T)

21. $V_{C1} = 8.77$ V; $V_{C2} = 3.24$ V; $V_{C3} = 3.24$ V; $V_{R1} = 2.1$ V; $V_{R2} = 1.14$ V

23. $I_T = 82.8$ mA; $I_{C2} = 15.3$ mA; $I_{C3} = 67.5$ mA; $I_{R1} = I_{R2} = 6.35$ mA

25. 4.03 VA

27. 0.91

29. (a) $0.056°$ (b) $5.60°$ (c) $44.4°$ (d) $84.2°$

31. (a) $90°$ (b) $86.4°$ (c) $57.8°$ (d) $9.04°$

33. See Figure P–11.

(a) For Figure 12–79   (b) For Figure 12–80

FIGURE P–11

35. Figure 12–79: 1.02 kHz; Figure 12–80: 1.59 kHz

37. No leakage: $V_{out} = 3.21$ V, $\theta = -18.7°$;   Leakage: $V_{out} = 2.83$ V, $\theta = -33.3°$

39. (a) 0 V   (b) 0.3 V   (c) 0.5 V   (d) 0 V

41. (a) $I_{LA} = 4.4$ A; $I_{LB} = 3.06$ A   (b) $P_{rA} = 509$ VAR; $P_{rB} = 211$ VAR
    (c) $P_{true(A)} = 823$ W; $P_{true(B)} = 641$ W   (d) $P_{aA} = 968$ VA; $P_{aB} = 674$ VA
    (e) Load A

43. 11.4 kΩ

45. $P_r = 1.32$ kVAR; $P_a = 2$ kVA

47. 0.1 μF

Chapter 13

1. 15 kHz

3. (a) 112 Ω   (b) 1.8 kΩ

5. (a) 17.4 Ω   (b) 64 Ω   (c) 127 Ω   (d) 251 Ω

7. (a) 89.3 mA   (b) 2.78 mA

9. 38.7°

11. See Figure P–12.

13. 7.69 Ω

FIGURE P–12

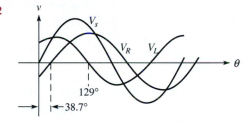

15. 2.39 kHz

17. (a) 274 Ω   (b) 89.3 mA   (c) 159 mA   (d) 183 mA   (e) 60.7° (I_T lagging V_s)

19. $V_{R1} = 7.92$ V; $V_{R2} = V_L = 20.8$ V

21. $I_T = 36$ mA; $I_L = 33.2$ mA; $I_{R2} = 13.9$ mA

23. 1.30 W; 1.04 VAR

25. $PF = 0.386$; $P_{true} = 347$ mW; $P_r = 692$ mVAR; $P_a = 900$ mVA

27. See Figure P–13.

29. See Figure P–14.

FIGURE P–13

FIGURE P–14

31. $V_{R1} = V_{L1} = 18$ V; $V_{R2} = V_{R3} = V_{L2} = 0$ V

33. 5.57 V

35. 88.7 mA

37. (a) 406 mA   (b) 228 mA   (c) 333 mA   (d) 336 mA

39. 0.133

41. See Figure P–15.

FIGURE P–15

Chapter 14

1. 480 Ω, 88.8° (V_s lagging I); 480 Ω capacitive

3. Impedance increases.

5. See Figure P–16.

FIGURE P–16

7. $X_L = 4.47$ kΩ; $X_C = 4.47$ kΩ, Z = 20 Ω; $I = 600$ mA

9. $f_r = 459$ kHz; $f_{C1} = 360$ kHz; $f_{C2} = 559$ kHz

11. (a) 14.5 kHz   (b) 25.2 kHz

13. (a) $f_r = 339$ kHz, $BW = 239$ kHz   (b) $f_r = 10.4$ kHz, $BW = 2.61$ kHz

15. Capacitive, $X_C < X_L$.

17. 758 Ω

19. 531 MΩ; 104 kHz

21. 62.5 Hz

23. 1.38 W

25. 200 Hz

27. $I_{R1} = I_C = 2.11$ mA; $I_{L1} = 1.33$ mA; $I_{L2} = 667$ μA; $I_{R2} = 667$ μA;
$V_{R1} = 6.96$ V; $V_C = 2.11$ V; $V_{L1} = V_{L2} = V_{R2} = 6.67$ V

29. $I_{R1} = I_{L1} = 41.5$ mA; $I_C = I_{L2} = 105$ mA; $I_T = 77.0$ mA

31. $L = 989$ μH; $C = 0.064$ μF

33. 8 MHz: $C = 40$ pF; 9 MHz: $C = 31$ pF; 10 MHz: $C = 25$ pF; 11 MHz: $C = 21$ pF

Chapter 15

1. 110 μs

3. 12.6 V

5. See Figure P–17.

7. See Figure P–18.

FIGURE P–17

FIGURE P–18

9. 15 V dc level with a very small charge/discharge fluctuation

11. Exchange positions of R and C. See Figure P–19 for the output voltage.
$5\tau = 5$ ms (repeating Problem 6)

FIGURE P–19

13. Approximately the same shape as input, but with an average value of 0 V

15. See Figure P–20.

FIGURE P–20

17. See Figure P–21.

FIGURE P–21

19. A 6 V dc level

21. **(b)** No fault   **(c)** Leaky capacitor   **(d)** Open C or shorted R

23. **(a)** 25 ms   **(b)** See Figure P–22.

FIGURE P–22

25. 1.44 s

GLOSSARY

Admittance (*Y*) A measure of the ability of a reactive circuit to permit current; the reciprocal of impedance. The unit is siemens (S).

Alternating current (ac) Current that reverses direction in response to a change in source voltage polarity.

Ammeter An electrical instrument used to measure current.

Ampere (A or amp) The unit of electrical current.

Ampere-hour rating A number given in ampere-hours determined by multiplying the current (A) times the length of time (h) a battery can deliver that current to a load.

Ampere-turn The unit of magnetomotive force (mmf).

Amplitude The maximum value of a voltage or current.

Apparent power (*P_a*) The product of the voltage times the current, expressed in volt-amperes (VA). In a purely resistive circuit it is the same as the true power.

Apparent power rating The method of rating transformers in which the power capability is expressed in volt-amperes (VA).

Atom The smallest particle of an element possessing the unique characteristics of that element.

Atomic number The number of electrons in a neutral atom.

Atomic weight The number of protons and neutrons in the nucleus of an atom.

Autotransformer A transformer in which the primary and secondary windings are in a single winding.

Average value The average of a sine wave over one half-cycle. It is 0.637 times the peak value.

AWG American wire gage. A standardization based on wire diameter.

Band-pass filter A filter that passes frequencies within a certain range and blocks all others.

Band-stop filter A filter that blocks frequencies within a certain range.

Bandwidth (*BW*) The characteristic of certain electronic circuits that specifies the usable range of frequencies for which signals pass from input to output without significant reduction in amplitude.

Baseline The normal level of a pulse waveform; the voltage level in the absence of a pulse.

Battery An energy source that uses a chemical reaction to convert chemical energy into electrical energy.

Bias The application of dc voltage to a diode or other electronic device to produce a desired mode of operation.

Bleeder current The current left after the total load current is subtracted from the total current into the circuit.

Branch One current path in a parallel circuit.

Capacitance The ability of a capacitor to store electrical charge.

Capacitive reactance The opposition of a capacitor to sinusoidal current. The unit is the ohm.

Capacitive susceptance (*B_C*) The ability of a capacitor to permit current; the reciprocal of capacitive reactance. The unit is the siemens (S).

Capacitor An electrical device consisting of two conductive plates separated by an insulating material and possessing the property of capacitance.

Cathode-ray tube (CRT) A vacuum tube device containing an electron gun that emits a narrow focused beam of electrons onto a phosphor-coated screen.

Center tap (CT) A connection at the midpoint of the secondary winding of a transformer.

Charge   An electrical property of matter that exists because of an excess or a deficiency of electrons. Charge can be either positive or negative.

Choke   An inductor. The term is used more commonly in connection with inductors used to block or choke off high frequencies.

Circuit   An interconnection of electrical components designed to produce a desired result. A basic circuit consists of a source, a load, and an interconnecting current path.

Circuit breaker   A resettable protective device used for interrupting excessive current in an electric circuit.

Circuit ground   A method of grounding whereby a large conducive area on a printed circuit board or the metal chassis that houses the assembly is used as the common or reference point; also called *chassis ground*.

Circular mil (CM)   The unit of the cross-sectional area of a wire.

Closed circuit   A circuit with a complete current path.

Coefficient of coupling (k)   A constant associated with transformers that is the ratio of secondary flux to primary flux. The ideal value of 1 indicates that all the flux in the primary winding is coupled into the secondary winding.

Coil   A common term for an inductor.

Conductance (G)   The ability of a circuit to allow current; the reciprocal of resistance. The unit is the siemens (S).

Conductor   A material in which electrical current can flow with relative ease. An example is copper.

Core   The structure around which the winding of an inductor is formed. The core material influences the electromagnetic characteristics of the inductor.

Coulomb (C)   The unit of electrical charge.

Coulomb's law   A physical law that states a force exists between two charged bodies that is directly proportional to the product of the two charges and inversely proportional to the square of the distance between them.

Covalent   Related to the bonding of two or more atoms by the interaction of their valence electrons.

CRT   See *Cathode-ray tube*.

Current   The rate of flow of electrons.

Current gain   The ratio of output current to input current.

Cutoff frequency (f_c)   The frequency at which the output voltage of a filter is 70.7% of the maximum output voltage.

Cycle   One repetition of a periodic waveform.

DC component   The average value of a pulse waveform.

Decibel (dB)   The unit of logarithmic expression of a ratio, such as a power ratio or a voltage ratio.

Degree   The unit of angular measure corresponding to 1/360 of a complete revolution.

Derivative   The instantaneous rate of change of a function determined mathematically.

Dielectric   The insulating material between the plates of a capacitor.

Dielectric constant   A measure of the ability of a dielectric material to establish an electric field.

Dielectric strength   A measure of the ability of a dielectric material to withstand voltage.

Differential amplifier   An amplifier that produces an output proportional to the difference of two inputs.

Differentiator   A circuit that produces an output which can approximate the instantaneous rate of change of the input function.

Duty cycle   A characteristic of a pulse waveform that indicates the percentage of time that a pulse is present during a cycle; the ratio of pulse width to period.

Earth ground   A method of grounding whereby one side of a power line is neutralized by connecting it to a water pipe or a metal rod driven into the ground.

Effective value   A measure of the heating effect of a sine wave; also known as the *rms* (root mean square) *value*.

Efficiency   The ratio of the output power to the input power, expressed as a ratio.

Electrical   Related to the use of electrical voltage and current to achieve desired results.

Electrical isolation   The condition that exists when two coils are magnetically linked but have no electrical connection between them.

Electromagnetic field   A formation of a group of magnetic lines of force surrounding a conductor created by electrical current in the conductor.

Electromagnetic induction   The phenomenon or process by which a voltage is produced in a conductor when there is relative motion between the conductor and a magnetic or electromagnetic field.

Electron   The basic particle of negative electrical charge in matter.

Electronic   Related to the movement and control of free electrons in semiconductors or vacuum devices.

Element   One of the unique substances that make up the known universe. Each element is characterized by a unique atomic structure.

Energy   The fundamental ability to do work and produce action or results. The unit is the joule (J).

Falling edge   The negative-going transition of a pulse.

Fall time (t_f)   The time interval required for a pulse to change from 90% to 10% of its full amplitude.

Farad (F)   The unit of capacitance.

Faraday's law   A law stating that the voltage induced across a coil of wire equals the number of turns in the coil times the rate of change of the magnetic flux.

Filter   A type of electrical circuit that passes certain frequencies and rejects all others.

Free electron   A valence electron that has broken away from its parent atom and is free to move from atom to atom within the atomic structure of a material.

Frequency   A measure of the rate of change of a periodic function; the number of cycles completed in 1 s. The unit of frequency is the hertz.

Frequency response   The way the output voltage of a circuit varies as the input frequency is changed.

Fundamental frequency   The repetition rate of a waveform.

Fuse   A protective device that burns open when excessive current flows in a circuit.

Generator   An energy source that produces electrical signals.

Ground   In electronic circuits, the common or reference point.

Half-power frequency   The frequency at which the output of a filter is 70.7% of maximum.

Harmonics   The frequencies contained in a composite waveform, which are integer multiples of the repetition frequency (fundamental).

Henry (H)   The unit of inductance.

Hertz (Hz)   The unit of frequency. One hertz equals one cycle per second.

High-pass filter   A certain type of filter in which higher frequencies are passed and lower frequencies are rejected.

Hypotenuse   The longest side of a right triangle.

Hysteresis   A lagging in the values of resulting magnetization in a magnetic material due to a changing magnetizing force.

Impedance (Z)   The total opposition to sinusoidal current expressed in ohms.

Impedance matching   A technique used to match a load resistance to an internal source resistance in order to achieve a maximum transfer of power.

Induced voltage   Voltage produced as a result of a changing magnetic field.

Inductance (L)   The property of an inductor that produces an opposition to any change in current.

Inductive reactance (X_L)   The opposition of an inductor to sinusoidal current. The unit is ohms.

Inductive susceptance (B_L)   The reciprocal of inductive reactance. The unit is siemens.

Inductor   An electrical device formed by a wire wound around a core having the property of inductance and the capability to store energy in its electromagnetic field; also known as a coil or a choke.

Input   The voltage, current, or power applied to an electrical circuit to produce a desired result.

Instantaneous value   The voltage or current value of a waveform at a given instant in time.

Insulator   A material that does not allow current under normal conditions.

Integral   The area under the curve of a function, determined mathematically.

Integrator   A circuit that produces an output which can approximate the area under the curve of the input function.

Interface   To make the output of one type of circuit compatible with the input of another so that they can operate properly together.

Ion   An atom that has gained or lost a valence electron resulting in a net positive or negative charge.

Joule (J)   The unit of energy.

Junction   A point at which two or more branches are connected together.

Kilowatt-hour (kWh)   A common unit of energy used mainly by utility companies.

Kirchhoff's current law   A law stating that the total current into a junction equals the total current out of the junction.

Kirchhoff's voltage law   A law stating that (1) the sum of the voltage drops around a closed loop equals the source voltage or (2) the sum of all the voltages (drops and sources) around a closed loop is zero.

Lag   To be behind; describes a condition of the phase or time relationship of waveforms in which one waveform is behind the other in phase or time.

Lead   To be ahead; describes a condition of the phase or time relationship of waveforms in which one waveform is ahead of the other in phase or time; also, a wire or cable connection to a device or instrument.

Leading edge   The first step or transition of a pulse.

Lenz's law   A physical law that states when the current through a coil changes, an induced voltage is created in a direction to oppose the change in current. The current cannot change instantaneously.

Linear   Characterized by a straight-line relationship.

Load   An element (resistor or other component) connected across the output terminals of a circuit that draws current from the circuit.

Low-pass filter   A certain type of filter in which lower frequencies are passed and higher frequencies are rejected.

Magnetic coupling   The magnetic connection between two coils as a result of the changing magnetic flux lines of one coil cutting through the second coil.

Magnetic flux   The lines of force between the north and south poles of a permanent magnet or an electromagnet.

Magnetic flux density   The number of lines of force per unit area in a magnetic field.

Magnetizing force   The amount of mmf per unit length of magnetic material.

Magnetomotive force   The force that produces a magnetic field.

Magnitude   The value of a quantity, such as the number of volts of voltage or the number of amperes of current.

Maximum power transfer theorem   A theorem that states the maximum power is transferred from a source to a load when the load resistance equals the internal source resistance.

Multimeter   An instrument that measures voltage, current, and resistance.

Mutual inductance (L_M)   The inductance between two separate coils, such as a transformer.

Neutron An atomic particle having no electrical charge.

Ohm (Ω) The unit of resistance.

Ohmmeter An instrument for measuring resistance.

Ohm's law A law stating that current is directly proportional to voltage and inversely proportional to resistance.

Open circuit A circuit in which the current path is broken.

Oscillator An electronic circuit consisting of an amplifier and a phase-shift network connected in a positive feedback loop that produces a time-varying output signal without an external input signal using positive feedback.

Oscilloscope A measurement instrument that displays signal waveforms on a screen.

Parallel The relationship in electric circuits in which two or more current paths are connected between the same two points.

Parallel resonance In a parallel *RLC* circuit, the condition where the impedance is maximum and the reactances are equal.

Pass band The range of frequencies passed by a filter.

Peak value The voltage or current value of a waveform at its maximum positive or negative points.

Peak-to-peak value The voltage or current value of a waveform measured from its minimum to its maximum points.

Period The time interval of one complete cycle of a given sine wave or any periodic waveform.

Periodic Characterized by a repetition at fixed time intervals.

Permeability The measure of ease with which a magnetic field can be established in a material.

Phase The relative displacement of a time-varying waveform in terms of its occurrence with respect to a reference.

Phasor A representation of a sine wave in terms of both magnitude and phase angle.

Photoconductive cell A type of variable resistor that is light-sensitive.

Potentiometer A three-terminal variable resistor.

Power The rate of energy consumption.

Power factor The relationship between volt-amperes and true power or watts. Volt-amperes multiplied by the power factor equals true power.

Power rating The maximum amount of power that a resistor can dissipate without being damaged by excessive heat buildup.

Power supply An electronic instrument that produces voltage, current, and power from the ac power line or batteries in a form suitable for use in various applications to power electronic equipment.

Primary winding The input winding of a transformer; also called *primary*.

Proton A positively charged atomic particle.

Pulse A type of waveform that consists of two equal and opposite steps in voltage or current separated by a time interval.

Pulse response The reaction of a circuit to a given input.

Pulse width The time interval between the opposite steps of an ideal pulse. For a nonideal pulse, the time between the 50% points on the leading and trailing edges.

Quality factor (*Q*) The ratio of reactive power to true power in a coil or a resonant circuit.

Radian A unit of angular measurement. There are 2π radians in one complete revolution. One radian equals 57.3°.

Ramp A type of waveform characterized by a linear increase or decrease in voltage or current.

Reactive power (*P_r*) The rate of which energy is stored and alternately returned to the source by a capacitor or inductor. The unit is the VAR.

Rectifier An electronic circuit that converts ac into pulsating dc; one part of a power supply.

Reflected load The secondary load as it appears to the source in the primary circuit.

Reflected resistance   The resistance in the secondary circuit reflected into the primary circuit.

Relay   An electromagnetically controlled mechanical device in which electrical contacts are open or closed by a magnetizing current.

Reluctance   The opposition to the establishment of a magnetic field.

Resistance   Opposition to current. The unit is the ohm (Ω).

Resistor   An electrical component designed specifically to provide resistance.

Resolution   The smallest increment of a quantity that a meter can measure.

Resonant frequency   The frequency at which a resonant condition occurs in a series or parallel *RLC* circuit.

Retentivity   The ability of a material, once magnetized, to maintain a magnetized state without the presence of a magnetizing force.

Rheostat   A two-terminal variable resistor.

Right triangle   A triangle with a 90° angle.

Ripple voltage   The variation in the dc voltage on the output of a filtered rectifier caused by the slight charging and discharging action of the filter capacitor.

Rise time (t_r)   The time interval required for a pulse to change from 10% to 90% of its amplitude.

Rising edge   The positive-going transition of a pulse.

Root mean square (rms)   The value of a sine wave that indicates its heating effect, also known as the *effective value*. It is equal to 0.707 times the peak value.

Sawtooth wave   A type of electrical waveform composed of ramps; a special case of a triangular wave.

Schematic   A symbolized diagram of an electrical or electronic circuit.

Secondary winding   The output winding of a transformer; also called *secondary*.

Selectivity   A measure of how well a filter passes certain frequencies and rejects others. The narrower the bandwidth, the greater the selectivity.

Semiconductor   A material that has a conductance value between that of a conductor and that of an insulator. Silicon and germanium are examples.

Sensitivity factor   The ohms-per-volt rating of a voltmeter.

Series   In an electrical circuit, a relationship of components in which the components are connected such that they provide a single current path between two points.

Series resonance   In a series *RLC* circuit, the condition where the impedance is minimum and the reactances are equal.

Shell   The orbit in which an electron revolves.

Short circuit   A zero or abnormally low resistance between two points; usually an inadvertent condition.

Solenoid   An electromagnetically controlled device in which the mechanical movement of a shaft or plunger is activated by a magnetizing current.

Source   Any device that produces energy; one of the three terminals of an FET.

Steady state   The equilibrium condition of a circuit that occurs after an initial transient time.

Step-down transformer   A transformer in which the secondary voltage is less than the primary voltage.

Step-up transformer   A transformer in which the secondary voltage is greater than the primary voltage.

Stop band   The range of frequencies between the upper and lower cutoff points.

Superposition   A method for analyzing circuits with two or more sources by examining the effects of each source by itself and then combining the effects.

Switch   An electrical or electronic device for opening and closing a current path.

Tank circuit   A parallel resonant circuit.

Tapered   Nonlinear, such as a tapered potentiometer.

Temperature coefficient   A constant specifying the amount of change in the value of a quantity for a given change in temperature.

Terminal   An external contact point on an electronic device.

Terminal equivalency   A condition that occurs when two circuits produce the same load voltage and load current where the same value of load resistance is connected to either circuit.

Tesla   The unit of flux density.

Termistor   A type of variable resistor that is temperature-sensitive.

Thevenin's theorem   A circuit theorem that provides for reducing any resistive circuit to a single equivalent voltage source in series with an equivalent resistance.

Time constant   A fixed time interval set by R, C, and/or L values that determines the time response of a circuit.

Tolerance   The limits of variation in the value of a component.

Trailing edge   The second step or transition of a pulse.

Transformer   A device formed by two or more windings that are magnetically coupled to each other and providing a transfer of power electromagnetically from one winding to another.

Transient   A temporary passing condition in a circuit; a sudden or temporary change in circuit conditions.

Transient time   An interval equal to approximately five time constants.

Triangular wave   A type of electrical waveform that consists of two ramps.

Trigger   The activating mechanism of some electronic devices or instruments.

Trimmer   A small variable capacitor.

Turns ratio   The ratio of turns in the secondary winding to turns in the primary winding.

Valence   Related to the outer shell or orbit of an atom.

Valence electron   An electron that is present in the outermost shell of an atom.

Volt   The unit of voltage or electromotive force.

Voltage   The amount of energy available to move a certain number of electrons from one point to another in an electrical circuit.

Voltage drop   The potential difference in voltage between two points when there is a drop in energy level due to reactance.

Volt-ampere reactive (VAR)   The unit of reactive power.

Voltmeter   An instrument used to measure voltage.

Watt (W)   The unit of power.

Watt's law   A law that states the relationships of power to current, voltage, and resistance.

Waveform   The pattern of variations of a voltage or current showing how the quantity changes with time.

Weber   The unit of magnetic flux.

Winding   The loops of wire in an inductor.

Wiper   The sliding contact in a potentiometer.

INDEX